Grand Challenges in Biology and Biotechnology

Series editor
Pabulo H. Rampelotto

More information about this series at http://www.springer.com/series/13485

Pabulo H. Rampelotto
Editor

Molecular Mechanisms of Microbial Evolution

Springer

Editor
Pabulo H. Rampelotto
Federal University of Rio Grande do Sul
Porto Alegre, Rio Grande do Sul, Brazil

ISSN 2367-1017 ISSN 2367-1025 (electronic)
Grand Challenges in Biology and Biotechnology
ISBN 978-3-030-09867-4 ISBN 978-3-319-69078-0 (eBook)
https://doi.org/10.1007/978-3-319-69078-0

This Springer imprint is published by the registered company Springer Nature Switzerland AG
The registered company address is: Gewerbestrasse 11, 6330 Cham, Switzerland

I dedicate this book to Werner Arber, whose remarkable career and personal life inspired me to develop this book project. I am blessed we share the same fraction of time in Earth's history.

Foreword

Experimental investigations made with relatively few kinds of microorganisms in the middle of the twentieth century rapidly revealed not only that genetic information is carried on nucleic acid molecules, but also how this information carries out its specific functions and how it evolves in the course of time. This fruitful development also led to the possibilities to determine the specific nucleotide sequences of genes. In recent decades, research has been more and more extended to yet little-studied kinds of microorganisms, which include bacteria, archaea, and small eukaryotes. For the selection of scientific reports included in this collection, particular attention was given to research contributions providing deeper insights into the remarkable inventiveness of nature in specific contributions to the slowly ongoing biological evolution by a kind of self-organization of nature. These investigations bring about highly valuable insights into the natural laws of biological evolution.

Biozentrum University of Basel
Basel, Switzerland

Werner Arber

Preface

This book is essentially the outcome of my intellectual cooperation with Werner Arber in the last couple of years. Our work toward the development of a previous journal I lead as EiC gave birth to discussions on different aspects of life, especially related to the microbial world, our mutual interest.

Through these enlightening discussions, I came to realize that one of the grand challenges in science nowadays is to understand how evolution continues to reshape life at a time when significant changes in biodiversity are happening all over the world. Moreover, the biggest current gap within this fundamental question is our limited understanding of the dynamics of microbial evolution in nature. If the past century and a half of studies has been mostly about the diversity and evolution of (macro) species (thanks to the landmark work of Darwin and the pioneer evolutionists), the twenty-first century will be mostly about the diversity and evolution of microbes (thanks to the next-generation sequencing technologies).

This gradual shift to a better understanding of the microbial world has also been driven by the fact that many of the ecological, environmental, and health challenges faced by our society are somehow linked to microbes. Although it is obvious for problems such as the rapid evolution of antibiotic resistance, it has, until recently, been less obvious for many other changes. The invasions of tree-boring insects devastating forests worldwide are not simple insect invasions. They are invasions by assemblages of species, including fungi and bacteria, of which insects are just the most visible members. On a different example, neuroscientists are convinced and supported by my many pieces of evidence that intestinal microbiota might influence brain development and behavior, which has caused a paradigm shift in neuroscience.

As such, it is possible to realize that one of the most profound paradigms that have transformed our understanding about life over the last decades was the acknowledgement that microorganisms play a central role in shaping the past and present environments on Earth and the nature of all life forms.

Stimulated by this paradigm shift, the field of microbial evolution is experiencing tremendous progress. Recent advances in DNA sequencing, high-throughput technologies, and genetic manipulation systems have enabled studies that directly characterize the molecular and genomic bases of evolution, producing data that are making us change our view of the microbial world. The notion that mutations in the coding regions of genomes are, in combination with selective forces, the main contributors to biodiversity needs to be reexamined as evidence accumulates, indicating that many noncoding regions that contain regulatory signals show a high rate of variation even among closely related organisms. Comparative analyses of an increasing number of closely related microbial genomes have yielded exciting insight into the sources of microbial genome variability with respect to gene content, gene order, and evolution of genes with unknown functions. Furthermore, laboratory studies (i.e., experimental microbial evolution) are providing fundamental biological insight through direct observation of the evolution process. They not only enable testing evolutionary theory and principles, but also have applications to metabolic engineering and human health. Overall, these studies ranging from viruses to bacteria to microbial eukaryotes are illuminating the mechanisms of evolution at a resolution that Darwin, Delbruck, and Dobzhansky could barely have imagined.

As a natural consequence, by late 2015 I found myself discussing with Werner the interesting possibility of exploring these topics in more detail through a concise and reliable reference featuring the current knowledge on the molecular mechanisms of microbial evolution with a collection of chapters written by the leading experts in the field. Werner gave full support to the project and was willing to write a chapter summarizing his decades of work in the field. That was the driving force to launch this unique book within the Grand Challenges Series. I hope our readers share the same enthusiasm I felt when working in this project. This landmark work will certainly have a special place in my library.

Porto Alegre, Brazil Pabulo H. Rampelotto

Contents

About the Editor

Pabulo Henrique Rampelotto is a molecular biologist and science editor with extensive editorial experience in the leading position. He is the founder and Editor-in-Chief of the Springer Book Series "Grand Challenges in Biology and Biotechnology." In addition, he is also Editor-in-Chief, Associate Editor, Senior Editor, Guest Editor, and member of the editorial board of several scientific journals in the field of life sciences and biotechnology. Furthermore, Pabulo is member of four Scientific Advisory Boards of Lifeboat Foundation, alongside several Nobel laureates and other distinguished scientists, philosophers, educators, engineers, and economists. Most of his recent work has been dedicated to the editorial process of several scientific journals in life science and biotechnology, as well as on the organization of special issues and books in his fields of expertise. In his special issues and books, some of the most distinguished team leaders in the field have published their work, ideas, and findings, including Nobel laureates and several of the highly cited scientists according to the ISI Institute.

"When he is not working, Pabulo enjoys spending time walking in the woods, in the mountains, and near the sea. . .thinking, always thinking." (Lifeboat Foundation, 2016).

Chapter 1
The Relevance and Challenges of Studying Microbial Evolution

Pabulo Henrique Rampelotto

1 Introduction

The pioneering work of Charles Darwin, unveiling key insights into how the evolutionary process works to generate different forms of life, has provided a cornerstone for understanding evolutionary relationships among different organisms. Nevertheless, Darwin's work primarily dealt with the macroscopic forms of life covering the most recent 1.0 billion years. The evolution of microbes, on the other hand, had been underway for about 3 billion years, covering much of the earlier evolutionary history of life. Thus, an understanding of the evolutionary relationships among microbes is of central importance for deciphering the origin and diversification of different forms of life on Earth.

Even now, most of the biodiversity of life on Earth is microbial. Many of the genes, molecular machines, regulatory, metabolic, and synthetic pathways found in all living organisms today evolved first in microorganisms. The great diversity of microbes allows them to synthesize or break down a vast range of chemical substrates and govern biogeochemical cycles that make Earth a habitable planet. As such, microbes are essential to Earth's functioning at every scale, and understanding them is imperative for a complete understanding of life.

Our own body is also home to a diverse assemblage of microbial cells. Bacteria that colonize our gastrointestinal tract help us maintain our health by extracting energy from undigested carbohydrates, synthesizing vitamins, and metabolizing xenobiotics. On a very practical level, understanding the mechanisms of microbial evolution will improve our ability to develop more effective antibiotics and vaccines,

P. H. Rampelotto (✉)
Center of Biotechnology and PPGBCM, Federal University of Rio Grande do Sul, Porto Alegre, RS, Brazil
e-mail: p.rampelotto@mail.ufsm.br

P. H. Rampelotto (ed.), *Molecular Mechanisms of Microbial Evolution*, Grand Challenges in Biology and Biotechnology,
https://doi.org/10.1007/978-3-319-69078-0_1

predict disease outbreaks and changes in virulence, and harness microorganisms' potential for rapid evolutionary change to create new products and processes. In this regards, basic research on microbial evolution has the potential to contribute across sectors and address applied problems in many fields, thereby leading to new approaches to treating disease, raising agricultural productivity, monitoring and addressing climate change, and producing clean energy.

Based on such tremendous potential, it is possible to realize that a better understanding of microbial evolution may give us more than just the ability to understand microbial diversity; it will help understand the world around us. The Darwinian revolution in the nineteenth century went far beyond the scientific domain and had the broadest philosophical and cultural implications. At the beginning of the twenty-first century, we are clearly moving toward similar direction in the microbial world.

As a brief introduction to this book, herein I highlight some of the reasons why understanding how microbes evolve is important to science and society and how challenging is to study microbial evolution.

2 Why Understanding Microbial Evolution Is Important

2.1 Major Events in Life's History Was, and Continuous to Be, Influenced by Microbes

Throughout the history of Earth, microorganisms have radically reshaped life on the planet, from creating the very air we breathe to wiping out almost all life on Earth.

2.1.1 How the Earth's Atmosphere Got Oxygen

For the first half of our planet's history, there was no oxygen in the atmosphere. This gas only started to appear about 2.4 billion years ago, in an episode known as the Great Oxidation Event (GOE), one of the most important events to ever happen on this planet. Such colossal event was triggered by cyanobacteria producing oxygen, which developed into multicellular forms as early as 2.3 billion years ago (approximately 200 million years before the GOE) (Schirrmeister et al. 2013). Interestingly, cyanobacteria are inferred to be ancestrally nonphototrophic and acquired the ability for photosynthesis by lateral gene transfer from another, unknown species, which further evolved within the group (Soo et al. 2017).

Before the GOE, any free oxygen produced by cyanobacteria was chemically captured by dissolved iron or organic matter. The GOE started when these oxygen sinks became saturated, at which point oxygen produced by cyanobacteria was free to escape into the atmosphere (Hamilton et al. 2016). The increased production of oxygen set Earth's original atmosphere off balance. Because oxygen was poisonous for large numbers of anaerobic organisms, most anaerobic types of bacteria were eliminated, opening up ecological "niches" for aerobic organisms to develop (Schirrmeister et al. 2015). Cyanobacteria were therefore responsible for one of the most significant mass extinctions in Earth's history. At the same time, these photosynthetic microbes were also responsible for a major turning point in the evolution of life on our planet.

2.1.2 Microbes May Have Caused Earth's Biggest Mass Extinction

About 252 million years ago, Earth suffered the biggest mass extinction event in its history, known as the Great Dying. The atmosphere filled with carbon, the planet baked in a warmer climate, and the oceans acidified. When it was over, approximately 95% of marine species and 70% of terrestrial species are becoming extinct (Sahney and Benton 2008). Scientists have long blamed volcanoes for triggering this catastrophe by pumping the atmosphere full of greenhouse gases. However, a recent study attributed the surge of carbon to the rapid evolution of a new microbe which developed a mechanism for the conversion of organic matter to methane.

According to the study, a group of microbes called *Methanosarcina* acquired two genes from an unrelated bacterium via gene transfer about 250 million years ago (Rothman et al. 2014). The new metabolic pathway allowed these microbes to rapidly consume large deposits of organic carbon in marine sediments while releasing vast amounts of methane, a greenhouse gas that warmed the atmosphere and acidified the oceans. Volcanoes could have still played in a role in spewing out nickel, which is necessary for the chemical reaction that allows microbes to make methane gas.

Although such implications remain speculative, the findings suggest that microbial evolution has important consequences for the evolution of the environment as a whole and indicates how a particular microbe may have played a crucial role in the evolution of life on Earth.

2.1.3 Microbes Control Critical Biogeochemical Processes

Through the evolution of oxygenic photosynthesis and continuous cycling of carbon, nitrogen, and other elements, microbes created and continue to sustain the conditions for life on Earth. This is true at all scales and in all environments, on land, in the ocean, and under the Earth's surface. Microbes directly modulate the amount of bioavailable nitrogen, carbon, phosphorous, sulfur, and many important metals (Long et al. 2016). The ability to transform nitrogen gas from the atmosphere into a form that organisms use to make critical biomolecules, like DNA and proteins (the process of nitrogen fixation), evolved in microbes very early in the history of life, and microbial nitrogen fixation continues to serve as a fundamental link in the nitrogen cycle.

While various microorganisms involved in carrying out biogeochemical processes have been identified, biogeochemical process rates are only rarely measured together with microbial growth, and one of the biggest challenges for advancing our understanding of biogeochemical processes is to systematically link biogeochemistry to the rate of specific metabolic processes (Rousk and Bååth 2011; Rousk and Bengtson 2014). We also need to identify the factors governing these activities and if it results in feedback mechanisms that alter the growth, activity, and interaction between primary producers and microorganisms (Treseder et al. 2012). By determining how different groups of microorganisms respond to individual environmental conditions by allocating, e.g., carbon to production of biomass, CO_2, and other products, a mechanistic and quantitative understanding of formation and decomposition of organic matter, and the production and consumption of greenhouse gases, can be achieved.

Human activities are changing the environment in which microbes evolve, creating innumerable new evolutionary pressures. How microbes will react to this new set of variables is critically important but difficult to study. Over the long term, there is no doubt that if microbial ecosystems are disturbed, their "evolutionary trajectories" will be affected in ways that we cannot still predict.

2.1.4 Climate Change and Microbial Evolution

Microbial chemical cycling also plays a critical role in the current status of the planet's greenhouse effect (Tian et al. 2016). Microbes can both absorb or release carbon, depending on their diets, so the direction of their influence is not so clear. However, altogether, they are huge players in the carbon cycle (Bardgett et al. 2008). Just the microbes that decompose dead plants in the soil, for example, release 55 billion tons of carbon dioxide a year, which is eight times what humans contribute through fossil fuels and deforestation (EPA 2018).

In addition, climate change is changing how these microbes function. In Arctic permafrost, for instance, where nearly half of the organic carbon stored in soil around the world is contained, there is normally not much microbial activity. In recent years, however, the permafrost is releasing more carbon dioxide than it absorbs, which scientists believe is due to rising temperatures allowing more microbes to feed in the tundra and release carbon dioxide (Schuur et al. 2015; Ward et al. 2017). Consequently, this feedback can accelerate climate change.

Moreover, denitrifiers and nitrifiers can generate nitric oxide and nitrous oxide, which are powerful greenhouse gases that have 280–320 times more potential for warming than carbon dioxide (Szukics et al. 2010). Also, phytoplankton produce dimethyl sulfide and dimethylsulfoniopropionate, and the cycling of these compounds produces sulfur gases that impact cloud formation and, hence, the water cycle and the global albedo (reflectivity). Archaeal methane production is the dominant natural source of methane, a gas that is over 20 times more powerful a greenhouse gas than carbon dioxide (Nazaries et al. 2013).

In essence, microbes have been changing the climate and have been changed by the climate, throughout Earth's history (Singh et al. 2010; Ladau et al. 2018). Although scientists have been studying microbial ecosystems for many years, there remains much more to learn and understood about complex microbial functions and their interactions with climate change.

2.2 *Models to Understand General Principles of Evolution*

Just as microbes have served as highly flexible model systems for molecular biology experimentation, microbes and microbial consortia can also be used for experiments on evolution (Adams and Rosenzweig 2014; O'Malley 2018). Microbes in experimental systems and in real-world situations like infectious disease offer the opportunity to test and observe microbial evolution in action (Koonin and Wolf 2012).

With larger organisms, until the advent of molecular biology, biologists were often forced to make inferences about evolution from observation. Genetic study has vastly enriched our understanding of the mechanisms of evolution, but the ability to carry out evolutionary experiments is still limited in long-lived organisms with large and complex genomes. By contrast, experimental evolution with microbes offers a rich alternative by providing a testable system based on hypothesis, experiment, and outcome (Elena and Lenski 2003). Control over selective pressures represents another advantage. Moreover, in microbiology it is possible to save and see the "mistakes" or evolutionary dead ends in an experiment; we don't necessarily "lose the losers" in an evolutionary microbiology experiment.

In practical terms, microbes are ideal models for understanding the effects of climate change on the diversity and evolution of biological systems. Because they have generation times as short as a few hours, they will do so at higher rates than most other organisms. Scientists can study the effects of climate change on microbes to both understand and hopefully predict the future effects of these environmental changes on all forms of life.

In a new frontier of science, the genomic and post-genomic studies of microorganisms living in extreme conditions on Earth are providing new insights about what it takes to life evolve in environments that were once thought uninhabitable, including potential habitable environments elsewhere in the universe (Rampelotto 2013; Bakermans 2015). Such fascinating studies are paving the way for the stablishment of astrobiology as a strong and vibrant field of research. Indeed, our increasing knowledge about the evolution of microbes in extreme habitats has led numerous scientists to raise the possibility of finding life in various planetary bodies within the Solar System (Rampelotto 2010).

2.3 *Animal Origins and Evolution*

Animal evolution traditionally has been viewed as the result of interactions between animals or with the physical environment. However, this understanding of evolution overlooks a huge missing piece of the puzzle: microbes. To put it bluntly, complex life-forms probably would never have evolved on planet Earth if it were not for microbes.

Bacteria have exerted critical influences on the evolution of eukaryotes and, ultimately, the origin and evolution of animals (Alegado and King 2014). Bacteria and archaea contributed to the cellular and genetic building blocks for the first eukaryotic cells, and bacteria formed stable associations with early eukaryotes in the form of mitochondria and plastids (McFall-Ngai et al. 2013). Moreover, bacteria were likely an important source of food for the progenitors of animals, as well as the first animals themselves (Rosenberg and Zilber-Rosenberg 2016).

After helping get animals started, bacteria also played an important role in helping them along their evolutionary path. While animal development is traditionally thought to be directed primarily by the animal's own genome in response to environmental factors, recent research has shown that animal development may be better thought of as an orchestration among the animal, the environment, and the coevolution of numerous microbial species (Bosch and Miller 2016).

Not only do animals share evolutionary history with microbes, but they also continue to interact with them on a daily basis—often in very profound ways. Bacteria living inside of animals can provide them with metabolic capabilities that the animal itself does not possess. Cows could not eat grass if it were not for the resident microbes that ferment it in the cow's rumen. Certainly, the evolution of the cow was heavily influenced by—if not largely dependent on—its microbial allies.

2.4 *Microbes and Humans*

Just like other animals, humans also share an interesting and deep evolutionary history with microbes (McFall-Ngai et al. 2013). Of the roughly 23,000 genes in the human genome, for instance, 37% are similar to genes in *Bacteria* and archaea. Another 28% are similar to genes in unicellular eukaryotes. Thus, a full 65% of human genes show similarity to microbes. Only 6% are found uniquely in primates.

Our bodies are made up of many more microbial cells than human cells. Thousands of species of bacteria, fungi, viruses, and other microbes live almost everywhere in and on our bodies, including the digestive system, nose, and skin, to name just a few (Gilbert et al. 2018). Some of the earliest research showed that the microbes living in our digestive systems help us digest food, make some of the vitamins we need, and balance the immune system (Arora and Bäckhed 2016; Foster et al. 2016). Since then, we've learned that these microbes, collectively called the microbiome, can affect body weight, susceptibility to cancer, and even behavior (Paun et al. 2017; Vuong et al. 2017; Goodman and Gardner 2018). The gut microbiome interacts with its host using signaling networks that employ the immune system, hormones, and the nervous system.

In short, microbes have a profound effect on our overall health and may become a key component of precision medicine (Kashyap et al. 2017; Petrosino 2018). As such, a better understanding on how microbes have adapted and evolved to colonize and influence our body will certainly revolutionize the way we view our health.

2.5 *Pathogen Evolution*

During evolution, humans developed many ways to protect themselves against bacterial pathogens. On the other hand, bacteria have developed strategies to evade, subvert, or circumvent these defenses. These microbial pathogens have a remarkable capacity for rapid evolution because they have large population sizes, short generation times, and high mutation rates. This capacity, combined with large dense human populations and rapid air travel, are leading to greatly increased risk of the evolution of novel pathogens.

As such, bacterial pathogens continue to cause problems for humans with the continuous evolution of known pathogens and the emergence of new ones (Martínez 2013). The immense social and economic impact of bacterial pathogens, from drug-resistant infections in hospitals to the devastation of agricultural resources, has resulted in major investment to understand the causes and consequences of pathogen evolution. Recent genome sequencing projects have provided insight into the evolution of bacterial genome structures, revealing the impact of mobile DNA on genome restructuring and pathogenicity (Jackson et al. 2012; Nuccio and Bäumler 2015; Tibayrenc 2017). Sequencing of multiple genomes of related strains has enabled the delineation of pathogen evolution and facilitated the tracking of bacterial pathogens globally (Bentley and Parkhill 2015). Other recent theoretical and empirical studies have shown that pathogen evolution is significantly influenced by ecological factors, such as the distribution of hosts within the environment and the effects of coinfection (Britton et al. 2015; Lloyd-Smith et al. 2015).

With a better knowledge on the molecular mechanisms of pathogen evolution, researchers can attempt to predict where disease outbreaks are likely to occur. Strategies can also be developed to control disease, for example, by promoting a lifestyle that maintains populations of beneficial microbes in our bodies and prevents the evolution or ingress of pathogens. In addition, researchers might be able to prevent the evolution of virulence by removing conditions that promote pathogen evolution and perhaps could even reverse adaptive changes.

2.6 *Microbial Evolution Can Be Used to Solve Global Problems*

Microorganisms can be used to solve some of the global problems through the generation of fuels, production of bio-based materials, improvement of crop productivity, remediation of pollution, and recyclement of wastes. These approaches show great promise for contributing to a transition to a sustainable human society. To improve microbial performance toward these aims, metabolic engineering is generally used. However, the remarkable complexity of dynamic interactions in cellular systems often prevents practical applications of metabolic engineering due to the requirement of extensive genetic and metabolic information on the organism of interest (Alkım et al. 2014). In contrast, evolutionary engineering follows the natural principles of evolution (i.e., variation and selection). The lack of need for prior genetic knowledge underlying the phenotypes of interest makes this a powerful approach for strain development for even species with minimal genotypic information (Cakar et al. 2012; Winkler and Kao 2014). Therefore, it is a complementary strategy that offers compelling scientific and applied advantages for strain development and process optimization.

With a more detailed and systematic understanding of microbial evolution, the manipulation of microbial communities can be applied to more complex problems. Whether the goal is to use microbial communities to produce desired materials, remedy environmental damage, or mitigate climate change, a comprehensive and predictive understanding of how microbial communities respond to change and stress in time is critical.

3 Challenges in the Study of Microbial Evolution

Despite their great relevance for the habitability of the planet and maintenance of our health, our understanding of the diversity, functioning, and evolution of microbial life is far from being complete.

This is partially due to our inability to bring microbial life into the lab for comprehensive and detailed investigations. In most cases, it is very challenging to successfully isolate individual microbes from their dynamic and complex environments and keep them functionally alive in a controlled setting. Even when isolation is possible, understanding how well the isolated members of microbial populations represent their environmental population is not necessarily always clear. Thankfully, in parallel to conventional approaches, we can use modern molecular and computational techniques to recover the genomic content of naturally occurring microbes directly from the environment and investigate some of the most fundamental aspects of their life and evolution.

Although tremendous progress has been achieved with the use of such advanced molecular techniques, the own complex nature of microbial systems imposes additional challenges for the study of microbial evolution. Microorganisms have several ways to generate far more dramatic and rapid genetic variation than plant and animals. They employ an impressive array of mechanisms to generate genetic variation, and this makes the study of their evolution a difficult task.

To tackle such challenging matter and discuss new insights on the molecular mechanisms of microbial evolution, some of the most distinguished team leaders in the field were invited to bring their interesting or provocative perspectives on topics of primary relevance for the theme. The outcome was a collection of 14 enlightening chapters that will drive you through the most interesting groundbreaking discoveries and emerging concepts on how microbes evolved and continue to evolve. The underlying goal of this volume was to span a range of topics and viewpoints to produce a timely and timeless work, one that would not become obsolete by the next generation of molecular data.

References

Adams J, Rosenzweig F (2014) Experimental microbial evolution: history and conceptual underpinnings. Genomics 104(6 Pt A):393–398

Alegado RA, King N (2014) Bacterial influences on animal origins. Cold Spring Harb Perspect Biol 6(11):a016162

Alkım C, Turanlı-Yıldız B, Cakar ZP (2014) Evolutionary engineering of yeast. Methods Mol Biol 1152:169–183

Arora T, Bäckhed F (2016) The gut microbiota and metabolic disease: current understanding and future perspectives. J Intern Med 280(4):339–349

Bakermans C (2015) Microbial evolution under extreme conditions. Walter de Gruyter, Berlin

Bardgett RD, Freeman C, Ostle NJ (2008) Microbial contributions to climate change through carbon cycle feedbacks. ISME J 2(8):805–814

Bentley SD, Parkhill J (2015) Genomic perspectives on the evolution and spread of bacterial pathogens. Proc Biol Sci 282(1821):20150488

Bosch TCG, Miller DJ (2016) The holobiont imperative: perspectives from early emerging animals. Springer, London

Britton T, House T, Lloyd AL, Mollison D, Riley S, Trapman P (2015) Five challenges for stochastic epidemic models involving global transmission. Epidemics 10:54–57

Cakar ZP, Turanli-Yildiz B, Alkim C, Yilmaz U (2012) Evolutionary engineering of *Saccharomyces cerevisiae* for improved industrially important properties. FEMS Yeast Res 12(2):171–182

Elena SF, Lenski RE (2003) Evolution experiments with microorganisms: the dynamics and genetic bases of adaptation. Nat Rev Genet 4(6):457–469

EPA (2018) Inventory of U.S. greenhouse gas. Emissions and sinks: 1990–2016. U.S. Environmental Protection Agency, Washington

Foster JA, Lyte M, Meyer E, Cryan JF (2016) Gut microbiota and brain function: an evolving field in neuroscience. Int J Neuropsychopharmacol 19(5):pyv114

Gilbert JA, Blaser MJ, Caporaso JG, Jansson JK, Lynch SV, Knight R (2018) Current understanding of the human microbiome. Nat Med 24(4):392–400

Goodman B, Gardner H (2018) The microbiome and cancer. J Pathol 244(5):667–676

Hamilton TL, Bryant DA, Macalady JL (2016) The role of biology in planetary evolution: cyanobacterial primary production in low-oxygen Proterozoic oceans. Environ Microbiol 18 (2):325–340

Jackson RW, Johnson LJ, Clarke SR, Arnold DL (2012) Bacterial pathogen evolution: breaking news. Trends Genet 27(1):32–40

Kashyap PC, Chia N, Nelson H, Segal E, Elinav E (2017) Microbiome at the frontier of personalized medicine. Mayo Clin Proc 92(12):1855–1864

Koonin EV, Wolf YI (2012) Evolution of microbes and viruses: a paradigm shift in evolutionary biology? Front Cell Infect Microbiol 2:119

Ladau J, Shi Y, Jing X, He J-S, Chen L, Lin X, Fierer N, Gilbert JA, Pollard KS, Chu H (2018) Climate change will lead to pronounced shifts in the diversity of soil microbial communities. bioRxiv 180174. https://doi.org/10.1101/180174

Lloyd-Smith JO, Funk S, McLean AR, Riley S, Wood JL (2015) Nine challenges in modelling the emergence of novel pathogens. Epidemics 10:35–39
Long PE, Williams KH, Hubbard SS, Banfield JF (2016) Microbial metagenomics reveals climate-relevant subsurface biogeochemical processes. Trends Microbiol 24(8):600–610
Martínez JL (2013) Bacterial pathogens: from natural ecosystems to human hosts. Environ Microbiol 15(2):325–333
McFall-Ngai M, Hadfield MG, Bosch TC et al (2013) Animals in a bacterial world, a new imperative for the life sciences. Proc Natl Acad Sci USA 110(9):3229–3236
Nazaries L, Pan Y, Bodrossy L, Baggs EM, Millard P, Murrell JC, Singh BK (2013) Evidence of microbial regulation of biogeochemical cycles from a study on methane flux and land use change. Appl Environ Microbiol 79(13):4031–4040
Nuccio SP, Bäumler AJ (2015) Reconstructing pathogen evolution from the ruins. Proc Natl Acad Sci USA 112(3):647–648
O'Malley MA (2018) The experimental study of bacterial evolution and its implications for the modern synthesis of evolutionary biology. J Hist Biol 51(2):319–354. https://doi.org/10.1007/s10739-017-9493-8
Paun A, Yau C, Danska JS (2017) The influence of the microbiome on type 1 diabetes. J Immunol 198(2):590–595
Petrosino JF (2018) The microbiome in precision medicine: the way forward. Genome Med 10 (1):12
Rampelotto PH (2010) Resistance of microorganisms to extreme environmental conditions and its contribution to astrobiology. Sustainability 2(6):1602–1623
Rampelotto PH (2013) Extremophiles and extreme environments. Life 3(3):482–485
Rosenberg E, Zilber-Rosenberg I (2016) Microbes drive evolution of animals and plants: the hologenome concept. MBio 7(2):e01395
Rothman DH, Fournier GP, French KL, Alm EJ, Boyle EA, Cao C, Summons RE (2014) Methanogenic burst in the end-Permian carbon cycle. Proc Natl Acad Sci USA 111 (15):5462–5467
Rousk J, Bååth E (2011) Growth of saprotrophic fungi and bacteria in soil. FEMS Microbiol Ecol 78:17–30
Rousk J, Bengtson P (2014) Microbial regulation of global biogeochemical cycles. Front Microbiol 5:103
Sahney S, Benton MJ (2008) Recovery from the most profound mass extinction of all time. Proc R Soc Lond B 275(1636):759–765
Schirrmeister BE, de Vos JM, Antonelli A, Bagheri HC (2013) Evolution of multicellularity coincided with increased diversification of cyanobacteria and the Great Oxidation Event. Proc Natl Acad Sci USA 110(5):1791–1796
Schirrmeister BE, Gugger M, Donoghue PC (2015) Cyanobacteria and the Great Oxidation Event: evidence from genes and fossils. Palaeontology 58(5):769–785
Schuur EA, McGuire AD, Schädel C et al (2015) Climate change and the permafrost carbon feedback. Nature 520(7546):171–179
Singh BK, Bardgett RD, Smith P, Reay DS (2010) Microorganisms and climate change: terrestrial feedbacks and mitigation options. Nat Rev Microbiol 8(11):779–790
Soo RM, Hemp J, Parks DH, Fischer WW, Hugenholtz P (2017) On the origins of oxygenic photosynthesis and aerobic respiration in Cyanobacteria. Science 355(6332):1436–1440
Szukics U, Abell GCJ, Hödl V et al (2010) Nitrifiers and denitrifiers respond rapidly to changed moisture and increasing temperature in a pristine forest soil. FEMS Microbiol Ecol 72:395–406
Tian H, Lu C, Ciais P et al (2016) The terrestrial biosphere as a net source of greenhouse gases to the atmosphere. Nature 531(7593):225–228
Tibayrenc M (2017) Genetics and evolution of infectious diseases. Elsevier, London
Treseder KK, Balser TC, Bradford MA et al (2012) Integrating microbial ecology into ecosystem models: challenges and priorities. Biogeochemistry 109(1–3):7–18

Vuong HE, Yano JM, Fung TC, Hsiao EY (2017) The microbiome and host behavior. Annu Rev Neurosci 40:21–49

Ward CP, Nalven SG, Crump BC, Kling GW, Cory RM (2017) Photochemical alteration of organic carbon draining permafrost soils shifts microbial metabolic pathways and stimulates respiration. Nat Commun 8(1):772

Winkler JD, Kao KC (2014) Recent advances in the evolutionary engineering of industrial biocatalysts. Genomics 104(6 Pt A):406–411

Chapter 2
Mayr Versus Woese: Akaryotes and Eukaryotes

Charles G. Kurland and Ajith Harish

1 Introduction

At the end of the last century, Ernst Mayr (1998) engaged with Carl Woese (1998) in an exchange that began as a straightforward query about the taxonomic status accorded to the Archaea. But that exchange stalled and remained an informative set piece illuminating a cultural conflict between what Woese characterized as pedantic biology and what Mayr characterized as a molecular approach lacking an adequate biological base. Resolving the issues brought out by this confrontation is more urgent today than it was 20 years ago because they involve the strategies of phylogenomics. Fundamental issues concerning these strategies have persisted since the Mayr-Woese confrontation (Mayr 1998; Woese 1998), an interim in which the number of sequenced genomes went from a few to many. Fortunately, innovative phylogenomic approaches have emerged along with the massive increase in genome sequence data.

These innovations as well as the thrust of new genomic data drive the present text, which describes a view of molecular evolution that has grown out of the Mayr-Woese confrontation. It will be no surprise that to follow Mayr's (1998) lead has meant loosening the grip of precisely those preconceptions most dear to Woese (1998). There, the greatest obstacle had been the restrictive view of species evolution supported by alignment-based gene trees.

C. G. Kurland
Department of Biology, Section of Microbial Ecology, Lund University, Lund, Sweden
e-mail: charles.kurland@biol.lu.se

A. Harish (✉)
Department of Cell and Molecular Biology, Section of Structural and Molecular Biology, Uppsala University, Uppsala, Sweden
e-mail: ajith.harish@icm.uu.se

P. H. Rampelotto (ed.), *Molecular Mechanisms of Microbial Evolution*, Grand Challenges in Biology and Biotechnology,
https://doi.org/10.1007/978-3-319-69078-0_2

In effect, we present the confrontation between the opposing briefs argued by Mayr and Woese (Mayr 1998; Woese 1998) as a springboard for an account of genome evolution that has reinvented itself since that confrontation. We begin with the exchange between Mayr and Woese because they were two highly regarded, iconic authors of pre-genomic versions of evolution theory. For this reason alone their opposition would have seismic repercussions. Also, the timing of their confrontation auspiciously punctuated an era of richly descriptive gene tree construction and opened on to the new vistas of the genomics era. Indeed, Mayr (1998) put his finger precisely on the weakest spot in contemporary molecular evolution: the view of evolution glimpsed through the narrow window of gene trees. That view of molecular evolution was an account so artificially refined that the descent of genomic novelty had been lost from genotypic reconstructions. In fact, evolution had been reduced to a description of the tempo of nucleotide or amino acid replacements in sanitized coding sequences.

The confrontation between Mayr and Woese was not wasteful of words. After acknowledging the enormity of Woese's discovery of the archaebacteria (now Archaea), which he likened to the discovery of "a new continent," Mayr frankly questioned, "where should one place the new group of organisms?" (Mayr 1998). Though acknowledging the enormous wealth of molecular data that were marshaled to distinguish Woese's three taxonomic domains (archaea, bacteria, and eukaryotes), Mayr (1998) suggested that it would be more sensible to divide biodiversity at the highest level into two empires made up of the anucleate archaea and bacteria on the one hand and the nucleated eukaryotes on the other.

Such a division would be second nature to anyone seasoned in the phenotypic view of Darwinian evolution. Unfortunately, that proposal was a red flag for Woese (1998), who for decades had been thundering against the conventional division of organisms into prokaryotes and eukaryotes. After all, this was the same division championed by Chatton (1938) as well as by Stanier and van Niel (1962) at a time when microorganisms as diverse as bacteria and fungi were classified as primitive or degenerate plants.

Woese (1998) categorically rejected Mayr's reasoning with a tart: "If there were ever an issue in biological classification that cannot be settled by pedantry, it is this one." As it happens, much data strongly suggest that Mayr's position was anything but a display of pedantry. Thus, Woese (1998) emphasized in his brief the sequence similarities between sampled genes in archaea and their homologues in eukaryotes. These similarities could be contrasted with a more distant relationship between the bacterial and eukaryote sequences. However, there were also by then clues suggesting that precisely the phylogenetic evidence that Woese (1998) marshaled to oppose Mayr's (1998) suggestion was quite possibly an artifact (Forterre et al. 1992; Philippe and Laurent 1998).

Subsequent work has verified these suspicions in detail by showing that Woese's (1998) phylogenetic inferences arise from biased mutation rates as well as from mutational saturation of ancient sequences (Brinkmann and Philippe 1999; Forterre and Philippe 1999; Philippe and Forterre 1999; Philippe et al. 2011). Furthermore, independent phylogenetic methods less vulnerable to distortion by biased mutation rates (Caetano-Anollés 2002; Harish et al. 2013) support Mayr's inference (Mayr 1998).

Though this conclusion has been ignored (Woese 1998), it is nevertheless confirmed by much data that emerged since the exchange between Mayr (1998) and Woese (1998). Some of Woese's signature assertions are in fact inconsistent with a wealth of more recent reliable data (Forterre et al. 1992; Philippe and Laurent 1998; Brinkmann and Philippe 1999; Forterre and Philippe 1999; Philippe and Forterre 1999; Caetano-Anollés 2002; Philippe et al. 2011; Harish et al. 2013). At a minimum, two of Woese's phylogenetic assertions must be corrected: First, the interpretation that archaea and eukaryotes are sister clades is not supportable. Instead, there is a very strong support for the identification of archaea and bacteria as sister clades that make up an akaryote empire, distinct from the eukaryote empire (Brinkmann and Philippe 1999; Caetano-Anollés 2002; Philippe et al. 2011; Harish et al. 2013). Second is the discovery that the most recent universal common ancestor (MRUCA) of the modern crown is not a bacterium (or an archaeon or a eukaryote). Rather, MRUCA has extensive phylogenetic affinities with eukaryotes as well as both bacteria and archaea, which could mean that MRUCA has matured beyond the progenote stage (Brinkmann and Philippe 1999; Caetano-Anollés 2002; Philippe et al. 2011; Harish et al. 2013). Furthermore, the taxonomic status (and phylogenetic relationships) of archaea and bacteria must be completely revised. In effect, we emphasize the reliability of more recent data based on overwhelming statistical support for both the rooting as well as the reconstruction of ancestral characters in MRUCA (Harish and Kurland 2017a). In contrast, the root inferred from the reciprocal rooting of paralogous genes (Iwabe et al. 1989) that was grafted onto the unrooted rRNA tree is highly ambiguous even after implementing sophisticated sequence evolution models (Gouy et al. 2015).

This reevaluation does not diminish our unqualified admiration for Woese's heroic efforts that transformed the landscape of biology and established archaea as separate and distinct from bacteria. The enormity of Woese's contributions is a fair measure of how much he left for us to assimilate and how much to criticize, in the normal way that science works. Though we challenge some of Woese's ideas in this review, we do so precisely because we honor the importance of those ideas as well as his body of work, which left its imprint on much of the molecular evolution and the microbiology of the past 50 years.

2 Unrooted Gene Trees

For decades, the paradigmatic method for gene tree construction was based on comparisons of sequence alignments to identify similarities among amino acid or nucleotide sequences, which then were recruited to reconstruct gene trees (Nei and Kumar 2000). Reconstruction methods surfaced during the heyday of neutral molecular evolution, which meant that they were developed under the immensely simplifying assumption of a universal molecular clock (Nei and Kumar 2000). This helped to fix the focus on DNA sequence-based phylogenies, that is to say, on genotypes (Woese 1998; Nei and Kumar 2000). The genotypic bias appears in Woese's polemic (Woese 1998) in which he seems to "accuse" Mayr of studying the evolution of phenotype. Up to then the genotypic approach favored by Woese (1998) seemed more natural to molecular biologists preoccupied with microorganisms and DNA. After all, microorganisms do not have feathers. Moreover, DNA sequences do provide very attractive alternative windows onto the world of microbial diversity. The methodological question we pursue in this review is "how should sequence information be annotated and implemented for species phylogeny?"

Though sequence alignments are a start, the sequence-based methods favored by molecular biologists are not direct sequence comparisons. Instead of following the sequences of taxa as such, averaged changes of composition within the sequence alignments are tracked. In other words, much of the sequence information in evolution is sacrificed to preserve the computational optimality rather than the phylogenetic optimality in sequence alignments used to construct gene tree (Mossel and Steel 2004; Penny and Collins 2010). It did not take long for problems created by the non-ideality of mutation rates to surface and to reveal the vulnerabilities of this method to mutational saturation and distortion, particularly for deep rooting of gene trees (Forterre et al. 1992; Brinkmann and Philippe 1999; Forterre and Philippe 1999; Philippe and Forterre 1999; Tourasse and Gouy 1999; Mossel and Steel 2004; Rokas and Carroll 2006; Penny and Collins 2010; Philippe et al. 2011). These together with recurrent alignment errors (Morrison 2009) make gene tree methodology much more challenging than was appreciated when Woese adapted his idiosyncratic version of rooted rRNA phylogeny (Fig. 2.1).

A stunning illustration of the variability of gene trees reconstructed from a common genome pool is provided by a recent study of 1070 trees reconstructed from 23 yeast genome sequences by Salichos and Rokas (2013). These are very closely related organisms with genomes that are not separated by exposure to especially long periods of mutational bias and saturation. Nevertheless, the reconstructions identified 1070 distinctive trees all differing from the tree produced by a concatenation analysis (Salichos and Rokas 2013). Equally discouraging results were obtained with other closely related eukaryote genome pools (Salichos and Rokas 2013). Accordingly, Salichos and Rokas argue in opposition to the current fashion to rely on concatenation "that selecting genes with strong phylogenetic signals and demonstrating the absence of significant incongruence are essential for accurately reconstructing ancient divergences" (Salichos and Rokas 2013).

Finally, it must be said that in spite of the many challenges they present (Degnan and Rosenberg 2006), unrooted protein and rRNA gene trees have done yeoman's service in establishing some broad contours of phylogeny. The most important of these contours is the robust identification of the three superkingdoms: archaea, bacteria, and eukaryotes (Fig. 2.1). Indeed, global phylogeny based on structural characters as diverse as secondary structure in rRNA (Caetano-Anollés 2002) or genome content of compact protein domains (Harish et al. 2013) confirms the broad outlines of Woese's three domains (Woese 1987; Woese et al. 1990).

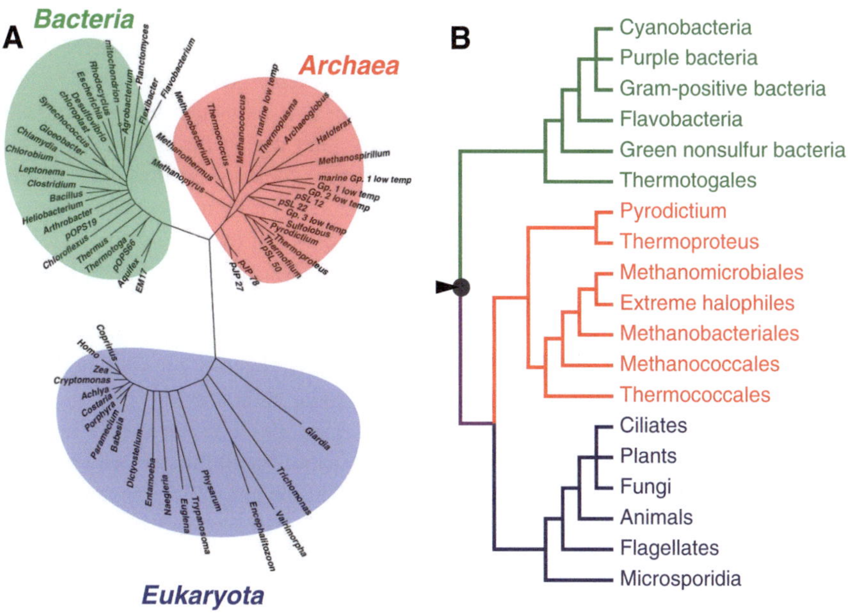

Fig. 2.1 Phylogeny of the small subunit rRNA gene inferred from sequence similarity. (**a**) Unrooted rRNA tree in which the root representing the common ancestor of extant lineages is unknown (adapted from Pace (1997)). (**b**) Rooted rRNA tree (adapted from Woese et al. (1990)). Here the position of the root was extrapolated from an independent analysis of gene duplication events based on the Dayhoff outgroup rooting method. Paralogous protein sequences for elongation factors were employed in this case (Iwabe et al. 1989)

3 Rooted rRNA and Protein Trees

Mayr (1998) challenged on phenotypic grounds the posited sister clade relationship between the archaea and the eukaryotes (Woese 1987; Woese et al. 1990; Pace 1997). Woese's response (Woese 1998) was to defend that sister relationship with sequence data, and more generally with an attack on evolutionary reasoning based on phenotypes. The latter was a red herring. The critical issue was how the three domains are rooted relative to each other and not whether genotypic or phenotypic methods are superior. In fact, that rooting ought to be the same in unbiased phylogenetic reconstructions whether they are genotypic or phenotypic.

The initial confusion about the phylogenetic relationship between archaea and bacteria arose because of a flawed approach to the rooting of rRNA gene trees (see Fig. 2.1). At first, rooting of rRNA sequences was deemed impossible, so an alternative strategy was chosen (Pace 1997). Here, a bacterial universal ancestor to the three superkingdoms identified by paralogous (outgroup) rooting of a pair of translation factors was recruited to root the "unrootable" rRNA tree (Woese 1987; Woese et al. 1990; Doolittle and Brown 1994; Baldauf et al. 1996; Pace 1997).

That a paralogous rooting of translation factors in the bacteria (Doolittle and Brown 1994; Baldauf et al. 1996) could be grafted onto an unrooted rRNA tree seemed to be an ingenious solution at the time. But in retrospect, it seems to be a less creative short-cut forced by the reliance on inappropriate sequence evolution models (Gouy et al. 2015; Harish and Kurland 2017b). One reason for skepticism is the textbook demonstration that rooted and unrooted trees for the same taxa may be very different, as shown in Fig. 5.1 (in Nei and Kumar 2000). Furthermore, on the basis of much prior experience (Nei and Kumar 2000), we would expect a priori that two different gene families will not present the same trees (Nei and Kumar 2000; Degnan and Rosenberg 2006; Salichos and Rokas 2013), let alone the canonical species tree, particularly, where one was an RNA gene tree and the other a protein gene tree. In such cases, it is highly unlikely that the mutational history of both could be commensurate. Furthermore, nothing in this protocol addresses the criticism that long branch attraction (LBA) (Forterre and Philippe 1999; Philippe and Forterre 1999) distorted the rooting of the translation factors (Doolittle and Brown 1994; Baldauf et al. 1996). Indeed, after correction for LBA, the tree of paralogs was found to root in a eukaryote-like ancestor not in the bacteria (Forterre and Philippe 1999; Philippe and Forterre 1999). Accordingly, the signature rooting of the three domains adopted by the Woese School (Woese 1987; Woese et al. 1990; Pace 1997) might be no more than an extremely influential phylogenomic artifact.

This regrettable conclusion was verified in great detail after Brinkmann and Philippe went on to study a pair of closely related paralogous translation proteins: one, the 54-kDa signal recognition protein (SRP54) and the other the signal receptor alpha protein (SRα) (Brinkmann and Philippe 1999). Paralogous reconstructions with sequences including the more rapidly mutating sites resulted in the bacterial rooting as expected for phylogeny perturbed by LBA (Brinkmann and Philippe 1999). In contrast, by focusing on the slowly mutating sites, which are presumably the most ancient and least susceptible to LBA, a phylogeny emerged that was rooted in a eukaryote-like ancestor with the archaea and bacteria emerging as sister groups within a clade as shown in Fig. 2.2a. This rooting (Brinkmann and Philippe 1999) provides another independent challenge to the bacterial rooting (Brinkmann and Philippe 1999), and it lends straightforward support for Mayr's (1998) inference that archaea and bacteria are sister clades.

Finally, G. Caetano-Anolles (2002) succeeded to root rRNA trees for both the small and the large subunit species by carrying out cladistic reconstructions based on both the conserved sequences and the evolving secondary structures in the rRNAs. The resulting rooted phylogenies (see Fig. 2.2b, c) describing the two-dimensional structural evolution of rRNAs confirm the association of archaea and bacteria as sister clades distinct from the eukaryote clade (Caetano-Anollés 2002) as noted earlier by Brinkman and Philippe (1999). Again, the universal common ancestor is identified as a taxon presenting rRNA structures more like those of eukaryotes than of bacteria (Caetano-Anollés 2002). Thus, the explicit rooting by an outgroup method identifies an ancestor with strong archaeal, bacterial, and eukaryote affinities for rooting of proteins (Brinkmann and Philippe 1999; Philippe et al. 2011) as well as independently of rRNAs (Caetano-Anollés 2002).

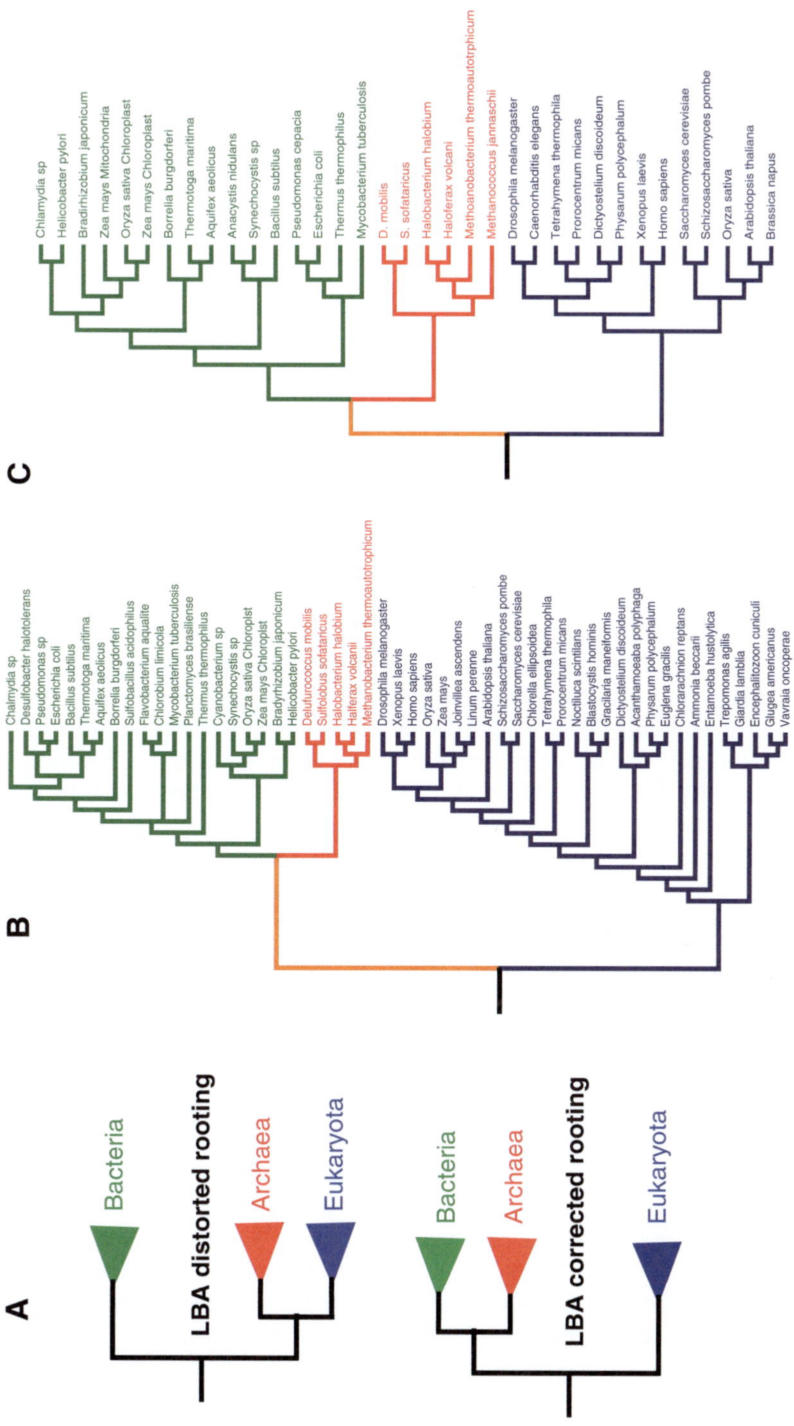

Fig. 2.2 Extrinsic and intrinsic rooting of the rRNA gene tree; (**a**) Ambiguous extrinsic rooting of the rRNA tree based on paralogous protein pairs due to long branch attraction (LBA) artifacts (Brinkmann and Philippe 1999). (**b**) Intrinsically rooted rRNA phylogeny inferred from homologous secondary structure elements in small subunit rRNA and (**c**) large subunit rRNA (Caetano-Anollés 2002)

4 The Unholy Sanctity of Genotype

Woese's concluding summation outlines the core of his dispute with Mayr (Woese 1998) "His [Mayr's] biology is centered on multicellular organisms and their evolutions; mine on the universal ancestor and its immediate descendants. His is the biology of visual experience, of direct observation. Mine cannot be directly seen or touched; it is the biology of molecules, of genes and their inferred histories. Evolution for Dr. Mayr's is an "affair of phenotypes".—For me, evolution is primarily the evolutionary process, not its outcomes" (Woese 1998). We interpret Woese's (1998) attempt to distance his phylogenetic views from those of Mayr (1998) as both his loyalty to his earlier work as well as a vivid statement of the prejudice common to molecular biologists who, following the lead of Kimura (1984), favor genotypic phylogeny. In all fairness, we also note that some of Mayr's supporters (Wheeler et al. 2013) are equally prejudiced partisans of phenotypic methods. Unfortunately, Woese (1998) chose to oppose the factual issue raised by Mayr (1998) concerning the phenotypic similarities of archaea and bacteria by appealing in part to the sanctity of the genotypic approach to phylogeny.

Indeed, Woese argues his brief in two ways that in our opinion are self-defeating. First, Woese did not accept the traditional evolutionary program, which for most biologists is about the evolution of phenotypes, i.e., form and function. Instead, Woese attacks Mayr's view as though that were self-evidently defective because it is "phenotypic" and, therefore, not applicable to microorganisms. Second, Woese argues that molecular studies provide the only option available to students of microbial evolution. The notion that there is an intrinsic opposition between the two approaches is a very common misunderstanding (Woese 1998; Wheeler et al. 2013). Nevertheless, the utility of annotating phenotypic structures from sequence data was already recognized in the literature at the time of the Mayr/Woese confrontation (Brunet et al. 1998; Fontana and Schuster 1998) and is much more commonplace now (Caetano-Anollés 2002; Harish et al. 2013).

The *On the Origin of Species* (Darwin 1859) and any standard work on evolutionary thought (Mayr 1982) recognize "descent with modification" and "natural selection" as the defining processes of Darwinian evolution. DNA replication and mutation are not on that short list. And, for Darwin, evolution was about phenotypes changing under selective pressures. Moreover, it seems that Woese (1998) could accept the Darwinian view of phylogeny as a genealogy of phenotypes for macroorganisms such as plants and animals, but not for the wee beasties, the archaea and the bacteria.

Now, such a division is oddly self-contradictory for a phylogenomic view that aspires to describe the universal common ancestor from which both the microscopic and macroscopic organisms descend. It is hard to believe that Woese would have wanted to argue that selection operates on the phenotypes of macroorganisms, while in contrast it is directly affecting the genotypes of the microorganisms. Instead, we suggest that Woese's dichotomy was a reflexive prejudice that stems from molecular biology's successful preoccupation with DNA sequences. That preoccupation may have suppressed the realization that though protein structures and functions may be more challenging to annotate as phylogenetic characters than are DNA sequences, they do provide access to microbial phenotypes and more generally, they allow the biologist to study the evolution of novelty more inclusively than is permitted by studies of alignment-based gene trees.

Though gene trees may under ideal circumstances or after much curating yield a rough ordering of the evolution of closely related species, that temporal order is not sufficient to describe the evolution of species, of genomes. We also need to account for "what" has evolved, not just when it evolved. But, alignment-based gene trees are too exclusive to report on what evolved. More precisely, alignment-based gene trees are unable to describe what evolves because they subject all the gains and losses, the novelties of genome content, to a ruthless curettage. There are no gene duplications, gene losses, or gene births in gene trees. Consequently, a genotypic model of evolution that is inspired solely by the mechanism of DNA replication for its dynamic is by its very nature an exclusive, conservative model, in which evolution from taxa to taxa can only be intuited from nucleotide substitutions. It is no surprise that there is strong mutual reinforcement between that preoccupation with phylogeny as a record of nucleotide substitutions and the view of evolution as mutation-driven, neutral evolution paced by a molecular clock (Zuckerkandl and Pauling 1965a, b; Kimura 1984; Nei and Kumar 2000).

In distinct contrast, much of what follows is informed by the search for inclusive alternatives to describe the evolution of novel characters in species and to root the corresponding trees. Here the obvious alternatives involve one or another genome content-based approximation to species phylogeny. In these, the structures and functions of expressed gene products become the phenotypic characters of phylogenomics. It may seem awkward to classically seasoned phylogeneticists and taxonomists to refer to a protein or a molecular domain of rRNA as a phenotypic character. However the discomfort with this terminology may fade as the understanding grows that for organisms such as bacteria, molecules and molecular domains are useful phenotypic characters of relevance to molecular evolutionists.

5 More Than One Way to Skin a Genome

Studies of genome content with proteome-implemented phylogenies provide a clear illustration of the differences between molecular phylogenies based on genotype and those based on phenotype (Gibbon and House 1999; Snel et al. 1999; Tekaia et al. 1999). In the latter, carried out around the time of the Mayr/Woese exchange, sequenced genomes were used to identify orthologous proteins that were encoded by the corresponding genomes. The presence or absence (occurrence) or the copy number of each ortholog (abundance) in each of the proteomes inferred from the genome sequences (taxa) was scored. The annotated data were used to implement an unrooted phylogenetic tree based on the corresponding proteomes. Here, the proteomes so defined are characteristic of each genomic taxon so that the resulting tree is an unrooted genome tree based on phenotypic (proteomic) characters, a so-called genome content tree.

The genome content trees (Fig. 2.5a) based on orthologous protein family content (Gibbon and House 1999; Snel et al. 1999; Tekaia et al. 1999) and the rRNA trees (Caetano-Anollés 2002) are in gratifying agreement with the genotypic results of Brinkmann and Philippe (Philippe et al. 2011). In particular, they show that different average evolutionary distances to eukaryotes do not distinguish archaea and bacteria. In effect (Fig. 2.2), the archaea and bacteria may be together considered a clade as convincingly emphasized by Dujon et al. (Tekaia et al. 1999).

In principle, a comparable genotypic approach to phylogenetic reconstructions might be to choose a very large multigene concatenation of the orthologous protein-coding sequences among the sequenced genomes and to use techniques such as those employed by Philippe et al. (Brinkmann and Philippe 1999; Forterre and Philippe 1999; Philippe and Forterre 1999; Philippe et al. 2011) to carry out an alignment-based comparison of compositional variation in the multigene concatenation of coding sequences to generate a phylogeny based on sites with the lowest mutation frequencies. All other things being equal, the congruence of the two phylogenies would probably depend very much on the choices of sites for the evaluation of mutation frequencies in the multigene concatenation (Brinkmann and Philippe 1999; Forterre and Philippe 1999; Philippe and Forterre 1999; Tourasse and Gouy 1999; Rokas and Carroll 2006; Philippe et al. 2011; Salichos and Rokas 2013). In principle, the same phylogenies should be obtainable in both cases if there are no selective distortions or no confusing discontinuities caused by duplicated paralogs or by gene loss in the evolutionary record of the alignments.

Still, there are several important ways that the results would be distinguishable for the concatenation-based gene trees and the genome content trees: one is that the three-dimensional structures of individual orthologous proteins would tend to be conserved over long evolutionary distances, while the corresponding coding sequences for those persistent structures would fluctuate significantly in an erratic, unrecognizable pattern. The consequence of those fluctuations is that only a minority of the structurally and functionally homologous coding sequences could be recognized by BLAST searches (Murzin et al. 1995; Gough et al. 2001; Illergård et al. 2009; Worth et al. 2009; Harish et al. 2013). Thus, for the homologous proteins selected within a lineage of proteomes, each taxon contains somewhat different coding sequences for the same conserved protein structure.

Another significant difference is that no gene acquisitions or gene losses are recorded in the augmented gene tree. In other words, the genomic changes that define the proteomes of species are absent from gene tree reconstructions. This means that at best only the mutational tempo of evolution of a minor fraction of a genome can be deduced from gene trees (Kimura 1984; Nei and Kumar 2000), while the species-defining phenotypic changes are left to genome content trees.

Such practical complications ought to balance the prejudices that molecular biologists typically mount against phenotypic approaches using structural characters to study phylogeny. We will take these up again, but first a key aspect of molecular "phenotypes" requires clarification. This is the relationship between the dimensionalities of structures and the degeneracy of DNA sequences that encode those structures, which is a clue to why many homologous protein structures are not identifiable in BLAST searches (Theobald and Wuttke 2005; Pethica et al. 2012).

5.1 The Dimensionality of Molecular Phenotypes

Much of the phenomenal progress that molecular biology has recorded in the latter half of the previous century was a consequence of pursuing the expression of DNA-encoded information in collinear RNA and protein molecules. This was the pursuit of life in one dimension, and of course that pursuit fixes attention on genotypes. In contrast, Darwin and Mayr (Darwin 1859; Mayr 1982) as well as many in between have had higher dimensional aspirations. After all, the construction of a genealogy of phenotypes involves, at the very least, the evolution of genetic information that is expressed in the real structures of macromolecules as well as in their dynamic interactions within the cellular context (Brunet et al. 1998; Fontana and Schuster 1998). Thus, evolutionary biologists must reckon with multiple dimensions of biological space that are encoded in genomes. This suggests that at a minimum the three-dimensional structure of coded gene products along with their functions will be of interest for a genealogy of phenotypes. After all it is the structures and functions that are selected and conserved.

The space of RNA secondary structures provides a simple illustration of a critical "informational" feature that arises in transcribing genetic information into the two-dimensional structures of say an rRNA (see Fig. 2.3). Imagine, if you will, that the phylogenetic character of interest is the length of helical domains in rRNA and that the evolutionary model tracks in time the increase or decrease in helical lengths of the rRNA, as discussed in Caetano-Anollés (2002). Here, the features of rRNA secondary structure are the phylogenetic characters, and the underlying evolution model describes the probabilities of changes between the different states of rRNA structure (Caetano-Anollés 2002).

Assume that the base pairs that can in principle stabilize helical domains are four in number (AU, UA, GC, CG) (Fig. 2.3). Then, the number of permutations of coding sequences in one dimension for N-base pairs is 4^N, which for any rRNA molecule is a very large number. This means that a priori there are potentially a very large number of genetic sequences that can code rRNA molecules with the same number and distributions of helical structures. Only by identifying the conserved distributions of annotated helical structures in rRNA molecules was Caetano-Anolles (2002) able to reconstruct phylogeny on the basis of the evolution of lengths of those structures.

The multiplicities of alternative amino acid sequences are considerably larger for the specification of alpha helices of aggregate length N. That means that in translating the nucleotide sequences encoding a protein characterized by N alpha helical residues there could be a very large number of alternative equivalent nucleotide sequences. Likewise, for beta sheet translation, there would be many alternative-coding sequences. Accordingly, without annotation for the positions of alpha helices and beta sheets as well as the tertiary structures they form, sequence alignments would not detect all of the similar structures that are encoded by the highly degenerate one-dimensional nucleotide coding sequences (Darwin 1859; Kimura 1984). In other words, comparisons of sequence alignments are expected to significantly underestimate the numbers of protein structure homologues encoded by degenerate genome sequences (see Fig. 2.3).

A library of Hidden Markov Models (HMMs) has been trained on atomic level structures of proteins obtained by X-ray crystallographic and NMR spectroscopic methods (Murzin et al. 1995; Gough et al. 2001; Worth et al. 2009). This library (the SCOP and SUPERFAMILY databases) can be used to recognize the folding propensities of amino acid sequences, so that the similarities among tertiary structures of protein domains become facilely recognizable though they are for the most part invisible in raw sequence alignments (Doolittle 1995; Murzin et al. 1995; Gough et al. 2001; Illergård et al. 2009; Worth et al. 2009; Harish et al. 2013). Thus, stable tertiary folds of specific groups of structurally equivalent superfamilies (SFs) are encoded by degenerate amino acid sequences that may be, in the extreme, related by reduced sequence similarities: as low as 15% sequence identity (Doolittle 1995, 2005; Murzin et al. 1995; Gough et al. 2001; Gough 2005; Yang et al. 2005; Illergård et al. 2009; Worth et al. 2009; Zmasek and Godzik 2011; Harish et al. 2013). Nevertheless, all the sequences that encode the same tertiary structure of a single SF are annotated as the coding sequences of structural homologues. Accordingly, the frequencies of annotatable structural similarities among compact protein domains at the level of SFs from the SCOP database are three- to tenfold higher than those recognizable in BLAST searches (Doolittle 1995; Murzin et al. 1995; Gough et al. 2001; Illergård et al. 2009; Worth et al. 2009; Harish et al. 2013)!

This very considerable discrepancy constitutes an important reason to favor compact protein domains as characters for genome content-based phylogeny: the network of homologous characters discernable by annotation of the tertiary folds of compact protein domains is much more extensive than the matrix of similar coding sequences detectable in sequence alignments. And, it is worth recalling that for the same reason, compact protein domains are phylogenetic characters that are robust to LBA or mutational saturation of ancient sequences precisely because they are supported by highly degenerate sequences. Thus, multiple mutational substitutions of an alignment sequence lower the recognition of mutant sequence homologies in BLAST searches and may also generate false similarities resulting in LBA. In contrast, the same substitutions have a smaller impact on the recognition of 3D homology encoded by a multiplicity of degenerate sequences that encode the same selected 3D fold structure (Doolittle 1995, 2005; Murzin et al. 1995; Gough et al. 2001; Gough 2005; Yang et al. 2005; Illergård et al. 2009; Worth et al. 2009; Wang et al. 2011; Zmasek and Godzik 2011).

Principal component analysis (PCA) projections as shown in Fig. 2.4 describe genome content of SFs in such a way that each genome is identifiable as a unique statistical distribution of SFs in a multidimensional statistical space (Harish et al. 2013). In such displays (Fig. 2.4), the proteomes of SFs characteristic of individual

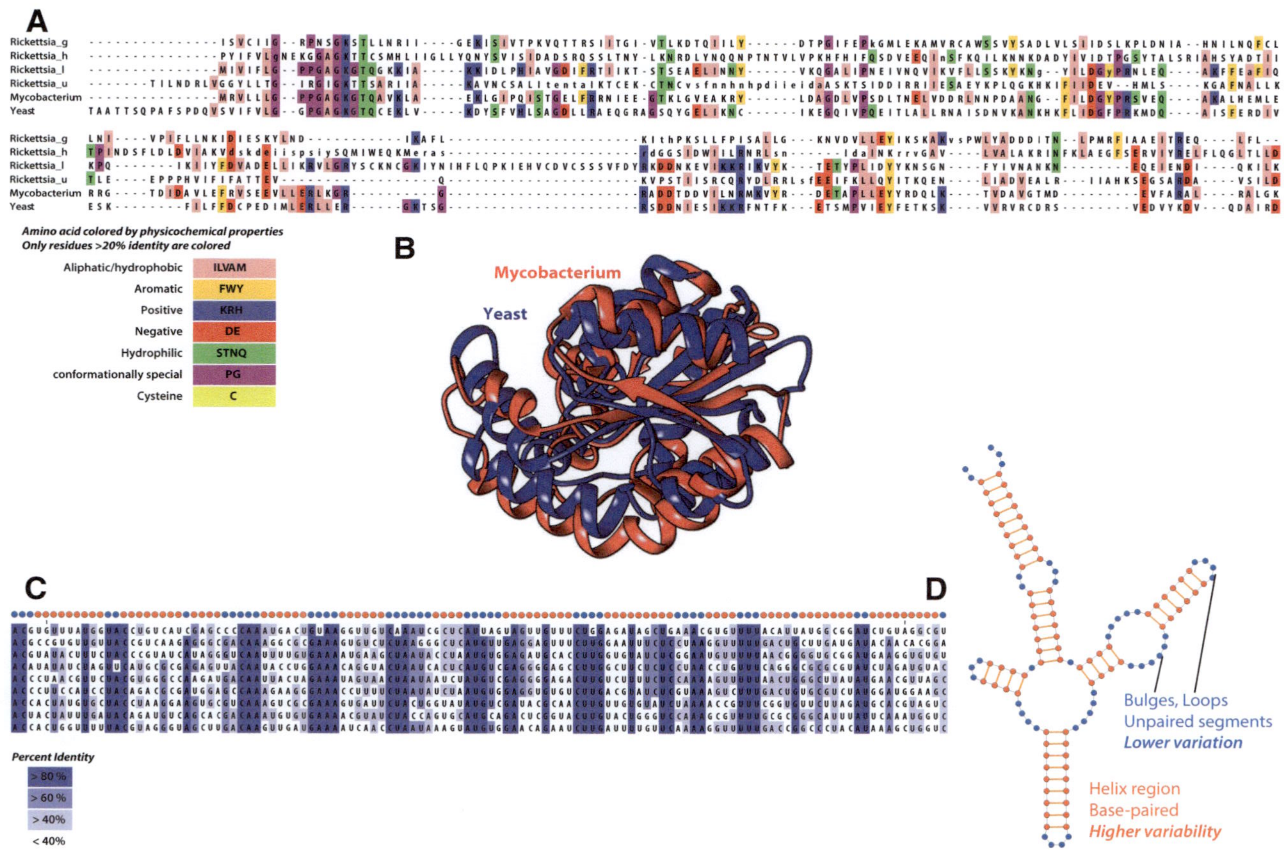

Fig. 2.3 Degeneracy of sequences that support a unique structural conformation. (**a**) Shows the variability of amino acid composition and length of protein domains corresponding to a unique 3D fold (**b**) in the P-loop containing the nucleoside triphosphate hydrolase domain. (**c**) Illustrates the degeneracy of rRNA sequences supporting the conserved secondary structure (**d**). Higher sequence variability is observed in the helical segments supported by the paired nucleotides. In contrast, there is lower sequence variability in the unpaired regions

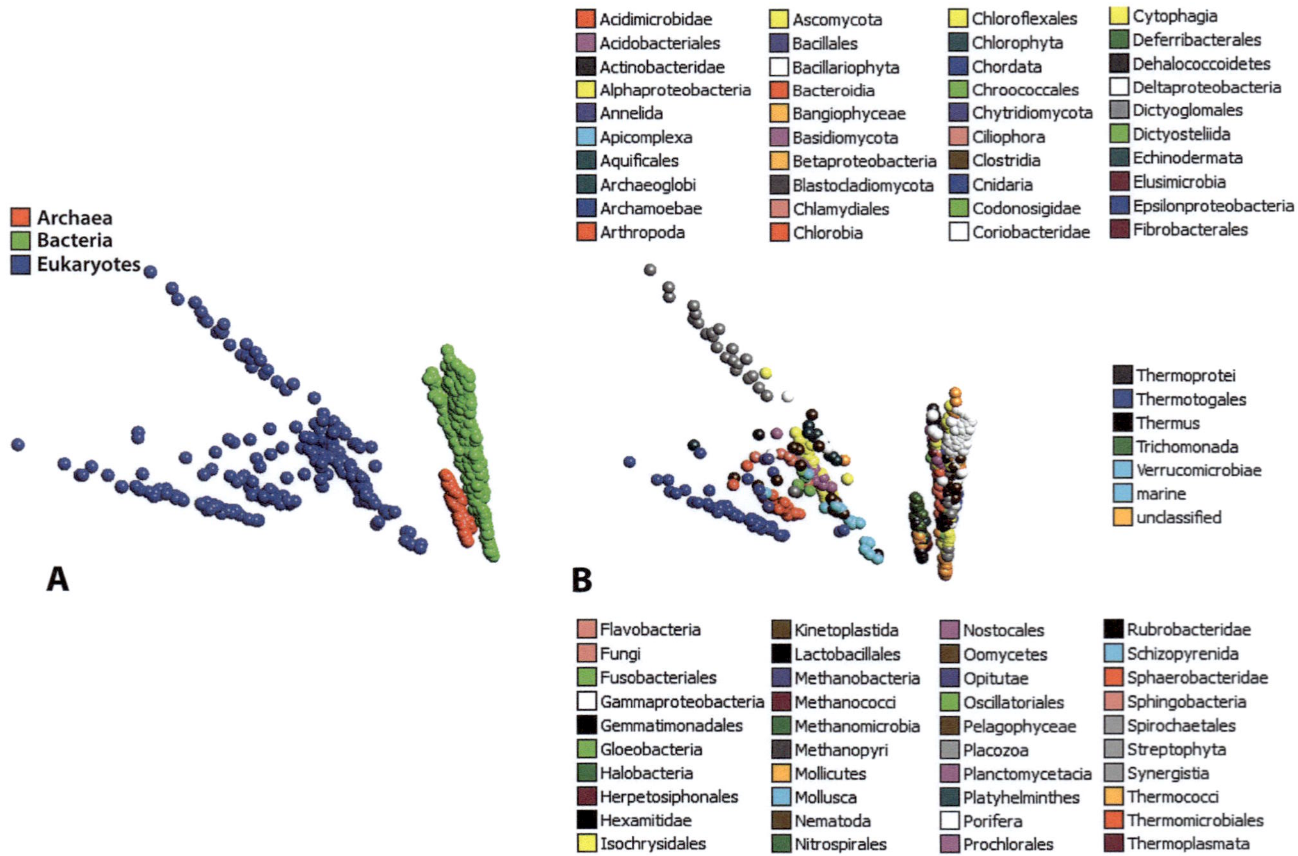

Fig. 2.4 In this principal component analysis (PCA) projection of SF domain distributions in sequenced genomes, each sphere represents a unique named species, 73 archaea, 682 bacteria, and 274 eukaryotes. (**a**) The groups of species are colored according the taxonomic superkingdoms and (**b**) according to phyla. The clustering of species groups at different taxonomic ranks shows that species can be described by their unique genomic SF domain content

genomes (taxa) are visualized as distinct phenotypic characters. The characteristic statistical distributions of SFs in PCA displays bring together individual genomes in neighborhoods of related taxonomic groups such as superkingdoms or phyla (see Fig. 2.4). Accordingly, the PCA projections illustrate the potential value of utilizing genome contents of SFs as phenotypic characters in phylogenetic reconstruction (Harish et al. 2013).

6 Phylogenies of Protein Domains

A steady stream of structural determinations accompanied the accumulation of genome sequences, both of which accelerated after the Mayr/Woese exchange (Mayr 1998; Woese 1998). Recently there were more than 80,000 protein structures at atomic resolution accessible in the Protein Data Bank. Each protein may present one or more phylogenetically exploitable compact 3D domain. These 3D domains recur both within the same and different genomes (Doolittle 1995; Murzin et al. 1995; Gough et al. 2001; Yang et al. 2005; Illergård et al. 2009; Worth et al. 2009; Harish et al. 2013). The 3D folding patterns of the domains are represented at several overlapping levels in the SCOP database (Doolittle 1995; Murzin et al. 1995; Gough et al. 2001; Yang et al. 2005; Illergård et al. 2009; Worth et al. 2009; Harish et al. 2013). Here, we focus on compact protein domains at the SF level of the SCOP hierarchy because these domains are demonstrably well behaved in phylogenetics, as shown in Fig. 2.5c (Doolittle 1995, 2005; Murzin et al. 1995; Gough et al. 2001; Yang et al. 2005; Illergård et al. 2009; Worth et al. 2009; Harish et al. 2013).

Very briefly, SFs alternate with disordered, non-domain regions of proteins that are more variable in structure and sequence length; the latter are called "linkers" (Wang et al. 2011). For the longer proteins of the eukaryotes, the SF to linker ratios are circa 1:1, while for the shorter length akaryote proteins, the ratios are closer to 3.5:1 (Kurland et al. 2007; Wang et al. 2011). However, we note that orthologous SFs of akaryotes have the same lengths as their homologues among the eukaryotes (Kurland et al. 2007; Wang et al. 2011). Evidently, the SFs are more conserved by structural constraints. In contrast, the linkers of homologous proteins present characteristic lengths that identify them as proteins of eukaryotes or akaryotes (Wang et al. 2011). Those recently discovered features of proteins resonate with the phylogenetic sorting of organisms into eukaryote and akaryote empires suggested by Mayr (1998). Further, the distinguishing linker lengths for the proteins of the two empires suggest that each has evolved under different average intensities of reductive evolution (Kurland et al. 2007; Wang et al. 2011; Harish et al. 2013), to which we return below.

Once it became possible to routinely annotate genome sequences to identify the structural elements that they encode, those same sequences that had previously been used for alignment-based gene families became phenotypic resources to describe the tertiary folds of SFs (Doolittle 1995, 2005; Murzin et al. 1995; Gough et al. 2001; Gough 2005; Illergård et al. 2009; Worth et al. 2009; Wang et al. 2011; Zmasek and

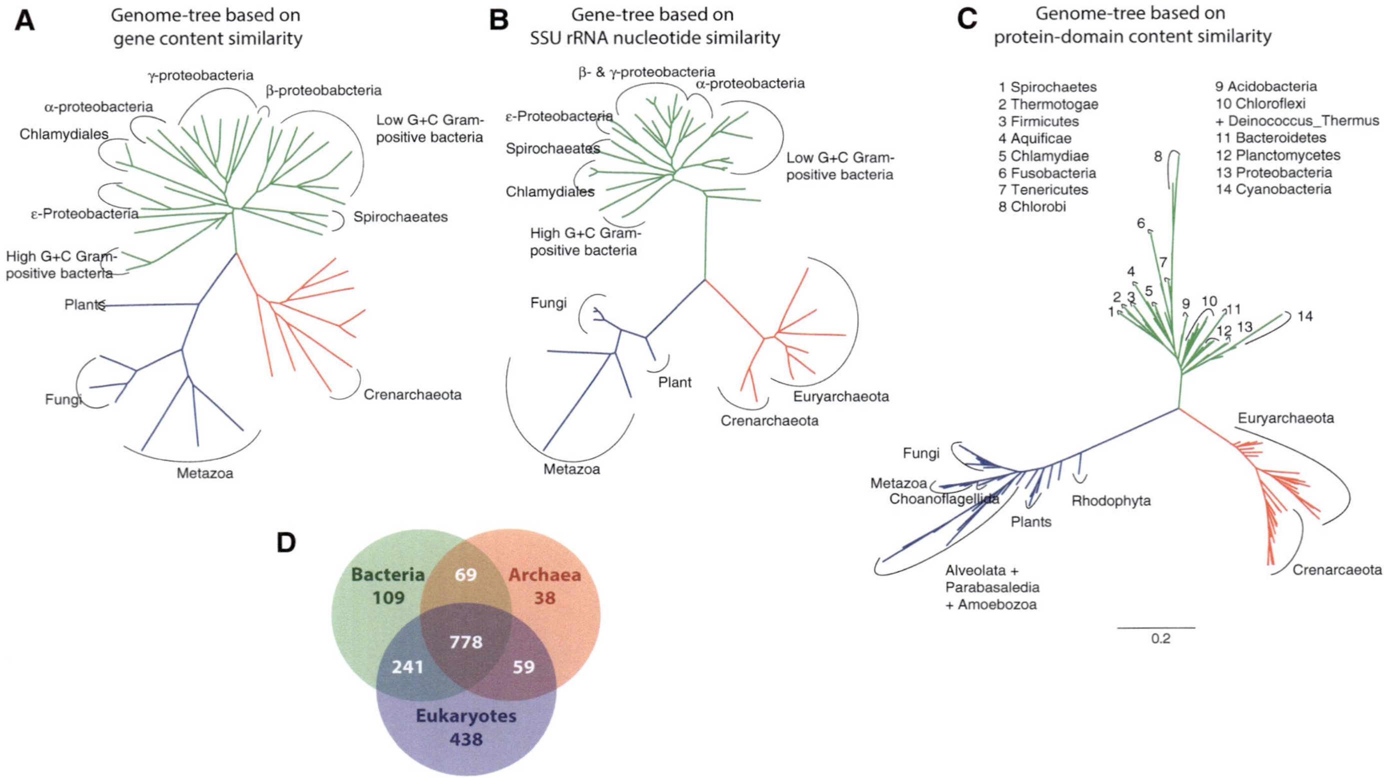

Fig. 2.5 Similarity of unrooted trees inferred from different data and different methods of tree reconstruction. (**a**) Neighbor-joining genome tree based on shared gene content in 50 species and (**b**) neighbor-joining gene tree based on rRNA sequence similarity of the species in (**a**). (**c**) Bayesian inference of a genome tree based on shared protein domain (SF) content in ~150 species sampled from ~150 unique genera. (**d**) Venn diagram depicts the extent of sharing of the 1732 homologous SFs among the 3 superkingdoms based on the species sampled in (**c**). All tree clusters of species are identified according to the phylum level classification. Individual species names for trees in A and C are described in Korbel et al. (2002) and for trees in (**c**) are described in Harish et al. (2013) as well as in Fig. 2.6

Godzik 2011; Harish et al. 2013). Thus, the annotated sequences from 141 sequenced genomes, 47 from each superkingdom, identify 1732 novel SFs of which 778 are annotated as shared by all three superkingdoms as shown in Fig. 2.5d. This degree of structural similarity between the proteomes of the three superkingdoms is invisible to BLAST searches. Furthermore, that half of the SFs that is shared by all three superkingdoms immediately suggests that a universal common ancestor of the three superkingdoms had a complex proteome with a minimum of half of all the unique SFs presented by modern crown organisms (Harish et al. 2013).

Comparisons of novel SFs shared between proteomes show that the cohort shared between archaeal and eukaryote proteomes is circa 0.55 (837/1516 SFs), while the bacterial cohort shared with eukaryotes is circa 0.67 (1019/1516 SFs) in modern organisms (Harish et al. 2013). These fractions are inconsistent with the idea that archaea and eukaryotes make up a sister clade that excludes bacteria (2, 16–20). Furthermore, rooted phylogenies of the SF-based proteomes from 141 broadly sampled species identify archaea and bacteria as sister clades distinct from the eukaryotes (Harish et al. 2013); see Fig. 2.6. Thus, the genome phylogenies confirm the previous results identifying distinct akaryote and eukaryote clades (Brinkmann and Philippe 1999; Tekaia et al. 1999; Caetano-Anollés 2002; Philippe et al. 2011), precisely as argued by Mayr (1998).

The phylogenies reconstructed from genome content of SF domains (Harish et al. 2013) are roughly but gratifyingly similar to those obtained by Caetano-Anolles for the structural evolution of rRNA (Caetano-Anollés 2002). The reconstructions identify three ancestral nodes: first, a most recent universal common ancestor (MRUCA) and then two others that emerge in parallel from MRUCA that are specific for akaryotes and eukaryotes, respectively. This characterization of the two ancillary ancestral nodes also supports Mayr's (1998) view that modern organisms may be divided meaningfully into two separate empires (Fig. 2.6).

Analysis of the proteomes encoded by these three ancestors as well as those of the crown clades in the relevant sampling suggests that MRUCA, though neither a bacterium nor a eukaryote, presents a very complex proteome that is the common ancestor to all three superkingdoms (Harish et al. 2013). That complexity is defined by a proteome of distinct SFs that corresponds to three fourths of all those found in all the sampled crown clades (Fig. 2.7). In addition, roughly 60% and 40% of those novel SFs are lost in the descent to the akaryote and eukaryote crown clades, respectively. However, in the same trajectories, the retained SFs are duplicated by between 4-fold and 20-fold, respectively. Thus, SF duplication and recombination are major molecular process in the acquisition of new genes by most genomes, but as seen in Fig. 2.7, these are more pronounced in the genomes of eukaryotes (Harish et al. 2013).

On the other hand, a bit less than one quarter of the crown SFs are unique characters not found in the ancestral proteomes (Harish et al. 2013). Thus, the birth of novel structures is a relatively minor pathway of genome evolution from MRUCA to the modern crown. In contrast, reductive evolution is quite evident in the trajectory to the modern crown. The loss of 60% and 40% of SFs from akaryotes and

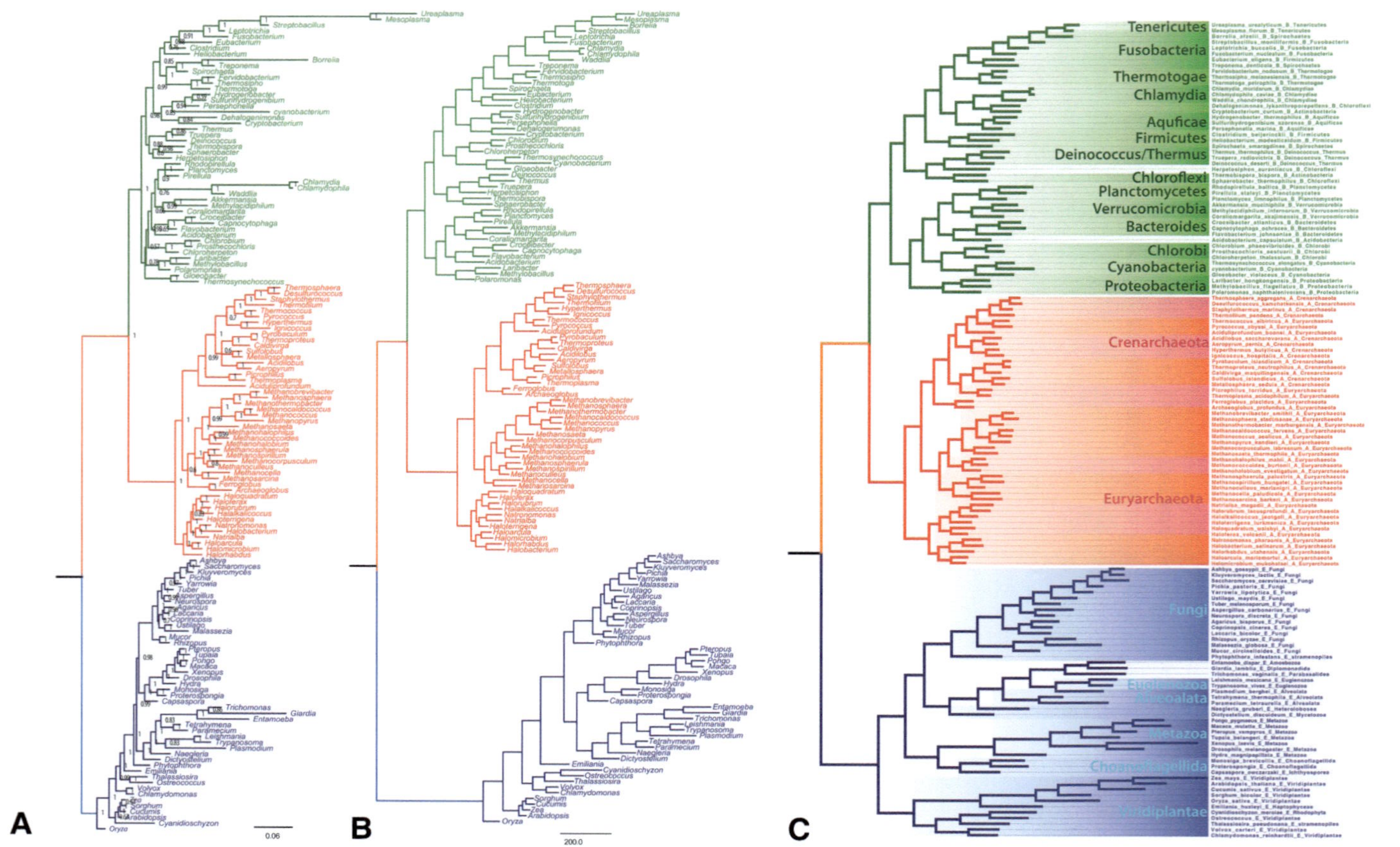

Fig. 2.6 Phylogeny based on genome content of SFs from ~150 species. Intrinsically rooted phylogeny inferred from (**a**) Bayesian analysis of unique SF content (SF occurrence), (**b**) empirical Sankoff parsimony. (**c**) Phylogeny inferred from SF abundance using the empirical Sankoff parsimony approach. Branch lengths correspond to the relative rate of gain and loss of SFs in (**a**), but in (**b**) and (**c**) branch lengths correspond to the sum of gains and losses of SFs

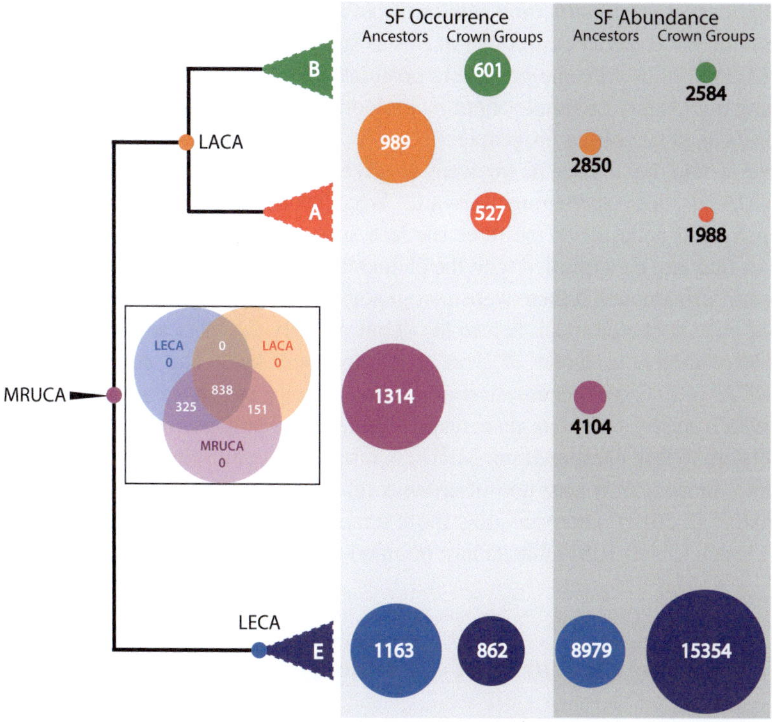

Fig. 2.7 SF domain content of the common ancestors based on the most parsimonious ancestral state reconstruction. Both SF occurrence (light gray panel) and SF abundance (dark gray panel) were reconstructed for MRUCA, LECA, and LACA. SF occurrences and abundances of ancestors are shown in the left side of each panel and that of the crown species in the right side of the panel. The numbers for the crown organisms are averages of occurrences/abundances for species in each superkingdom analyzed in Fig. 2.6. *MRUCA* most recent universal common ancestor, *LACA* last akaryote common ancestor, *LECA* last eukaryote common ancestor

eukaryotes, respectively, would be inconceivable without a universal common ancestor that was as complex as the MRUCA.

The complex pattern of gain and loss of SFs attending genome evolution of the crown organisms would completely elude molecular phylogeneticists exclusively fixated on gene tree constructions from sequence alignments. That is to say, contrary to Woese's (1998) assertion, a pure genotypic approach to phylogeny provides at best a bare bones temporal scheme for evolution and even that is not too reliable in detail (Degnan and Rosenberg 2006; Salichos and Rokas 2013).

Finally, for want of space as well as for lack of direct relevance, we cannot devote much attention to the concept of the progenote that so engaged Woese (1998). As we understand it, the progenote was the primitive launching pad for the evolution of the structural, replicative, and metabolic molecular repertoires eventually shared by crown organisms. In general, the characteristic cellular phenotypes of modern organisms would not prefigure in the proteome of the progenote, out of which these eventually evolved. In contrast, MRUCA is so complex that the functional profiles as well as the structural similarities of its proteome approach an aggregate of 75% of the total modern ensemble of SFs as summarized in Fig. 2.7. It is difficult to identify MRUCA with the progenote postulated by Woese (1998) because there is nothing elementary or simple about its proteome, but then there is nothing simple in Woese's sketches of the progenote.

We would be remiss if we were to give the impression that the descent of MRUCA is clear and straightforward. What seems clear is that MRUCA, the reconstructed ancestor of the three modern superkingdoms or of the two empires, may or may not be identified with the elemental first cell that Woese (1998) referred to as the "progenote." If there were a progenote in the early evolution of organisms, it would have appeared much before the debut of MRUCA. It is anyone's guess how long the progenote mode of gene exchange persisted. The suggestion is that MRUCA is a bottlenecked survivor of a previous crown of organisms, which survived a nearly complete planetary environmental collapse (Harish et al. 2013). According to this interpretation, MRUCA reseeded the biosphere and, by so doing, became the root of a new tree of crown organisms when the biosphere recovered (Harish et al. 2013). There are other more sensational scenarios such as seeding from outer space (Crick 1981) that cannot be ruled out at present.

7 Preserving the Purity of the Lineage

A defining feature of Woese's progenote was a hypothetical inclination to indulge in horizontal gene transfer (HGT) (Woese 1998, 2000). Woese suggests that such an inclination would qualify progenote evolution for the special status of "non-Darwinian" evolution (Woese 1998, 2000). So, what characteristics identify HGT as non-Darwinian? The short answer is nothing does! If Darwinian evolution is taken to be "descent with modifications and natural selection" (Darwin 1859; Mayr 1982; Penny 2011), HGT is most certainly Darwinian (Penny 2011) because it is evident that selective forces control fixation of HGT in populations (Berg and Kurland 2002; Kurland and Berg 2010). Indeed, HGT is just one sort of novel sequence acquisition event that is a normal but admittedly a non-Mendelian component of genome evolution. It may be unappealing to some, but HGT seems to make little difference to the phylogeny of the three superkingdoms (Gibbon and House 1999; Snel et al. 1999; Tekaia et al. 1999; Harish et al. 2013) probably because HGT is often transient and patchy as well as infrequently fixed in populations (Berg and Kurland 2002; Kurland and Berg 2010; Harish et al. 2013).

Shown in Fig. 2.4 is a valuable reality check on the phylogenomic disorder that has been attributed to HGT (Doolittle 1999). These PCA projections of SF proteomes show that the genome content of SFs from a broad diversity of species is well resolved into unique distributions of superfamilies that define each sequenced genome as a unique taxon suitable for the reconstructions of phylogeny. In light of such data, we find that it is difficult to translate an appreciation of the sterility of gene trees into the rhetoric of "How Bacterial Species Form and Why They Don't Exist" (Doolittle 2012).

On the other hand, a transferred gene undeniably creates a discontinuity in gene trees, which do not accommodate the gain or loss of any particular sequence in the evolution of genomes (Kurland and Berg 2010; Harish et al. 2013). In effect, gene trees present a quasi-static antithesis to genome evolution because they do not describe any evolutionary novelties more complex than nucleotide substitutions. Nucleotide substitutions are potentially useful to define the mutational tempo of evolution, but they cannot describe the novel modes that truly define species (genome) evolution such as gene duplication, gene loss, and gene birth.

Indeed, that limitation was nearly grasped by W. Ford Doolittle in his hortatory essay on the "paradigm shift" (Doolittle 1999). There, he flirts with a critical reader: "Perhaps it would be easier, and in the long run more productive, to abandon the attempt to force the data that Zuckerkandl and Pauling stimulated biologists to collect into the mold provided by Darwin." (Doolittle 1999). Regrettably, these prescient thoughts faded from the literature on HGT, but the false image of Zuckerkandl and Pauling (1965a) as solely interested in genotypes remained. Their early preoccupation with protein structure and its influence on mutation rates was conveniently erased from the record.

The rampant HGT paradigm shift (Doolittle 1999) replaces an inclusive respect for the realities of Darwinian evolution with a molecular fetish: Here, evolution is depicted as a model of DNA replication, and the possibilities of studying the phylogenomics of species evolution are limited to what can be gleaned from nucleotide substitutions. This limitation arises because as Doolittle asserts "After Zuckerkandl and Pauling (1965b) biologists came to think that the universal tree could be reduced to a tree based on sequences of orthologous genes, any of which (practical considerations aside) could serve as a marker for an entire genome, organism, or species." In fact, the only biologists that may have held that extreme opinion were dyed in the wool neutralists who were committed to an evolutionary biology in which nothing other than nucleotide substitutions happens. And, it would not be an exaggeration to suggest that the thrust of work on gene tree phylogeny since 1999 (Degnan and Rosenberg 2006), which culminates in Salichos and Rokas (2013), has been to illustrate the proposition that a gene tree is rarely a faithful description of species phylogeny (Salichos and Rokas 2013).

The preoccupation with HGT may have seemed fitting just before the millennium shift because then every glitch and hiccup in gene trees was reflexively identified with HGT (Doolittle 1999). But, in truth, it is no mean feat to distinguish paralogous duplications, or parallel evolution, from HGT particularly when mutation rates are biased and variable (Kurland 2000; Kurland and Berg 2010). Recall that duplications make up the majority of compact protein domains (e.g., 10). Accordingly a combination of high duplication rates with irregular mutation rates might generate a large fraction of HGT look-alikes.

In effect, a misunderstanding about what is and what is not Darwinian evolution coupled to the failure to recognize the limited scope of genotypic phylogenies combined to make acceptable the assertion that HGT "compromises the definition of taxa at all ranks" (Doolittle 1999). Likewise, Woese's suggestion that for the progenote "organismal genealogies probably had no meaning at the time when the domains formed" (Woese 1998) is incomprehensible. Unfortunately, such forced misinterpretations of HGT were just the beginning of the phylogenetic razzle-dazzle that became a preoccupation with the "new paradigm," rampant HGT (Doolittle 1999; Doolittle and Zhaxybayeva 2013). It might have been more beneficial all around had the new paradigm been informed by the new perspectives that genome sequence data opened as well as the obvious differences between a gene tree and a species tree (Harish et al. 2013).

It is worth emphasizing that it is in fact alignment-based gene trees that are not Darwinian! In contrast, HGT is as Darwinian as any process that involves descent with modifications (Nei and Kumar 2000; Berg and Kurland 2002; Penny 2011; Wheeler et al. 2013). Of course, inclusive genome content trees are needed to account for novel evolution in the form of gene duplications, gene loss, and novel sequence acquisitions, including HGT and parallel (convergent) evolution. Generally the frequencies of the last two classes of novel acquisitions are overestimated by confusion with gene duplication combined with paralogy in alignment-based analyses. Indeed, it was understood from the start of the rampant HGT fashion that reticulate phylogenies provide a particularly friendly window through which to track novel sequence acquisitions in detail (Kurland 2000). Indeed, novel gene acquisitions or losses become useful phylogenetic markers if they persist (Abby et al. 2012). Furthermore, they are valuable identifiers of patches or ecotypes within a global population (Berg and Kurland 2002). In our view, HGT is just one of a number of gene tree glitches that emerge from genome content trees as a natural (Darwinian) dimension of genome evolution (Berg and Kurland 2002; Penny 2011; Abby et al. 2012; Harish et al. 2013). HGT happens, as do duplications, gene loss as well as gene birth and parallel evolution.

8 Novel Protein Discovery

Woese never abandoned his disdain for phenotypic phylogeny or explained why he feared that phenotypic phylogeny might place archaea in a phylogenetic relationship different from that supported by gene trees (Woese 1998). But there is an influential, unwritten mythology that gene tree *aficionados* recruit nowadays to undermine confidence in genome content trees: This suggests that compact domains are unstable and continually morph from one tertiary fold to another in response to simple nucleotide substitutions in their coding sequences. In effect, this urban myth suggests that protein domains are not good phylogenetic characters, while a nucleotide sequence alignment is the gold standard. Remarkably, this prejudice has been shielded from the realities of erratic mutation rates and gene tree incongruences (Forterre et al. 1992; Philippe and Laurent 1998; Brinkmann and Philippe 1999; Forterre and Philippe 1999; Philippe and Forterre 1999; Tourasse and Gouy 1999; Mossel and Steel 2004; Degnan and Rosenberg 2006; Rokas and Carroll 2006; Penny and Collins 2010; Philippe et al. 2011; Salichos and Rokas 2013). Furthermore, it is inconsistent with the fact that well-curated gene trees (Brinkmann and

Philippe 1999; Philippe et al. 2011) and trees based on evolving secondary structures of rRNAs (Caetano-Anollés 2002), genome content trees based on protein families (Gibbon and House 1999; Snel et al. 1999; Tekaia et al. 1999), and genome content trees based on SFs (Harish et al. 2013) all identify archaea and bacteria as sister clades within the same three superkingdom phylogeny. Quite a coincidence for unsuitable phylogenetic characters!

The amino acid sequences that spontaneously fold into the stable structures of compact protein domains have been selected through their influence on the folding process itself (Grishin 2001; Oliveberg and Wolynes 2005; Haglund et al. 2008; Danielsson et al. 2013). Experimental studies in which amino acid sequences that encode compact protein domains are altered reveal that those sequences are selected by their capacity to support an intramolecular phase transition in which a hydrophobic core is stabilized by specific secondary structures that restrict the backbone configurations of the amino acid sequences (Grishin 2001; Oliveberg and Wolynes 2005; Haglund et al. 2008; Danielsson et al. 2013). This folding process is described as a "minimally frustrated folding funnel" to illustrate the smooth energetically favored phase transition to the folded state. The resulting structure is a unique one because the sequences so selected, though degenerate, simply do not support alternative folds (Grishin 2001; Oliveberg and Wolynes 2005; Haglund et al. 2008; Danielsson et al. 2013).

According to this conception, it is the physical chemistry of compact protein domains that guarantees that they are stable phylogenetic characters (Fig. 2.3b). Thus, while the sequences that encode a given compact domain are multiple, the tertiary fold of a domain is within limits unique and rarely morphs into another tertiary fold (Doolittle 1995, 2005; Gough et al. 2001; Gough 2005; Illergård et al. 2009). Exceptions are identifiable as instances of convergent evolution (parallel evolution), but they are rare (Gough 2005; Theobald and Wuttke 2005; Castoe et al. 2010; Pethica et al. 2012). The observed stability of domain tertiary fold is an expression of the rigors of the posttranslational folding process: when this process is interrupted, for example, by mutations or experimental alteration of the sequences, the folding process fails (Grishin 2001; Oliveberg and Wolynes 2005; Haglund et al. 2008; Danielsson et al. 2013). Failure to adopt a canonical domain fold results in polypeptide entanglement, precipitation, and proteolysis by ubiquitous, dedicated chaperonin *cum* protease systems (Kurland 2010; Kurland and Berg 2010). Thus, cells normally do not tolerate unfolded and misfolded polypeptides. This means that mutant structures that might otherwise provide a transition from one tertiary fold to another are rarely tolerated.

How then do new activities and functions arise? The invention of novel compact protein domains de novo is not very frequent (Harish et al. 2013) probably for the same reasons that compact domains are stable characters and do not easily morph from one tertiary fold to another (Grishin 2001; Oliveberg and Wolynes 2005; Haglund et al. 2008; Kurland 2010; Danielsson et al. 2013). An alternative that was popular shortly after the Mayr/Woese confrontation was the appeal to HGT as the source of novel activities (Ochman et al. 2000). But from the evolutionary perspective relying on exchanged activities begs the question of how these activities arose in the first place. There is, however, a quite serviceable mechanism that has recently been directly verified by experiment: this is an updated version of Ohno's venerable idea that novel protein functions arise through gene duplications and paralogy (Ohno 1970; Bergthorsson et al. 2007; Näsvall et al. 2012; Andersson and Näsvall 2013).

This selection mechanism for novel functions takes off from the recognition that enzymes normally mediate at slow rates a spectrum of dissipative reactions that represent novel (noncanonical) departures from their normal functions (Copley 2003). Under a selective regime for such a dissipative reaction, duplications will be favored to meet the novel selective demands on the cell for the novel product. Once multiple copies of the enzyme are accumulated, mutation of these may create more kinetically effective paralogs that are eventually selected as independent alleles. This well-defined selection mechanism, the innovation-amplification-divergence (IAD) model, has been verified for two amino acid biosynthetic reactions (Bergthorsson et al. 2007; Näsvall et al. 2012) as well as for nonenzymatic functions (Deng et al. 2010).

The IAD model and especially its supporting data provide convincing verification of Ohno's suggestion that gene duplications are the starting point for the selection of paralogous structures to support the selective evolution of novel functions (Bergthorsson et al. 2007; Deng et al. 2010; Näsvall et al. 2012). Of course, neutralists who have argued that duplications are neutral events will want to suggest that the IAD model does not exclude a role for neutral duplications. That may be a valid observation. Nevertheless, that objection does not establish by default a role for neutral duplications in genome evolution of novel functions. Data is needed to support the neutral assertion. That is to say something more than logically consistent population genetic theory. As we note in the next section, an active genetic sequence that supports transcription and translation is not a neutral sequence. Accordingly, we cannot take for granted that gene duplications are neutral.

It remains to point out that gene duplications are supplemented by recombination events in which different domains can be brought together in novel arrangements to create novel activities. According to the tabulations of Wang et al. (2011), there are circa 30% of unique proteins that present multiple domains in all three superkingdoms. Accordingly, novel combinations of multiple compact domains have provided a substantial pathway through which novel proteins may evolve. The frequency of neutral contributions to such recombination events remains an article of faith.

In the Darwinian scheme of things, molecular characters such as secondary structures and tertiary folds of protein domains are historical features that persist as long as they continue to maintain the fitness of cells. Studies of reductive evolution, which are discussed next, as well as the IAD model for the origin of novel activities put a decidedly Darwinian spin on evolution (Bergthorsson et al. 2007; Deng et al. 2010; Näsvall et al. 2012). They suggest that persistent biological structures are not simply accidental in the way that mutations are accidental. Phenotypic characters at all levels persist because they are selected, and when they are not maintained by selection, they are lost (Kimura 1984; Berg and Kurland 2002).

9 Akaryotes: Less Is More

The notion that structures not supported by selection can be lost from genome repertoires has a more positive connotation when it is rephrased in the form: loss of structures may have selective advantages. This aspect of architectural design, captured in Mies van der Rohe's aphoristic "less is more," is also a fundamental aspect of the evolution of cellular architectures (Ehrenberg and Kurland 1984; Kurland 1992; Kurland et al. 2007; Wang et al. 2011; Koskiniemi et al. 2012).

Orgel and Crick (Kurland 1992) recognized that any sequence of DNA, even if it is not transcribed or translated, is at the very minimum a small drag on the growth rate of a cell because it "costs" something to reproduce sequences. That is to say there is no such thing as a truly neutral DNA sequence. Of course, Orgel and Crick (1980) stipulated that if there is transcription and translation of a DNA sequence, the costs go up proportionately. Consequently, Orgel and Crick likened the spread of selfish DNA sequences within genomes to "the spread of a not too harmful parasite within its host" (Orgel and Crick 1980). Here, the architectural principle for cells is that if the cost of copying, transcribing, and translating a sequence is greater than the benefit it confers on the cell, growth rate competition in cell populations will favor loss of the coding sequence because random lethal mutations will not be counterselected (Ehrenberg and Kurland 1984; Kurland 1992).

This reasoning also applies to any neutral or growth inhibitory DNA sequence propagating through a population of cells as long as it is associated with an adequately active infective vehicle such as a virus or a transposon (Orgel and Crick 1980; Berg and Kurland 2002). However, if that sequence is not by itself contributing to the growth rate of the cell, it will be degraded by random mutation even while being actively propagated to the cell's progeny (Berg and Kurland 2002). So, infectivity per se does not protect neutral or deleterious sequences from mutational degradation. This conclusion suggests that the ultimate fate of both neutral and deleterious sequences in growing cells is mutational meltdown (Kimura 1984; Berg and Kurland 2002).

We may couple this conclusion to another more basic view of cell architectures. Here, the idea is that there is an optimal arrangement of working cellular parts under any particular steady state growth condition. The maximum rate of growth is a function of the dynamics of individual component functions as well as the relative amounts of each component in the cell. The maximized arrangement for each cellular component is defined conditionally: at this optimum an incremental increase or decrease of the mass or the rate of function of that component leads to a decrease of the rate of cellular growth (Ehrenberg and Kurland 1984). In other words, the maximum rate of function per mass of component is a condition of optimality. Hence small is good. Optimal arrangements can be stabilized by mutations that enhance dynamic properties. Thus, mutations that reduce the kinetic load on cell growth rates or that favor the deletion of destructive coding sequences are selected (Kurland 1992; Koskiniemi et al. 2012). By the same token, selective loss of coding sequences, reductive evolution, has also been identified as a principal genomic adaptation of endosymbionts and endocellular parasites (Orgel and Crick 1980; Andersson and Kurland 1998; Silva et al. 2001; Moran 2003). Nevertheless, reductive evolution is much more general, as is reflected in the growing attention it has received recently in studies of genome evolution (Lynch 2007; Maeso et al. 2012; Harish et al. 2013; Wolf and Koonin 2013).

We interpret the ubiquity of the reductive mode of genome evolution (Fig. 2.7) as an expression of the general tendency to purge sequences that do not support maximum fitness of cells (Maeso et al. 2012; Harish et al. 2013; Wolf and Koonin 2013). However, we distinguish the suspension of selection for a fitness-enhancing sequence from the purging of neutral and deleterious sequences. The purging mechanism can be a mutation-driven one such as that expected when there is a greater frequency of deletions than insertions impacting genomes (Lynch 2007; Kuo and Ochman 2009). However, the destructive impact of deletions and insertions is in general the same for genes that are fitness enhancing: in such cases both sorts of mutation are counterselected by purifying selection. It is only when the fitness contribution of a coding sequence has been compromised that reductive evolution is initiated. The purging of the unselected sequence is then set in motion. Accordingly, purging is "selective" to the extent that it reduces the drag on cellular growth rates (Ehrenberg and Kurland 1984; Kurland 1992).

The requirement for architectural optimality for cells can be summarized in another way: a cell can have too little or too much of a good thing (Ehrenberg and Kurland 1984; Kurland 1992). As an obvious consequence, regulation of the expression levels of cellular components must be a major aspect of genome evolution. Decades of molecular biological research certainly attest to the importance and the sophistication of the control functions operating to regulate gene expression. In addition, there is yet another dimension of architectural optimality: the sizes of molecules.

Since one of the determinants of growth efficiency is the mass-normalized kinetic efficiency of proteins, the optimal sizes of proteins as well as their expression levels are relevant to their contribution to growth rate efficiency (Ehrenberg and Kurland 1984; Kurland 1992). Comparisons of the lengths of orthologous proteins reveal that lengths are systematically distributed differently in akaryotes and in eukaryotes. Thus, eukaryote proteins have mean lengths of 508 amino acids, while archaea and bacteria present mean lengths of 309 and 311, respectively (Kurland et al. 2007). In addition, the standard deviations of length normalized to mean lengths of proteins vary between 0.099, 0.091, and 0.21 in archaea, bacteria, and eukaryotes, respectively (Wang et al. 2011). This variation suggests that the selective pressure on length minimization is twice as high for akaryotes as it is for eukaryote proteins (Kurland et al. 2007).

The reductive pressure on amino acid sequences seems to be directed preferentially to the N- and C-terminal sequences of proteins, more specifically to the linkers

(Kurland et al. 2007; Wang et al. 2011). In contrast, the compact protein domains are composed of virtually identical amino acid lengths in all three superkingdoms (Wang et al. 2011). The length conservation of SF coding sequences across the superkingdoms is an impressive reminder of the selective structural constraints that guide the descent of SFs (Wang et al. 2011). There is little or no room for mutation-driven length variation within any nominal SF, though there is significant sequence degeneracy among the coding sequences of individual SFs (Chothia and Gough 2009; Pethica et al. 2012). On the other hand, the eukaryotes have more than threefold greater linker lengths than do archaea and bacteria with mean lengths of 250, 73, and 86 amino acids, respectively (Wang et al. 2011).

We interpret this evidence for persistent selection for minimum lengths of proteins in the akaryotes as a reflection of the general tendency of akaryotes to exploit a reductive evolutionary mode. Thus the abundance of SFs encoded by a representative eukaryote genome is sixfold greater than a representative akaryote (Harish et al. 2013). So, the descent from a common ancestor has involved a greater systematic reduction of coding in akaryotes than in eukaryotes. It is worth emphasizing the selective nature of that reduction. Mutation-driven purging mechanisms provided by different relative rates of insertion and deletion mutations (Chothia and Gough 2009; Pethica et al. 2012) are selective only to the extent that they are favored by purifying selection for maximum growth rates. Where there is no such effect, the reduction is mutation driven as for neutral or deleterious sequences.

On the other hand, the greater diversity of eukaryote proteomes means that selective pressure on individual eukaryote proteins is lower than it is in akaryotes. Here, the greater diversity of coding sequences is translated to lower mass fractions invested in any representative protein. Since the mass fraction of an expressed protein is proportional to the selective pressure exerted on that protein, individual eukaryote proteins in general are subject to lower reductive pressure than are akaryote proteins (Kurland et al. 2007). The influence of effective population sizes also conspires to vitiate selective pressure in eukaryotes, while enormous population sizes tend to maximize these selective forces in akaryotes (Kurland et al. 2007). The sum of both these effects is to support more stringent reductive pressure in akaryotes (Kurland et al. 2006, 2007). Of course, there are stunning exceptions to these generalizations such as the yeasts (Dujon 2010).

The donut charts that display distributions of SF functions among the three superkingdoms reveal a remarkably conserved distribution of different functions among organisms (Fig. 2.8). However, that uniformity is deceptive. It is in part a reflection of the fact that different SFs may mediate similar functions. Nevertheless, there are at present only circa 2000 superfamilies known to structural biologists. Consequently, much of the species specificity of genomes must reflect the combinatorial possibilities of a limited number of unique SFs. Here, every taxon has a unique genome determined by its capacity to encode SFs in characteristic quantitative combinations, as shown in Fig. 2.4. There are not enough unique proteins in the protein database to identify each proteome with a collection of unique SFs. Accordingly, the reductive evolution of akaryote proteomes determines that the SFs shared with eukaryotes are lightly scattered among the genomes of akaryotes but more

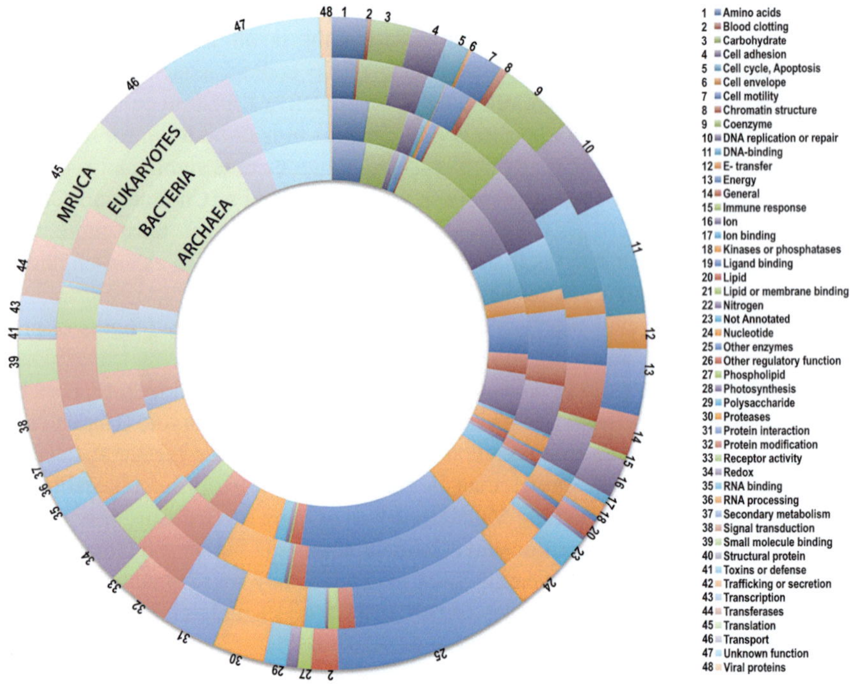

Fig. 2.8 Conservation of functions between MRUCA and crown clades. Despite the large variation in proteome sizes, the fractions of proteomes invested in different functions are remarkably conserved. In general, similar fractions of the proteomes in the each superkingdom are invested in different functions. There are however small differences in the abundances of SFs in certain functions, which presumably relate to specific ecological adaptations

uniformly distributed among the eukaryote genomes (Chothia and Gough 2009; Pethica et al. 2012; Harish et al. 2013). This of course puts a rather different spin on the consequences of an occasional HGT event. Thus, the transfer of a SF will alter the distributions of at a minimum several hundred SFs by one SF. Little wonder that the influences of HGT are demonstrably minimal and so fall short of a paradigm-shifting phenomenon (Kurland and Berg 2010; Harish et al. 2013).

10 No Genome Is an Island

There are by now two traditional ways to account for the evolution of molecular structures: one is a modern extension of the Darwinian view of descent with modification by natural selection, and the other is neutral evolution (Ehrenberg and Kurland 1984; Kimura 1984; Nei and Kumar 2000; Lynch 2007). Molecular evolution theory has been permeated by neutral theory because of the seductive simplicity of population genetic calculations based on the neutral view (Kimura 1984). However, the elephant in the room is the challenging deficiency of molecular data from coding sequences that supports such neutralist claims (Lind et al. 2010a). For example, W. Ford Doolittle echoed Zuckerkandl and Pauling's molecular clock model expectation (Zuckerkandl and Pauling 1965b) that any gene tree would be expected to provide a good phylogenetic replica of the corresponding species tree (Doolittle 1999). However, this is certainly not what is observed in general (Mossel and Steel 2004). Indeed, no gene tree from any of a thousand different genes within a pool of related yeast genomes yields the same phylogeny as that from a concatenation analysis (Salichos and Rokas 2013). Of course, this is not to say that neutral mutations never go to fixation in populations. The problem is finding them.

There is another way of viewing the conflict between neutral theory and selection theory. Genome evolution can be described theoretically as an autonomous process if it is assumed that genomes evolve in essence through mutation-driven diffusion processes (Zuckerkandl and Pauling 1965b; Kimura 1984; Lynch 2007). In this extreme it would be immaterial whether or not members of the same species or different species were sharing the environments of organisms.

However, what is consistently observed is a prevalence of competition between related organisms in which purifying selection determines the fate of mutant alleles even when the discovery of neutral mutants might be favored by the experimental design (Lind et al. 2010a). Thus, in spite of protestations to the contrary (Lynch 2007), there is no empirical molecular support for the dominance of neutral autonomy in the evolution of coding sequences. Though neutral theory has haunted molecular genetics for decades (Zuckerkandl and Pauling 1965b; Kimura 1984; Lynch 2007), any textbook on evolutionary genetics suggests that individuals within the same species or population must be viewed most often as competitors to make sense of genetic data (Ehrenberg and Kurland 1984; Lynch 2007). Of course, at the extremes of very small population numbers, mutation-driven diffusion processes may play a transient role (Kimura 1984; Lynch 2007). Nevertheless, purifying selection introduces a dimension of community to the evolution of genomes that is expanded to include members of different species by the incidence of selective HGT as well as by symbiosis, parasitism, and predation (Ehrenberg and Kurland 1984; Berg and Kurland 2002; Lynch 2007; Kurland and Berg 2010). In other words, though it is seldom viewed as such, genome evolution seems to be primarily an ecological process.

The ecology underlying the distinctive strategies that akaryotes and eukaryotes have favored in genome evolution is to some extent apparent (Kurland et al. 2006). Eukaryote genomes are much more complex than those of akaryotes, and eukaryote cells tend to be phagotrophic (Kurland et al. 2006). We know of no phagotrophic akaryotes. Akaryotes favor small efficient cellular architectures that support growth rates within a given environment that are faster than their larger eukaryote neighbors (Kurland et al. 2006). This generalization is not meant to imply that there is no slowly growing akaryote or no rapidly growing eukaryote. However, it is a generalization that describes the normal constraints on growth rates for the optimal architectures of growth rate-maximized cells growing in a given medium. The only way for such cells to increase their growth rates is to lose cellular mass (Ehrenberg and Kurland 1984; Kurland 1992). Essentially, this is the reality of "less is more" for cellular architectures in an ecological context.

Initially, similarity in size and shape of eukaryote organelles and certain bacteria observed under the light microscope were interpreted as evidence for Sagan's conjecture of an endosymbiotic origin for mitochondria and other organelles (Sagan 1967). However, the model was rejected for the specific case of flagella in eukaryotes because no resident genomes were observed in association with flagella (Gray 2017). In contrast, the presence of [circular] genomes in mitochondria and chloroplasts was taken to be a strong evidence in favor of the Sagan's conjecture (Gray 2017). Further, when rRNA was employed as a "phylogenetic marker" gene, sequence similarity between bacterial and organellar rRNAs depicted in unrooted gene trees was considered to be a strong evidence in support of the endosymbiotic model (Gray and Doolittle 1982; Yang et al. 1985).

Nevertheless, an unrooted tree is not an evolutionary tree per se, and "phenetic" similarity without regard to any ancestor-descendant polarity is not evidence of specific evolutionary relationships (Morrison 2006; Wiley and Lieberman 2011). Furthermore, conclusions based on the observed similarities of a single gene family as opposed to similarities of gross cellular phenotypes seemed suspect to Mayr (1998). Instead, a comparative analysis of the archaeal proteomes available at the time (that of *Methanococcus jannaschii* and *Methanobacterium thermoautotrophicum*) suggested to Mayr that for the "proteins or sequences that can be definitely assigned, about 77% are more eubacterial and less than a quarter (23%) are eukaryotic. These percentages reinforce the conclusion that the two kinds of bacteria [eubacteria and archaebacteria] are far more similar to each other than are the archaebacteria to the fully evolved eukaryotes." (Mayr 1998).

Woese dismissed Mayr's comparative genomic analysis (Woese 1998). Rather, Woese favored the interpretation that the migrant genes do not define the ancestors of the three domains. Instead, he insisted that it was the novel genes that were confined to a single domain and were transmitted vertically that defined the ancestors of the domains. Thus, he insisted that a core of unique, vertically inherited genes defined and distinguished each of the primary lines of descent, and it was this core that must define "the highest level taxa in a biological classification theory" (Woese 1998).

In sharp contrast, a systematic analysis of the origins of proteomes based on empirical, nonstationary and nonreversible genome evolution models subsequently demonstrated that the primary lines of descent are defined by the descent from MRUCA of three quarters of the circa 2000 known SFs in these samplings (Harish et al. 2013; Harish and Kurland 2017a). Here, phylogenies inferred by both parsimony and Bayesian methods yield the same genomic "tree of life" (ToL): this is a global phylogeny that features the separate divergence of akaryote (archaea and bacteria) and eukaryote lineages from their common ancestor MRUCA (Harish et al. 2013; Harish and Kurland 2017a). We find no indication of massive inter-lineage exchange of coding sequences for SFs during the descent of the two lineages.

Independent analyses of three different ancestral proteome reconstructions based on 141, 228, and 336 genomes indicate that 97% of modern mitochondrial SFs as well their homologues in bacteria and archaea were present in MRUCA. Accordingly, we conclude that the evolution of the mitochondrial proteome was autogenic (endogenic) and not endosymbiotic (exogenic) (Harish and Kurland 2017a, c). In addition, an explicit Bayesian test of the competing hypotheses indicates that the independent and parallel descent of akaryotes and eukaryotes as in Mayr's interpretation is at least 10^{87} times more probable than Woese's three-domain hypothesis. In contrast the so-the eocyte hypothesis (aka fusion of prokaryote lineages) is demonstrably highly improbable on the basis of the same data (Harish and Kurland 2017a).

Results and conclusions from our initial study based on 141 genomes (from 141 unique genera) (Harish et al. 2013) were further confirmed with expanded samplings of 228 and 336 species. In addition, exhaustive tests of the robustness of the phylogenetic analyses when challenged by potential artifacts due to sampling biases and genome sizes biases also confirmed the conclusions. In each case equal numbers of unique archaea, bacteria, and eukaryote species were sampled (Harish and Kurland 2017a, b), and they yielded comparable phylogenies. The statistical consistency of the genome evolution models as well as the empirical support for the grounding assumptions of the evolution models were explicitly verified with Bayesian posterior predictions (Harish and Kurland 2017a, c). This broad taxonomic sampling of genome sequences from more than 300 species is the most comprehensive sampling of species from archaea, bacteria, and eukaryotes analyzed to date.

Our conclusions are unlikely to be affected by the recent discovery of many novel phyla of archaea, such as *lokiarchaea*, *asgard*, *archaea*, and the DPANN archaea (Rinke et al. 2013; Zaremba-Niedzwiedzka et al. 2017) as well as the bacterial candidate phyla radiation (Hug et al. 2016). Our confidence comes from the fact that despite the relatively larger proteome of *lokiarchaea* with an estimated ~5400 proteins, which is ~1000 genes more than next largest archaeal proteome, there is only a modest increase in the number of SFs identified in the archaeal cohort: this increase amounts to a mere 10 SFs. The *lokiarchaea-specific* genes include homologues of many eukaryote-specific proteins. Nevertheless, none of the 10 SFs are unique to *lokiarchaea*: all 10 SFs are shared with either bacteria and eukaryotes or viruses (Nasir et al. 2015). Accordingly, the addition of *lokiarchaea* does not affect the phylogeny inferred from SF composition (Fig. 2.6) even though *lokiarchaea* is an early-diverging archaeal lineage (unpublished data).

Our recent empirical proteome evolution models (Figs. 2.6 and 2.7; (Harish and Kurland 2017a, c) are consistent with the earlier kinetic models of growth rate efficiency that are constrained by protein synthesis efficiency (Ehrenberg and Kurland 1984; Kurland 1992). Together, these analyses suggest that akaryote-like cells with their large population sizes and their relatively small proteomes can select characters with exquisitely small fitness differences. In addition, akaryotes can rely on linkage effects that promote the hitchhiking of multiple mutations, each with modest fitness enhancement linked to a single genomic unit (Smith and Haigh 1974; Berg and Kurland 1997). Here the linkage effects would be augmented in genomes that are not undergoing persistent recombination that separates favorable mutant alleles. Indeed, hitchhiking provides a realistic mechanism to enhance the fitness of cells by promoting multiple reductive changes in akaryote progeny (Smith and Haigh 1974; Berg and Kurland 1997).

In contrast, the eukaryotes in their descent from the common ancestor evolved ever more complex genomic and cellular structures, primarily by sequence duplications. Here, reductive losses may also have relevance (see Fig. 2.7). Eukaryotes may have been driven in part by the pressure to become more effective raptors by evolving into multicellular organisms with elaborate structures for which fitness is not reflected in individual cellular growth rates. In contrast, akaryotes descend primarily as highly reduced cells that are specialized in small scale, rapidly growing cellular economies (Kurland et al. 2007; Wang et al. 2011; Harish et al. 2013). Here, the differences between archaea and bacteria are thought to have arisen from their differentially specialized outer membranes and proton pumps (Valentine 2007). And, most provocatively, the akaryotic proteomes seem to have descended to a significant extent from the nominal "mitochondrial" proteome of MRUCA (Harish et al. 2013).

11 Darwin Rules

Rereading 50-year-old Zuckerkandl and Pauling papers (Zuckerkandl and Pauling 1965a, b) is informative about the more recent Mayr/Woese confrontation (Mayr 1998; Woese 1998) and its aftermath (Doolittle 1999). In the first place, Mayr emerges as the eternal gadfly even then at the dawning of molecular evolution. But there is a significant difference between the earlier (Zuckerkandl and Pauling 1965b) and later (Woese 1998) responses to his challenges: earlier when Mayr made pointed comments or posed a difficult question, Zuckerkandl and Pauling respond with answers based on relevant calculations (Zuckerkandl and Pauling 1965b). They (Zuckerkandl and Pauling 1965b) did not respond with unsupported assertion spiced in pique and extoling the virtues of "this" over "that" (Woese 1998). In addition, it emerges very clearly that the phylogenetic method employed by Zuckerkandl and Pauling (1965b) was hijacked and translated into something that might be misinterpreted by readers of genotype diatribes (Woese 1998; Doolittle 1999, 2012; Doolittle and Zhaxybayeva 2013). So, that there is no mistake in this matter, we rephrase that clarification: it is simply not the case that Zuckerkandl and Pauling (1965a, b) had encouraged molecular biologists to collect sequence information from randomly selected genes as the starting point for studies of mutational histories as might be incorrectly inferred from later accounts (Woese 1998; Doolittle 1999, 2012; Doolittle and Zhaxybayeva 2013).

In fact, Zuckerkandl and Pauling began their studies of mutational histories with sets of proteins that were considered to be structural homologues such as the globins (Zuckerkandl and Pauling 1965a, b). Furthermore, the structural constraints were often the context for their analyses of amino acid substitutions (Zuckerkandl and Pauling 1965a, b). Of course, their focus was on quantitating mutational changes in the corresponding sequences to infer historical relationships along with and homology guided by the workings of the molecular clock. However, by the time of the Mayr/Woese exchange, that methodology had been turned on its head. By then, as now, alignment-based gene trees begin with coding sequences, then alignments were made, from which sequence similarity was statistically inferred and interpreted as structural similarity. After all this, mutational histories are used to construct phylogeny.

The sticking point is that sequence similarity is not the same thing as structural similarity. W. R. Pearson, the creator of FASTA, emphasizes in a particularly lucid presentation that similar sequences in alignment-based studies are inferred from the statistics of sequence similarity, but structural similarities are inferred from their three-dimensional structures as such (Pearson 1995). Pearson cautions us "even when homologous sequences have different functions, they share a common three dimensional fold," but "absence of significant similarity does not guarantee nonhomology" (Pearson 1995). As we have seen, the sequence similarity search BLAST tends to underestimate sequence similarity to a significant degree. BLAST can also generate false similarities (Murzin et al. 1995; Gough et al. 2001; Illergård et al. 2009; Harish et al. 2013). For example, the phenomenon of long branch attraction illustrates the disjunction between sequence similarity and homology in deep phylogeny (Brinkmann and Philippe 1999; Forterre and Philippe 1999; Philippe and Forterre 1999; Tourasse and Gouy 1999; Mossel and Steel 2004; Rokas and Carroll 2006; Penny and Collins 2010; Philippe et al. 2011). Likewise, extreme codon preferences can provide false indications of putative HGT between organisms that do not share sequence exchanges (Singer and Hickey 2003).

By using the SCOP database to identify homologous structures from genome sequence data, we return to a methodology more like that of Zuckerkandl and Pauling (1965b), but it is operating at a different hierarchical level. Thus, it is clear that genome content tree construction from sequence data is an innovation clearly distinguishable from the approach favored by Zuckerkandl and Pauling (1965b).

The choice between implementing alignment-based gene trees or phylogeny based on genome content of structural characters is straightforward: That choice depends on whether the objective is a specific gene tree or a species tree. Had it not been obvious before (Nei and Kumar 2000), it is now demonstrably clear (Salichos and Rokas 2013) that implementing a gene tree in order to create a species tree is not a valid procedure. Furthermore, we emphasize the conclusion that genome content cladistics using structural homologues is unlike gene trees. Gene trees and genome trees obviously describe different things. Genome content trees do transcend the hierarchical limitations of gene trees based on sequence alignments. In effect, the two methods explicitly exploit different levels of the genic and the genomic hierarchies.

Molecular biologists inspired by Woese's (1998) passion for understanding the molecular processes underlying the evolution of organisms have been the beneficiaries of a marvelous gift with which to implement their passion. This is the SCOP database, at present the annotated key to the atomic dimension structures of more than 85,000 unique proteins, which together with appropriate Hidden Markov models can transform the nucleotide sequences of genomes into the three-dimensional structures of altogether circa 2000 compact protein domains at the level of superfamily (Murzin et al. 1995; Gough et al. 2001; Worth et al. 2009). This gift was created by hundreds of structural biologists and informaticists working for decades, inspired by the goal of acquiring a comprehensive, detailed structural atlas of all the proteins that have been fashioned by the evolution of modern organisms. Other databases are under construction, and much more is required before we have at hand a complete structural atlas that includes the disordered regions of proteins. But, the success of the SCOP database is already an inspiration and a workable tool with which to further Woese's inspiring quest to understand evolution at the molecular level.

The reason that the SCOP database is such a treasure trove for evolutionary biologists is that it transcends some of the ambiguity about homology that plagues alignment-based gene trees. In that sense, SCOP returns us to Zuckerkandl and Pauling's (1965b) secure starting point in studying structural homologues. Thus, SCOP is a database of homologous protein domain structures, for example, homologous SFs (Murzin et al. 1995; Gough et al. 2001; Gough 2005; Chothia and Gough 2009; Illergård et al. 2009; Zmasek and Godzik 2011). SCOP is not a database of homologous coding sequences or of homologous proteins. Those distinctions are important. As W. R. Pearson asserts (Pearson 1995), homologous sequences are ones that share "a common ancestor" just as do homologous structures, which mean that they are homologous in the Darwinian sense.

We do not mean to imply that phylogeny based on genome content of SFs is the last word on species phylogeny. At the very least, it remains to account for the universe of disordered protein regions, for which the Database of Disordered Protein Predictions, D^2P^2 (Oates et al. 2013) is a hopeful development. Also high on the list of things to do is an integrated account of the nominally noncoding genome sequences that dominate eukaryote genomes. Here, we have only documented in some detail a coarse-grained phylogenetic consensus between four different sorts of phylogenetic reconstructions for coding sequences. These are based on selected but well-curated gene trees (Brinkmann and Philippe 1999; Philippe et al. 2011), on secondary structures of rRNA (Caetano-Anollés 2002), on genome content of orthologous proteins (Tekaia et al. 1999; Snel et al. 1999), and on genome content of compact protein domains (Yang et al. 2005; Harish et al. 2013).

That consensus tells us that the phylogeny of the three superkingdoms is not the tree favored by Woese and his cohort (Woese 1987, 1998; Woese et al. 1990; Doolittle and Brown 1994; Baldauf et al. 1996; Pace 1997). In fact, the Woese tree with its sister clades of archaea and eukaryotes as well as the so-called bacterial root (Fig. 2.1) has long been considered a textbook illustration of the destructive influences of long branch attractions and mutational saturation on gene trees (Forterre et al. 1992; Philippe and Laurent 1998; Forterre and Philippe 1999; Philippe and Forterre 1999; Mossel and Steel 2004; Penny and Collins 2010). In distinct contrast, the consensus recognizes archaea and bacteria as sister clades within a monophyletic akaryote empire (Forterre 1992; Harish et al. 2013). Here, the akaryote empire is phylogenetically paired with the eukaryote empire, while the common root of the modern crown is thought to be a complex universal ancestor, MRUCA (Brinkmann and Philippe 1999; Caetano-Anollés 2002; Kurland et al. 2007; Philippe et al. 2011; Harish et al. 2013). Though not a bacterium nor a eukaryote, MRUCA presents phylogenetic characters shared by both akaryotes and eukaryotes (Brinkmann and Philippe 1999; Caetano-Anollés 2002; Kurland et al. 2007; Philippe et al. 2011; Harish et al. 2013).

The evolutionary status of the akaryotes in the new consensus phylogeny, not just their names, is substantially different from that in Woese's phylogeny. That new status reflects more than a different tempo for the emergence of the three superkingdoms, which is all that one might expect from a new gene tree. Rather, explicit, systematic changes in the proteomes of organisms flesh out the differences between the superkingdoms. Just the inference that MRUCA and the crown groups shared three quarters of all the superfamilies (Fig. 2.7) informs us that it is primarily combinatorial rearrangements that provide the species specificity reflected in PCA displays of SF proteomes (Fig. 2.4), and it is not de novo protein evolution that paces the speciation of the modern crown clades. There is no indication in genome content phylogeny that akaryotes are ancestors of eukaryotes. Rather, it seems that the akaryotes and eukaryotes diverged from their common ancestor and that their modern descendants subsequently evolved independently.

Indeed, the donut charts (Fig. 2.8) that summarize the distributed activities of the proteomes of the superkingdoms and MRUCA reinforce that view of functional similarities primarily derived from the common ancestor. What is invisible in gene trees is precisely the quantitative variation of individual protein domains that is the substance of genome content phylogeny as shown in Figs. 2.6 and 2.7. Furthermore, it is the characteristic mix of gain and loss of superfamilies that defines the divergence of the akaryotes and eukaryotes: Reductive evolution dominates over gene duplication for akaryotes, while the duplication of superfamilies and to a lesser extent novel sequence acquisition are quantitatively dominant over loss in eukaryotes (Fig. 2.7). Accordingly, the number of novel superfamilies decreases by more than half in akaryotes and by roughly one third in eukaryotes. However, the crown eukaryotes have in abundance terms nearly four times as many superfamilies as inferred for MRUCA, while that ratio is roughly two to one for akaryotes. Evidently there is a pronounced dominance of duplication in the decent of eukaryote proteomes and a predominance of reductive evolution for the akaryotes (Harish et al. 2013).

The consensus phylogeny obtained with both gene trees as well as cladistic methods (Brinkmann and Philippe 1999; Caetano-Anollés 2002; Kurland et al. 2007; Philippe et al. 2011; Harish et al. 2013) is suggestive of a heuristic resolution for the rhetorical confrontation between Mayr and Woese concerning the relative merits of genotypic and phenotypic phylogeny (Mayr 1998; Woese 1998). Thus, each of the four methods begins with DNA sequences that are annotated to produce characters that are further exploited for phylogeny. So far we have identified secondary structural characters and tertiary structural characters as phenotypic. In addition, we have viewed some aspects of primary sequences as phenotypic characters but in a veiled way. In effect, mutational bias due to LBA and mutational saturation of ancient sequences may be interpreted as one-dimensional phenotypic annotation of DNA sequences. Accordingly, if there is in fact a continuum of primary, secondary, and tertiary structural characters encoded in genome sequences, the barrier between genotypic and phenotypic phylogeny raised by Woese (1998) is a semantic one lacking in substance (Mayr 1998; Woese 1998; Doolittle and Zhaxybayeva 2013; Wheeler et al. 2013). In other words that aspect of the Mayr/Woese confrontation is eminently forgettable.

Likewise, the influence of HGT on phylogeny has been distorted beyond recognition (Doolittle 1999, 2012; Doolittle and Zhaxybayeva 2013; Wheeler et al. 2013). In the first place, there is no reason to believe that HGT is non-Darwinian (Penny 2011). Likewise, the identification of unique bacterial proteomes in PCA displays, as shown in Fig. 2.4, clearly beggar the claim that HGT subverts the recognition of bacterial species (Doolittle 2012). Thus, bacterial genomes present proteomes that are species-specific assemblages of superfamilies readily distinguished from each other (Harish et al. 2013). Finally, HGT happens sufficiently often to be taken as a normal non-Mendelian mode of sequence acquisition in bacteria as elsewhere.

On the other hand, it is true that gene trees do not accommodate HGT but neither do they accommodate duplications, gene birth, or gene loss. Does this mean that novel gene duplications, losses, and births are non-Darwinian genomic events that destroy the possibility of meaningful species phylogeny? Yes, according to some views W. F. Doolittle, because when "different genes give different trees, and there is no fair way to suppress this disagreement, then a species (or phylum) can 'belong' to many genera (or kingdoms)" (Doolittle 1999). We can agree that gene trees provide a very cloudy window on genome evolution (Salichos and Rokas 2013). So, perhaps molecular evolutionists ought to stop insisting that gene trees provide the only reliable method to describe the species trees (Doolittle 1999; Kurland et al. 2007). After all, phylogeny based on alignments from diploid organisms featuring intense meiotic recombination does not permit the resolution of a single haploid genotypic lineage in gene trees. So, what's a little HGT between friends?

Gene trees are often unable to distinguish HGT from other novel gene acquisition events such as gene duplications and paralogy. At a minimum, this leads to serious overestimations of HGT frequencies (Berg and Kurland 2002; Kurland and Berg 2010). Fixation of HGT in species is routinely overestimated because transferred sequences are often purged from cells in the absence of specific selective factors, and counterselection limits the persistence of HGT to adaptive patches within global populations of species (Berg and Kurland 2002; Kurland and Berg 2010; Lind et al. 2010b). As a consequence, just the numbers of HGT events recognizable in different samplings or strains of a given species is not informative about the actual numbers of HGT events that are fixed in that species. Indeed, global fixation rates of HGT may not be far from zero because of the pronounced tendency of gene transfers to be transiently limited to patchy distributions (Berg and Kurland 2002; Kurland and Berg 2010; Lind et al. 2010b). Future studies of HGT as well as efforts to track individual novel sequence acquisitions seem most appropriately carried out in sequence-based reticulate reconstructions (Kurland 2000).

A disappointing feature of rooted phylogeny is that reconstruction of the most recent universal common ancestor seems to exhaust the possibilities of travel further back in evolutionary time (Harish et al. 2013). In other words, it seems that MRUCA is a singularity beyond which phylogeny cannot go. At present, we suspect that MRUCA represents relatively "recent organisms," perhaps as young as 1000–2000 MYA. It is not altogether clear how molecular biologists can reach back to study urancestors to test some of Woese's ideas about the progenote (Woese 1998) or other conceptions of the first cellular chemistries (Woese 1998; Kurland 2010). However, it is inconceivable that a cell as complex as MRUCA could have been a progenote – or so it seems now.

The idea of reductive genome evolution as a selective adaptation has a long history. To our knowledge the idea originated along with the first genome sequence determinations of mitochondria (Attardi 1985). From there it was an obvious step to infer a reductive descent for the genomes of endosymbionts as well as of endocellular parasites (Andersson and Kurland 1991, 1998; Andersson et al. 1998; Moran 2003). But, extreme reductive scenarios are not reserved for the genomes of intracellular residents. There are numerous free-living fungi and bacteria as well as archaea that have descended through genome reduction, but none so marked as certain pelagic free-living *Alphaproteobacteria* (Giovannoni et al. 2005; Viklund et al. 2012).

There are two extreme modes of adaptation for the most extreme shrinking genomes: one of these would be to abandon the alternative metabolic pathways that support a broad range of utilizable substrates. This would fix the organism to a very narrow choice and concentration of a useable substrate that supports very slow growth rates for intermittent periods of time. That arrangement will be stable as long as the limited availability of foodstuffs is also in the long term at least intermittently reliable.

Another more classic reductive scenario would be reductive genome evolution that leads to a commensal dependence on cross feeding of the metabolic products of another genome. If that commensal relationship is stable, it allows one genome to reduce its cellular investment in the corresponding genes and proteins, with an attendant growth rate enhancement. Such complex commensality has been observed in nature. Thus, the Black Queen Hypothesis (Morris et al. 2012) describes a communal dimension of evolution in the reductive mode. In such a case, loss of gene function in one cell is supported by the compensatory activities of another cell. Accordingly, the Black Queen Hypothesis (Morris et al. 2012) points to a novel mode of phylogenomics that may transcend genome content-based phylogeny by explicitly accounting for the ecological connectivity between genomes.

Ideally the tracking of such connectivity would also facilitate integration of HGT events into the normality of genome ecology (Kurland 2000). For example, there is in mealybugs a complex symbiotic interaction between the insect and one of the smallest known bacterial endosymbiont genomes that is associated with yet another more normal bacterial endosymbiont (Husnik et al. 2013). Part of the adaptation of this three-genome system has been to accommodate dozens of exchanges of amino acid biosynthetic genes into the insect host, though none are from the minute genome of the smallest endosymbiont (Husnik et al. 2013).

We suggest that such communal exchanges would be more transparent in genome content constructions than in networks of alignments. The reason for this expectation is that multiplicities of genome adaptations including duplications, loss, and HGT would be incompatible with gene tree networks that do not accommodate multiple genomic events. A new genome content methodology will be needed for the next generation of phylogenetic methods that takes us from genome phylogeny to the phylogeny of genome communities. This upgrade will bring us a long way from Zuckerkandl and Pauling (1965b) and the roots of gene trees, but a long way is where we want to be.

In spite of Woese's (1998) attachment to a deeply flawed rooting of the three superkingdoms (Iwabe et al. 1989; Doolittle and Brown 1994; Baldauf et al. 1996), no one did more than he to clear the path to the acquisition of genome sequences and a reliable view of their phylogenies. But now we need to move on.

Acknowledgments We are indebted to Siv Andersson, Otto Berg, Dan Dykhuizen, Måns Ehrenberg, Julian Gough, Diarmaid Hughes, Bruce Levin, Michael Levitt, Mikael Oliveberg, David Penny, Antonis Rokas, Anders Tunlid, Richard Villems, and Irmgard Winkler for essential criticism and help in preparing this paper.

References

Abby SS, Tannier E, Gouy M, Daubin V (2012) Lateral gene transfer as a support for the tree of life. Proc Natl Acad Sci 109(13):4962–4967

Andersson G, Kurland C (1991) An extreme codon preference strategy: codon reassignment. Mol Biol Evol 8(4):530–544

Andersson SG, Kurland CG (1998) Reductive evolution of resident genomes. Trends Microbiol 6 (7):263–268

Andersson DI, Näsvall J (2013) New genes arise via innovation, amplification, divergence. Microbe 8(4):166–170

Andersson SG, Zomorodipour A, Andersson JO, Sicheritz-Pontén T, Alsmark UCM, Podowski RM, Näslund AK, Eriksson A-S, Winkler HH, Kurland CG (1998) The genome sequence of Rickettsia prowazekii and the origin of mitochondria. Nature 396(6707):133–140

Attardi G (1985) Animal mitochondrial DNA: an extreme example of genetic economy. Int Rev Cytol 93:93–145

Baldauf SL, Palmer JD, Doolittle WF (1996) The root of the universal tree and the origin of eukaryotes based on elongation factor phylogeny. Proc Natl Acad Sci 93(15):7749–7754

Berg OG, Kurland C (1997) Growth rate-optimised tRNA abundance and codon usage. J Mol Biol 270(4):544–550

Berg OG, Kurland C (2002) Evolution of microbial genomes: sequence acquisition and loss. Mol Biol Evol 19(12):2265–2276

Bergthorsson U, Andersson DI, Roth JR (2007) Ohno's dilemma: evolution of new genes under continuous selection. Proc Natl Acad Sci 104(43):17004–17009

Brinkmann H, Philippe H (1999) Archaea sister group of Bacteria? Indications from tree reconstruction artifacts in ancient phylogenies. Mol Biol Evol 16(6):817–825

Brunet LJ, McMahon JA, McMahon AP, Harland RM (1998) Noggin, cartilage morphogenesis, and joint formation in the mammalian skeleton. Science 280(5368):1455–1457

Caetano-Anollés G (2002) Evolved RNA secondary structure and the rooting of the universal tree of life. J Mol Evol 54(3):333–345

Castoe TA, de Koning AP, Pollock DD (2010) Adaptive molecular convergence: molecular evolution versus molecular phylogenetics. Commun Integr Biol 3(1):67–69

Chatton ÉPL (1938) Titres et travaux scientifiques (1906–1937) de Edouard Chatton. Impr. E. Sottano, Sète

Chothia C, Gough J (2009) Genomic and structural aspects of protein evolution. Biochem J 419:15–28

Copley SD (2003) Enzymes with extra talents: moonlighting functions and catalytic promiscuity. Curr Opin Chem Biol 7(2):265–272

Crick F (1981) Life itself: its origin and nature. Simon and Schuster, New York, NY. 192 p

Danielsson J, Awad W, Saraboji K, Kurnik M, Lang L, Leinartaitė L, Marklund SL, Logan DT, Oliveberg M (2013) Global structural motions from the strain of a single hydrogen bond. Proc Natl Acad Sci 110(10):3829–3834

Darwin C (1859) On the origin of species by means of natural selection, or the preservation of favoured races in the struggle for life. Murray, London

Degnan JH, Rosenberg NA (2006) Discordance of species trees with their most likely gene trees. PLoS Genet 2(5):e68

Deng C, Cheng C-HC, Ye H, He X, Chen L (2010) Evolution of an antifreeze protein by neofunctionalization under escape from adaptive conflict. Proc Natl Acad Sci 107 (50):21593–21598

Doolittle RF (1995) The multiplicity of domains in proteins. Annu Rev Biochem 64(1):287–314

Doolittle WF (1999) Phylogenetic classification and the universal tree. Science 284 (5423):2124–2128

Doolittle RF (2005) Evolutionary aspects of whole-genome biology. Curr Opin Struct Biol 15 (3):248–253

Doolittle WF (2012) Population genomics: how bacterial species form and why they don't exist. Curr Biol 22(11):R451–R453

Doolittle WF, Brown JR (1994) Tempo, mode, the progenote, and the universal root. Proc Natl Acad Sci 91(15):6721–6728

Doolittle WF, Zhaxybayeva O (2013) What is a prokaryote? In: Rosenberg E, DeLong EF, Lory S, Stackebrandt E, Thompson F (eds) The prokaryotes: prokaryotic biology and symbiotic associations. Springer, Berlin

Dujon B (2010) Yeast evolutionary genomics. Nat Rev Genet 11(7):512–524

Ehrenberg M, Kurland CG (1984) Costs of accuracy determined by a maximal growth rate constraint. Q Rev Biophys 17(1):45–82

Fontana W, Schuster P (1998) Continuity in evolution: on the nature of transitions. Science 280 (5368):1451–1455

Forterre P (1992) Neutral terms [14]. Nature 355(6358):305

Forterre P, Philippe H (1999) Where is the root of the universal tree of life? Bioessays 21 (10):871–879

Forterre P, Benachenhou-Lahfa N, Confalonieri F, Duguet M, Elie C, Labedan B (1992) The nature of the last universal ancestor and the root of the tree of life, still open questions. Biosystems 28 (1):15–32

Gibbon STF, House CH (1999) Whole genome-based phylogenetic analysis of free-living microorganisms. Nucleic Acids Res 27(21):4218–4222

Giovannoni SJ, Tripp HJ, Givan S, Podar M, Vergin KL, Baptista D, Bibbs L, Eads J, Richardson TH, Noordewier M (2005) Genome streamlining in a cosmopolitan oceanic bacterium. Science 309(5738):1242–1245

Gough J (2005) Convergent evolution of domain architectures (is rare). Bioinformatics 21 (8):1464–1471

Gough J, Karplus K, Hughey R, Chothia C (2001) Assignment of homology to genome sequences using a library of hidden Markov models that represent all proteins of known structure. J Mol Biol 313(4):903–919
Gouy R, Baurain D, Philippe H (2015) Rooting the tree of life: the phylogenetic jury is still out. Philos Trans R Soc B 370(1678):20140329
Gray MW (2017) Lynn Margulis and the endosymbiont hypothesis: 50 years later. Mol Biol Cell 28 (10):1285–1287
Gray MW, Doolittle WF (1982) Has the endosymbiont hypothesis been proven? Microbiol Rev 46 (1):1
Grishin NV (2001) Fold change in evolution of protein structures. J Struct Biol 134(2):167–185
Haglund E, Lindberg MO, Oliveberg M (2008) Changes of protein folding pathways by circular permutation overlapping nuclei promote global cooperativity. J Biol Chem 283 (41):27904–27915
Harish A, Kurland CG (2017a) Akaryotes and eukaryotes are independent descendants of a universal common ancestor. Biochimie 138:168–183
Harish A, Kurland CG (2017b) Empirical genome evolution models root the tree of life. Biochimie 138:137–155
Harish A, Kurland CG (2017c) Mitochondria are not captive bacteria. J Theor Biol 434:88–98
Harish A, Tunlid A, Kurland CG (2013) Rooted phylogeny of the three superkingdoms. Biochimie 95(8):1593–1604
Hug LA, Baker BJ, Anantharaman K, Brown CT, Probst AJ, Castelle CJ, Butterfield CN, Hernsdorf AW, Amano Y, Ise K (2016) A new view of the tree of life. Nat Microbiol 1:16048
Husnik F, Nikoh N, Koga R, Ross L, Duncan RP, Fujie M, Tanaka M, Satoh N, Bachtrog D, Wilson AC (2013) Horizontal gene transfer from diverse bacteria to an insect genome enables a tripartite nested mealybug symbiosis. Cell 153(7):1567–1578
Illergård K, Ardell DH, Elofsson A (2009) Structure is three to ten times more conserved than sequence—a study of structural response in protein cores. Proteins 77(3):499–508
Iwabe N, Kuma K, Hasegawa M, Osawa S, Miyata T (1989) Evolutionary relationship of archaebacteria, eubacteria, and eukaryotes inferred from phylogenetic trees of duplicated genes. Proc Natl Acad Sci 86(23):9355–9359
Kimura M (1984) The neutral theory of molecular evolution. Cambridge University Press, Cambridge
Koskiniemi S, Sun S, Berg OG, Andersson DI (2012) Selection-driven gene loss in bacteria. PLoS Genet 8(6):e1002787
Korbel JO et al (2002) SHOT: a web server for the construction of genome phylogenies. Trends Genet 18(3):158–162
Kuo C-H, Ochman H (2009) Deletional bias across the three domains of life. Genome Biol Evol 1:145
Kurland C (1992) Translational accuracy and the fitness of bacteria. Annu Rev Genet 26(1):29–50
Kurland CG (2000) Something for everyone: horizontal gene transfer in evolution. EMBO Rep 1 (2):92
Kurland CG (2010) The RNA dreamtime. Bioessays 32(10):866–871
Kurland CG, Berg OG (2010) A hitchhikers guide to evolving networks. In: Caetano-Anollés G (ed) Evolutionary genomics and systems biology. Wiley, Hoboken, NJ
Kurland C, Collins L, Penny D (2006) Genomics and the irreducible nature of eukaryote cells. Science 312(5776):1011–1014
Kurland CG, Canbäck B, Berg OG (2007) The origins of modern proteomes. Biochimie 89 (12):1454–1463
Lind PA, Berg OG, Andersson DI (2010a) Mutational robustness of ribosomal protein genes. Science 330(6005):825–827
Lind PA, Tobin C, Berg OG, Kurland CG, Andersson DI (2010b) Compensatory gene amplification restores fitness after inter-species gene replacements. Mol Microbiol 75(5):1078–1089
Lynch M (2007) The frailty of adaptive hypotheses for the origins of organismal complexity. Proc Natl Acad Sci 104(Suppl 1):8597–8604

Maeso I, Roy SW, Irimia M (2012) Widespread recurrent evolution of genomic features. Genome Biol Evol 4(4):486–500

Mayr E (1982) The growth of biological thought: diversity, evolution and inheritance. Harvard University Press, Cambridge, MA

Mayr E (1998) Two empires or three? Proc Natl Acad Sci U S A 95(17):9720–9723

Moran NA (2003) Tracing the evolution of gene loss in obligate bacterial symbionts. Curr Opin Microbiol 6(5):512–518

Morris JJ, Lenski RE, Zinser ER (2012) The Black Queen hypothesis: evolution of dependencies through adaptive gene loss. MBio 3(2):e00036-12

Morrison DA (2006) Phylogenetic analyses of parasites in the new millennium. Adv Parasitol 63:1–124

Morrison DA (2009) Why would phylogeneticists ignore computerized sequence alignment? Syst Biol 58(1):150–158

Mossel E, Steel M (2004) A phase transition for a random cluster model on phylogenetic trees. Math Biosci 187(2):189–203

Murzin AG, Brenner SE, Hubbard T, Chothia C (1995) SCOP: a structural classification of proteins database for the investigation of sequences and structures. J Mol Biol 247(4):536–540

Nasir A, Kim KM, Caetano-Anollés G (2015) Lokiarchaeota: eukaryote-like missing links from microbial dark matter? Trends Microbiol 23(8):448–450

Näsvall J, Sun L, Roth JR, Andersson DI (2012) Real-time evolution of new genes by innovation, amplification, and divergence. Science 338(6105):384–387

Nei M, Kumar S (2000) Molecular evolution and phylogenetics. Oxford University Press, Oxford

Oates ME, Romero P, Ishida T, Ghalwash M, Mizianty MJ, Xue B, Dosztányi Z, Uversky VN, Obradovic Z, Kurgan L, Dunker AK, Gough J (2013) D2P2: Database of disordered protein predictions. Nucleic Acids Res 41(D1):D508–D516

Ochman H, Lawrence JG, Groisman EA (2000) Lateral gene transfer and the nature of bacterial innovation. Nature 405(6784):299–304

Ohno S (1970) Evolution by gene duplication. George Alien & Unwin/Springer, London/Berlin

Oliveberg M, Wolynes PG (2005) The experimental survey of protein-folding energy landscapes. Q Rev Biophys 38(03):245–288

Orgel LE, Crick FH (1980) Selfish DNA: the ultimate parasite. Nature 284(5757):604–607

Pace NR (1997) A molecular view of microbial diversity and the biosphere. Science 276 (5313):734–740

Pearson WR (1995) Effective protein sequence comparison. Methods Enzymol 266:227–258

Penny D (2011) Darwin's theory of descent with modification, versus the biblical tree of life. PLoS Biol 9(7):e1001096

Penny D, Collins L (2010) Evolutionary genomics leads the way. In: Caetano-Anollés G (ed) Evolutionary genomics and systems biology. Wiley, Hoboken, NJ

Pethica R, Levitt M, Gough J (2012) Evolutionarily consistent families in SCOP: sequence, structure and function. BMC Struct Biol 12(1):27

Philippe H, Forterre P (1999) The rooting of the universal tree of life is not reliable. J Mol Evol 49 (4):509–523

Philippe H, Laurent J (1998) How good are deep phylogenetic trees? Curr Opin Genet Dev 8 (6):616–623

Philippe H, Brinkmann H, Lavrov DV, Littlewood DTJ, Manuel M, Wörheide G, Baurain D (2011) Resolving difficult phylogenetic questions: why more sequences are not enough. PLoS Biol 9 (3):e1000602

Rinke C, Schwientek P, Sczyrba A, Ivanova NN, Anderson IJ, Cheng J-F, Darling A, Malfatti S, Swan BK, Gies EA, Dodsworth JA, Hedlund BP, Tsiamis G, Sievert SM, Liu W-T, Eisen JA, Hallam SJ, Kyrpides NC, Stepanauskas R, Rubin EM, Hugenholtz P, Woyke T (2013) Insights into the phylogeny and coding potential of microbial dark matter. Nature 499(7459):431–437

Rokas A, Carroll SB (2006) Bushes in the tree of life. PLoS Biol 4(11):e352

Sagan L (1967) On the origin of mitosing cells. J Theor Biol 14(3):225–275

Salichos L, Rokas A (2013) Inferring ancient divergences requires genes with strong phylogenetic signals. Nature 497(7449):327–331

Silva FJ, Latorre A, Moya A (2001) Genome size reduction through multiple events of gene disintegration in Buchnera APS. Trends Genet 17(11):615–618

Singer GA, Hickey DA (2003) Thermophilic prokaryotes have characteristic patterns of codon usage, amino acid composition and nucleotide content. Gene 317:39–47

Smith JM, Haigh J (1974) The hitch-hiking effect of a favourable gene. Genet Res 23(1):23–35

Snel B, Bork P, Huynen MA (1999) Genome phylogeny based on gene content. Nat Genet 21 (1):108–110

Stanier RY, van Niel C (1962) The concept of a bacterium. Arch Microbiol 42(1):17–35

Tekaia F, Lazcano A, Dujon B (1999) The genomic tree as revealed from whole proteome comparisons. Genome Res 9(6):550–557

Theobald DL, Wuttke DS (2005) Divergent evolution within protein superfolds inferred from profile-based phylogenetics. J Mol Biol 354(3):722–737

Tourasse NJ, Gouy M (1999) Accounting for evolutionary rate variation among sequence sites consistently changes universal phylogenies deduced from rRNA and protein-coding genes. Mol Phylogenet Evol 13(1):159–168

Valentine DL (2007) Adaptations to energy stress dictate the ecology and evolution of the Archaea. Nat Rev Microbiol 5(4):316–323

Viklund J, Ettema TJ, Andersson SG (2012) Independent genome reduction and phylogenetic reclassification of the oceanic SAR11 clade. Mol Biol Evol 29(2):599–615

Wang M, Kurland CG, Caetano-Anollés G (2011) Reductive evolution of proteomes and protein structures. Proc Natl Acad Sci 108(29):11954–11958

Wheeler Q, Assis L, Rieppel O (2013) Phylogenetics: heed the father of cladistics. Nature 496 (7445):295–296

Wiley EO, Lieberman BS (2011) Phylogenetics: theory and practice of phylogenetic systematics. Wiley, New York

Woese CR (1987) Bacterial evolution. Microbiol Rev 51(2):221

Woese CR (1998) Default taxonomy: Ernst Mayr's view of the microbial world. Proc Natl Acad Sci U S A 95(19):11043–11046

Woese CR (2000) Interpreting the universal phylogenetic tree. Proc Natl Acad Sci 97 (15):8392–8396

Woese CR, Kandler O, Wheelis ML (1990) Towards a natural system of organisms: proposal for the domains Archaea, Bacteria, and Eucarya. Proc Natl Acad Sci 87(12):4576–4579

Wolf YI, Koonin EV (2013) Genome reduction as the dominant mode of evolution. Bioessays 35 (9):829–837

Worth CL, Gong S, Blundell TL (2009) Structural and functional constraints in the evolution of protein families. Nat Rev Mol Cell Biol 10(10):709–720

Yang D, Oyaizu Y, Oyaizu H, Olsen GJ, Woese CR (1985) Mitochondrial origins. Proc Natl Acad Sci U S A 82(13):4443–4447

Yang S, Doolittle RF, Bourne PE (2005) Phylogeny determined by protein domain content. Proc Natl Acad Sci U S A 102(2):373–378

Zaremba-Niedzwiedzka K, Caceres EF, Saw JH, Bäckström D, Juzokaite L, Vancaester E, Seitz KW, Anantharaman K, Starnawski P, Kjeldsen KU, Stott MB, Nunoura T, Banfield JF, Schramm A, Baker BJ, Spang A, Ettema TJG (2017) Asgard archaea illuminate the origin of eukaryotic cellular complexity. Nature 541(7637):353–358

Zmasek CM, Godzik A (2011) Strong functional patterns in the evolution of eukaryotic genomes revealed by the reconstruction of ancestral protein domain repertoires. Genome Biol 12(1):R4

Zuckerkandl E, Pauling L (1965a) Evolutionary divergence and convergence in proteins. In: Vogel VBH (ed) Evolving gene and proteins. Academic Press, New York, pp 97–166

Zuckerkandl E, Pauling L (1965b) Molecules as documents of evolutionary history. J Theor Biol 8 (2):357–366

Chapter 3
The Tree of Life

Morgan Gaia, Violette Da Cunha, and Patrick Forterre

1 Introduction

Darwin wrote in 1857 that he foresaw "the time [. . .] when we shall have very fairly true genealogical trees of each great kingdom of nature" (Woese 2000). Fulfilling Darwin's dream by deciphering the topology of the universal tree of life (TOL) is undoubtedly one of the most important and rewarding tasks for a biologist. For a long time, trees of life were only based on subjective assumptions about the relative evolutionary relationships between organisms, tentatively deduced from their phenotypes (autotrophs evolved from heterotrophs, aerobes from anaerobes, diderm from monoderm, and so on). A major assumption was that evolution always goes from simpler to more complex organisms, reminiscent of Aristotle's *Scala Naturae*. For instance, mycoplasma were considered to be the most "primitive" living organisms because they lack peptidoglycan and were the smallest known bacteria (Wallace and Morowitz 1973). At the turn of the last century, the advances in cytology led to the view that life's main division was between organisms featuring a nucleus and those lacking it (i.e., the eukaryotes and the prokaryotes, respectively) (Stanier and Van Niel 1962). This division of life based on two different types of cellular organization was rapidly interpreted as an evolutionary division assuming that

M. Gaia
Département de Microbiologie, Institut Pasteur, Unité de Biologie Moléculaire du Gène chez les Extrêmophiles (BMGE), Paris, France

V. Da Cunha · P. Forterre (✉)
Département de Microbiologie, Institut Pasteur, Unité de Biologie Moléculaire du Gène chez les Extrêmophiles (BMGE), Paris, France

Institute for Integrative Biology of the Cell (I2BC), CEA, CNRS, Univ. Paris-Sud, Univ. Paris-Saclay, Gif-sur-Yvette, France
e-mail: forterre@pasteur.fr

P. H. Rampelotto (ed.), *Molecular Mechanisms of Microbial Evolution*, Grand Challenges in Biology and Biotechnology,
https://doi.org/10.1007/978-3-319-69078-0_3

simpler "pro"karyotes (meaning before karyotes), assimilated at that time to Bacteria, preceded more complex eukaryotes in the history of life. In particular, reviving the endosymbiotic theory for the origin of mitochondria and chloroplasts (see Box 3.1), Lynn Margulis suggested that eukaryotes emerged from a collection of diverse prokaryotes assimilated to Bacteria (the flagella from *Spirochaeta*, the cytoplasm from a wall-less bacterium, the nucleus from a bacterium containing histones) (Sagan 1967) (Fig. 3.1a).

Box 3.1 The Origin of Organelles

For a long time, an important part of the discussions about the topology of the tree of life turned around the position of cellular organelles, such as mitochondria and chloroplasts. As soon as the early twentieth century, a few biologists suggested that these organelles could be enslaved bacteria (Mereschkowski 1905, 1910; Martin and Kowallik 1999; López-García et al. 2017). The idea that mitochondria and chloroplasts originated from within eukaryotic cell (the autogenous theory) remained the dominant one for some time. However, the discovery of DNA and bacterial-like ribosomes (in size) within mitochondria and chloroplasts in the early 1960s restarted the debate (Gray and Doolittle 1982). A seminal paper by Lynn (Sagan) Margulis made the endosymbiosis hypothesis again popular, extending the idea of a prokaryotic origin to all eukaryotic organelles, including the nucleus (Sagan 1967). She suggested that mitochondria originated from a heterotrophic Gram-negative bacterium, chloroplast from a cyanobacterium, the flagella from spirochetes, and the nucleus from a wall-less bacterium (or later on a wall-less archaea). The debate remained speculative until Carl Woese and his colleagues solved this question through their strategy based on the analysis of ribosomal rRNA sequences. In 1975, they demonstrated that the 16S RNA oligonucleotide catalogue of a chloroplast was much more similar to the 16S rRNA of *Cyanobacteria* than to the 18S rRNA of the plant containing this chloroplast, not only in terms of size but also in sequence (Zablen et al. 1975). The demonstration turned out to be more difficult for mitochondria because of the extent to which the mitochondrial rRNA has evolved compared to its bacterial ancestor, but it was finally proven in studying plant mitochondria that these organelles too originated from bacteria, more precisely from alphaproteobacteria (Yang et al. 1985). In contrast, Lynn Margulis's idea turned out to be wrong in the case of the eukaryotic flagellum, and it is still controversial in the case of the nucleus. In the last four decades, several authors have regularly suggested a prokaryotic ancestor (an archaeon) for the nucleus (for review, see Forterre 2011). The idea of an endosymbiotic origin of the nucleus has now been abandoned for the more drastic hypothesis that the entire eukaryotic cells (both the nucleus and the cytoplasm) originated from an archaeon (the eocyte hypothesis) that engulfed the bacterial ancestor of mitochondria. It is important in this

(continued)

Box 3.1 (continued)
discussion to keep in mind a striking difference between the eukaryotic universal proteins of mitochondrial origin compared to those with archaeal affinity. Whereas the eukaryotic universal proteins are clearly distinct (even if closely related) from the archaeal ones, the eukaryotic universal proteins of mitochondrial origin (most of them nuclear encoded) are identical to their bacterial ancestors, often branching within bacteria and close to alphaproteobacteria in phylogenetic trees. This helps identifying these proteins that are often annotated as bona fide eukaryotic proteins in public database.

Addressing the TOL question through an experimental program became only possible in the 1960s, thanks to the molecular biology revolution. As soon as 1957, Crick predicted that we would once study evolution by comparing sequences of informational macromolecules, proteins, and nucleic acids (Cobb 2017). This possibility was formally discussed in the seminal paper of Zuckerkandl and Pauling (1965), based on their early work on hemoglobin sequences comparison. Their studies were limited to animals, but they already allowed them to show a rather good correlation between phylogenies based on macromolecular sequences (semantides sensu Zuckerkandl and Pauling) and those obtained by paleontologists based on the fossil record. The molecular revolution in biology was about to change forever our comprehension of the living world and its history, the TOL finally becoming a valid object for sound scientific approaches. Deciphering the topology of the TOL is still an ongoing and controversial quest. This is the history of this quest, as well as the present state of the art, that we will try to describe in this chapter.

2 The Three-Domain Concept: A Revolution in Biology

2.1 The Discovery of Archaea and the Tripartite Division of Life

The first scientific depiction of a complete universal TOL based on sequences comparison was obtained in the late 1970s and early 1980s by Carl Woese and his colleagues of the "Urbana School," using 16/18S rRNA as molecular chronometers (for historical accounts of this scientific saga, see Sapp and Fox 2013; Forterre 2016a). From their painstaking work analyzing catalogues of oligonucleotides obtained after digestion of rRNA with ribonuclease T1 (a method previously designed by Sanger et al. 1965), Woese and Fox concluded 40 years ago that life should not be divided into two realms, eukaryotes and prokaryotes, but into three "urkingdoms": archaebacteria (methanogens only at that time), eubacteria, and eukaryotes (Woese and Fox 1977a) (Fig. 3.1b). Woese and Fox could propose this challenging view to the traditional prokaryote/eukaryote dichotomy because

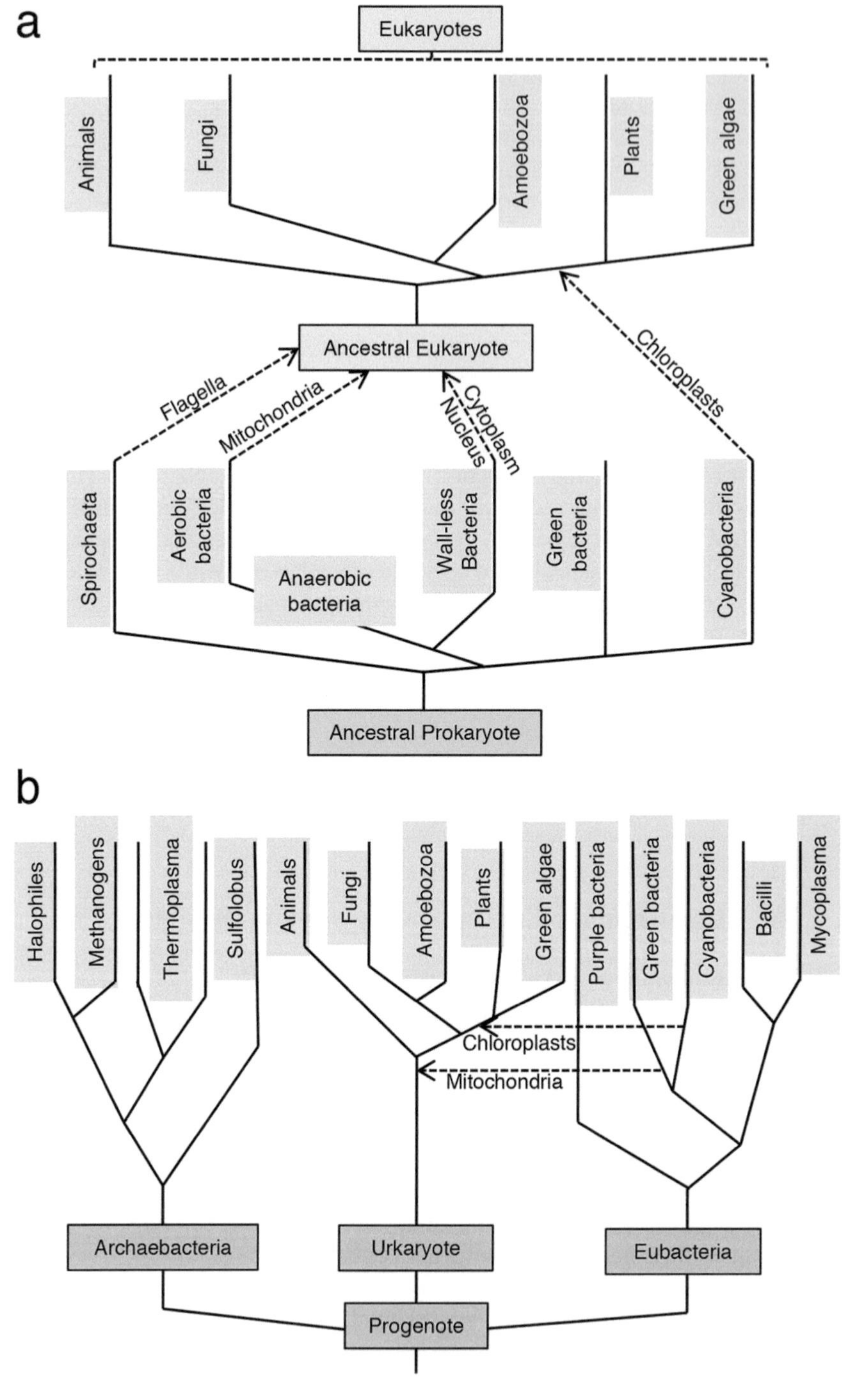

Fig. 3.1 Illustrations showing the transition in the views on life evolution. In the 1960s, a popular view was that eukaryotes emerged from prokaryotes (**a**), but in 1977, Woese and Fox proposed that

archaebacteria—although prokaryotes—were not more closely related to eubacteria than to eukaryotes in the rRNA comparison analysis. They further suggested that the three domains emerged "independently" from a common ancestor that was much simpler than modern organisms, a progenote, an organism that *had not yet completed evolving the link between genotype and phenotype*. This idea was in fact discussed earlier in another historical paper they published the same year (Woese and Fox 1977b). Woese and Fox thus proposed a new vision of the evolution of life in which the one-step scheme (from prokaryotes to eukaryotes) was replaced by a two-step scheme: first from the progenote to prokaryotes and then to eukaryotes (Fig. 3.1b). In this scenario, the prokaryotic phenotype supposedly originated three times independently, the ancestor of modern eukaryotes being an extinct prokaryote (as "an organizational, not a phylogenetic distinction," Woese and Fox 1977b), an "urkaryote" sensu Woese and Fox.

When DNA sequencing techniques became available and widespread in the 1980s, it became possible to build TOLs based on the alignments of orthologous nucleotides from complete rRNA sequences (Pace et al. 1986). Archaebacteria (later on renamed Archaea, the term we will use thereafter in this chapter; see further) were also clearly distinct from eubacteria (later on renamed Bacteria) in the first rRNA trees built using distance-based methods (Pace et al. 1986). However, the branch leading to eukaryotes was much longer than the two others in these trees. This could have suggested that the division between prokaryotes and eukaryotes was meaningful after all, with Archaea and Bacteria on one side and Eukaryotes on the other (Fig. 3.2a). Nevertheless, Woese and his co-workers maintained the three "urkingdoms" concept, considering that both Archaea and Bacteria were monophyletic groups in the 16/18S rRNA trees (Pace et al. 1986; Olsen and Woese 1989).

At the end of the 1980s and in the 1990s, the first published TOLs based on universal proteins (elongations factors, RNA polymerase, and ATP synthases) confirmed the tripartite division of the living world but revealed a striking difference with the rRNA trees. Indeed, whereas Archaea were more closely related to Bacteria in rRNA trees (this is still the case in recent analyses; see Furukawa et al. 2017), they were much more closely related to eukaryotes in all universal protein trees (Iwabe et al. 1989, 1991; Gogarten et al. 1989; Pühler et al. 1989; Linkkila and Gogarten 1991; Klenk et al. 1991; Klenk and Zillig 1994; reviewed in Brown and Doolittle 1997) (see Fig. 3.2b for an example). This finally convinced many biologists of the necessity to replace the classic prokaryote/eukaryote dichotomy by the three-domain concept of Carl Woese. Interestingly, there is no clear explanation yet to interpret this difference between the eukaryotic branch lengths in the rRNA and protein TOLs.

Fig. 3.1 (continued) three kingdoms emerged independently from the progenote, an organizational prokaryote (**b**). Illustrations adapted from *Microbes from Hell*, Patrick Forterre, University of Chicago Press (2016)

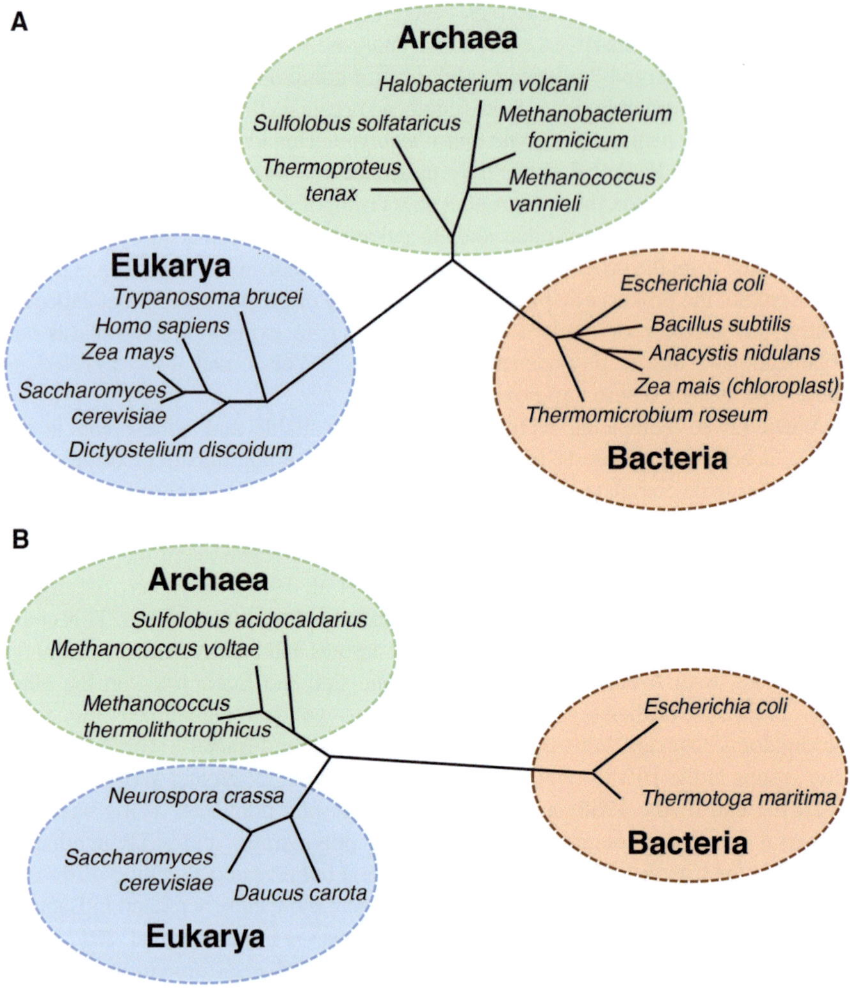

Fig. 3.2 Illustrations of three-domain trees. In rRNA trees, Archaea (Archaebacteria) were usually more closely related to Bacteria (Eubacteria) than to eukaryotes, since the branch leading to the latter was much longer than the two others (**a**) (adapted from Pace et al. 1986). However, Archaea are much more closely related to Eukarya in universal protein trees (**b**) (adapted from Linkkila and Gogarten 1991)

The three-domain concept turned out to be heuristic in that many biologists became inspired to study Archaea, which at that time were poorly known microorganisms. Some started working on various aspects of their molecular biology and biochemistry, while others started looking for new archaea in various biotopes. Both approaches were rapidly successful: biochemists confirmed the uniqueness and unity of Archaea by showing that organisms as diverse as methanogens, halophiles, and thermoacidophiles shared atypical phospholipids that differ from those of Bacteria and Eukarya, both in terms of composition and stereospecificity (Tornabene and Langworthy 1979). Wolfram Zillig and Karl-Otto Stetter soon discovered new groups of Archaea living at very high temperatures in anoxic environments (reviews in Stetter 2013; Albers et al. 2013; Forterre 2016a). Stetter designated the term "hyperthermophile" for those organisms with optimal growth temperatures of 80 °C or higher (Stetter 1989). Zillig was the first to detect striking similarities between the molecular mechanisms of eukaryotes and archaea in studying their RNA polymerases (Zillig et al. 1979; Huet et al. 1983). He also discovered in Archaea new families of viruses that were dramatically different from viruses known to infect bacteria or eukaryotes (reviewed in Zillig et al. 1996; Prangishvili 2013). These findings gave support to the notion of Archaea being a kingdom on their own at the same taxonomic level as Bacteria and Eukarya.

The work of Carl Woese and his colleagues was also important to redefine the status of Bacteria. Bacteria were (and still are) often confused with prokaryotes and even more with primitive organisms. Comparative analyses of rRNA from mycoplasma revealed, for instance, that these minute Bacteria were not, as previously supposed, the remnants of primordial cells but of highly derived *Firmicutes* having loss their cell wall (Woese et al. 1980). The classic Gram-positive/Gram-negative division used to classify Bacteria appeared meaningless on an evolutionary point of view, since most bacterial phyla were populated with Gram-negative species (Woese 1987). In fact, Bacteria could probably be considered as the most successful of the three domains, being present everywhere on the planet and colonizing all Eukarya. Bacteria share unique features, most likely inherited from their last common ancestor, such as rigid cell wall made of peptidoglycan or else DNA gyrase allowing them to directly connect their physiological state to their gene expression pattern (Forterre and Gadelle 2009).

Our view of the domain Eukarya was also impacted by the rRNA comparative analyses, revealing, for example, that fungi are more closely related to animals than to plants. In that case, however, rRNA phylogenies were partly misleading because of a long branch attraction (LBA) artifact (see Box 3.2) that clusters together several groups of parasitic eukaryotes lacking mitochondria at the base of the eukaryotic tree (Vossbrinck et al. 1987). This suggested that these organisms, tentatively called Archezoa (Cavalier-Smith 1989), diverged from other eukaryotes before the mitochondrial endosymbiosis. Refuting the archezoa hypothesis, later studies uncovered nuclear genes of mitochondrial origin and/or mitochondria-derived organelles in all putative Archezoa, revealing that all modern Eukarya actually originated from a last eukaryotic common ancestor (LECA) that already had mitochondria (Embley and Hirt 1998; Hirt et al. 1999; Philippe et al. 2000a). The so-called Archezoa are actually fast-evolving eukaryotic parasitic species that have lost mitochondria (Embley and Hirt 1998). Their long branches in phylogenetic analyses based on rRNA and some fast-evolving proteins, such as the elongation factor EF2, were systematically attracted at the base of the eukaryotic tree by the long branches of the outgroup sequences from another domain (Kamaichi et al. 1996; Philippe and Germot 2000; Philippe et al. 2000b). Notably, phylogenies of slowly evolving proteins allowed recovering the correct position of these organisms such as the grouping of *Microsporidia* with *Fungi* (Keeling and Doolittle 1996; Edlind et al. 1996; Hirt et al. 1999). This was a major alert in the field about the pitfalls associated to the interpretation of molecular phylogenetic analyses, especially LBA (Gribaldo and Philippe 2002; Inagaki et al. 2004). The "Archezoa" story revealed how unequal rate of evolution could bias interpretations and how cautious one should be in introducing fast-evolving organisms in species datasets. This marked the start of a continuous search for a model and framework that could correctly deal with any sort of bias possibly embedded in a dataset. In particular, distance-based methods were progressively abandoned and replaced by more elaborated methods that try to take into account the complex history of sequence evolution (see Box 3.2).

Box 3.2 The Evolution of Phylogenetics

Until the mid-twentieth century, evolutionary relationships between species were essentially estimated by taxonomists through the comparison of limited, observed traits. Starting in the 1950s, a desire to bring more rigorous mathematics into the field of evolution emerged, pushed by the combined rise of the molecular revolution that was offering access to large data with unambiguous states for characters (DNA or protein sequences) and of the accessibility of computers. This ambition of using mathematics approaches on large and objective data was initially transcribed by phenetics (Michener and Sokal 1957), for which the main principle is to compare the overall similarity between organisms without any subjective assumption. This led to the development of distance methods (initiated by Fitch and Margoliash 1967), which score the data based on the comparative identity of related sequences (i.e., from a multiple sequence alignment) and translate it into genetic distances. In opposition to phenetics, cladistics (Hennig 1966) posed that characters should be weighted differently in order to differentiate the signal inherited from a shared ancestor from homoplasy (convergent evolution, transfers, so on). If the phenetic vision was progressively abandoned because it only reflected apparent global similarity instead of evolutionary histories of related sequences (orthologs), its intensive use of statistical approaches made its way into cladistics to deal with the very large dataset of DNA and protein sequences. Parsimony approaches (Edwards and Cavalli-Sforza 1963; Camin and Sokal 1965) hence proposed to differently score the states of characters in a comparison of sequences and to compute the trees (building phylogenies) explaining the distribution of data with the fewest evolutionary changes. With time, more variable parameters and approaches were introduced, to the point that it became complex to really tell if a method is phenetic or cladistic by nature; "computational phylogenetics" now refers to this statistical aspect of cladistics. The rise of probabilistic methods (maximum likelihood, Felsenstein 1981; and Bayesian inferences, Li 1996) allowed integrating more sophistication, with the use of models designed to simulate the

(continued)

Box 3.2 (continued)

evolutionary behavior of sequences. The most recent models can deal with different complex issues, either by their design based on very numerous alignments (empirical models, such as LG; Le and Gascuel 2008) or by encompassing complex variables to notably model heterogeneity across sites and sequences (such as CAT-GTR, the most heterogeneity-aware model to date; Lartillot and Philippe 2004). Obviously, the more variables to compute, the more computationally demanding the analysis will be: this can restrict the choice of methods, since, for instance, the CAT-GTR model can only be used in Bayesian frameworks so far, less resource-consuming than the maximum likelihood approaches by their different mathematical computations. The aspiration to develop more sophisticated approaches was not only motivated by the need to integrate or combine more data but also to limit the impact of known artifacts (Tavare 1986; Lake 1987; Gouy and Li 1989; Bruno and Halpern 1999; Swofford et al. 2001; Huelsenbeck et al. 2002). For instance, a well-known artifact is the trend of sequences with rates of evolution too much divergent than the average in the dataset to falsely locate together in the resulting tree, sometimes with strong support: accordingly, long branches in phylogenetic trees can be attracted to the root by the outgroup (long branch attraction or LBA). Parsimony approaches are recognized to be quite sensitive to this artifact (the assumption that evolutionary changes are rare event cannot indeed be systematically hold true for large data), and probabilistic approaches and their different models were designed in part to deal with this issue (Felsenstein 1981). Yet, and despite clear improvements in the resolution of trees (notably through computations of supports), it has been shown that even the most recent approaches to date cannot compellingly deal with too strong LBA artifacts (Gouy et al. 2015). Particularly, some were designed to deal with this issue under specific conditions, such as the length of the sequences to analyze (this is the case for the CAT-GTR model in Bayesian framework, essentially made for long concatenations). The composition of the dataset to analyze will also severely impact the resulting trees and can generate other issues, such as amino acid composition biases or uninformative sites. Accordingly, the ability of a method to deal with artifacts has to be considered in relative terms, and choices of approaches and models have to be made with caution to fit the data. For example, a too simplistic model will probably not faithfully reflect the evolutionary scenario of proteins; in opposition, an overfitting model will also generate artifacts.

Regardless of the approach used, the global concept of computational phylogenetics however stayed the same: aligning comparable sequences (i.e., orthologs), removing ambiguous sites, and reconstructing a putative evolutionary tree explaining the distribution of data and accordingly reflecting a hypothesis about the evolution of the gene(s)/protein(s)/species (Yang and Rannala 2012). Each step, which can be performed through different approaches, is hence critical and will impact the downstream analyses.

2.2 The Root of the Trees and the Last Universal Common Ancestor

The first TOLs published by Woese and his co-workers based on rRNA were unrooted since, by definition, there is no organism living today that can be considered as outgroup to modern organisms. In 1989, two research groups independently proposed to root the TOLs using paralogous proteins that originated from a gene duplication event that occurred before the emergence of the last universal common ancestor (LUCA) (Iwabe et al. 1989; Gogarten et al. 1989). They identified two couples of proteins fulfilling this criterion: the two translation elongation factors (called EFTu and EFG in Bacteria, EF1, and EF2 in Archaea and Eukarya) and the two major subunits of the ATP synthases (catalytic and regulatory). The similarities between paralogous proteins in each couple allow building two TOLs linked to each other and to root each of these TOLs using the other as the outgroup. This strategy provided independently four roots for the TOLs, each corresponding to the root of one of the four paralogous protein trees rooted using its paralogous counterpart. Iwabe, Gogarten, and their colleagues found that these four roots were all located between Bacteria and a lineage leading to the common ancestor of Archaea and Eukarya. Using a few other couples of paralogous proteins (carbamoyl synthetase subunits, signal recognition particles, and a few tRNA synthetases), Doolittle and colleagues also recovered the same rooting (often dubbed the bacterial root) in the following decade (Brown and Doolittle 1997).

Some authors, including one of us, debated about the validity of the bacterial root in the 1990s, especially concerned by the possibility that the very long branch of Bacteria in all couples of paralogous proteins used could be artificially attracted by the long branches of the outgroup paralogous sequences (Forterre and Philippe 1999). Another concern was the mutational saturation of all protein sequences in these protein couples (Philippe and Forterre 1999). Nonetheless, most biologists rapidly accepted this so-called bacterial rooting, possibly because it suggested at first sight that the universal ancestor, sharing traits common to Archaea and Bacteria, was a prokaryote after all. The rRNA TOL could thus be interpreted in the classic evolutionary framework, prokaryotes being the ancestors of eukaryotes. Carl Woese himself supported this rooting because the two prokaryotic domains are not sister groups in this TOL, definitely justifying to unlink archaea(bacteria) and (eu) bacteria in a proper taxonomic system. In their seminal 1990 paper, Woese, Kandler, and Wheelis thus rooted the universal rRNA TOL in the bacterial branch, drawing what is now referred as the classic Woese tree of life (Woese et al. 1990). Considering that the "urkingdoms" formerly called "archaebacteria" and "eubacteria" were separated in this tree, they proposed the nomenclature now commonly used, classifying cellular organisms into three "domains" (instead of urkingdoms): Archaea, Bacteria, and Eukarya. In the same paper, Woese and his colleagues proposed dividing the domain Archaea into two major kingdoms, Crenarchaeota and Euryarchaeota, corresponding to the two main branches of Archaea previously observed in 16S rRNA trees. The Crenarchaeota were named from the Greek word *crenos* (origin) because they included only thermophiles and Woese was a strong proponent of hypotheses suggesting a hot origin of life on our planet (Woese 1979). The Euryarchaeota were named from the Greek word *euryos* (diversity) because they included archaea with very diverse phenotypes: halophiles, methanogens, and various hyperthermophiles.

During the last two decades, the accumulation of observations showing that the bacterial informational molecular machineries are very different from those of archaea and eukaryotes eventually convinced most biologists of the bacterial rooting. It is now widely accepted that this root corresponds to the deepest bifurcation of the TOL. Indeed, despite the absence of robust phylogenetic support in the phylogenies of paralogous proteins that diverged before LUCA (Philippe and Forterre 1999), the bacterial rooting appears to be the most parsimonious one to explain the distribution pattern of molecular features between the three domains (Olsen and Woese 1997; Forterre 2015). For example, Archaea and Eukarya share 33 ribosomal proteins (r-proteins) that have no homologue in Bacteria, whereas there is not a single r-protein only shared by Archaea and Bacteria or by Bacteria and Eukarya (Fig. 3.3a). These 33 proteins are localized in the ribosomes at about the same position than 23 nonhomologous r-proteins specific of bacteria (Lecompte et al. 2002). The bacterial rooting easily explains this observation by the independent addition of these 33 and 23 r-proteins in the branches leading from LUCA to the ancestor of Bacteria on one side and to Archaea/Eukarya on the other, respectively (Forterre 2015) (Fig. 3.3b). More generally, comparative biochemistry and later on comparative genomics have shown that Archaea and Eukarya share many unique traits in their informational mechanisms (replication, transcription, translation) and operational ones (secretion proteins, membrane-bound ATPase, components of the cytoskeleton, and vesicular systems) that might have emerged in the branch leading to the common ancestor of these two domains. This implies (as first proposed by Woese) that modern forms of transcription, translation, and genome replication were only settled after the divergence of two main branches of the TOL (Woese and Fox 1977b).

The bacterial rooting implies that only homologous traits present in Bacteria, and at least one of the two other domains can be safely attributed to LUCA, provided they were not laterally transferred between domains. This suggests that LUCA was much simpler than modern cells, with smaller ribosomes (Lecompte et al. 2002), a RNA genome (Leipe et al. 1999; Forterre 2006), no ATP synthase (Mulkidjanian et al. 2007), and no initiation transcription factors (Werner and Grohmann 2011), since many proteins involved in these processes are not homologous between Bacteria and Archaea. LUCA was therefore clearly different from modern prokaryotes, and all superficial resemblances observed between Archaea and Bacteria

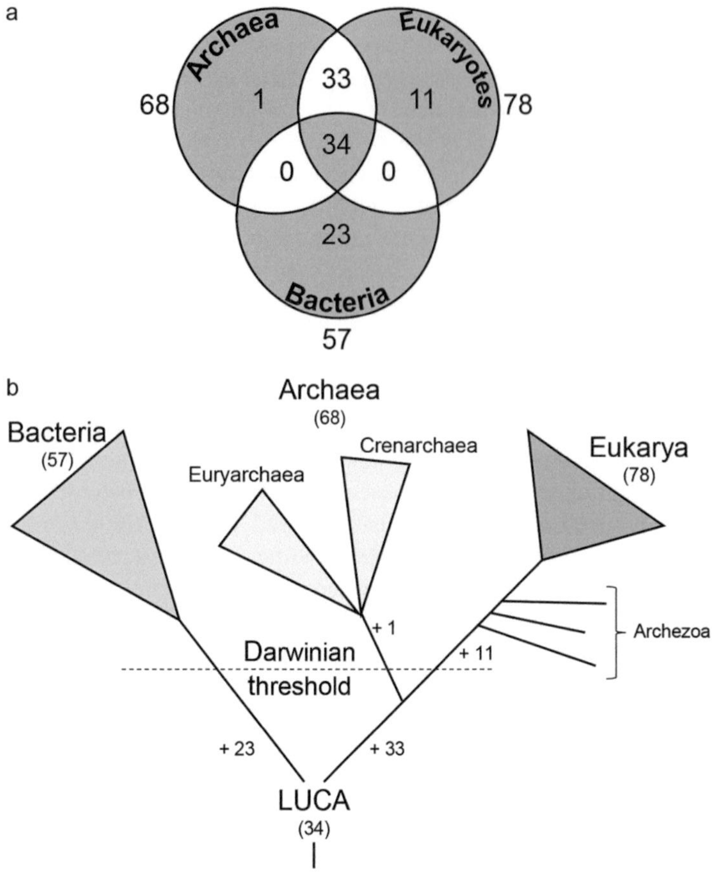

Fig. 3.3 The ribosomal proteins in the three domains. (**a**) Venn diagram showing the distribution of ribosomal proteins among the three domains of life. (**b**) Illustration of the most parsimonious scenario of gains of these proteins in the different domains when rooting the tree in the Bacteria

(the prokaryotic phenotype per say) could hence be due to convergent evolution (Forterre 2015). However, LUCA was probably not a progenote in its original definition, i.e., an organism so primitive that it was not able to synthesize proteins accurately (Woese and Fox 1977b). This is unlikely, because some tRNA modification enzymes critical for faithful translation are known to be universal, so probably present in LUCA (Grosjean et al. 2007; Forterre 2015).

2.3 *The Tree of Life and the Tempo of Biological Evolution*

A striking feature of the TOLs, in both their ancient and more recent versions, is that the branches leading from the last common ancestors of each of the three domains (thereafter called the three ancestors) to modern organisms are similar in length or even sometimes longer than the branches leading from LUCA to the three ancestors. This is especially noticeable for the ancestors of Archaea and of Bacteria, which could have emerged more than 3 billion years ago, possibly only a few hundred million years after LUCA. These two ancestors should have nonetheless been quite similar to modern Archaea and Bacteria, since they shared all characters present today in all members of these domains. If LUCA was an organism much simpler than modern organisms, this means that the tempo of evolution was drastically higher between LUCA and the three ancestors than it has been since the three domains were established and has slowed down dramatically thereafter. During the last 3 billion years, the organisms continued evolving, but within the borders of their domain, bacteria remained bacteria, archaea remained archaea, and eukaryotes remained eukaryotes. Notably, all members of a given domain have very similar ribosomes, whereas ribosomes are very different from one domain to the other. Woese already noticed this drastic reduction in the tempo of life evolution in his early papers and suggested first that it was due to the establishment of a firm link between the genotype and the phenotype that marked the end of the "progenote" stage of evolution (Woese and Fox 1977a). Later on, he suggested that this reduction was linked to a dramatic decrease in the extent of lateral gene transfers (LGTs) that he supposed to have taken place between LUCA and the three ancestors (Woese 1998, 2000). He suggested that LGTs were so prevalent in early times after LUCA that speciation was impossible and that all organisms evolved in concert ("communal evolution"). In this scenario, the formation of the domains took place progressively when the LGT rate cooled down, opening the possibility for speciation (branching) to occur. In Woese words: "This tree [the TOL] became an organismal tree only as it grew, only as its more superficial branches emerged" (Woese 1998). Woese suggested that the Darwinian model of evolution was only valid after this emergence and called the transition period between the communal and Darwinian types of evolution the "Darwinian threshold" (Woese 2002). The formulation by Woese of the "Darwinian threshold" could explain why he never proposed a name to the clade grouping Archaea and Eukarya in his TOL, considering that branching patterns that occurred before this threshold should not be used to determine taxonomic divisions.

The Darwinian threshold concept was criticized by authors who noticed that there is no evidence that LGTs were much more frequent before the emergence of the three ancestors (Poole 2009) and argued that LGT or symbiosis events are actually bona fide "Darwinian" variations that do not alter the tree-like structure of the overall evolutionary pattern (Forterre 2012). Forterre thus suggested more recently naming "Arkarya" the clade grouping Archaea and Eukarya (Forterre 2015). As an alternative to explain the reduction of evolutionary tempo that occurred between LUCA and the three ancestors, Forterre also suggested that the dramatic slowdown in the evolutionary tempo could have corresponded to the transition from organisms with RNA genomes toward organisms with DNA genomes (Forterre 2006).

3 The Eocyte Controversy: New Methods and New Trees

3.1 The Eocyte Hypothesis

From the very beginning, the work by Carl Woese and his colleagues triggered intense debates about the different visions of the TOL. The three-domain concept was rapidly challenged by James Lake based on structural features of the ribosome observed by electron microscopy. Lake and his co-workers reported in 1984 that the ribosomes of thermophilic archaea of the genera *Sulfolobus*, *Thermoproteus*, *Desulfurococcus*, *Thermococcus*, and *Thermoplasma* exhibit some structural features that are also present in the eukaryotic ribosomes, but not in the ribosomes of bacteria and of other archaea (mostly halophiles and methanogens at that time) (Lake et al. 1984). Lake interpreted these features as synapomorphies testifying for a clade grouping eukaryotes and most thermophilic archaea. He proposed naming these organisms "eocytes" (from the Greek words "dawn" and "cells"), according to the prevalent view at the time that thermophiles were the first living forms that appeared on our planet. Lake and his co-workers thus proposed a new TOL in which the living world was not divided in three but in four primary kingdoms: Bacteria, Archaea (reduced to methanogens and halophiles), Eocytes, and Eukaryotes (Lake et al. 1984). Once rooted between Bacteria and Archaea, such tree produces a topology in which the Archaea (including Eocyte) are paraphyletic, with eukaryotes emerging from them, as sister group to Eocytes (Fig. 3.4a). The TOL suggested by Lake in 1984 became the ancestor of all recent so-called eocyte trees, in which eukaryotes emerge from a specific archaeal lineage. To explain why the rRNA tree of Woese and colleagues did not recover the eocyte topology, Lake suggested that the phylogenetic tree reconstruction algorithms used by these authors (distance-based) were biased by artifacts, notably unequal rates of nucleotide substitutions among branches, that produced an artificial grouping of the Archaea together (including Eocytes) due to the attraction between the long branches of Eukarya and Bacteria (Lake 1988).

The specific eocyte ribosomal features described by Lake and colleagues were soon disputed by other ribosome experts who observed similar features in the ribosome of *Methanococcus vannielii*, a mesophilic euryarchaeon (Stöffler-Meilicke et al. 1986). Also worth to be mentioned, the first described Eocytes based on ribosome structure included organisms that later turned out to be both Crenarchaea (*Sulfolobus*, *Thermoproteus*, *Desulfurococcus*) and Euryarchaea (*Thermococcus* and *Thermoplasma*). However, when Lake published few years later further papers supporting an "eocyte tree," this time based on rRNA sequence comparison, Eocytes were limited to Crenarchaea (Lake 1988). Lake obtained this result using a new algorithm called "evolutionary parsimony" (Lake 1987) designed to correct the putative LBA effect that would be at the origin of the Woese TOL in previous analyses.

In the following years, different authors discussed Lake's conclusions and published various TOLs, first based on rRNA analyses and later on universal protein

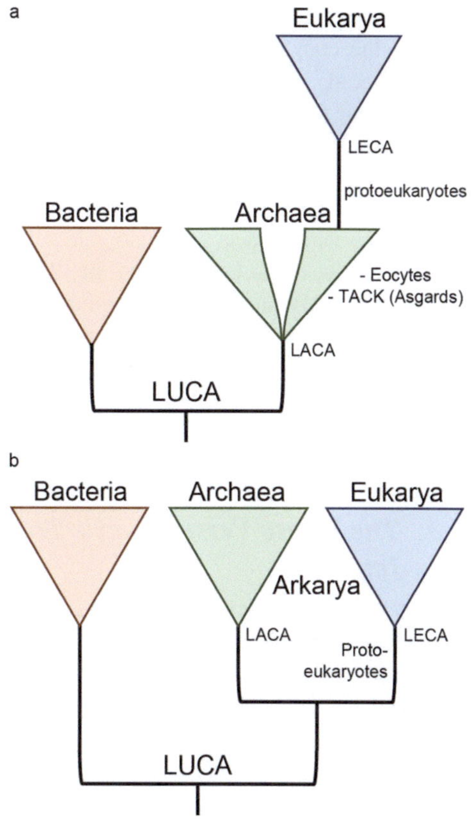

Fig. 3.4 Schematic representations of the eocyte and the Woese TOL. In the eocyte TOL (**a**), Eukarya emerged from within the Archaea, sister group to the Eocytes in the early versions of this scenario, and to Asgard archaea more recently (part of the putative TACK superphylum). In the Woese TOL (**b**), Archaea and Eukarya are sister groups and share a common ancestor. It has been proposed to name this grouping "Arkarya" (Forterre 2015). In both scenarios, we do not know much about proto-eukaryotes that existed before the last eukaryotic common ancestor (LECA) and that emerged from Archaea in the eocyte TOLs or diverged from the common ancestor shared with Archaea in the Woese TOLs

sequences. Gouy and Li found that analyses of 16/18S rRNA data by the neighbor joining (distance method) but also by maximum parsimony strongly supported the Woese's tree and that the probability of recovering the eocyte tree was relatively low for the evolutionary parsimony method designed by Lake (Gouy and Li 1989). In the same paper, their analyses of 23/28S rRNA sequences with the three methods strongly favored the Woese TOL. The Woese TOL was also supported by the first published universal protein phylogenies based on the translation elongation factors (Gribaldo and Cammarano 1998) or based on the RNA polymerases (Klenk et al. 1991; Iwabe et al. 1991) (Fig. 3.4b). Surprisingly, the RNA polymerase was progressively abandoned in the following years as universal TOL marker, and later works in the next decade were more focused on the elongation factors. However, the use of the elongation factors to build the TOL rapidly appeared to be a challenging exercise. For instance, Cammarano and co-worker reported that EF1(Tu) and EF2(G) produce different topologies, with EF2 strongly supporting the Woese tree with any method of tree reconstruction available at that time, whereas EF1(Tu) weakly supported an eocyte tree (Cammarano et al. 1992; Creti et al. 1994). Lake and co-workers also obtained an eocyte tree with EF1(Tu), using a newly designed algorithm, paralinear distances (Lake 1994). In 1996, Doolittle and co-workers published a paper in which a combination of EF1(Tu) and EF2(G) favored the eocyte tree with various distance and maximum likelihood methods (Baldauf et al. 1996). Later on, from an in-depth analysis of the EF2(G) sequence aligned using structural data, Cammarano and colleagues finally concluded that both maximum likelihood and maximum parsimony methods do not allow discrimination between the eocyte and Woese trees for this particular protein (Cammarano et al. 1999). This was possibly because the EF2(G) dataset was already heavily saturated at that time with respect to amino acid substitutions (Philippe and Forterre 1999). More recently, it has been shown that these proteins have a very complex evolutionary history, with many duplications that occurred in the branches leading to Bacteria and to Eukaryotes, and a possible duplication before LUCA (Atkinson 2015). All these observations indicate that EF2(G) is not a reliable marker for deep phylogeny.

3.2 The Woese Versus Eocyte Trees of Life: A Phylogenetic Impasse?

In the first years of this century, new sequencing data were produced at an accelerating pace, revealing first more and more microbial protein sequences, then complete microbial genomes, and finally metagenomes. This led to the description of new microbial lineages from in silico reconstructions from environmental DNA of partial or nearly complete genomes of uncultivated microbes. This progressively increased the number of candidate proteins for universal markers, allowing the comparison of many new trees based on different universal proteins and the production of integrated trees, either based on the concatenation of universal protein sequences or on the combination of individual trees to produce "supertrees." The first published papers analyzing either the concatenation of universal proteins or of the topologies of individual universal proteins recovered (or favored) the Woese's tree with all methods available at that time (Brown et al. 2001; Harris et al. 2003; Ciccarelli et al. 2006). The Woese TOL was also obtained from the building of "whole genome trees" (for reviews, see Wolf et al. 2002; House and Fitz-Gibbon 2002) or from statistical analyses of many single-protein trees (Yutin et al. 2008). In parallel, other groups started recovering "eocyte trees" using gene presence/absence analyses (Rivera and Lake 2004), supertree methods (Pisani et al. 2007), and finally universal protein concatenations (Cox et al. 2008; Foster et al. 2009). All these studies were plagued by important drawbacks (for their detailed analyses, see Gribaldo et al. 2010). They were either based on limited datasets with few representatives of each domain (Brown et al. 2001; Harris et al. 2003) and/or on unbalanced datasets with very different number of species and/or proteins per domain (Ciccarelli et al. 2006). Some of them were performed with methods very sensitive to lateral gene transfer (LGT) (Rivera and Lake 2004; Pisani et al. 2007) producing odd results. For instance, the eukaryotes were sister group to *Thermoplasma* (a Euryarchaeon) in the eocyte tree obtained by McInerney and colleagues (Pisani et al. 2007).

For a few years now, the Woese and eocyte TOL have been described as the 3D (three domains) versus 2D (two domains) TOL, respectively (Gribaldo et al. 2010), considering that the eocyte tree is characterized by two primary domains (Archaea and Bacteria) that merged later on to produce a secondary domain: Eukarya. Strictly speaking, the Woese TOL can however be also viewed as a 2D TOL if the clade grouping Archaea and Eukarya (Arkarya sensu Forterre 2015) is considered itself as a primary domain. This is the reason why we have preferred in this chapter the nomenclature, "Woese" versus "eocyte" TOLs (Fig. 3.4).

Some authors wondered if the difference between the Woese (3D) and eocyte (2D) TOLs was really meaningful, considering that the specific archaeal branch was anyway always short in Woese TOLs (Rochette et al. 2014). They interpreted this short stem as suggesting that the last common ancestor of Archaea and Eukarya was probably very similar to the last archaeal common ancestor (LACA). However, one should not forget that the construction of TOLs from trimmed alignments artificially reduces the evolutionary distance between organisms by eliminating regions that are too variable in terms of sequence such as regions containing indels. The branch lengths thus necessarily underestimate the extent of divergence between the organisms at the various nodes for deep branches. As an example, the branches connecting modern organisms to the three-domain ancestors are sometimes longer than the branches connecting LUCA to these ancestors, although these ancestors were probably very similar to modern organisms but certainly very different from LUCA. If the Woese TOL is correct, the last common ancestor of Archaea and Eukarya may have been thus quite different from LACA, despite the short length of the archaeal branch.

3.3 *The Revival of the Eocyte Tree*

From this succession of contradictory results obtained at the beginning of this century by scientists using mostly overlapping protein datasets, it was concluded that we were in a *phylogenomic impasse* regarding the evolutionary relationships between Archaea and Eukarya (Gribaldo et al. 2010). However, in the following years, the work of Embley's group (Cox et al. 2008; Foster et al. 2009) started to shift progressively the balance in favor of eocyte trees in the literature. These authors argued that their work, supporting the emergence of eukaryotes from within Archaea, was of better quality than previous ones because they used newer models of evolution, more adapted to TOL reconstruction, in a Bayesian framework. As previously suggested by Lake, Embley and colleagues claimed that the Woese topology was an LBA artifact due to the stationary models often used. However, it should be noticed that Embley and colleagues also recovered the eocyte topology with less sophisticated homogeneous models (Cox et al. 2008), suggesting that "their result may be linked more to the dataset used (the gene and the taxonomic sampling) than to the new evolutionary models" (Gribaldo et al. 2010).

In the meantime, a third major phylum of Archaea, the Thaumarchaeota, was proposed by Forterre and colleagues based on the concatenation of ribosomal proteins conserved between Archaea and Eukarya (Brochier-Armanet et al. 2008a). In their tree, Archaea were rooted between Thaumarchaeota and all other Archaea. Interestingly, the genomes of Thaumarchaeota turned out to encode several new eukaryotic signature proteins (ESPs), i.e., proteins highly similar to their eukaryotic homologues that were never previously detected in Archaea (Brochier-Armanet et al. 2008a, b; Spang et al. 2010). This suggested that these ESPs were already present in the last common ancestor of Archaea and Eukarya and were later on lost in Euryarchaeota and Crenarchaeota (Brochier-Armanet et al. 2008b). However, in the framework of the eocyte TOL, these ESPs might have also been synapomorphies testifying for the grouping of Eukarya with Thaumarchaeota. In 2011, Guy and Ettema obtained an eocyte tree based on 26 universal proteins in which eukaryotes were sister group to a clade including Thaumarchaeota, Aigarchaeota (metagenomics reconstructed genome of a candidate archaeal phylum close to Thaumarchaeota), Crenarchaeota, and *Candidatus Korarchaeum cryptofilum* (Guy and Ettema 2011). They proposed grouping these four phyla into a putative TACK superphylum. A year after, Embley and colleagues again obtained the sisterhood of eukaryotes and the putative TACK superphylum in a tree based on the concatenation of 29 universal proteins (Williams et al. 2012). The TACK nomenclature was rapidly accepted by most authors in the community of evolutionists and scientists working on Archaea, a clear indication of the eocyte revival (Fig. 3.4a).

In parallel to phylogenomic studies performed to determine the topology of the TOL, the validity of the TOL itself as a metaphor has been criticized in the first decade of this century by authors who proposed replacing the tree by a web or rhizome of life (Dagan and Martin 2009; Koonin 2009; Doolittle 2009; Raoult 2010). The TOL cannot indeed take into account all aspects of life evolution but is a necessary framework to understand events such as LGT, endosymbiosis, hybridization, and so on. However, all these events can also be considered as specific cases of variations that do not change the overall tree-like structure of the TOL (Forterre 2012). Notably, the trend suggesting that microbial evolution cannot be described within the classical Darwinian framework has been vanishing in recent years, possibly because most proponents of the "web of life" became advocates of the recent eocyte trees (Baross and Martin 2015; Koonin 2015; Sousa et al. 2016).

Besides technical and philosophical disputes among evolutionists concerning the topology of the TOL, several authors had previously discussed the relative merit of the Woese versus eocyte trees in terms of biological credibility. For instance, the eocyte hypothesis was strongly criticized by Woese who argued that "modern cells are fully evolved entities [...] sufficiently complex, integrated and 'individualized' that further major change in their designs does not appear possible" (Woese 2000). In other words, it was not possible for an archaeon to become a eukaryote: they should have originated from a common ancestor that was different from both of them. Many biological arguments against the eocyte TOL were published supporting the idea that eukaryotes evolved from a lineage of "proto-eukaryotes," corresponding to the "urkaryotes" of Woese and Fox (de Duve 2007; Kurland et al. 2006; Poole and Penny 2001; Forterre 2011, 2013, 2015). Proponents of eocyte scenarios counterargued that eukaryogenesis was a very special and unique event triggered by the endosymbiosis of one or several bacteria into an archaeal host (López-García and Moreira 2006; Koonin 2006; Williams et al. 2013; Lane and Martin 2015; López-García et al. 2017; and many references therein). Based on this assumption, many hypotheses have been proposed to explain the origin of all modern eukaryotic features in the descendants of the archaeal host. For instance, William Martin and colleagues recently suggested that replacement of archaeal lipids by bacterial ones in modern eukaryotes was triggered by membrane vesicles produced by the bacterial ancestor of mitochondria (Gould et al. 2016). These membrane vesicles would have both accumulated in the cytoplasm to form the intracellular membrane system typical of eukaryotic cells and fused with the archaeal host cytoplasmic membrane. Martin and Lane suggested that all specific eukaryotic features only emerged after the mitochondrial endosymbiosis because this event was necessary to provide the amount of energy required for the complexification of the eukaryotic cell (Lane and Martin 2010). This "mitochondria early" scenario has been refuted by others, including some proponents of the eocyte scenarios, who advocated a late acquisition for mitochondria (Martijn and Ettema 2013; Booth and Doolittle 2015; and references therein). More generally, Forterre argued that it is unclear how the mitochondrial endosymbiosis could have dramatically changed the fundamental mechanisms of the archaeal host molecular biology and the rate of evolution of universal proteins, especially considering that the endosymbiosis at the origin of the chloroplast did not produce such dramatic transformation in the eukaryotic fabric of plants (Forterre 2011, 2013, 2015).

4 The Tree of Life and the Metagenomics Challenges

4.1 The Discovery of Asgards

One of the arguments against the scenario in which eukaryotes originated from the merging of an archaeon and a bacterium is that there is no known example of bacterial endosymbiont living within an archaeon. To answer this point, Martijn and Ettema predicted the existence of a still unknown complex phagocytosing archaeon, member of the TACK superphylum, at the origin of Eukaryotes (the phagocytosing archaeon theory) (Martijn and Ettema 2013). Two years later, Ettema and colleagues announced that they had discovered such an archaeon from sediments located close to the vent site Loki's Castle (Spang et al. 2015). They reconstituted in silico, from environmental DNA, the genomes of three new archaeal lineages, Loki 1, 2, and 3, members of a new tentative phylum that they called Lokiarchaeota. Strikingly, Eukaryotes emerged next to the Lokiarchaea among the Archaea, as sister group to Loki 3 in a concatenation of 36 universal markers in a

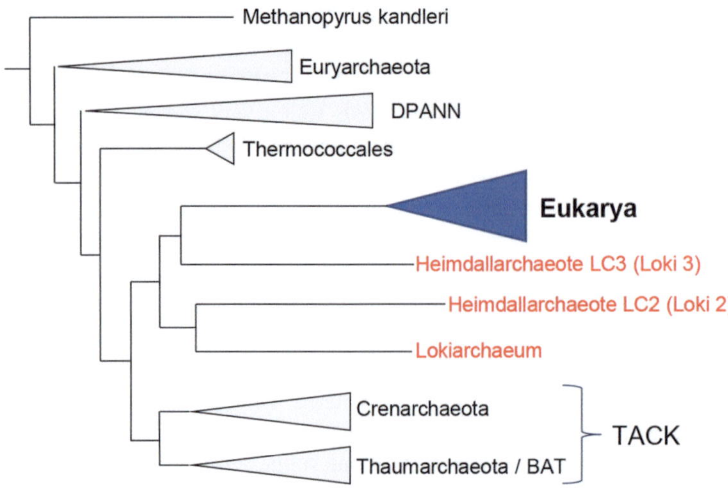

Fig. 3.5 Schematic representation of the Bayesian phylogeny obtained by Spang and co-workers (Fig. 3.2b in Spang et al. 2015) from the concatenation of 36 universal markers. Since its publication, Loki 2 and 3 have been renamed Heimdallarchaeote LC2 and LC3, respectively

Bayesian phylogeny (CAT-GTR model; Fig. 3.2b in Spang et al. 2015) (Fig. 3.5). Moreover, Ettema and colleagues identified in the proteomes of the three Loki new ESPs that could testify for the presence in these organisms of a primitive phagocytosis apparatus, such as actin much more similar to eukaryotic actins than those previously detected in Archaea, new proteins of the ESCRT-III vesicular transport system, and multiple G-proteins. Two years later, Ettema and colleagues published in collaboration with few other laboratories the description of more genomes from new archaeal lineages related to Lokiarchaea, reconstructed from DNA from different site environments worldwide (Zaremba-Niedzwiedzka et al. 2017). These cosmopolitan new archaea again contain many ESPs and branch together with eukaryotes in their phylogenies, supporting their first observations. These lineages were very diverse, justifying and proposing several new phyla: Thorarchaeota (Seitz et al. 2016), Odinarchaeota, and Heimdallarchaeota (Zaremba-Niedzwiedzka et al. 2017). The Loki 2 and 3 initially described in 2015 appeared to be actually members of Heimdallarchaeota (Heimdallarchaeote LC2 and LC3, respectively). Loki, Thor, Odin, and Heimdall, being gods of the ancient Nordic mythology, were grouped by Ettema and colleagues in a putative superphylum, called Asgard (the world of gods).

The discovery of Asgard archaea was considered by some in the scientific community as the final proof validating the eocyte scenario (López-García and Moreira 2015; Koonin 2015; Embley and Williams 2015; Dacks et al. 2016; López-García et al. 2017). The first critics came from Caetano-Anollés and colleagues who noticed that the lokiarchaeal proteomes added only 10 new members (0.1%) to the archaeal protein fold superfamilies (Nasir et al. 2015). A low number for organisms supposed to bridge the gap between Eukarya and other Archaea. These authors also criticized the imbalanced number of species in the dataset studied by Ettema and colleagues (10 Bacteria, 10 Eukarya, and 87 Archaea) (Spang et al. 2015). Such imbalance can affect downstream analyses such as alignment, trimming, and selection of phylogenetically informative regions for tree reconstruction (Nasir et al. 2016). In simulation experiments, Nasir and colleagues recovered seven Woese TOLs out of ten TOLs obtained from subsets of species randomly picked from the Spang et al. (2015) species dataset after the removing of metagenomics reconstructed genomes, each subset containing ten members for each domain (Fig. 3.2 in Nasir et al. 2016). Remarkably, removing the fast-evolving archaeal species *Methanopyrus kandleri* and *Candidatus Korarchaeum cryptofilum* from the three subsets favoring the eocyte topology switched these trees into Woese TOLs, revealing that fast-evolving species (FES) favor the eocyte topology. This confirmed that both an imbalance dataset and the presence of FES in the dataset could influence the TOL reconstruction. Notably, the dataset used more recently by Ettema and colleagues (Zaremba-Niedzwiedzka et al. 2017) was still imbalanced (13 Bacteria, 26 Eukarya, and 116 Archaea) and contained a high number (around 30) of FES and/or genomes reconstructed from metagenomics data. The phylogenetic analyses of Ettema and colleagues were also criticized by Forterre and colleagues who suggested that the grouping of Asgards and Eukaryotes in their trees was due to a combination of contamination of some Asgard proteins by patches of eukaryotic sequences, the choice of universal markers used, and LBA artifacts due to the presence of FES in the dataset (Da Cunha et al. 2017). We will discuss these particular critics in more details in the following paragraphs.

The discovery of many new ESPs in various lineages of Archaea and in other deep branching archaeal lineages provides important information to try reconstructing the history of eukaryogenesis. In the eocyte framework, these ESPs were present in the last common ancestor of Asgard archaea and Eukarya, and all new specific eukaryotic traits originated de novo in the lineage leading from the first eukaryotic common ancestor (FECA) to the last eukaryotic common ancestor, LECA (Eme et al. 2017). In the Woese framework, these ESPs were present in the last common ancestor of Archaea and Eukaryotes, and some specific eukaryotic traits might have originated de novo in the eukaryotic branch (from FECA to LECA), but others might have originated before the split between Archaea and Eukarya and later on lost in Archaea. In both scenarios, one should explain the origin of specific eukaryotic traits. Some could have arisen de novo, but others can have been introduced into proto-eukaryotic lineages by LGTs from ancient bacteria (via predation or endosymbiosis) or from ancient viral lineages that coevolved with proto-eukaryotes.

It remains to be known if all ESPs detected in Asgards are real. A troublesome observation is that Heimdallarchaea LC3 (the most contaminated Asgard according to Da Cunha et al. 2017) is by far the one containing more ESPs (22 out of 31) (Zaremba-Niedzwiedzka et al. 2017). Therefore, "one cannot exclude that genes encoding some ESP were reconstructed from small patches of eukaryotic sequences that were combined with the homologous archaeal sequences present in the sample" (Da Cunha et al. 2017). Alternatively, some of these ESPs could also correspond to ancient gene transfers between Asgards and proto-eukaryotes. This emphasizes the importance to cultivate members of the various Asgard lineages in order to get complete and clean genomes to analyze. At the moment, the only physiological data on Asgard cells have been obtained for Lokiarchaea, which in fact correspond to an archaeal lineage previously detected via rRNA analysis and known as the Deep Sea Archaeal Group (DSAG). Kittel and co-workers, using FISH analysis with a DSAG probe, detected coccoid-like cells with size around 0.2 to 0.4 microns (see panel 4 in Fig. 3.2 in Knittel et al. 2005). A priori, such a small size does not seem really compatible with a phagocytosing organism.

4.2 *The Pandora Box of New Archaeal and Bacterial Lineages*

Independent of the controversy about the position of Asgards in the TOL, the reconstruction by Ettema and colleagues of so many genomes from archaeal lineages previously only known from rRNA sequence analyses was a real tour de force. In the same vein, several groups described during the same period a bunch of new lineages of archaea from metagenomic analyses, which branch as sister groups to Crenarchaeota or Thaumarchaeota and were previously only known from environmental rRNA sequences (for reviews, see Adam et al. 2017; Spang et al. 2017). The groups of Jillian Banfield, Tanja Woyke, and their colleagues further described an explosion of new uncultivated archaeal and bacterial lineages corresponding to organisms with small genomes (between 0.5 and 1 Mb) (Rinke et al. 2013; Brown et al. 2015; Castelle et al. 2015). These organisms have also small sizes <500 nm since they were filtered through 0.22 μm membrane filters. This was confirmed for some of them by electron microscopic observations (Baker et al. 2006, 2010; Comolli et al. 2009). Notably, their genomes systematically lack many genes encoding several essential metabolic pathways, strongly suggesting that these nanosized organisms are ectosymbionts (Brown et al. 2015; Castelle et al. 2015; Golyshina et al. 2017), similar to the extensively studied archaeon *Nanoarchaeum equitans*, that can only grow in symbiosis with its crenarchaeal host, *Ignicoccus hospitalis* (Forterre et al. 2009). Nanosized archaea were grouped into the candidate superphylum DPANN (Diapherotrites, Parvarchaeota, Aenigmarchaeota, Nanoarchaeota, and Nanohaloarchaeota) (Rinke et al. 2013), whereas nanosized bacteria are known either as the *Patescibacteria* superphylum (Rinke et al. 2013) or the candidate phyla radiation (CPR) (Brown et al. 2015). Many sequences from DPANN were progressively added in recent TOL phylogenies and were supposed to profoundly modify our view of the TOL (Spang et al. 2015; Hug et al. 2016; Zaremba-Niedzwiedzka et al. 2017; Parks et al. 2017). Banfield and colleagues suggested that these organisms represent the most part of life diversity and proposed a "new view of the TOL" to emphasize this point (Hug et al. 2016). In their new eocyte TOL, based on the concatenation of 16 ribosomal proteins, Archaea were rooted within DPANN, whereas Bacteria were rooted between CPR and all other

Bacteria (Hug et al. 2016). Similarly, Williams, Ettema, and colleagues rooted the archaeal tree between DPANN and all other archaea from a "consensus unrooted archaeal topology" (Williams et al. 2017). This root was selected because it corresponded to the most parsimonious scenario explaining the pattern of LGTs, gene duplications, and gene losses in a collection of 3242 single-gene trees.

4.3 The Conundrum of Fast-Evolving Species

A major concern in using nanosized organisms in phylogenetic analysis is the risk of LBA, as previously discussed in the case of "Archezoa." Indeed, and as mentioned before, nanosized archaea with reduced genomes are most likely fast-evolving species (FES), as shown in the case of *N. equitans* (Brochier et al. 2005b). In agreement with this hypothesis, Forterre and colleagues identified in the universal protein Kae1 a region that is strictly conserved in length in all cellular organisms except for the presence of various insertions and/or deletions in DPANN, *C. Korarchaeum cryptofilum* and the well-known FES *M. kandleri* (Fig. S38b, in Da Cunha et al. 2017). FES are very difficult to position in phylogenetic analyses, frequently occupying either very deep or very late positions (long branches). Indeed, whereas *Parcubacteria* are early branching in the ribosomal tree of Banfield and colleagues (Hug et al. 2016), they are late branching in the rRNA tree of Woyke and colleagues (Schulz et al. 2017).

The addition of FES in universal protein datasets is often justified by the assumption that adding these sequences is essential to take into account the whole diversity of each domain and to break some long branches of the TOL. Bayesian analyses with nonhomogeneous models such as CAT-GTR (see Box 3.2) have been precisely designed to prevent LBA after the Archezoa conundrum (Lartillot et al. 2007). However, using simulation experiment, Philippe and colleagues have shown in a case study that this correction is critically dependent on the length of the outgroup branch (Gouy et al. 2015). These authors concluded that: "even recent Bayesian methods of tree reconstruction cannot eliminate LBA when the outgroup is very distant." Accordingly, many evolutionists thus prefer avoiding these sequences in their TOL analyses (Petitjean et al. 2014; Raymann et al. 2015).

In reanalyzing the data of Spang and co-workers (Spang et al. 2015), Forterre and colleagues observed that the high number of DPANN and other fast-evolving archaea, such as *Methanopyrus kandleri* and *C. Korarchaeum cryptofilum*, in the species dataset strongly affects single-gene phylogenies (Table 1 in Da Cunha et al. 2017). The individual trees were systematically less robust and recovered less frequently the monophyly of the major archaeal phyla in the presence of FES sequences. This raises legitimate concern about the validity of studies based on global analyses of LGTs in multiple single trees, such as those used by Williams and colleague in order to root the archaeal tree (Williams et al. 2017), since the systematic misplacements of various lineages in single trees can be confused with LGTs. Forterre and colleagues also observed that the number of Woese TOLs jumped from

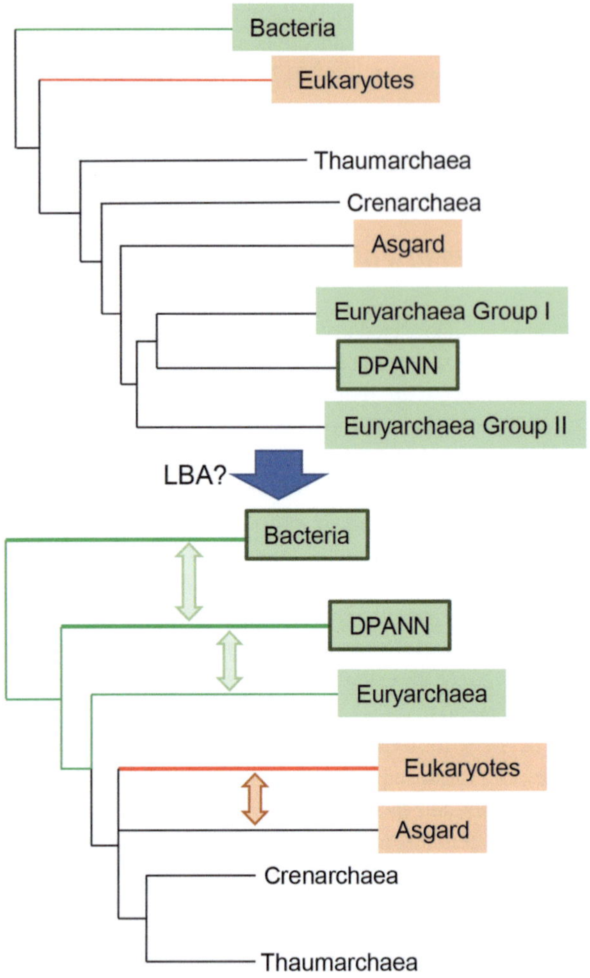

Fig. 3.6 Illustrations of the putative effects of LBA artifacts in the TOL reconstruction. The DPANN would be attracted toward Bacteria due to their long branches and attract the Euryarchaea to which they are closely related. This could result in the artificial grouping of Eukarya with Asgard archaea

1 to 11 (out of 36 universal protein TOLs) when FES were removed from the dataset of Spang et al. (2015). This confirmed that FES tend to favor the eocyte versus the Woese TOL, as shown previously by Nasir and colleagues (Nasir et al. 2016). It is possible that FES attracted Euryarchaeota toward the long branch of Bacteria because *M. kandleri* and DPANN are themselves Euryarchaeota (Brochier et al. 2004, 2005a, b; Brochier-Armanet et al. 2011). This attraction could artificially break the monophyly of Archaea, transforming the actual Woese TOL into an eocyte TOL (Fig. 3.6). The presence of an LBA effect in the first tree that supported the grouping of Asgard archaea and Eukarya (Fig. 3.2b in Spang et al. 2015) is visible,

since Archaea and Eukarya are both rooted in the branch leading to FES. The Eukarya are rooted in the fast-evolving parasitic species, *Trichomonas*, *Entamoeba*, *Leishmania, and Plasmodium*, whereas the archaea are rooted in *M. kandleri* (Fig. 3.5). This again confirms that even the most sophisticated model in a Bayesian framework cannot always correct LBA. Strikingly, Forterre and colleagues observed that *M. kandleri* is located at its correct position (sister group to *Methanobacteriales*; see Brochier et al. 2004) in a maximum likelihood analysis obtained with the same dataset (Fig. S19 in Da Cunha et al. 2017), compared to the original tree (Fig. 3.2b in Spang et al. 2015). This indicates that a Bayesian analysis with the CAT-GTR model can be sometimes more sensitive to LBA than a maximum likelihood analysis with a more stationary model.

4.4 The Problem of Contamination in TOL Analyses

The increasing number of genomes reconstructed from metagenomes in datasets used to build TOLs raises the problem of possible contamination. These genomes can include contaminating patches of sequences from other organisms or from artificially built sequences. These phenomena can increase sequence variations that can mimic the behavior of sequences of FES. It is very difficult to avoid all possible contamination during in silico genome reconstruction. Several tools, such as CheckM and Anvi'o, have been designed to estimate heterogeneity and contamination (Parks et al. 2015; Eren et al. 2015), and they usually consider that a genome is safe to work with when contamination is below 5%. Such low level of contamination can still confuse some phylogenetic analyses. An Anvi'o analysis performed by Forterre and colleagues suggests that the contigs composing the lokiarchaeal genomes reconstructed by Ettema and colleagues can be separated into different sets, with one of them adding only 2% of completeness but increasing the contamination index from 14% to 56% (Fig. S12 in Da Cunha et al. 2017). This suggests that they did not use an optimal strategy for selecting contigs for the final version of the genome and that some contamination cannot be excluded.

Contamination can be sometimes tentatively deduced from the presence in protein sequences of regions that cannot be aligned with sequences from the evolutionarily related organisms (introducing insertions in the alignment) but with distantly related ones. Screening the untrimmed alignments of the 36 proteins analyzed by Spang and colleagues (2015), Forterre and colleagues noticed in the three Asgard elongation factor EF2 sequences several large insertions (up to 31 amino acids) strikingly similar either to sequences of eukaryotic EF2 or to sequences of eukaryotic EF2 paralogues, suggesting that Asgard EF2 sequences could be contaminated by patches of DNA from eukaryotes present in the same environment. In parallel, examination of the 36 individual trees revealed that the EF2/EFG tree was among the few robust ones and the only one strictly identical to the tree obtained after the concatenation of the 36 protein sequences (Fig. 3.2 in Da Cunha et al. 2017). This suggested that EF2 harbors indeed a strong signal that influences the overall topology, a concerning signal considering the potential contamination abovementioned. Indeed, after removing EF2 from the dataset, Forterre and colleagues obtained a tree in which Eukarya were no longer sister group to Asgard archaea (Fig. 3.4 in Da Cunha et al. 2017).

The eukaryotic-like insertions were especially abundant in Heimdallarchaea LC3, the one being sister group to eukaryotes in Spang et al. (2015). The presence of short contaminating sequences in the genomes of Heimdallarchaea could be explained by the fact that they were obtained using multiple displacement amplification (MDA), a method known to introduce chimeric DNA sequences (Lasken and Stockwell 2007; Nurk et al. 2013). Significantly, Ettema and colleagues initially reported that the presence of heimdallarchaeal sequences (corresponding to Loki 2 and Loki 3) was essential to recover the sisterhood of Eukarya and Lokiarchaea (Loki 1) (Fig. S13b in Spang et al. 2015), whereas Forterre and colleagues noticed that removing EF2 from Heimdallarchaea LC3 was sufficient to break the sisterhood of Eukarya and Asgard archaea in maximum likelihood analyses (Fig. 3.2 in Da Cunha et al. 2017).

Ettema and colleagues did not include EF2 in their new dataset of 48 universal proteins (Zaremba-Niedzwiedzka et al. 2017) and again recovered the emergence of Eukarya within Asgard archaea. However, they could not observe convergence in their Bayesian analysis with the CAT-GTR model since all maxdiff values were higher than the recommended values for an acceptable run (<0.3) according to the manual, let alone values for good runs (<0.1) (Table S5 in Zaremba-Niedzwiedzka et al. 2017). This could confirm that the strong signal they obtained with their previous set of 36 universal proteins was indeed deeply influenced by EF2. Ettema and colleagues only obtained a robust grouping of eukaryotes and Asgards in an r-proteins tree without bacterial sequences (not a TOL) arbitrarily rooted in Euryarchaeota and in maximum-likelihood analyses of an rRNA TOL, after removing DPANN. However, they could not observe any convergence of their trees in a Bayesian analysis of the same dataset with the CAT-GTR model (maxdiff values at 1) (Table S5 in Zaremba-Niedzwiedzka et al. 2017).

4.5 Contradictory Signals in Protein Datasets

Although individual protein trees generally carry a weak signal, their analysis can be very informative. In particular, examination of single-gene trees can suggest the presence of contradictory signals in protein sequence datasets; such signal being hidden once the sequences have been concatenated, since by definition, a concatenation acts like an artificial supergene for which only one tree, reflecting the strongest signal, will be inferred. For instance, in a phylogeny obtained after the concatenation of 50 ribosomal proteins, *Nanoarchaeum equitans* branched between Crenarchaeota and Euryarchaeota (Brochier et al. 2005b). However, examining single-protein phylogenies, Forterre and colleagues noticed that single-protein trees rarely recovered this position but that *N. equitans* branched either within Crenarchaeota (9 r-proteins) or within Euryarchaeota (33 r-proteins, 13 as sister group to Thermococcales). It is indeed possible that some original *N. equitans* r-proteins have been replaced by r-proteins from the crenarchaeal host (*Ignicoccus hospitalis*) of this nanoarchaeon. After the removal of the nine crenarchaeal-like r-proteins from the dataset, Forterre and colleagues obtained a tree with *N. equitans* branching as sister group to Thermococcales at the base of Euryarchaeota with strong support (Brochier et al. 2005b). This position was also obtained in single-protein phylogenies of several important informational proteins (Brochier et al. 2005a) and later on supported by a unique synapomorphy (Urbonavicius et al. 2008) and a new phylogeny of archaeal r-proteins with an increased species dataset (Brochier-Armanet et al. 2011). Hence, in-depth analysis of single-gene phylogenies of archaeal ribosomal proteins (r-proteins) was essential to correctly position this nanoarchaeon in the archaeal tree. This example shows that it is sometimes important to restrict the analysis to less data in order to avoid important artifacts that present tools cannot confidently resolve yet.

Examination of single-protein phylogenies of the 36 universal proteins of the Spang et al. (2015) dataset after the removal of FES also revealed two types of contradictory signals: one concerning the position of Asgards in single-gene trees and the other concerning the TOL topology (Da Cunha et al. 2017). Forterre and colleagues noticed that 19 out of 90 Asgard proteins (including EF2) branched as sister group to Eukarya, while all the others branched within Archaea. They suggested that, as in the case of EF2, some of the Asgard proteins with eukaryotic affinity could be contamination by hidden patches of eukaryotic sequences. Another possibility is that their atypical position results from an LBA attraction between Asgard archaea and Eukarya.

Regarding the TOL topology, Forterre and colleagues also identified a set of 25 proteins displaying the eocyte topology and another set of 11 proteins supporting the Woese topology (hereafter called the eocyte proteins and the Woese proteins, respectively) (Fig. 3.1 in Da Cunha et al. 2017). Using a topology test, they could identify in each set the proteins that were the most statistically robust, supporting one topology and rejecting the other (6 Woese proteins and 11 eocyte proteins). Independent concatenations of these two sets produced robust trees with opposite topologies, in both ML and Bayesian frameworks, confirming that the topology of the TOL is strongly dependent on the universal markers used (Figs. S9, S10 in Da Cunha et al. 2017) (Fig. 3.7). The 6 Woese proteins are larger on the average than the 11 eocyte proteins (EF2 is the longest of these) (Fig. 3.7a). Larger proteins usually contain stronger phylogenetic signals, and the authors indeed noticed that they produce more robust TOLs where the monophyly of major archaeal phyla is recovered with higher bootstrap values (Table S1 in Da Cunha et al. 2017) (Figure).

Small r-proteins (less than 150 amino acids) produced poorly resolved trees, even after removing FES. In fact, Forterre, Philippe, and colleagues concluded 15 years ago that r-proteins were useful as markers for archaeal phylogeny but probably not for the TOL because "alignment between ribosomal proteins of different domains (even between Archaea and Eukarya) turned out to be difficult, preventing a meaningful use of bacterial/eukaryal ribosomal proteins to root the archaeal tree or to test the monophyly of Archaea" (Matte-Taillez et al. 2002). Notably, most

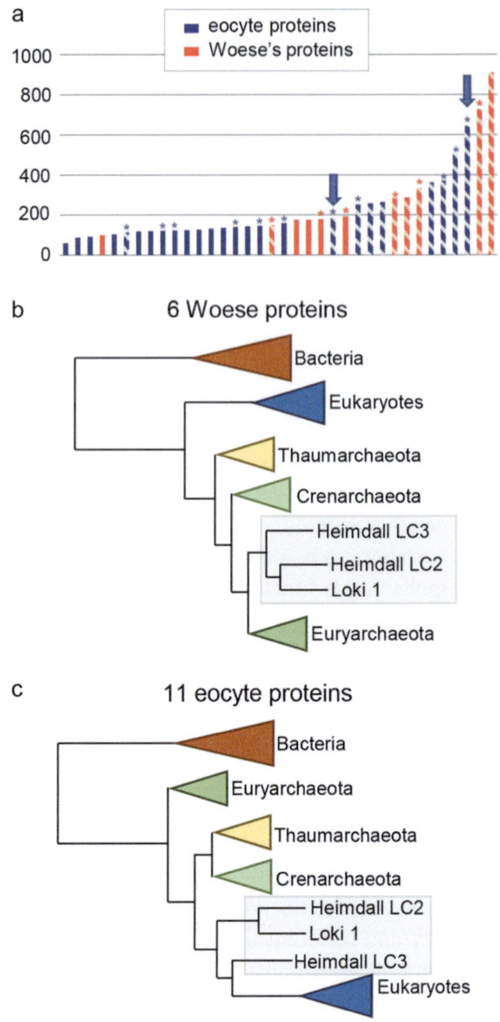

Fig. 3.7 Schematic representations of the results obtained by Da Cunha and colleagues (Da Cunha et al. 2017) from their reanalyses of the individual markers used in Spang et al. (2015). (**a**) The proteins displaying a Woese topology (red) are on average larger than the proteins displaying an eocyte topology (blue). Ribosomal and non-ribosomal proteins are indicated in solid and hashed bars, respectively. Proteins that were statistically robust according to topology tests are shown with the asterisk symbol. The arrows indicate the proteins that should probably be considered with caution: EF2, because of its likely contamination, and a Zn protease, because the alignment had only three eukaryotes to represent the entire domain. Independent concatenations of the 6 Woese (**b**) and 11 eocyte (**c**) proteins that were statistically robust produce highly supported trees, in both ML and Bayesian frameworks, with opposite topologies

r-proteins from the dataset of Spang et al. (2015), except the three largest ones, favored eocyte TOL topologies in the analysis of Forterre and colleagues, in line with the many TOLs based on r-proteins that systematically support the eocyte scenario. This is concerning because r-proteins, especially those encoded by conserved operons, are more and more used as the only universal markers in recent analyses (Lasek-Nesselquist and Gogarten 2013; Hug et al. 2016) or represent the large majority of universal proteins used (20 out of 26 in Guy and Ettema 2011 or 31 out of 48 in Zaremba-Niedzwiedzka et al. 2017). Also concerning, some of the most robust Woese proteins according to Da Cunha et al. (2017) were missing in several recent studies that support the eocyte scenario (Raymann et al. 2015;

Zaremba-Niedzwiedzka et al. 2017). A general conclusion is that exhaustive analyses of all possible universal proteins remain to be done.

4.6 TOLs Based on RNA Polymerase

RNA polymerase large subunits are the two largest universal proteins. As previously discussed, several TOLs based on RNA polymerases, producing the Woese topology, were published at the end of the 1980s and beginning of the 1990s (Klenk et al. 1991; Iwabe et al. 1991). The authors used either RNA polymerase II alone or the three RNA polymerases (I, II, III) as representatives of the eukaryotes. Later on, the RNA polymerase was abandoned as single-protein marker for the TOLs, possibly because the existence of three homologous RNA polymerases in eukaryotes complicated the design of proper datasets. A few authors nonetheless still used RNA polymerases as a single molecular marker in some studies: Embley and colleagues found that RNA polymerase II was a good marker to retrieve the position of microsporidia in the eukaryotic tree (Hirt et al. 1999), whereas Forterre and colleagues reported that RNA polymerase (a concatenation of all subunits) was also a good marker for archaeal phylogeny (Brochier et al. 2005a).

Recently, Forterre and colleagues selected RNA polymerase (concatenation of the two largest subunits) to test a new species dataset in building the TOL. They argued that RNA polymerase could be a good marker because "the rate of substitution should be relatively homogeneous for the two large subunits that are both involved in the catalytic activity and both important to conserve the global structure and the interaction with DNA and RNA. These characteristics stand well compared to ribosomal proteins that are much smaller and occupy external positions on the ribosome" (Da Cunha et al. 2017). Avoiding FES and balancing as much as possible their species dataset (39 species for each domain), Forterre and colleagues performed phylogenetic analyses of the two RNA polymerase large subunits in a concatenation with different probabilistic approaches (notably ML and Bayesian with CAT-GTR) and obtained highly congruent and supported Woese TOLs, with all Asgards branching as a monophyletic group at the base of Euryarchaeota (Da Cunha et al. 2017) (Fig. 3.8). Their trees recovered the monophyly and consensus phylogeny of Archaea but also the monophyly of some clade a priori difficult to recover in TOL phylogenies, such as the *Proteobacteria* in Bacteria and the Amorpha in Eukarya, suggesting that the RNA polymerase tree is a good representation of the species tree.

In their RNA polymerase trees, the Archaea were rooted between the Thaumarchaeota and other archaea (Fig. 3.7 in Da Cunha et al. 2017). After adding the recently identified Bathyarchaea in their analysis, Forterre and colleagues recovered an archaeal tree rooted in a similar manner. Bathyarchaea form a monophyletic group with Aigarchaeota (*Candidatus Caldiarchaeum subterraneum*) and Thaumarchaeota in this tree. If correct, this rooting suggests proposing a new candidate archaeal superphylum (thereafter dubbed BAT) that would not be compatible with the existence of the putative TACK superphylum. Notably, the same

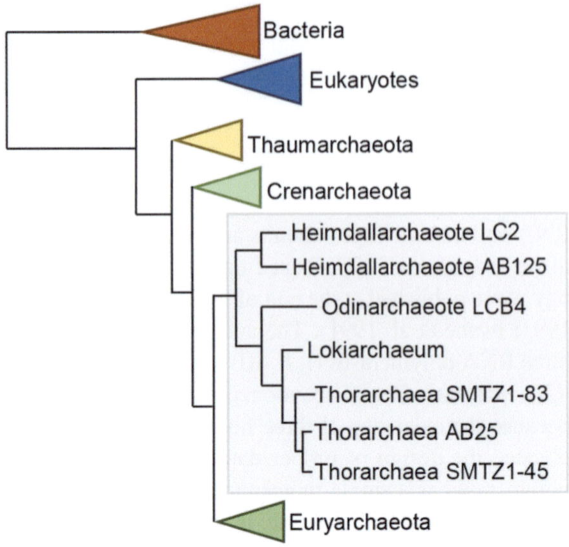

Fig. 3.8 Schematic representation of the phylogeny of the concatenated two largest RNA polymerase subunits obtained by Da Cunha and colleagues (Fig. 3.8 in Da Cunha et al. 2017). The phylogeny, highly supported in both ML and Bayesian frameworks (LG and CAT-GTR models, respectively), displays the monophyly of the Asgard archaea, close to the Euryarchaea

rooting was obtained by these authors in their TOL based on the concatenation of the six Woese proteins statistically robust (Da Cunha et al. 2017). As previously discussed, the BAT root was first observed when sequences of Thaumarchaeota were included in archaeal phylogenies (Brochier-Armanet et al. 2008a; Spang et al. 2010). These phylogenies were based on the concatenation of 53 archaeal r-proteins and rooted with eukaryotic sequences. It was thus remarkable that the same rooting was obtained with completely different datasets (RNA polymerase) that not only include Eukarya but also Bacteria (Da Cunha et al. 2017).

Forterre and colleagues also argued that the BAT root is supported by the distribution pattern of RNA polymerase A subunit structures in the three domains (Da Cunha et al. 2017). This subunit is indeed split into two subunits A' and A'', corresponding to the N- and C-terminal moieties, in most archaea, but, never in BAT, Eukaryotes, nor Bacteria. Such split appears to be a rare event since it is only observed in some *Mimivirus* and *Cyanobacteria* but at different positions. The BAT root easily explains the evolution of RNA polymerase structure since it requires only one split; this would have been followed in a few species by a secondary fusion between the two subunits encoded by neighboring genes. In contrast, rooting between Euryarchaeota and other Archaea requires two or more splits (depending of the precise tree topology) that should have occurred exactly at the same position, to explain the evolution of RNA polymerase structure.

4.7 The Archaeal Root

The BAT root contradicts other recent results. For instance, Gribaldo and colleagues published a TOL in which the eocyte topology is linked to a new archaeal root located within Euryarchaeota (Raymann et al. 2015). They obtained the same root when they only used Bacteria as an outgroup. Using the same strategy (removing Eukarya and DPANN from their dataset) but slightly different species and protein markers, Moreira and colleagues rooted the archaeal tree between Euryarchaeota and the putative TACK superphylum (renamed Proteoarchaeota by these authors) (Petitjean et al. 2014). As mentioned before, Williams, Ettema, and colleagues obtained a root between DPANN and all other archaea using an approach which does not require an outgroup but the search for the most parsimonious scenario explaining gene and species distribution in a myriad of single-gene trees (Williams et al. 2017). In summary, the Woese TOLs seem to be linked to the BAT root, whereas the eocyte TOLs are linked to roots located within Euryarchaeota (considering that DPANN probably belongs to this phylum).

These two roots imply very different scenarios for the evolution of Archaea and the origin of eukaryotic signature proteins (ESPs) (Fig. 3.9). The euryarchaeal root suggests that ESPs originated progressively within archaea and accumulated in the final archaeal lineage that became the host of the mitochondrial ancestor (Fig. 3.9a). In that scenario, LACA had "a relatively small-genomed archaeal ancestor that subsequently increased in complexity" (Williams et al. 2017). It remains to understand in that model why the increase in molecular biology complexity that took place during archaeal evolution never occurred in Euryarchaeota and why some ESPs kept accumulating in a few archaeal phyla. In contrast to the euryarchaeal root, the BAT root suggests that all ESPs originated in the lineage common to Archaea and Eukarya and were present in LACA (Fig. 3.9b). These ESPs were later on selectively lost in various archaeal phyla. This implies that LACA was somehow more complex than modern Archaea (Csurös and Miklós 2009; Wolf et al. 2012).

If the last common ancestor of Archaea and Eukarya was indeed more complex than modern archaea, it means that evolution from this ancestor to modern organisms should have undertaken divergent paths in the lineages that led to these two domains: Archaea evolving by reduction, whereas Eukarya probably continued evolving toward more complexity. It has been suggested that reductive evolution leading to the prokaryotic phenotypes of Archaea was due to their adaptation to high-temperature biotopes (the thermoreduction hypothesis, Forterre 2013) and/or to predatory proto-eukaryotes (the phagotrophic unicellular raptor scenario) (Kurland et al. 2006). In contrast, Eukarya evolution toward more complexity could have been facilitated by their interaction with complex viruses (in terms of gene content or replication mechanisms) that are not known in Archaea, such as large DNA viruses and/or retroviruses (Forterre 2011, 2013). For instance, it has been proposed that many bacterial genes were introduced in proto-eukaryotes by the integration of ancient NCLDVs (nucleocytoplasmic large DNA viruses) since their genomes often encode a rather large proportion (up to 10%) of bacterial genes (Forterre 2013). It has even been speculated that NCLDVs could have played an important role in the origin of the eukaryotic nucleus since some of them produce viral factories the size of this organelle (Takemura 2001; Bell 2001, review in Forterre and Gaia 2016). The recent discovery of a bacteriophage that induces the formation of a nucleus in the infected bacteria makes such speculations reasonable (Chaikeeratisak

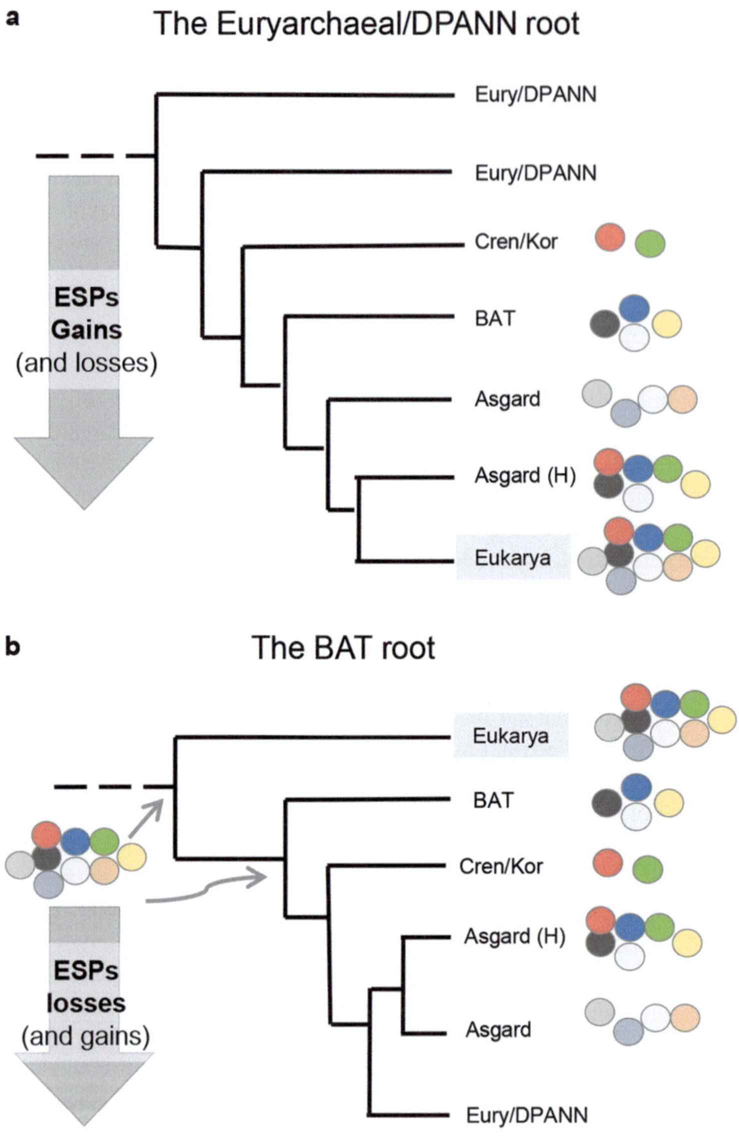

Fig. 3.9 Illustrations of scenarios explaining the presence of ESPs in Archaea. (**a**) If the Archaea are rooted in the Euryarchaeota/DPANN (eocyte TOLs), ESPs would have progressively accumulated in some archaeal phyla. In the opposite, (**b**) if the Archaea are rooted in the BAT with the Archaea being monophyletic (Woese TOLs), most ESPs would have already been present in the common ancestor of Archaea and Eukarya, before being lost progressively in the archaeal phyla (through reduction)

et al. 2017). Notably, hypotheses suggesting an important role for viruses during eukaryogenesis can be also valid in the eocyte scenario to explain the evolution toward more and more complexity from FECA to LECA.

4.8 Where Are the Viruses in the TOL?

Considering the importance that viruses and related elements (plasmids, transposons, and so on) have played in the history of life (Forterre and Prangishvili 2013; Koonin and Dolja 2013), it seems desirable to include viruses in the TOL (Brüssow 2009). However, this is not technically possible because there is no equivalent of rRNA or universal proteins in the viral world to place all viruses in the TOL. Some scientists have argued that, in any case, viruses do not have their place in the TOL because they are not alive (Moreira and López-García 2009), but this is disputable (Forterre 2010); the definition of life is a difficult exercise, not only for viruses but also for cells (Forterre 2016b). One way to introduce viruses in the TOL is to consider that viruses are "here there and everywhere" (as sung by the Beatles) from the trunk to the leaves of the TOL. Another possibility is to focus on viruses encoding their own RNA polymerase, since they can technically be positioned in a TOL based on this universal protein. RNA polymerases are encoded by the genomes of a few *Caudovirales* (head and tail viruses) infecting bacteria, *Baculoviridae*, and most viruses of the NCLDV superfamily. All these RNA polymerases are homologous to cellular ones, but they are very divergent, and only NCLDVs' RNA polymerases sequences can be aligned with those of their cellular counterparts.

Several TOLs including both NCLDVs' and cellular RNA polymerase sequences have been published in the last 10 years. Didier Raoult, Pierre Pontarotti, and colleagues published ML TOLs in which NCLDVs form a fourth monophyletic clade (Boyer et al. 2010; Sharma et al. 2014, 2015a, b). From their results, Raoult and colleagues suggested that viruses of the NCLDV superfamily should be considered as a fourth domain of life. In contrast, using Bayesian analysis, Williams, Embley, and Heinz obtained a tree in which NCLDVs were mixed with Eukarya and concluded that NCLDVs' RNA polymerases originated from eukaryotic RNA polymerases, refuting the fourth domain hypothesis (Williams et al. 2011). Lopez-Garcia and Moreira added up to the criticism by comparing ML and Bayesian TOLs of the RNA polymerase B subunit from the species dataset previously used by Boyer and colleagues (Moreira and López-García 2015). They concluded that the monophyly of NCLDVs was an LBA artifact because they were monophyletic in the ML tree but branch at different positions within eukaryotes in the Bayesian tree. They proposed that NCLDV RNA polymerases were independently recruited several times from different eukaryotes, in agreement with the view that viruses are mainly pickpockets of cellular genes (Moreira and López-García 2009).

All these analyses were however plagued by various pitfalls such as the small size and/or imbalance of the dataset or the use of a very short alignment from a single RNA polymerase subunit (272 amino acids in Boyer et al. 2010; 272 in Williams et al. 2011; 427 in Moreira and López-García 2015). The data of Lopez-Garcia and Moreira also included many sequences retrieved from metagenomic data. Surprisingly, focusing on the "fourth domain" question, neither Williams and colleagues nor Lopez-Garcia and Moreira discussed the fact that the archaeal and eukaryal domains were completely unresolved in their Bayesian trees, raising doubt on their interpretation (for instance, Bacteria branches within Archaea in the tree of Williams and colleagues). The fact that the topology of these domains was fully resolved in the RNA polymerase tree of Forterre and colleague (using a concatenation of both subunits for a total of more than 1600 positions) in both ML and Bayesian analyses with nonhomogeneous models again suggests that short sequence alignments as well as the presence of FES can be major problems in Bayesian frameworks. In any case, it is probably not a good idea to focus on the "fourth domain" question when analyzing the origin and evolution of NCLDVs. Forterre, Krupovic, and Prangishvili proposed that the term domain should only be applied to organisms encoding ribosomes (in accordance with the domain definition of Woese based on rRNA sequences comparison) (Forterre et al. 2014). They suggested that viruses, defined as "capsid-encoding organisms" (Raoult and Forterre 2008), should be classified into lineages, based on the history of their virions (Forterre et al. 2014).

The controversy between the Woese and eocyte topologies has actually important implications considering the origin of eukaryotic viruses (Fig. 3.10). In the eocyte TOL, all eukaryotic viruses should have originated either de novo in the specific eukaryotic branch of the tree or from archaeal and/or bacterial viruses that infected the first eukaryotic common ancestor (FECA) shortly after the merging of the archaeal and bacterial ancestors of eukaryotes (Koonin et al. 2015). In the Woese TOL hypothesis, some specific eukaryotic viruses might also have originated in the eukaryotic branch of the tree, but others might have originated from viruses that predated the divergence between Archaea and Eukaryote and even from viruses that predated LUCA (Forterre and Gaia 2016). In both cases, the major divergence observed between eukaryotic and archaeal viruses remains enigmatic, considering the close similarity between these two domains in terms of molecular biology (Forterre et al. 2014). A major paradox of the TOL, still unexplained, is indeed the clear-cut similarities between the archaeal and bacterial mobilomes, whereas their molecular biology is profoundly different. It is possible that these similarities arose from the greater possibility of LGT between Archaea and Bacteria and/or from convergence in the cellular organization of these two "prokaryotic" domains.

5 Perspectives

Forty years after the discovery of the large evolutionary gap that split "prokaryotes" into Archaea and Bacteria, and 30 years after the publication of the first TOLs based on rRNA and protein sequences comparisons, the debate between the eocyte and Woese TOLs is still ongoing. As seen in this chapter, the choice between the two topologies strongly depends on the universal markers and species dataset used in a

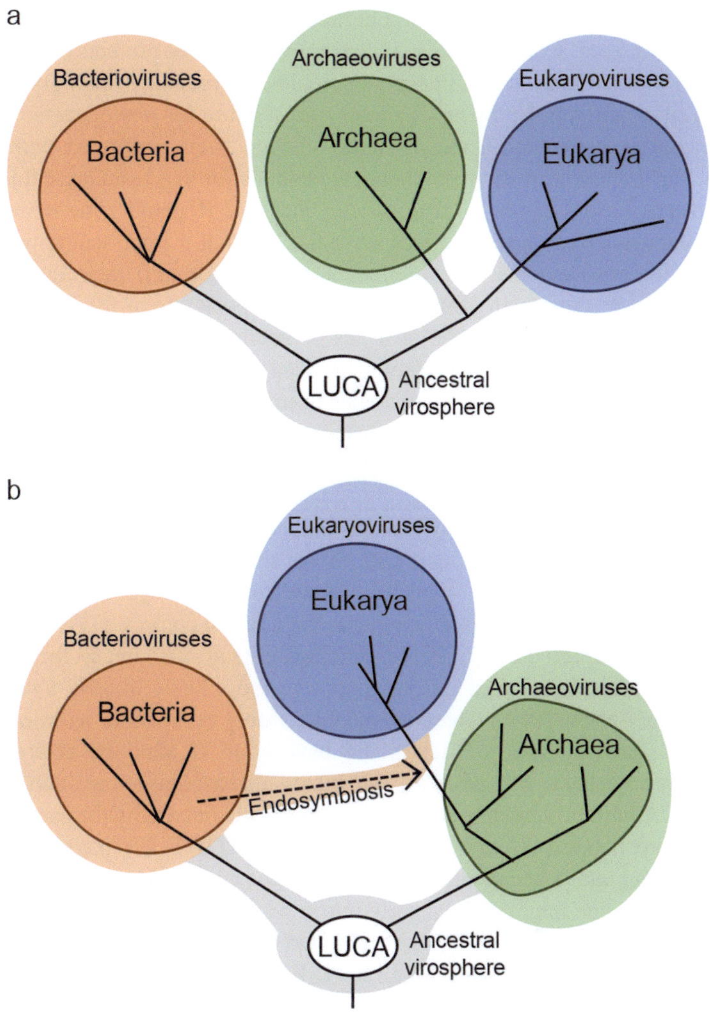

Fig. 3.10 Illustrations of the possible scenarios for the origin of eukaryotic viruses. (**a**) In the Woese scenario, the three virospheres are derived from the ancestral virosphere, and thus the modern eukaryotic and archaeal virospheres are derived from the virosphere infecting their common ancestor. (**b**) In the eocyte scenario, it has been proposed that eukaryotic viruses come from the bacterioviruses related to the bacteria that were internalized in the archaeon at the origin of eukaryotes (fusion hypothesis)

particular study. It remains to determine why different proteins produce different topologies; proponents of the eocyte TOL could argue that proteins favoring the Woese TOLs are evolving faster, amplifying an LBA attraction between Bacteria and Eukarya. Proponents of the Woese TOL could argue instead that proteins favoring the eocyte TOLs do not contain enough signal to detect the short archaeal branch. The latter hypothesis would explain why longer proteins tend to favor the

Woese TOL. However, this correlation between protein size and the TOL topology is not absolute, and it would be interesting to know why some proteins of similar size sometimes produce trees with opposite topologies. Surprisingly, the rigorous identification of all universal proteins available and suitable for phylogenetic analyses still remains to be done, all recent studies having been performed with partial datasets. Finally, one can still hope that new methods for tree building will improve the recovery of valid signal in phylogenetic analyses. Regarding the recent introduction of nanosized bacteria and archaea in the TOLs, it is important now to study in detail their mechanisms of evolution to confirm (or not) their fast-evolving phenotype and to determine with confidence their positions in the TOL. Besides classic phylogenetic analyses, the analyses of indels could be helpful to address some of these issues. We have already mentioned the possibility to detect contaminations in metagenome-assembled genomes from the detection of anomalous insertions. Rare indels, especially significant insertions, could be also useful to determine the position of nanosized organisms in the TOL by identifying structural synapomorphies with their closest relatives.

Regardless of the TOL topology considered, understanding what kind of dramatic evolutionary event led to the emergence of the three domains would be a major advance. Until now, most discussions have focused on eukaryogenesis, especially on the timing of the acquisition of mitochondria (for a recent controversy, see Pittis and Gabaldón 2016; Degli Esposti 2016). Another critical question is the origin of spliceosome and spliceosomal introns that were most likely already abundant in LECA (Csurös et al. 2011). The eocyte TOL is only compatible with intron-late hypotheses since the spliceosome should have evolved after the emergence of Eukarya from Archaea. It is often assumed that spliceosomal introns derived from group II RNA introns were introduced in the eukaryotic lineage by the bacteria at the origin of mitochondria (Koonin 2006). The Woese TOL is compatible with both intron-early and intron-late theories. For instance, the analogy between the ribosome and the spliceosome might suggest early intron scenarios in which the spliceosome was an ancient RNA molecular machine that was present in LUCA and disappeared in Archaea and Bacteria (Penny et al. 2009; Forterre 2013). In contrast with the myriad of papers that discussed eukaryogenesis, very few have discussed bacteriogenesis and archaeogenesis. The “invention” of peptidoglycan and of DNA gyrase, as well as the establishment of very efficient mechanisms for DNA replication (with long Okazaki fragments and very fast rate of DNA synthesis), has probably played a major role in the emergence of Bacteria. In the case of Archaea, adaptation to high temperatures might have been a key factor in the emergence of this domain, because archaeal lipids seem especially well suited for proper membrane fluidity and impermeability in hot environments (Glansdorff et al. 2008). Forterre suggested that adaptation to high-temperature environments could have triggered bacteriogenesis and archaeogenesis, because reduction of cell size, coupling between transcription and translation, and high macromolecular turnover are beneficial for organisms with a “prokaryotic” phenotype (Forterre 1995, 2013). Convergent evolution toward this phenotype could have also been favored by the necessity for proto-bacteria and proto-archaea to escape predation by proto-eukaryotes (Kurland et al. 2006).

6 Conclusion

Trees have been extensively used as symbols, notably of eternity over humans' relative short lives. As a metaphor of life history, the tree of life is however a very special tree. Far from a strong oak standing against the wind, the final leaves are here the most stable part of the tree; the major ramifications are underground, along with the trunk and the roots. Over the last decades, visions of the tree of life have gathered around a rather limited number of hypotheses. Nowadays, there are essentially two major schools of thoughts: those adhering to Woese's vision of a tripartite life and those proposing that one of the modern domain emerged by a fusion of members of the two others. Both concepts have their proponents and critics, and debates can sometimes be somewhat heated. Nonetheless, this has created a dynamic that pushes forward faster than ever. Dreams are sometimes unreachable fantasies. Yet, if Darwin's dream of fairly true trees of each domain is not really fulfilled for now, evolutionists are now in the race to make it happen.

After redaction of this chapter, two comments were published discussing the work of Da Cunha et al. (2017) on the tree of life topology. These papers illustrate the controversy surrounding the analyses of the universal proteins and the problem of fast-evolving species or contamination. Spang and colleagues (2018) suggested that the Woese trees obtained by Da Cunha et al. (2017) were due to long branch attractions between Bacteria and Eukaryotes. Da Cunha and colleagues (2018) answered that this is not the case and published trees without bacteria showing that the archaeal branches are similar or even sometimes longer than eukaryotic branches in universal trees.

Acknowledgments MG, VDC, and PF are supported by an ERC grant from the European Union's Seventh Framework Program (FP/2007–2013)/Project EVOMOBIL-ERC Grant Agreement no. 340440. We thank Pabulo Henrique Rampelotto for the opportunity to write this chapter.

References

Adam PS, Borrel G, Brochier-Armanet C, Gribaldo S (2017) The growing tree of Archaea: new perspectives on their diversity, evolution and ecology. ISME J 11(11):2407–2425

Albers SV, Forterre P, Prangishvili D, Schleper C (2013) The legacy of Carl Woese and Wolfram Zillig: from phylogeny to landmark discoveries. Nat Rev Microbiol 11(10):713–719

Atkinson GC (2015) The evolutionary and functional diversity of classical and lesser-known cytoplasmic and organellar translational GTPases across the tree of life. BMC Genomics 16:78

Baker BJ, Tyson GW, Webb RI et al (2006) Lineages of acidophilic archaea revealed by community genomic analysis. Science 314(5807):1933–1935

Baker BJ, Comolli LR, Dick GJ et al (2010) Enigmatic, ultrasmall, uncultivated Archaea. Proc Natl Acad Sci USA 107(19):8806–8811

Baldauf SL, Palmer JD, Doolittle WF (1996) The root of the universal tree and the origin of eukaryotes based on elongation factor phylogeny. Proc Natl Acad Sci USA 93(15):7749–7754

Baross JA, Martin WF (2015) The ribofilm as a concept for life's origins. Cell 162(1):15–15

Bell PJL (2001) Viral eukaryogenesis: was the ancestor of the nucleus a complex DNA virus? J Mol Evol 53:251–256

Booth A, Doolittle WF (2015) Reply to lane and Martin: being and becoming eukaryotes. Proc Natl Acad Sci USA 112(35):E4824

Boyer M, Madoui M-A, Gimenez G et al (2010) Phylogenetic and phyletic studies of informational genes in genomes highlight existence of a 4th domain of life including giant viruses. PLoS One 5(12):e15530

Brochier C, Forterre P, Gribaldo S (2004) Archaeal phylogeny based on proteins of the transcription and translation machineries: tackling the *Methanopyrus kandleri* paradox. Genome Biol 5(3): R17

Brochier C, Forterre P, Gribaldo S (2005a) An emerging phylogenetic core of Archaea: phylogenies of transcription and translation machineries converge following addition of new genome sequences. BMC Evol Biol 5:36

Brochier C, Gribaldo S, Zivanovic Y et al (2005b) Nanoarchaea: representatives of a novel archaeal phylum or a fast-evolving euryarchaeal lineage related to Thermococcales? Genome Biol 6(5): R42

Brochier-Armanet C, Boussau B, Gribaldo S, Forterre P (2008a) Mesophilic Crenarchaeota: proposal for a third archaeal phylum, the Thaumarchaeota. Nat Rev Microbiol 6(3):245–252

Brochier-Armanet C, Gribaldo S, Forterre P (2008b) A DNA topoisomerase IB in Thaumarchaeota testifies for the presence of this enzyme in the last common ancestor of Archaea and Eucarya. Biol Direct 3:54

Brochier-Armanet C, Forterre P, Gribaldo S (2011) Phylogeny and evolution of the Archaea: one hundred genomes later. Curr Opin Microbiol 14(3):274–281

Brown JR, Doolittle WF (1997) Archaea and the prokaryote-to-eukaryote transition. Microbiol Mol Biol Rev 61(4):456–502

Brown JR, Douady CJ, Italia MJ et al (2001) Universal trees based on large combined protein sequence data sets. Nat Genet 28(3):281–285

Brown CT, Hug LA, Thomas BC et al (2015) Unusual biology across a group comprising more than 15% of domain bacteria. Nature 523(7559):208–211

Bruno WJ, Halpern AL (1999) Topological bias and inconsistency of maximum-likelihood using wrong models. Mol Biol Evol 16(4):564–566

Brüssow H (2009) The not so universal tree of life or the place of viruses in the living world. Philos Trans R Soc Lond Ser B Biol Sci 364(1527):2263–2274

Camin JH, Sokal RR (1965) A method for deducing branching sequences in phylogeny. Evolution 19:311–323

Cammarano P, Palm P, Creti R et al (1992) Early evolutionary relationships among known life forms inferred from elongation factor EF-2/EF-G sequences: phylogenetic coherence and structure of the archaeal domain. J Mol Evol 34(5):396–405

Cammarano P, Creti R, Sanangelantoni AM, Palm P (1999) The archaea monophyly issue: a phylogeny of translational elongation factor G(2) sequences inferred from an optimized selection of alignment positions. J Mol Evol 49(4):524–537

Castelle CJ, Wrighton KC, Thomas BC et al (2015) Genomic expansion of domain archaea highlights roles for organisms from new phyla in anaerobic carbon cycling. Curr Biol 25 (6):690–701

Cavalier-Smith T (1989) Molecular phylogeny. Archaebacteria and Archezoa. Nature 339 (6220):100–101

Chaikeeratisak V, Nguyen K, Khanna K et al (2017) Assembly of a nucleus-like structure during viral replication in bacteria. Science 355(6321):194–197

Ciccarelli FD, Doerks T, von Mering C et al. (2006) Toward automatic reconstruction of a highly resolved tree of life. Science 311 (5765):1283–7. Erratum in: (2006) Science 312 (5774):697
Cobb M (2017) 60 years ago, Francis Crick changed the logic of biology. PLoS Biol 15(9): e2003243
Comolli LR, Baker JB, Downing KH et al (2009) Three-dimensional analysis of the structure and ecology of a novel, ultra-small archaeon. ISME J 3(2):159–167
Cox CJ, Foster PG, Hirt RP et al (2008) The archaebacterial origin of eukaryotes. Proc Natl Acad Sci USA 105(51):20356–20361
Creti R, Ceccarelli E, Bocchetta M et al (1994) Evolution of translational elongation factor (EF) sequences: reliability of global phylogenies inferred from EF-1 alpha(Tu) and EF-2(G) proteins. Proc Natl Acad Sci USA 91(8):3255–3259
Csurös M, Miklós I (2009) Streamlining and large ancestral genomes in Archaea inferred with a phylogenetic birth-and-death model. Mol Biol Evol 26(9):2087–2095
Csurös M, Rogozin IB, Koonin EV (2011) A detailed history of intron-rich eukaryotic ancestors inferred from a global survey of 100 complete genomes. PLoS Comput Biol 7(9):e1002150
Da Cunha V, Gaia M, Gadelle D et al (2017) Lokiarchaea are close relatives of Euryarchaeota, not bridging the gap between prokaryotes and eukaryotes. PLoS Genet 13(6):e1006810
Da Cunha V, Gaia M, Nasir A, Forterre P (2018) Asgard archaea do not close the debate about the universal tree of life topology. PLoS Genet 14(3):e1007215
Dacks JB, Field MC, Buick R et al (2016) The changing view of eukaryogenesis – fossils, cells, lineages and how they all come together. J Cell Sci 129(20):3695–3703
Dagan T, Martin W (2009) Getting a better picture of microbial evolution en route to a network of genomes. Philos Trans R Soc Lond Ser B Biol Sci 364(1527):2187–2196
de Duve C (2007) The origin of eukaryotes: a reappraisal. Nat Rev Genet 8(5):395–403
Degli Esposti M (2016) Late mitochondrial acquisition, really? Genome Biol Evol 8(6):2031–2035
Doolittle WF (2009) The practice of classification and the theory of evolution, and what the demise of Charles Darwin's tree of life hypothesis means for both of them. Philos Trans R Soc Lond Ser B Biol Sci 364(1527):2221–2228
Edlind TD, Li J, Visvesvara GS et al (1996) Phylogenetic analysis of beta-tubulin sequences from amitochondrial protozoa. Mol Phylogenet Evol 5(2):359–367
Edwards AWF, Cavalli-Sforza LL (1963) The reconstruction of evolution. Heredity 18:553
Embley TM, Hirt RP (1998) Early branching eukaryotes? Curr Opin Genet Dev 8(6):624–629
Embley TM, Williams TA (2015) Evolution: steps on the road to eukaryotes. Nature 521 (7551):169–170
Eme L, Spang A, Lombard J et al (2017) Archaea and the origin of eukaryotes. Nat Rev Microbiol 15(12):711–723
Eren AM, Esen ÖC, Quince C et al (2015) Anvi'o: an advanced analysis and visualization platform for 'omics data. PeerJ 3:e1319
Felsenstein J (1981) Evolutionary trees from DNA sequences: a maximum likelihood approach. J Mol Evol 16(6):368–376
Fitch WM, Margoliash E (1967) Construction of phylogenetic trees. Science 155(3760):279–284
Forterre P (1995) Thermoreduction, a hypothesis for the origin of prokaryotes. C R Acad Sci III 318 (4):415–422
Forterre P (2006) Three RNA cells for ribosomal lineages and three DNA viruses to replicate their genomes: a hypothesis for the origin of cellular domain. Proc Natl Acad Sci USA 103 (10):3669–3674
Forterre P (2010) Defining life: the virus viewpoint. Orig Life Evol Biosph 40(2):151–160
Forterre P (2011) A new fusion hypothesis for the origin of Eukarya: better that previous ones, but probably also wrong. Res Microbiol 62(1):77–91
Forterre P (2012) Darwin's goldmine is still open: variation and selection run the world. Front Cell Infect Microbiol 2:106
Forterre P (2013) The common ancestor of Archaea and Eukarya was not an archaeon. Archaea 2013:372396

Forterre P (2015) The universal tree of life: an update. Front Microbiol 6:717

Forterre P (2016a) Microbes from Hell. Chicago University Press, Chicago

Forterre P (2016b) To be or not to be alive: how recent discoveries challenge the traditional definitions of viruses and life. Stud Hist Phil Biol Biomed Sci 59:100–108

Forterre P, Gadelle D (2009) Phylogenomics of DNA topoisomerases: their origin and putative roles in the emergence of modern organisms. Nucleic Acids Res 37(3):679–692

Forterre P, Gaia M (2016) Giant viruses and the origin of modern eukaryotes. Curr Opin Microbiol 31:44–49

Forterre P, Philippe H (1999) Where is the root of the universal tree of life? BioEssays 21 (10):871–879

Forterre P, Gribaldo S, Brochier-Armanet C (2009) Happy together: genomic insights into the unique *Nanoarchaeum/Ignicoccus* association. J Biol 8(1):7. https://doi.org/10.1186/jbiol110

Forterre P, Prangishvili D (2013) The major role of viruses in cellular evolution: facts and hypotheses. Curr Opin Virol 3(5):558–565

Forterre P, Krupovic M, Prangishvili D (2014) Cellular domains and viral lineages. Trends Microbiol 22(10):554–558

Foster PG, Cox CJ, Embley TM (2009) The primary divisions of life: a phylogenomic approach employing composition-heterogeneous methods. Philos Trans R Soc Lond Ser B Biol Sci 364 (1527):2197–2207

Furukawa R, Nakagawa M, Kuroyanagi T et al (2017) Quest for ancestors of eukaryal cells based on phylogenetic analyses of aminoacyl-tRNA synthetases. J Mol Evol 84(1):51–66

Glansdorff N, Xu Y, Labedan B (2008) The last universal common ancestor: emergence, constitution and genetic legacy of an elusive forerunner. Biol Direct 3:29

Gogarten JP, Kibak H, Dittrich P et al (1989) Evolution of the vacuolar H+-ATPase: implications for the origin of eukaryotes. Proc Natl Acad Sci USA 86(17):6661–6665

Golyshina OV, Toshchakov SV, Makarova KS et al (2017) 'ARMAN' archaea depend on association with euryarchaeal host in culture and in situ. Nat Commun 8(1):60

Gould SB, Garg SG, Martin WF (2016) Bacterial vesicles secretion and the evolutionary origin of the eukaryotic endomembrane system. Trends Microbiol 24(7):525–534

Gouy M, Li WH (1989) Phylogenetic analysis based on rRNA sequences supports the archaebacterial rather than the eocyte tree. Nature 339(6220):145–147

Gouy M, Baurain D, Philippe H (2015) Rooting the tree of life: the phylogenetic jury is still out. Philos Trans R Soc Lond Ser B Biol Sci 370(1678):20140329

Gray MW, Doolittle WF (1982) Has the endosymbiont hypothesis been proven? Microbiol Rev 46 (1):1–42

Gribaldo S, Cammarano P (1998) The root of the universal tree of life inferred from anciently duplicated genes encoding components of the protein-targeting machinery. J Mol Evol 47 (5):508–516

Gribaldo S, Philippe H (2002) Ancient phylogenetic relationships. Theor Popul Biol 61(4):391–408

Gribaldo S, Poole AM, Daubin V et al (2010) The origin of eukaryotes and their relationship with the Archaea: are we at a phylogenomic impasse? Nat Rev Microbiol 8(10):743–752

Grosjean H, Marck C, de Crécy-Lagard V (2007) The various strategies of codon decoding in organisms of the three domains of life: evolutionary implications. Nucleic Acids Symp Ser (Oxf) 51:15–16

Guy L, Ettema TJG (2011) The archaeal 'TACK' superphylum and the origin of eukaryotes. Trends Microbiol 19(12):580–587

Harris JK, Kelley ST, Spiegelman GB, Pace NR (2003) The genetic core of the universal ancestor. Genome Res 13(3):407–412

Hennig W (1966) Phylogenetic systematics. University of Illinois Press, Urbana, IL

Hirt RP, Logsdon JM Jr, Healy B et al (1999) Microsporidia are related to Fungi: evidence from the largest subunit of RNA polymerase II and other proteins. Proc Natl Acad Sci USA 96 (2):580–585

House CH, Fitz-Gibbon ST (2002) Using homolog groups to create a whole-genomic tree of free-living organisms: an update. J Mol Evol 54(4):539–547

Huelsenbeck JP, Larget B, Miller RE, Ronquist F (2002) Potential applications and pitfalls of Bayesian inference of phylogeny. Syst Biol 51(5):673–688
Huet J, Schnabel R, Sentenac A, Zillig W (1983) Archaebacteria and eukaryotes possess DNA-dependent RNA polymerases of a common type. EMBO J 2(8):1291–1294
Hug LA, Baker BJ, Anantharaman K et al (2016) A new view of the tree of life. Nat Microbiol 1:16048
Inagaki Y, Susko E, Fast NM et al (2004) Covarion shifts cause a long-branch attraction artifact that unites microsporidia and archaeabacteria in EF-1alpha phylogenies. Mol Biol Evol 21 (7):1340–1349
Iwabe N, Kuma K, Hasegawa M et al (1989) Evolutionary relationship of archaebacteria, eubacteria, and eukaryotes inferred from phylogenetic trees of duplicated genes. Proc Natl Acad Sci USA 86(23):9355–9359
Iwabe N, Kuma K, Kishino H et al (1991) Evolution of RNA polymerases and branching patterns of the three major groups of Archaeabacteria. J Mol Evol 32(1):70–78
Kamaichi T, Hashimoto T, Nakamura Y et al (1996) Complete nucleotide sequences of the genes encoding translation elongation factors 1alpha and 2 from a microsporidian parasite, Glugea plecoglossi: implications for the deepest branching of eukaryotes. J Biochem 20(6):1095–1103
Keeling PJ, Doolittle WF (1996) Alpha-tubulin from early-diverging eukaryotic lineages and the evolution of the tubulin family. Mol Biol Evol 13(10):1297–1305
Klenk HP, Zillig W (1994) DNA-dependent RNA polymerase subunit B as a tool for phylogenetic reconstructions: branching topology of the archaeal domain. J Mol Evol 38(4):420–432
Klenk HP, Palm P, Zillig W (1991) A monophyletic holophyletic archaeal domain versus the 'eocyte tree'. Trends Biochem Sci 16(8):288–290
Knittel K, Lösekman T, Boetus A et al (2005) Diversity and distribution of methanotrophic archaea at cold seeps. Appl Environ Microbiol 71(1):467–479
Koonin EV (2006) The origin of introns and their role in eukaryogenesis: a compromise solution to the introns-early versus introns-late debate? Biol Direct 1:22
Koonin EV (2009) Darwinian evolution in the light of genomics. Nucleic Acids Res 37 (4):1011–1034
Koonin EV (2015) Archaeal ancestors of eukaryotes: not so elusive any more. BMC Biol 13:84
Koonin EV, Dolja VV (2013) A virocentric perspective on the evolution of life. Curr Opin Virol 3 (5):546–557
Koonin EV, Dolja VV, Krupovic M (2015) Origins and evolution of viruses of eukaryotes: the ultimate modularity. Virology 479–480:2–25
Kurland CG, Collins LJ, Penny D (2006) Genomics and the irreducible nature of eukaryote cells. Science 312(5776):1011–1014
Lake JA (1987) A rate-independent technique for analysis of nucleic acid sequences: evolutionary parsimony. Mol Biol Evol 4(2):167–191
Lake JA (1988) Origin of the eukaryotic nucleus determined by rate-invariant analysis of rRNA sequences. Nature 331(6152):184–186
Lake JA (1994) Reconstructing evolutionary trees from DNA and protein sequences: paralinear distances. Proc Natl Acad Sci USA 91(4):1455–1459
Lake JA, Henderson E, Oakes M, Clark MW (1984) Eocytes: a new ribosome structure indicates a kingdom with a close relationship to eukaryotes. Proc Natl Acad Sci USA 81(12):3786–3790
Lane N, Martin W (2010) The energetics of genome complexity. Nature 467(7318):929–934
Lane N, Martin WF (2015) Eukaryotes really are special, and mitochondria are why. Proc Natl Acad Sci USA 112(35):E4823
Lartillot N, Philippe H (2004) A Bayesian mixture model for across-site heterogeneities in the amino-acid replacement process. Mol Biol Evol 21(6):1095–1109
Lartillot N, Brinkmann H, Philippe H (2007) Suppression of long-branch attraction artefacts in the animal phylogeny using a site-heterogeneous model. BMC Evol Biol 7(Suppl 1):S4
Lasek-Nesselquist E, Gogarten JP (2013) The effects of model choice and mitigating bias on the ribosomal tree of life. Mol Phylogenet Evol 69(1):17–38

Lasken RS, Stockwell TB (2007) Mechanism of chimera formation during the multiple displacement amplification reaction. BMC Biotechnol 7:19

Le SQ, Gascuel O (2008) An improved general amino acid replacement matrix. Mol Biol Evol 25 (7):1307–1320

Lecompte O, Ripp R, Thierry JC et al (2002) Comparative analysis of ribosomal proteins in complete genomes: an exemple of reductive evolution at the domain scale. Nucleic Acids Res 30(24):5382–5390

Leipe DD, Araving L, Koonin EV (1999) Did DNA replication evolve twice independently? Nucleic Acids Res 27(17):3389–3401

Li S (1996) Phylogenetic tree construction using Markov chain Monte Carlo. PhD dissertation, Ohio State University

Linkkila TP, Gogarten JP (1991) Tracing origins with molecular sequences: rooting the universal tree of life. Trends Biochem Sci 16(8):287–288

López-García P, Moreira D (2006) Selective forces for the origin of the eukaryotic nucleus. BioEssays 28(5):525–533

López-García P, Moreira D (2015) Open questions on the origin of Eukaryotes. Trends Ecol Evol 30(11):697–708

López-García P, Eme L, Moreira D (2017) Symbiosis in eukaryotic evolution. J Theor Biol 434:20–33

Martijn J, Ettema TJ (2013) From archeon to eukaryote: the evolutionary dark ages of the eukaryotic cell. Biochem Soc Trans 41(1):451–457

Martin W, Kowallik KV (1999) Annotated English translation of Mereschkowsky's 1905 paper "Über Natur und Ursprung der Chromatophoren im Pflanzenreiche". Eur J Phycol 34:287–295

Matte-Taillez O, Brochier C, Forterre P, Philippe H (2002) Archaeal phylogeny based on ribosomal proteins. Mol Biol Evol 19(5):631–639

Mereschkowski C (1905) Über Natur und Ursprung der Chromatophoren im Pflanzenreiche. Biol Centralbl 25:593–604

Mereschkowski K (1910) Theorie der zwei Plasmaarten als Grundlage der Symbiogenesis, einer neuen Lehre von der Ent-stehung der Organismen. Biol Centralbl 30:353–367

Michener CD, Sokal RR (1957) A quantitative approach to a problem in classification. Evolution 11 (2):130–162

Moreira D, López-García P (2009) Ten reasons to exclude viruses from the tree of life. Nat Rev Microbiol 7(4):306–311

Moreira D, López-García P (2015) Evolution of viruses and cells: do we need a fourth domain of life to explain the origin of eukaryotes? Philos Trans R Soc Lond Ser B Biol Sci 370 (1678):20140327

Mulkidjanian AY, Makarova KS, Galperin MY, Koonin EV (2007) Inventing the dynamo machine: the evolution of the F-type and V-type ATPases. Nat Rev Microbiol 5(11):892–899

Nasir A, Kim KM, Caetano-Anollés G (2015) Lokiarchaeota: eukaryote-like missing links from microbial dark matter? Trends Microbiol 23(8):448–450

Nasir A, Kim KM, Da Cunha V, Caetano-Anollés G (2016) Arguments reinforcing the three-domain view of diversified cellular life. Archaea 2016:1851865

Nurk S, Bankevich A, Antipov D et al (2013) Assembling single-cell genomes and mini-metagenomes from chimeric MDA products. J Comput Biol 20(10):714–737

Olsen GJ, Woese CR (1989) A brief note concerning archaebacterial phylogeny. Can J Microbiol 35(1):119–123

Olsen GJ, Woese CR (1997) Archaeal genomics: an overview. Cell 89(7):991–994

Pace NR, Olsen GJ, Woese CR (1986) Ribosomal RNA phylogeny and the primary lines of evolutionary descent. Cell 45(3):325–326

Parks DH, Imelfort M, Skennerton CT et al (2015) CheckM: assessing the quality of microbial genomes recovered from isolates, single cells, and metagenomes. Genome Res 25 (7):1043–1055

Parks DH, Rinke C, Chuvochina M et al (2017) Recovery of nearly 8,000 metagenome-assembled genomes substantially expands the tree of life. Nat Microbiol 2(11):1533–1542

Penny D, Hoeppner MP, Poole AM, Jeffares DC (2009) An overview of the introns-first theory. J Mol Evol 69(5):527–540
Petitjean C, Deschamps P, López-García P, Moreira D (2014) Rooting the domain archaea by phylogenomic analysis supports the foundation of the new kingdom Proteoarchaeota. Genome Biol Evol 7(1):191–204
Philippe H, Forterre P (1999) The rooting of the universal tree of life is not reliable. J Mol Evol 49 (4):509–523
Philippe H, Germot A (2000) Phylogeny of eukaryotes based on ribosomal RNA: long-branch attraction and models of sequence evolution. Mol Biol Evol 17(5):830–834
Philippe H, Germot A, Moreira A (2000a) The new phylogeny of eukaryotes. Curr Opin Genet Dev 10(6):596–601
Philippe H, Lopez P, Brinkmann H et al (2000b) Early-branching or fast-evolving eukaryotes? An answer based on slowly evolving positions. Proc Biol Sci 267(1449):1213–1221
Pisani D, Cotton JA, McInerney JO (2007) Supertrees disentangle the chimerical origin of eukaryotic genomes. Mol Biol Evol 24(8):1752–1760
Pittis AA, Gabaldón T (2016) Late acquisition of mitochondria by a host with chimaeric prokaryotic ancestry. Nature 531(7592):101–104
Poole AM (2009) Horizontal gene transfer and the earliest stages of the evolution of life. Res Microbiol 160(7):473–480
Poole A, Penny D (2001) Does endo-symbiosis explain the origin of the nucleus? Nat Cell Biol 3 (8):E173–E174
Prangishvili D (2013) The wonderful world of archaeal viruses. Annu Rev Microbiol 67:565–585
Pühler G, Leffers H, Gropp G et al (1989) Archaeabacterial DNA-dependent RNA polymerases testify to the evolution of the eukaryotic nuclear genome. Proc Natl Acad Sci USA 86 (12):4569–4573
Raoult D (2010) The post-Darwinian rhizome of life. Lancet 375(9709):104–105
Raoult D, Forterre P (2008) Redefining viruses: lessons from Mimivirus. Nat Rev Microbiol 6 (4):315–319
Raymann K, Brochier-Armanet C, Gribaldo S (2015) The two-domain tree of life is linked to a new root for the Archaea. Proc Natl Acad Sci USA 112(21):6670–6675
Rinke C, Schwientek P, Sczyrba A et al (2013) Insights into the phylogeny and coding potential of microbial dark matter. Nature 499(7459):431–437
Rivera MC, Lake JA (2004) The ring of life provides evidence for a genome fusion origin of eukaryotes. Nature 431(7005):152–155
Rochette NC, Brochier-Armanet C, Gouy M (2014) Phylogenomic test of the hypotheses for the evolutionary origin of eukaryotes. Mol Biol Evol 31(4):832–845
Sagan L (1967) On the origin of mitosing cells. J Theor Biol 14(3):255–274
Sanger F, Brownlee GG, Barrell BG (1965) A two-dimensional fractionation procedure for radioactive nucleotides. J Mol Biol 13(2):373–398
Sapp J, Fox GE (2013) The singular quest for a universal tree of life. Microbiol Mol Biol Rev 77 (4):541–550
Schulz F, Eloe-Fadrosh EA, Bowers RM et al (2017) Towards a balanced view of the bacterial tree of life. Microbiome 5(1):140
Seitz KW, Lazar CS, Hinrichs KU et al (2016) Genomic reconstruction of a novel, deeply branched sediment archaeal phylum with pathways for acetogenesis and sulfur reduction. ISME J 10 (7):1696–1705
Sharma V, Colson P, Giorgi R et al (2014) DNA-dependent RNA polymerase detects hidden giant viruses in published databanks. Genome Biol Evol 6(7):1603–1610
Sharma V, Colson P, Chabrol O et al (2015a) Welcome to pandoraviruses at the 'Fourth TRUC' club. Front Microbiol 6:423
Sharma V, Colson P, Chabrol O et al (2015b) Pithovirus sibericus, a new bona fide member of the 'Fourth TRUC' club. Front Microbiol 6:722

Sousa FL, Neukirchen S, Allen JF et al (2016) Lokiarchaeon is hydrogen dependent. Nat Microbiol 1:16034

Spang A, Hatzenpichler R, Brochier-Armanet C et al (2010) Distinct gene set in two different lineages of ammonia-oxidizing archaea supports the phylum Thaumarchaeota. Trends Microbiol 18(8):331–340

Spang A, Saw JH, Jørgensen S et al (2015) Complex archaea that bridge the gap between prokaryotes and eukaryotes. Nature 521(7551):173–179

Spang A, Caceres EF, Ettema TJG (2017) Genomic exploration of the diversity, ecology, and evolution of the archaeal domain of life. Science 357(6351):eaaf3883

Spang A, Eme L, Saw JH et al (2018) Asgard archaea are the closest prokaryotic relatives of eukaryotes. PLoS Genet 14(3):e1007080

Stanier RY, Van Niel CB (1962) The concept of a Bacterium. Arch Mikrobiol 42:17–35

Stetter KO (1989) Extremely thermophilic chemolithoautotrophic archaebacteria. In: Schlegel HG, Brown B (eds) Autotrophic bacteria. Science Tech Publishers and Springer, Berlin, pp 167–171

Stetter KO (2013) A brief history of the discovery of hyperthermophilic life. Biochem Soc Trans 41 (1):416–420

Stöffler-Meilicke M, Böhme C, Strobel O et al (1986) Structure of ribosomal subunits of *M. vannielii*: ribosomal morphology as a phylogenetic marker. Science 231(4743):1306–1308

Swofford DL, Waddell PJ, Huelsenbeck JP et al (2001) Bias in phylogenetic estimation and its relevance to the choice between parsimony and likelihood methods. Syst Biol 50(4):525–539

Takemura M (2001) Poxviruses and the origin of the eukaryotic nucleus. J Mol Evol 52:419–425

Tavare S (1986) Some probabilistic and statistical problems on the analysis of DNA sequences? Lect Math Life Sci 17(2):57–86

Tornabene TG, Langworthy TA (1979) Diphytanyl and dibiphytanyl glycerol ether lipids of methanogenic archaeabacteria. Science 203(4375):51–53

Urbonavicius J, Auxilien S, Walbott H et al (2008) Acquisition of a bacterial RumA-type tRNA (uracil-54,C5)-methyltransferase by Archaea through an ancient horizontal gene transfer. Mol Microbiol 67(2):323–335

Vossbrinck CR, Maddox JV, Friedman S et al (1987) Ribosomal RNA sequence suggests microsporidia are extremely ancient eukaryotes. Nature 326(6111):411–414

Wallace DC, Morowitz HJ (1973) Genome size and evolution. Chromosoma 40:121–122

Werner F, Grohmann D (2011) Evolution of multisubunit RNA polymerases in the three domains of life. Nat Rev Microbiol 9(2):85–98

Williams TA, Embley TM, Heinz E (2011) Informational gene phylogenies do not support a fourth domain of life for nucleocytoplasmic large DNA viruses. PLoS One 6(6):e21080

Williams TA, Foster PG, Nye TM et al (2012) A congruent phylogenomic signal places eukaryotes within the Archaea. Proc Biol Sci 279(1749):4870–4879

Williams TA, Foster PG, Cox CJ, Embley TM (2013) An archaeal origin of eukaryotes supports only two primary domains of life. Nature 504(7479):231–236

Williams TA, Szöllősi GJ, Spang A et al (2017) Integrative modeling of gene and genome evolution roots the archaeal tree of life. Proc Natl Acad Sci USA 114(23):E4602–E4611

Woese CR (1979) A proposal concerning the origin of life on the planet earth. J Mol Evol 13 (2):95–101

Woese CR (1987) Bacterial evolution. Microbiol Rev 51(2):221–271

Woese CR (1998) The universal ancestor. Proc Natl Acad Sci USA 95(12):6854–6859

Woese CR (2000) Interpreting the universal phylogenetic tree. Proc Natl Acad Sci USA 97 (15):8392–8396

Woese CR (2002) On the evolution of cells. Proc Natl Acad Sci USA 99(13):8742–8747

Woese CR, Fox GE (1977a) Phylogenetic structure of the prokaryotic domain: the primary kingdoms. Proc Natl Acad Sci USA 74(11):5088–5090

Woese CR, Fox GE (1977b) The concept of cellular evolution. J Mol Evol 10(1):1–6

Woese CR, Maniloff J, Zablen LB (1980) Phylogenetic analysis of the mycoplasmas. Proc Natl Acad Sci USA 77(1):494–498

Woese CR, Kandler O, Wheelis ML (1990) Towards a natural system of organisms: proposal for the domains Archaea, Bacteria, and Eucarya. Proc Natl Acad Sci USA 87(12):4576–4579

Wolf YI, Rogozin IB, Grishin NV, Koonin EV (2002) Genome trees and the tree of life. Trends Genet 18(9):472–479

Wolf YI, Makarova KS, Yutin N, Koonin EV (2012) Updated clusters of orthologous genes for Archaea: a complex ancestor of the Archaea and the byways of horizontal gene transfer. Biol Direct 7:46

Yang Z, Rannala B (2012) Molecular phylogenetics: principles and practice. Nat Rev Microbiol 13 (5):303–314

Yang D, Oyaizu Y, Oyaizu H et al (1985) Mitochondrial origins. Proc Natl Acad Sci USA 82 (13):4443–4447

Yutin N, Makarova KS, Mekhedov SL et al (2008) The deep archaeal roots of eukaryotes. Mol Biol Evol 25(8):1619–1630

Zablen LB, Kissil MS, Woese CR, Buetow DE (1975) Phylogenetic origin of the chloroplast and prokaryotic nature of its ribosomal RNA. Proc Natl Acad Sci USA 72(6):2418–2422

Zaremba-Niedzwiedzka K, Caceres EF, Saw JH et al (2017) Asgard archaea illuminate the origin of eukaryotic cellular complexity. Nature 541(7637):353–358

Zillig W, Stetter KO, Janeković D (1979) DNA-dependent RNA polymerase from the archaebacterium Sulfolobus acidocaldarius. Eur J Biochem 96(3):596–604

Zillig W, Prangishvili D, Schleper C et al (1996) Viruses, plasmids and other genetic elements of thermophilic and hyperthermophilic Archaea. FEMS Microbiol Rev 18(2–3):225–236

Zuckerkandl E, Pauling L (1965) Molecules as documents of evolutionary history. J Theor Biol 8 (2):357–366

Chapter 4
Multiple Clocks in the Evolution of Living Organisms

Antoine Danchin

1 Introduction

Textbook knowledge anticipates that bacteria are immortal. Yet, for example, experiments that glued a single bacterium in a nanowell, where it could be supplied by ample nutrients, showed that it initiated and sustained its natural multiplication cycle, shedding daughter cells on a more or less regular basis. In due time, the bacterium slowly shrank and eventually died (Rusconi et al. 2014; Taheri-Araghi et al. 2015). Also, it is usually accepted that prokaryotes (bacteria and archaea) are "simple" organisms, and therefore the witnesses of what may have developed at the origin of life. This assumes both that the first cells were prokaryotes and that all extant living organisms arose from a common origin after plenty of mutation events kept accumulating. Furthermore it is still believed, more often than not, that there must exist some common molecular clock ticking in parallel for all organisms, based on the steady evolution of genes submitted to a more or less constant flux of mutations (Kumar 2005). This clock is then used to explore backward the evolution of living forms and to construct phylogenetic trees. This ends up in a tree of life—where prokaryotes are always perceived as primitive organisms—often considered as the ultimate account of the history of extant living forms. Alas, these common assumptions (that actually took long to be widely accepted, showing indeed that they are not as straightforward as they seem to be, a debate that we do not discuss here) fail to fully take into account an obvious fact: living organisms do not keep multiplying. They live most of their lifespan in a state where they are undoubtedly

A. Danchin (✉)
Institute of Cardiometabolism and Nutrition, Paris, France

School of Biological Sciences, Li KaShing Faculty of Medicine, Hong Kong University, Pokfulam, SAR Hong Kong, China
e-mail: antoine.danchin@normalesup.org

P. H. Rampelotto (ed.), *Molecular Mechanisms of Microbial Evolution*, Grand Challenges in Biology and Biotechnology,
https://doi.org/10.1007/978-3-319-69078-0_4

alive but do not multiply yet still metabolise. The study of bacteria in natural environments revealed, early on, that a large proportion of those were viable but nonculturable, substantiating the idea that being alive is a separate state from multiplying [(VBNC) (Roszak and Colwell 1987)]. This meant that VBNC cells could not be isolated by standard procedures although it was clear that they were still in a metabolising state (Jannasch 1967). During this "animated" but not multiplying stage, cells age while expressing only part of their genome programme. In parallel, it is generally young organisms that produce a progeny. However, only senescent organisms have no descent, so that mature organisms, including VBNCs, may contribute to evolution.

Variations in the time when a progeny is produced have consequences that are described here. In this short reflection, I show that ageing matters and that in the same way as important genes must differ between the different stages of life, the clocks associated to the resting phase are intrinsically different from the one that ticks when organisms multiply and which is used as the "universal" molecular evolutionary clock. As a consequence, the phylogeny of different genes may differ, explaining why we observe considerable discrepancies between related studies when history is taken into account. To support this view, I end up emphasising the consequences of an unexpected general drive of evolution, transcription of specific genes associated with resting cells.

2 Prologue: Against Any Singular Origin in General Physical Processes

We tend to be "adamists". We assume that there exists a single origin for all emerging worlds. Fred Hoyle's joke about the Big Bang became widely accepted, and rare are those who challenge the idea (Afshordi and Magueijo 2016). Most biologists will speak about LUCA, our elusive last universal common ancestor (Ouzounis and Kyrpides 1996), and we generally surmise that man came to being suddenly, within a particular group of primates that were the ancestors of apes and human beings. The quest for a well-identified origin is so deeply ingrained in our minds that I have chosen to depict the origin of modern man as a telling example of fuzzy origins. No doubt that the same is true for the origin of cells (Danchin 2017a, b), with many clocks at work, as I shall summarise in the course of this chapter.

2.1 No Eve, No Adam

Let us dig a little more into the origin of man, which may also apply to the origins of the genus *Homo*. We already know that modern Europeans are a mixture of *Homo sapiens* sapiens, out of Africa, and some *Homo neanderthalensis* (Stringer 2016).

People native from Asia and, above all, Oceania have also some *Homo denisova* in their genome (Paixão-Cortes et al. 2012). Tibetans extracted a gene that allowed them to thrive at 4500 m and above from members of that particular *Homo* clade (Huerta-Sanchez and Casey 2015). This already tells us that our origin must be intricate. We could perhaps, with a twist, still accept that *H. neanderthalensis* belongs to the *H. sapiens* clade. The fact that we got a fertile descent from a cross between members of both groups argues in favour of a quite similar chromosome organisation. These ancestors belonging to the genus *Homo* most likely had 46 chromosomes, in contrast to the number found in apes, 48. Now, this opens up a straightforward question: we know that our chromosome 2 results from the head to head fusion of two chromosomes still separate in apes (Stankiewicz 2016). This implies that a gamete from a member (male or female is indifferent at this stage) of our ancestral ape group had, following an accidental fusion, 23 chromosomes instead of 24, begetting offsprings with an unbalanced number of chromosomes, 47, after standard intercourse. As a consequence, when mating (and this assumes that this did not affect its fertility), this individual with a chromosome aberration produced a progeny, half of which with the right number of its normal parent, 48, and half with 47 chromosomes, again. In a social group, especially if the unbalanced parent was a male, this may lead to a small but significant descent of individuals with 47 chromosomes, of both sexes. At some point 2 individuals with 47 chromosomes mated, and this created a progeny that carried chromosomes in the proportions ¼ 48, ½ 47 and ¼ 46 (assuming Mendelian standard inheritance). Now, the individuals carrying 46 chromosomes could mate with any of those members, and, after several generations, a stable colony of members with 46 chromosomes could start developing, adapting to this new genome, presumably across many generations. This is far from a smooth well-defined origin, QED.

2.2 *Multiple Origins of the First Cells*

There are reasons to assume that the same is true for the origin of the first cells. Indeed, Freeman Dyson, in his *Origins of Life*, established without ambiguity that life required at least *two* origins, a first one where reproduction of metabolism (making similar copies) was established, before emergence of a process of replication (making identical copies) (Dyson 1985). A complex scenario of the further development of cell-like entities led to the first cells, as structures that maintained themselves by splitting, fusing together while engulfing each other, preying on their neighbours [a primitive phagocyte-like behaviour (Harish et al. 2013; Danchin 2014, 2017a, b; Harish and Kurland 2017)]. This implies that evolution trees of individual components of cells are not condemned to be trees but, rather, intricate networks composed of cell communities (Woese 2002; Ragan et al. 2009) and, in any event, should not always follow the same treelike dichotomies.

To understand more in-depth the nature of the functions that allow life to develop in time, we now need to get an idea of the master functions that drive the general behaviour of living things. What is it to be alive?

3 The Master Functions of Life

The success of genome transplantation into recipient hosts (Lartigue et al. 2007) allowed us to look into biology with new eyes. We would now be right in considering cells as computers making computers (Danchin 2009a, b). To go further, I apply here the agenda of functional analysis (Cole 1998) to cells considered as nonmechanical "machines" or "automata", where a programme can be explicitly (physically) told from the machine that runs it. Trying to understand how an organism can be fit for a particular niche, I first split its biological functions into two functional categories, *master functions* and their associated *helper functions* meant to achieve the target of the master functions. Illustrated with a human artefact, the master function of a printer would be to print documents. The associated helper functions would be supplying ink, paper, electric power, etc. Other helper functions would correspond to the design of the printer's chassis.

When cells are pictured as factories rather than computers, their designed master function is production of some compounds, possibly via metabolic engineering. However, this human design is entirely dependent on the ability of cells to build up a large number of tiny similar factories. They must multiply while replicating both their own programme and the designed programme construct, thus yoking the cell factory to a hidden inevitable biological master function (cell multiplication). Making a young progeny is certainly essential to understand, but it is not realistic to consider it as the only master function of cells. Indeed this would require that a proper environment with relevant nutrients came to the organism (i.e. provided it with all basic building blocks that are needed for life, as, indeed, in human intelligent designs). By contrast organisms have to run in the opposite way, with cells explored their environment to get these nutrients. Cells must be autonomous, and to multiply they must find a niche for multiplication, which entails the ability to explore it, and develop a progeny there. This implies that, associated with the master function "make a progeny", cells are more often than not in the process of exploration.

Interestingly, involvement of exploration is not a mystifying attribute of life. Exploration is a necessary consequence of the second principle of thermodynamics. The imperative increase of entropy implies the inevitable exploration of available space and energy states by any physical entity. For most (this is the common view), "propagating life" is the destination of life. However, we may propose an alternative position, where "exploration" would be the master function, with "propagating life" as the immediately downstream helper function to that particular master function. Life would thus be a particular physico-chemical process conveying farther, and in a somewhat ordered way, the intrinsic propensity for exploration carried over by all entities present in the universe (following the second law of thermodynamics that tells that physical systems will tend to occupy as many space and energy states as they can). This allows us, with Jacques Monod, to recognise that life does not fight against a major constraint of physics (inevitable increase of entropy), but, rather, uses it as a driver for one of its key master functions (Monod 1972). A myriad of helper functions have evolved to enable these master functions, as reflected in the huge variety of living organisms. Because they span widely different niches, there is no reason why they should have evolved following similar trajectories, using clocks ticking at the same rhythm. This difference should be reflected in the nature and organisation of the relevant programmes (defining specific genes and gene products). The main question we must answer, then, is to understand the details of the way the progeny is constructed, being aware that the building blocks of cells, macromolecules in particular, are doomed to age and will need at some point to be replaced by young counterparts.

4 Horizontal Gene Transfer

Cells combine two elements, a programme on a physical support and a machine that runs the programme. Multiplication requires reproduction of the machine and replication of the programme. Exploration must also be tied to both. To this aim, the machine has a variety of means to travel in the environment while probing its makeup, exchanging matter, energy and information. For a long time, it was accepted that genomes, which carried the programme, were fairly stable and evolved vertically via mutations. The dictionary of mutation processes was fairly large, based on chemical, biochemical and biomechanical constraints. It comprised an important component, identified very early on by Susumu Ohno, gene duplication (Ohno 1987) that allowed innovation via variation upon a structural/functional theme. It was also known, especially in bacteria, that genes could be exchanged via propagation of viruses (prophages and general transduction) as well as in the processes of bacterial sexuality (conjugation) and more generally DNA transformation. All these processes introduced genes in genomes via recombination. Yet, transfer of foreign genes was long seen as more or less anecdotal (see early references Alacevic 1963; Sukhodolets 1988; Landan et al. 1990).

The situation changed dramatically in 1991 with the first large sequence data sets collected in the *Escherichia coli* genome [more than one third at the time (Médigue et al. 1991a, b)]. Analysing these data, it appeared that a significant part of the genome differed from the bulk in terms of protein expression, as witnessed by consistent biases in codon usage. This was interpreted as the signature of universally spread horizontal gene transfer (Médigue et al. 1991a, b). Unforeseen by then, this discovery was concomitant with what can be considered as another key discovery of genomics, presented at Elounda in Crete the same year: The sequence of chromosome III of *Saccharomyces cerevisiae* and a 100 kb chunk of the genome of *Bacillus subtilis*, simultaneously, showed—in such widely distant organisms—that more than half of the genes newly identified from their sequence did not look like anything known. These elusive, esoteric, conspicuous genes, as Piotr Slonimski named them (Goujon 2001), suddenly revealed that sequencing genomes was not simply a technological feat, but a rich mine for the discovery of totally unanticipated biological functions. The "exploration" function of cells was not limited to finding proper niches for multiplication but also comprised discovery of new genes, expressed in other cell types, in other environments.

To be sure, since the very early times when cells fused, split and engulfed one another (Danchin 2017a, b), cells must have kept developing a major process of exploration via primitive gene exchanges. This has considerable consequences in terms of maintaining or evolving a general machinery that allows cells to read pieces of genome programmes coming from anywhere. However, the outcome is not always positive for the cell's fate. Because there are a variety of advantages to transfer innovation horizontally, there is a positive selection pressure to maintain a universal "gene reading" process (this is a general explanation of the quasi-universality of the genetic code). By contrast, because foreign DNA may carry over harmful genes (viruses are cases in point, but they are certainly not the only possible vectors of harm), there is a negative selection pressure that will tend to split the gene reading machinery into components that would not be compatible with each other. The negative impact of foreign DNA is therefore a driving force for the emergence of functions such as DNA restriction/modification, a domain which is far from fully explored [see, e.g. synthesis of phosphorothioates (Cao et al. 2014; Wu et al. 2017) or ingress of 7-deazaguanine as a guanine substitute in DNA (Hutinet et al. 2017)].

As a consequence, one may expect that the descent of the early cells will result in a partition of clades that tend to be, in terms of genes and gene expression, isolated from one another (Harish and Kurland 2017). We have further well-identified examples of this situation. Lack of foreign gene expression in a distant recipient can therefore be used for the replication of foreign DNA in a variety of hosts (Karas et al. 2013). The experiment of whole genome construction and gene transplantation, for example, was made easy using *S. cerevisiae* as an intermediate, especially (but not only) because the genomes of *Mollicutes* (*Mycoplasma*, *Spiroplasma* and the like) code for proteins with a variant genetic code where codon UGA codes for tryptophan, in contrast to the majority of organisms (Ohama et al. 2008). This makes these genomes more or less inert in most other hosts. In the same way, the genomes of A + T-rich *Firmicutes* and that of cyanobacteria are mutually exclusive in terms of gene expression. It has been demonstrated that *B. subtilis* could carry over in its progeny a whole genome of cyanobacteria, without expressing it (Watanabe et al. 2012). The consequence is that, despite common practice, while it may be relatively easy to reconstruct root history within specific clades, this may become extremely difficult, if not impossible, between clades.

5 Microbial Genomes Organisation

The very existence of horizontal gene transfer suggests that genomes may be a patchwork of genes made of functionally consistent islands (Acevedo-Rocha et al. 2013). Using again functional analysis, genomes can be split into three functional domains, driving major processes: to live (coded by the paleome), to live in a specific environmental context (coded by the cenome) and to live as a multicellular organism (Fig. 4.1). This latter part is coded by what could be named the "histome" (from ἰστος, tissue, as in "histology"). It comprises the functions that allow cell differentiation and all the cell types of the organism (ca 250 tissues in an animal, not counting the many types of neurons and glial cells). This latter class makes a significant contribution to the general organisation of genomes. It even exists in microbes, where it codes for processes such as formation of biofilms (Flemming et al. 2016), heterocysts (Muro-Pastor and Hess 2012), mycelium (McCormick and Flärdh 2012), fruiting bodies [see *Myxococcales* (Huntley et al. 2010)], sporulation and any other types of differentiated cell. I shall not discuss further here the contribution of this part of a functional genome as this likely corresponds to a more recent part in the evolution of genomes. It certainly contributes to lateral connections between phylogenetic clades as co-evolution of genes belonging to these different classes will adjust them to one another.

At this point, it is important to remark that building up a genome made of linear DNA, within a cell that is three-dimensional, creates a space constraint that has to be worked out when the cell grows. Indeed, making a linear structure will ask for considerably less material than the building up of a three-dimensional structure, especially a sphere. Filling up the cytoplasm with proteins as the cell grows requires an increase in biosynthesis as the cube of the cell's size (if the cell is spherical, less when it is of another shape, and this may account for the cylindrical shape of many bacteria), while placing proteins in the membrane would go as the square of the cell's size. In fact, overall, there is a trade-off between metabolism kinetics (including transport of metabolites), growth rate and the overall shape of cells, which can be

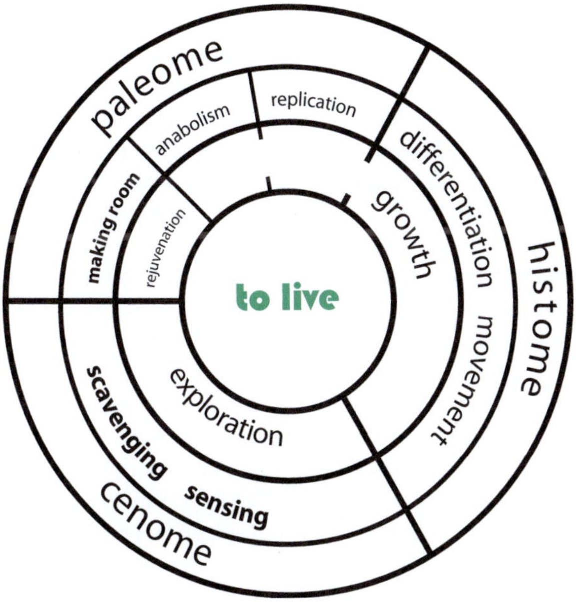

Fig. 4.1 Functional organisation of genomes. The three parts of the genome will evolve independently from one another. Standard phylogenetic trees reflect the evolution of some of the paleome

extremely varied to answer this lack of fit between space dimensions (Kysela et al. 2016).

This discrepancy also introduces a considerable constraint on the length of the genome, which is built as a one-dimensional molecule. A first metabolic solution lies in the biosynthesis of its building blocks. It has been recognised—and found as a puzzling metabolic feature (Danchin 1997)—that deoxyribonucleotides are almost universally made from ribonucleoside *di*phosphates, not *tri*phosphates. This way, the amount of available NDP building blocks, which is two orders of magnitude lower than that of their NTP parents, considerably curtails DNA biosynthesis. Yet, even with this limitation in place, a genome cannot be too short. This implies that despite a selective tendency to streamline the genome sequence because of the cost to maintain functional genes, there is an opposite tendency to fill it in with extra DNA sequences (hence the misleading notion that genomes comprise "junk" DNA). Gene duplication, amplification of insertion sequences or similar structures and horizontal gene transfer maintain a fair size for the genome, better matching the size of the cell. Overall, insertions and deletions create an equilibrium that results in an optimum length, where the DNA length is considerably longer than that of the cell.

In summary, DAPI staining shows that the genome in itself occupies a fair portion of the cell's volume, folding into a variety of shapes when streamlined (Hashimoto et al. 2005). The overall size of cells is also modulated by membrane growth and metabolic rates (with, again, a natural tendency to become bigger or to create membrane compartments, because membranes are two-dimensional) while shaping its genome into more like a three-dimensional structure (Fig. 4.2), via folding into a Peano curve-like space-filling setup (Danchin et al. 2000). This constraint is likely to be important in the gene flow that maintains a particular genome length (Fang et al. 2008). It also maintains a positive selective pressure for the generation of genomes as a patchwork associating a paleome, for making a progeny, complemented by a cenome and sometimes a histome, as just discussed.

Investigating the cenome's composition should help us tell the niche where the organism is thriving. This is how we could confirm (this was known from the time when this bacterium was discovered but curiously overlooked) that *B. subtilis* is an epiphyte (Belda et al. 2013), widely escorting plants (probably herbaceous plants in particular, but different strains might have variants of that particular niche (Nongkhlaw 2014), and this should be told from their genome sequence). However, a large number of genes present in this category are coding for unknown functions and often of unknown history. This is not unexpected, as exploration and colonisation of the environment result from an immense variety of possible functions. Among those are metabolic pathways that construct molecules allowing the organism to settle in a particular place (Moonens and Remaut 2017), as well as those which code for degradation of these same molecules, which become natural food sources as soon as they exist. The large variety of carbohydrates in the environment allows, for example, for a considerable panel of such activities (Cockburn and Koropatkin 2016). In a few words, the cenome of bacteria remains an unlimited source of genes coding for proteins of considerable interest, for metabolic

Fig. 4.2 The genome DNA fills up a significant amount of the cell's volume. The genetic programme has a material support that must be synthesised in parallel with other cytoplasmic components. Tight ad hoc regulation would be difficult to implement securely. A way out selected by evolution is to make a long molecule that fills in. Confining it to a nucleus is a complementary way to perform this task when cells are large

engineering in particular. While many of those genes could be from ancient descent, it is not the genomic domain that we can easily use to go backward in time. Its contribution to evolution is therefore still a matter of investigation.

By contrast, the paleome codes for the small number of functions that are essential for perpetuation of life. In niches that supply all the building blocks required to build up a cell, we find some 400–500 conserved functional macromolecules. This figure corresponds to the minimal genome of a *Mollicutes* that grows without iron, *Mycoplasma mycoides* JCVI-syn3.0 (Hutchison et al. 2016), in media supplemented by all essential building blocks. A fair number of the corresponding functions have been proposed using in silico studies (Danchin and Fang 2016). The share of the genome that codes for building biomass is fairly well understood (it comprises the cell's biosynthesis machinery, replication, transcription and, above all, translation, as well as genes involved in transport and coupling of core metabolism to macromolecule biosynthesis processes). The genes coding for cleaning, maintenance and repair, while clearly identified, still ask for a detailed analysis of the corresponding functions. In particular, because cells continuously age, the functions required to cherry-pick the components that must be placed in the young daughter cells are still far from well understood.

6 The Making of a Progeny

Indeed, among the many definitions that identify the operations tied to life, a specific process is always present: life perpetuates itself by generating a progeny. While it is still relevant to say that a sterile organism is alive, such an organism misses a key property of life, in that it does not have a progeny. Indeed, its very existence is simply borrowing time: maintenance of a machine linked to a programme, both doomed to age, can hardly allow long-term survival in an ever-changing environment (for a discussion see Danchin (2009a, b) and references therein). Some animal societies have classes of sterile individuals, but they are always directly connected to a fertile lineage. If life were only comprising infertile individuals, it would be unable to persist as rocks do, for example. It would already be extinct, unless there existed a steady and speedy process of spontaneous generation with a creation time shorter than the lifespan of individual organisms. This is an unlikely feat that cannot, anyhow, fit with the chemistry of life as we know it.

Considering the construction of a progeny, we can see that an ultimate destination of the genetic programme is to make a copy of itself within a copy of the machine. "Copy" here is poorly defined. How are the processes of programme copying with that of cell copying linked together? Remarkably, the actual concrete copying process differs whether dealing with the programme or with the machine (Dyson 1985; Danchin 2012): again, the programme is replicated in most of the cell's progeny, while the machine's future is much sloppier, and it is only reproduced (i.e. made of similar components, not an exact replica). To this dichotomy two time scales are associated: in general, replication is trustworthy for many generations, while reproduction makes copies that vary rapidly over time (this is a basis for epigenetic heredity). The genome transplantation experiment gives us a vivid illustration of this functional dichotomy. Extracted at the end of the experiment and sequenced, the genome of the bacteria in the recovered colonies is identical to that which has been transplanted in the host. By contrast, the machinery and even the cell's shape differ in the initial host and in the cells making the final colonies (Fig. 4.3). In terms of engineering, this is somewhat unusual, although we all know of man-made devices that have been progressively modified, as was Theseus's boat [that did not keep a single original of its boards after some time, but was still considered the same (Danchin 2003)]. The parent machine has aged, and its components have been replaced by new ones. In the transplantation experiment, this regeneration process required the use of a new programme, differing from the parent one that had been destroyed, thanks to an astute genetic design (Lartigue et al. 2007). As a consequence, during multiplication, the programme that was used is that of the transplanted genome, directing the synthesis of entities that differ from those of the initial host machine.

This state of affairs is far more general than that in the transplantation experiment: in any living form, the components age and are replaced. In parallel, the environment changes, and some components are no longer required and are diluted out, while others are expressed. In short, while the programme may remain the same, the

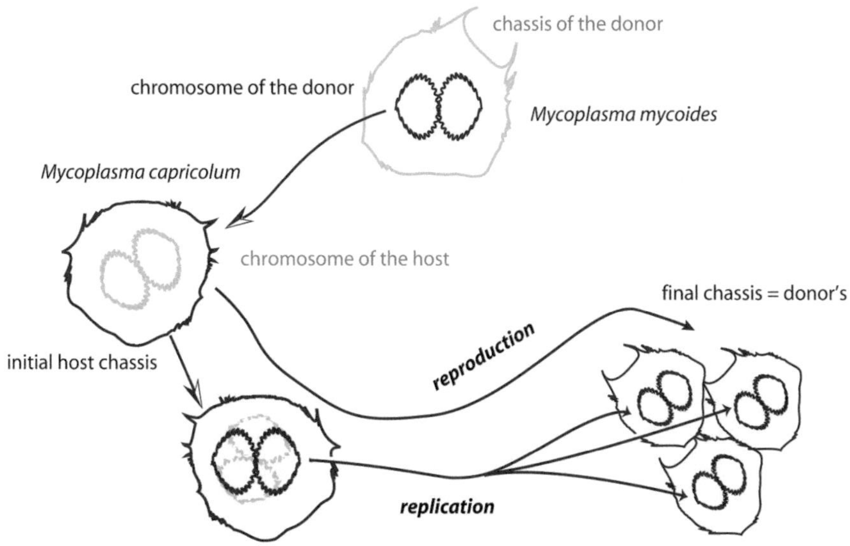

Fig. 4.3 The programme replicates, while the cell reproduces. Genome transplantation is a vivid illustration of the difference between reproduction (making a similar copy) and replication (making an identical copy). For example, the ribosomes of the host cell will differ from those of the members of the final transplanted colony, but they will display the same function

machine that runs it is quite variable. It keeps however its main functional (abstract) properties: reading and expressing the programme and directing the construction of a progeny while monitoring and exploring the environment, extracting proper resources and discarding useless or worn out components. The relationship between the machine and the programme is central to this essential interaction. This situation is also common in contemporary computers, which remember our past actions and do not behave today as they did some time ago, improving their adaptation to our wishes as time elapses. In cells, this corresponds to exploiting information that is not directly that present in the genes, but, rather, contextual information present in the way genes are placed (and sometimes tagged by specific biochemical processes) in the genome and its disposition within the cell as well as in the ultimate matter making the genome.

It follows that, when looking for the core functional genome, we have to understand how the cell succeeds in telling what is young from what is old, and puts the young entities all within the progeny, getting rid of the old ones. Construction of a young progeny from aged cells highlights that there is a specific management of information by cells, in a way that is highly reminiscent of the way the so-called Maxwell's demons operate (Binder and Danchin 2011). One hundred and fifty years ago, James Clerk Maxwell investigated the way he could separate moving gas molecules according to their speed, within an enclosure split into two compartments by a thin wall with an opening trap that could be opened or closed at will. He proposed that an intelligent being could measure the speed of incoming molecules and either open or close the trap, according to their measured speed. This process retained all fast molecules on one side (making it hot) and the slow one on the other side (making it cold). If this were possible, this would allow him to create a steam machine, and hence a perpetual movement, as it appeared that it could be possible to use such a demon without energy. This was discussed for decades until Rolf Landauer and Charles Bennett showed that if acquiring memory (computing) indeed does not require energy, erasing memory will, so that the process is indeed energy consuming, precluding perpetual movement (Landauer 1961; Bennett 1988).

Apart from the trap mechanics, many other processes would allow separating old and young things. Here are two examples. Besides separating young and aged components according to their age into two compartments, another way would be, for example, to evolve specific devices (other types of Maxwell's demons) that patrol within the cell compartment, consistently interacting with properly tagged molecules there (via a selective process of complementarity), leading to destruction of those that have aged and then using ATP or GTP hydrolysis to reset their memory (i.e. restore their original shape that has been modified during interaction with relevant substrates) for another fruitful interaction. This is illustrated, for example, by the newly discovered hydrolysis process mediated by the universal molecular chaperone Hsp90. This protein actively induces conformational changes in specific protein clients. To this aim it has both to recognise proper substrates, to induce functional folding, while changing its own shape, and then to restore its original state to start again the process. This involves hydrolysis of two ATP molecules, sequentially. The first hydrolysis is used for client remodelling, while the second one is used to reset the ATP-dependent cycle (Elnatan et al. 2017). Misfolded proteins escaping this function will aggregate and be disposed of. In another process, proteins such as septins prevent aged proteins from going from the mother cell to the daughter cells (Budovsky et al. 2010), or organise cell division (Li et al. 2012), using energy to reset their state to ground level (Binder and Danchin 2011; Danchin 2012). In general, ATP- or GTP-dependent chaperones and proteases identify specific targets within the cytoplasm and in membranes (Bittner et al. 2016, 2017; Pearl 2016) and use energy to manage information about the correct nature of their substrates. Finally, time-dependent protein clocks, based on isomerisation of asparagine or aspartate, may be repaired or degraded (Danchin et al. 2011). All these devices provide a general way to cope with information pertaining to age and distribute it in a non-random way between growing cells, ending up in a young progeny.

7 Evolution of Resting Cells Progeny

The time when a progeny is built up is also important. When cells rest, they are not metabolically inert. They still need to manage energy, import building blocks and export waste. This is essential because their components are submitted to inevitable accidents triggered by reactive chemical species (Danchin 2017a, b) and spontaneously age. As a matter of fact, the protein polypeptide backbone is never entirely stable. As I previously brought up, some amino acid residues, aspartate and asparagine in particular, tend to cyclise into L-succinimide (Robinson and Robinson 2004), which subsequently evolves spontaneously to the D-enantiomer, and may finally lead to D-aspartate. This aged or maturation product is often not functional, of a different function or less active than the original polypeptide. It follows that the cell has to replace it, on a routine basis. In some cases, the protein may be repaired (using *S*-adenosylmethionine), but this is only an exact repair for aspartate, as asparagine would cycle to aspartate and not to the original residue. Otherwise the protein has to be resynthesised.

The cycle of ageing/repair/replacement operates on a subfamily of proteins, those that are expressed during the resting stage, in a particular environment. This requires that specific genes are transcribed and then translated. Remarkably, as emphasised by Barbara Wright in a series of remarkable studies (Longacre et al. 1999; Wright 2004), because transcription opens up the DNA double helix, it is doomed to be a mutagenic process. Isolated nucleic acid bases will become accessible to reactive molecules, and cytosine will have a considerably enhance probability to tautomerise, making it prone to deamination (Jinks-Robertson and Bhagwat 2014). Until recently, in particular because of fairly efficient repair processes [including a transcription-repair coupling system (Adebali et al. 2017)], this inevitable source of accidents was perceived as pervasive but anecdotal, despite the fact that it could be readily identified (Sakofsky and Grogan 2013; Gaillard and Aguilera 2016). To explore how bacteria may evolve by generating adaptive mutants during ageing, we constructed "intelligent" bacteria that would be able to extract information from their environment while resting. Subsequently, using genome sequencing of the mutants we obtained, we observed that mutagenesis was widespread under these ageing conditions. However, mutagenesis was not at all random, but restricted to hotspots in the genome. Remarkably, these hotspots highlighted a consistent pattern of metabolic pathways.

The experimental setup is straightforward. We used *E. coli* cells deleted for their gene coding for adenylate cyclase. The consequence is that these cells are unable to use a considerable number of carbon sources. Yet they form colonies on a rich medium and then stop growing when they have used all metabolites that do not depend on cyclic AMP regulation to be used. The chosen medium was close to the situation in the human gut at the exit of the second duodenum, where *E. coli* thrives normally, in the presence of the host-rich diet and bile salts. In this medium, we added a carbon source (typically maltose) that these mutated parent bacteria could not use, and we waited. Nothing changed after 24 h. Colonies remained of the same size and appearance. However, after a few days (typically 5 days), papillae began to outgrow on some colonies, and this process lasted up to 2 months, with papillae keeping appearing on resting colonies (most did not display visible changes). We sequenced the genome of 96 of those and explored the corresponding mutations (Sekowska et al. 2016). Beside the expected mutations, which rendered the activator of carbohydrate catabolism constitutively active, we found hotspots that linked together genes common to specific metabolic pathways. As an example, we found mutations in the *cmk* gene, coding for cytidylate kinase, together with mutations in *pyrG* (CTP synthesis), *pnpA* [polynucleotide phosphorylase, making CDP upon phosphorolysis of RNA (Danchin 1997)] and *udk* (pyrimidine kinase). A corollary of all these mutations is that they reduced CTP availability (Fig. 4.4). In the same way, genes involved in cAMP-dependent carbon metabolism were also affected by mutations (often leading to functional pathways in this case), indicative again of concerted evolution.

This may be interpreted as follows. During the resting phase, some metabolic pathways are expressed and running. The proteins involved (transporters, regulators, enzymes) are ageing. Bacteria of a given species have evolved in specific niches where they tend to collect genes that allow them to display a fitting metabolism. If the same environment is regularly met by the bacteria, this had on previous occasions induced a selective pressure that allowed the genes of such pathways to act in concert. Interestingly, because every protein has a specific in-built ageing clock (via its aspartate and asparagine residues, which make a kind of hourglass), a subset of proteins may mature and age at the same rate, more or less simultaneously. Subsequently, they will break down more or less together and will be resynthesised at the same time. This process will begin by a transcription step, and we saw that transcription is mutagenic (Longacre et al. 1999). The most likely outcome will be simultaneous inactivation. This might account for the previously puzzling concerted disappearance of some pathways—such as the molybdopterin-dependent pathways in *Pseudoalteromonas haloplanktis* (Médigue et al. 2005). In contrast, this mutagenic process may also result in a concerted positive evolution, in particular when the proteins of interest interact together. The major consequence of the fact that a subset of genes is expressed when cells are ageing is that a subset of the genome will evolve in a way that may extract specific information from the environment where

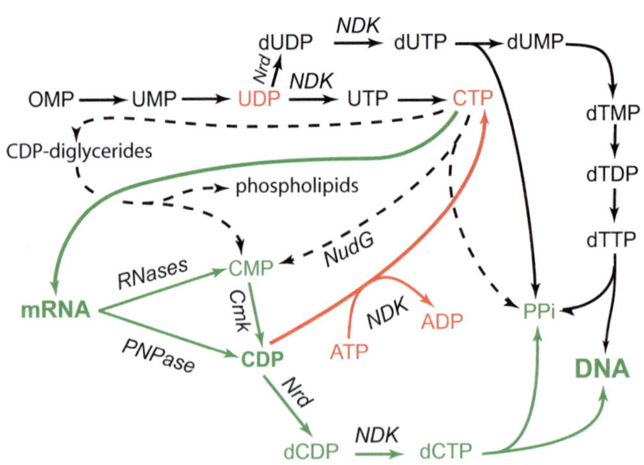

Fig. 4.4 Pyrimidine metabolism, with emphasis on DNA synthesis [adapted from Fig. 2 in Danchin et al. (2018)]. Note that de novo dCTP synthesis requires turnover of RNA because de novo CTP synthesis does not go through a CDP step. Concerted mutations found in the aged-induced mutagenesis of a strain of *E. coli* deleted for its *cya* gene were in *cmk*, *pyrG*, *pnpA* and *udk* (Sekowska et al. 2016). This suggests a selection pressure for CTP limitation

cells are located, a selective pressure that does not follow strict neo-Darwinian interpretation of evolution. Another outcome would be a strong selection pressure to refrain from evolving when the environment keeps being the same for a long time. This is consistent with recruitment of antimutator genes by horizontal gene transfer, as indeed witnessed in *E. coli* (Médigue et al. 1991a, b). It will be of interest to explore the antimutator genotypes of microbiota, depending on the type of environment they thrive in.

8 Provisional Conclusion: No Universal Molecular Clock

The functional analysis of the way genomes are organised shows that they comprise a set of genes driving cell multiplication (the paleome) and a set of genes (the cenome) for life in context (managing exploration of environment and its metabolic consequences). The genes of the paleome are expressed in a process that is by definition concerted. For this reason it is expected that its elements will evolve with a common clock. This fits observations when considering the evolution rate of the translation machinery based on ribosomal RNA in parallel with that of related highly conserved genes (Hug et al. 2016). As a matter of fact, the widespread tree of life used in discussions of the origins of life is based after this extremely limited set of genes [this is well illustrated by the way multilocus typing is achieved (Glaeser and Kämpfer 2015)]. Recent work suggests that this would in fact define a clock that ticked for a specific compartment, presumably nuclear (Staley and Fuerst 2017). Yet, no genome [except for the majority of the genes present in the new synthetic genome of a *M. mycoides* derivative JCVI-syn3.0 (Hutchison et al. 2016)] is only coding for genes allowing multiplication, as no organism is continuously multiplying. This means that a considerable part of genomes must be made of genes managing context and evolving in parallel with the context as we saw with the *E. coli* bacteria producing adaptive mutations during stationary phase. We may accept that a more or less constant molecular evolutionary clock, corresponding to the key biosynthetic machinery genes, might have kept running with constant speed. It is however extremely unlikely that the vast majority of the genes that make the cenome of individual species would evolve with exactly the same speed. In that case we may expect niche-dependent clocks, tying individual species to their environment. As a special situation, we should finally take into account those organisms that remain in a resting phase during much of their life cycle and do not make spores or seeds (which age only very slowly). This resting stage, which is certainly much more sensitive to catastrophic changes, will, by contrast, provide the organism with a way to evolve rapidly while adapting to new environments. This general evolution scheme might explain why some organisms appear as unchanged for millions of years, while others have evolved very fast.

Acknowledgements This work benefited from discussions held in the E-seminar Stanislas Noria.

References

Acevedo-Rocha CG, Fang G et al (2013) From essential to persistent genes: a functional approach to constructing synthetic life. Trends Genet 29(5):273–279

Adebali O, Chiou Y-Y et al (2017) Genome-wide transcription-coupled repair in *Escherichia coli* is mediated by the Mfd translocase. Proc Natl Acad Sci USA 114(11):E2116–E2125

Afshordi N, Magueijo J (2016) Critical geometry of a thermal big bang. Phys Rev D 94(10):101301

Alacevic M (1963) Interspecific recombination in Streptomyces. Nature 197(4874):1323–1323

Belda E, Sekowska A et al (2013) An updated metabolic view of the *Bacillus subtilis* 168 genome. Microbiology 159(Pt 4):757–770

Bennett C (1988) Notes on the history of reversible computation. IBM J Res Dev 44:270–277

Binder PM, Danchin A (2011) Life's demons: information and order in biology. What subcellular machines gather and process the information necessary to sustain life? EMBO Rep 12 (6):495–499

Bittner L-M, Arends J et al (2016) Mini review: ATP-dependent proteases in bacteria. Biopolymers 105(8):505–517

Bittner L-M, Arends J et al (2017) When, how and why? Regulated proteolysis by the essential FtsH protease in *Escherichia coli*. Biol Chem 398(5–6):625–635

Budovsky A, Fraifeld VE et al (2010) Linking cell polarity, aging and rejuvenation. Biogerontology 12(2):167–175

Cao B, Chen C et al (2014) Genomic mapping of phosphorothioates reveals partial modification of short consensus sequences. Nat Commun 5:3951

Cockburn DW, Koropatkin NM (2016) Polysaccharide degradation by the intestinal microbiota and its influence on human health and disease. J Mol Biol 428(16):3230–3252

Cole EL Jr (1998) Functional analysis: a system conceptual design tool [and application to ATC system]. IEEE Trans Aerosp Electron Syst 34(2):354–365

Danchin A (1997) Comparison between the *Escherichia coli* and *Bacillus subtilis* genomes suggests that a major function of polynucleotide phosphorylase is to synthesize CDP. DNA Res 4 (1):9–18

Danchin A (2003) The Delphic boat. What genomes tell us. Harvard University Press, Cambridge

Danchin A (2009a) Bacteria as computers making computers. FEMS Microbiol Rev 33:3–26

Danchin A (2009b) Natural selection and immortality. Biogerontology 10(4):503–516

Danchin A (2012) Scaling up synthetic biology: do not forget the chassis. FEBS Lett 586 (15):2129–2137

Danchin A (2014) The emergence of the first cells. Rev Cell Biol Mol Med:https://doi.org/10.1002/3527600906.mcb.20130025

Danchin A (2017a) Coping with inevitable accidents in metabolism. Microb Biotechnol 10 (1):57–72

Danchin A (2017b) From chemical metabolism to life: the origin of the genetic coding process. Beilstein J Org Chem 13:1119–1135

Danchin A, Fang G (2016) Unknown unknowns: essential genes in quest for function. Microb Biotechnol 9(5):530–540

Danchin A, Guerdoux-Jamet P et al (2000) Mapping the bacterial cell architecture into the chromosome. Philos Trans R Soc Lond Ser B Biol Sci 355(1394):179–190

Danchin A, Binder PM et al (2011) Antifragility and tinkering in biology (and in business); flexibility provides an efficient epigenetic way to manage risk. Genes 2(4):998–1016

Danchin A, Ouzounis C et al (2018) No wisdom in the crowd: genome annotation in the era of big data - current status and future prospects. Microb Biotechnol 11(4):588–605

Dyson FJ (1985) Origins of life. Cambridge University Press, Cambridge

Elnatan D, Betegon M, et al (2017) Symmetry broken and rebroken during the ATP hydrolysis cycle of the mitochondrial Hsp90 TRAP1. Elife 6

Fang G, Rocha EPC et al (2008) Persistence drives gene clustering in bacterial genomes. BMC Genomics 9:4

Flemming H-C, Wingender J et al (2016) Biofilms: an emergent form of bacterial life. Nat Rev Microbiol 14(9):563–575

Gaillard H, Aguilera A (2016) Transcription as a threat to genome integrity. Annu Rev Biochem 85 (1):291–317

Glaeser SP, Kämpfer P (2015) Multilocus sequence analysis (MLSA) in prokaryotic taxonomy. Syst Appl Microbiol 38(4):237–245

Goujon P (2001) From biotechnology to genomes: the meaning of the double helix. World Scientific, Singapore

Harish A, Kurland CG (2017) Akaryotes and Eukaryotes are independent descendants of a universal common ancestor. Biochimie 138:168–183

Harish A, Tunlid A et al (2013) Rooted phylogeny of the three superkingdoms. Biochimie 95 (8):1593–1604

Hashimoto M, Ichimura T et al (2005) Cell size and nucleoid organization of engineered *Escherichia coli* cells with a reduced genome. Mol Microbiol 55(1):137–149

Huerta-Sanchez E, Casey FP (2015) Archaic inheritance: supporting high-altitude life in Tibet. J Appl Physiol 119(10):1129

Hug LA, Baker BJ et al (2016) A new view of the tree of life. Nat Microbiol 1:16048

Huntley S, Hamann N et al (2010) Comparative genomic analysis of fruiting body formation in Myxococcales. Mol Biol Evol 28(2):1083–1097

Hutchison CA 3rd, Chuang RY et al (2016) Design and synthesis of a minimal bacterial genome. Science 351(6280):aad6253

Hutinet G, Swarjo MA et al (2017) Deazaguanine derivatives, examples of crosstalk between RNA and DNA modification pathways. RNA Biol 14(9):1175–1184

Jannasch H (1967) Growth of marine bacteria at limiting concentrations of organic carbon in seawater. Limnol Oceanogr 12(2):264–271

Jinks-Robertson S, Bhagwat AS (2014) Transcription-associated mutagenesis. Annu Rev Genet 48 (1):341–359

Karas BJ, Jablanovic J et al (2013) Direct transfer of whole genomes from bacteria to yeast. Nat Methods 10(5):410–412

Kumar S (2005) Molecular clocks: four decades of evolution. Nat Rev Genet 6(8):654–662

Kysela DT, Randich AM et al (2016) Diversity takes shape: understanding the mechanistic and adaptive basis of bacterial morphology. PLoS Biol 14(10):e1002565

Landan G, Cohen G et al (1990) Evolution of isopenicillin N synthase genes may have involved horizontal gene transfer. Mol Biol Evol 7(5):399–406

Landauer R (1961) Irreversibility and heat generation in the computing process. IBM J Res Dev 3:184–191

Lartigue C, Glass JI et al (2007) Genome transplantation in bacteria: changing one species to another. Science 317(5838):632–638

Li S, Ou XH et al (2012) Septin 7 is required for orderly meiosis in mouse oocytes. Cell Cycle 11 (17):3211–3218

Longacre A, Reimers JM et al (1999) Specificity of transcription-enhanced mutations. Ann N Y Acad Sci 870:383–385

McCormick JR, Flärdh K (2012) Signals and regulators that govern *Streptomyces* development. FEMS Microbiol Rev 36(1):206–231

Médigue C, Rouxel T et al (1991a) Evidence for horizontal gene transfer in *Escherichia coli* speciation. J Mol Biol 222(4):851–856

Médigue C, Viari A et al (1991b) *Escherichia coli* molecular genetic map (1500 kbp): update II. Mol Microbiol 5(11):2629–2640

Médigue C, Krin E et al (2005) Coping with cold: the genome of the versatile marine Antarctica bacterium *Pseudoalteromonas haloplanktis* TAC125. Genome Res 15(10):1325–1335

Monod J (1972) Chance and necessity: an essay on the natural philosophy of modern biology. Vintage Books, New York
Moonens K, Remaut H (2017) Evolution and structural dynamics of bacterial glycan binding adhesins. Curr Opin Struct Biol 44:48–58
Muro-Pastor AM, Hess WR (2012) Heterocyst differentiation: from single mutants to global approaches. Trends Microbiol 20(11):548–557
Nongkhlaw FMW (2014) Epiphytic and endophytic bacteria that promote growth of ethnomedicinal plants in the subtropical forests of Meghalaya, India. Universidad de Costa Rica
Ohama T, Inagaki Y et al (2008) Evolving genetic code. Proc Jpn Acad Ser B Phys Biol Sci 84 (2):58–74
Ohno S (1987) Repetition as the essence of life on this Earth: music and genes. Modern trends in human Leukemia VII: new results in clinical and biological research including pediatric oncology. Springer, Berlin, pp 511–519
Ouzounis C, Kyrpides N (1996) The emergence of major cellular processes in evolution. FEBS Lett 390(2):119–123
Paixão-Cortes VR, Viscardi LH et al (2012) Homo sapiens, Homo neanderthalensis and the Denisova specimen: new insights on their evolutionary histories using whole-genome comparisons. Genet Mol Biol 35(4 Suppl):904–911
Pearl LH (2016) Review: the HSP90 molecular chaperone – an enigmatic ATPase. Biopolymers 105(8):594–607
Ragan MA, McInerney JO et al (2009) The network of life: genome beginnings and evolution. Introduction. Philos Trans R Soc Lond B Biol Sci 364(1527):2169–2175
Robinson N, Robinson A (2004) Molecular clocks deamidation of asparaginyl and glutaminyl residues in peptides and proteins. Althouse Press, Cave Junction, OR
Roszak DB, Colwell RR (1987) Survival strategies of bacteria in the natural environment. Microbiol Rev 51(3):365–379
Rusconi R, Garren M et al (2014) Microfluidics expanding the frontiers of microbial ecology. Annu Rev Biophys 43:65–91
Sakofsky CJ, Grogan DW (2013) Endogenous mutagenesis in recombinant *Sulfolobus* plasmids. J Bacteriol 195(12):2776–2785
Sekowska A, Wendel S et al (2016) Generation of mutation hotspots in ageing bacterial colonies. Sci Rep 6(1):2
Staley JT, Fuerst JA (2017) Ancient, highly conserved proteins from a LUCA with complex cell biology provide evidence in support of the nuclear compartment commonality (NuCom) hypothesis. Res Microbiol 168(5):395–412
Stankiewicz P (2016) One pedigree we all may have come from – did Adam and Eve have the chromosome 2 fusion? Mol Cytogenet 9:72
Stringer C (2016) The origin and evolution of *Homo sapiens*. Philos Trans R Soc B Biol Sci 371 (1698)
Sukhodolets VV (1988) Organization and evolution of the bacterial genome. Microbiol Sci 5 (7):202–206
Taheri-Araghi S, Brown SD et al (2015) Single-cell physiology. Annu Rev Biophys 44(1):123–142
Watanabe S, Shiwa Y et al (2012) Complete sequence of the first chimera genome constructed by cloning the hwole genome of *Synechocystis* strain PCC6803 into the *Bacillus subtilis* 168 genome. J Bacteriol 194(24):7007–7007
Woese CR (2002) On the evolution of cells. Proc Natl Acad Sci USA 99(13):8742–8747
Wright BE (2004) Stress-directed adaptive mutations and evolution. Mol Microbiol 52(3):643–650
Wu T, Huang Q et al (2017) Mechanistic investigation on ROS resistance of phosphorothioated DNA. Sci Rep 7:42823

Chapter 5
Natural Strategies of Spontaneous Genetic Variation: The Driving Force of Biological Evolution

Werner Arber

Mainly based on experimental investigations carried out with *Escherichia coli* bacteria and their viruses (bacteriophages), we discuss here the variety of specific molecular mechanisms that occasionally contribute to the spontaneous formation of genetic variants, the drivers of biological evolution. The identified mechanisms are assigned to three natural strategies of genetic variation. These are local changes in the nucleotide sequences of DNA, intra-genomic rearrangements of DNA segments, and acquisition of a foreign DNA segment by horizontal gene transfer. A number of evolution genes contribute by their in part rarely available products as variation generators and as modulators of the frequencies of genetic variation. Various nongenetic elements can also be involved in spontaneous mutagenesis. A conceptual conclusion is the duality of the genome which carries besides the essential genes for the individual living organism also a number of genes that insure the very slow but steady biological evolution at the population level including, besides adaptation to evolving habitats, also the building up of a rich biodiversity on Earth.

About 150 years ago, the scientific branches of evolutionary biology (Ch. Darwin, 1859), classical genetics (G. Mendel, 1866), and nucleic acid biochemistry (F. Miescher, 1874) took their origin, mainly based on work with different traits of plants and of animals. In the course of time, phenotypic traits became assigned to depend on activities of chromosomes. It remained long unknown if this was due to chromosomal proteins or to functions determined by the nucleic acids. Work carried out with microorganisms (mainly bacteria and bacterial viruses, i.e., bacteriophages) succeeded in the 1940s to trace the way to molecular genetics. Purified *Pneumococcal* DNA preparations (Avery et al. 1944) transferred traits from a donor bacterium into an appropriate receptor strain (transformation). In addition, sexual transfer of traits between two related bacterial *Escherichia coli* strains was

W. Arber (✉)
Biozentrum University of Basel, Basel, Switzerland
e-mail: Werner.Arber@unibas.ch

P. H. Rampelotto (ed.), *Molecular Mechanisms of Microbial Evolution*, Grand Challenges in Biology and Biotechnology,
https://doi.org/10.1007/978-3-319-69078-0_5

seen to occur linearly upon conjugation, pointing to a linear order of the transferred genetic information (Lederberg 1947). Finally, some bacteriophages could be shown to carry along genetic information picked up in their bacterial host and to give rise by transduction to horizontal gene transfer upon infection of another host (Zinder and Lederberg 1952).

Structural investigations by Watson and Crick revealed in 1953 the filamentous double-helical structure of DNA molecules, which strongly suggested how the nucleotide sequences of DNA might carry the biological information (Watson and Crick 1953a). Further experimental work revealed detailed structural aspects of natural gene vectors (small auto-replicating fertility plasmids and bacteriophage particles and their genomes). These fundamental insights pointed to various microbial abilities to occasionally transfer genetic information horizontally between different bacterial strains (see references Hayes 1964; Goodenough and Levine 1974).

Enzymatic restriction-modification systems (Arber 1965a) can enable bacteria to destroy invading foreign DNA molecules, but they protect the cells' own genome by site-specific nucleotide methylation from their restriction endonuclease activity (Arber 1965b; Kühnlein et al. 1969). Several other factors can also impede the acquisition of foreign genetic information (Arber 2012). Nevertheless, occasional horizontal gene transfer represents an effective contribution to biological evolution.

In view of the linearity of the genetic information carried in DNA molecules, a comparison with our written texts can be helpful for a broad understanding. The genome of the intensively studied *Escherichia coli* bacteria is contained in just one large circular DNA molecule, and its information content roughly corresponds to the size of the Bible. Some other bacterial strains are known to have somewhat smaller genomes. But we can say that, in general, the genetic content of bacteria has the size of a book, whereas the genomes of plants and animals are much larger and may correspond to encyclopedias of hundreds up to a thousand volumes of the size of the Bible. The encyclopedia of the human genome contains about 700 volumes.

Bacteria are haploid unicellular organisms. Under appropriate growth conditions in liquid media, their generation time is often less than an hour. This leads to large populations within a day, which facilitates exploring the various mechanisms of spontaneous mutagenesis. It was an early general finding that, by far, not all novel genetic variants are favorable and provide a selective advantage. One has thus no evidence for a directive of spontaneous mutagenesis in response to an identified need; sites and times of spontaneous mutagenesis are more random. In addition, only one in several hundred bacterial cells per generation becomes affected by spontaneous mutagenesis in a growing population.

The different so far identified molecular mechanisms contributing occasionally to the production of novel genetic variants testify of a rich inventiveness of Mother Nature to provide DNA sequence alterations. These are the drivers of biological evolution, whereas natural selection directs evolution.

Based on the available specific knowledge, we can attribute individual mutagenesis events to three natural strategies of spontaneous production of novel genetic variants. These strategies are:

(a) Local nucleotide sequence variations affecting one or a few adjacent nucleotides
(b) Intra-genomic rearrangements of particular DNA segments
(c) The acquisition of a segment of foreign DNA by horizontal (also called lateral) gene transfer

Short-living isomeric forms of nucleotides are a natural source for a local mutagenesis leading to a nucleotide substitution (Watson and Crick 1953b). For example, the imino-form of adenine does not pair with thymine, but it does so with cytosine. This leads to a mispairing when adenine reassumes its stable standard form. Enzymatic repair systems succeed to correct part of nascent mispairings, but some nucleotide substitutions can escape repair.

A number of enzymatic activities can contribute to intra-genomic DNA rearrangements (Arber 2014). Mobile genetic elements can occasionally transpose to an alternative site in the genome or in a resident plasmid (Shapiro 1983). Other DNA rearrangements can be brought about by general recombination between sequence homologies at different locations in the genome. A source for novel gene fusions and operon fusions can be site-specific recombination occurring at so-called secondary crossover sites or quasi sites (Iida and Hiestand Nauer 1987). Still other so-called illegitimate recombination processes can also contribute to rare DNA rearrangements. In many of these processes of DNA rearrangements, low tolerable levels of mutagenesis are reached by a low availability of relevant enzymes and by the activities of repair systems.

It is clear that some spontaneously occurring genetic variations can be lethal for the organism in question. This is obviously practically impossible to be studied. However, a relevant experimental approach to identify lethal mutants in prophage P1 maintained as a plasmid in its lysogenic host revealed that 95% of lethal mutations corresponded to the insertion of a mobile genetic element out of the host genome into one of the essential phage genes, and only about 5% were due to a local mutagenesis (Sengstag and Arber 1983). Lethality was thereby identified by the absence of production of active phage particles upon induction of phage replication.

Evolutionary emergence of novel properties can be caused by intra-genomic DNA rearrangements bringing about a novel fusion of functional domains already carried on the genome.

Another important source for novel properties is the acquisition of foreign genetic information by horizontal DNA transfer; this allows the recipient cell to share in successful developments made by others. Intensive fundamental studies of relevance were made with genes for antibiotic resistance, but it rapidly became obvious that many other gene functions can also be welcome acquisitions for microorganisms.

In view of the evolutionary role of horizontal gene transfer, randomly placed horizontal connectors can be drawn between branches of the tree of evolution originally drawn by Darwin. Whereas this author intended to symbolically show a common origin of forms of life, horizontal connectors in addition can symbolize that branches of the tree of evolution have also a common future by profiting of the occasional acquisition of a foreign genetic potential (Arber 1991). Note that the universality of the genetic code, i.e., the common genetic language, contributes to the effectiveness of gene acquisition (Arber 2006). In addition, close cohabitation can favor occasional horizontal gene transfer: think on microbiomes in eukaryotic organisms (Blaser et al. 2013).

According to the theory of molecular evolution, novel genetic variants are not due to errors and accidents. They rather depend, on the one hand, on enzymatic activities such as variation generators and modulators of the frequencies of genetic variation and, on the other hand, on several nongenetic elements which can also contribute to mutagenesis. As we have already mentioned for rare isomeric forms of nucleotides, structural flexibilities of biologically active molecules can become involved in the production of genetic variants. Other contributions can be assigned to environmental mutagens and to random encounter including the chance to undergo horizontal gene transfer. We can conclude that the natural reality actively takes care of biological evolution.

For obvious reasons, genes for products acting as variation generators and as modulators of the frequencies of genetic variation must be carried in the genome. Some of their gene products are known to be inessential for the life of the concerned individual, whereas some other gene products serve both, for the individuals' life and for the capacity of the population to undergo biological evolution. If we call "evolution genes" the genes that are essential for insuring biological evolution, we can conclude that genomes generally show a duality (Arber 2014): A majority of the genes carried in the genome are essential for the fulfillment of the life of the concerned individual, whereas a minority of genes may uniquely contribute to the slow evolution of the concerned population, i.e., to the expansion of life and to enrich biodiversity. Products of some of the genes can obviously serve for both purposes.

References

Arber W (1965a) Host-controlled modification of bacteriophages. Annu Rev Microbiol 19:365–378

Arber W (1965b) Host specificity of DNA produced by *Escherichia coli,* V. The role of methionine in the production of host specificity. J Mol Biol 11:247–256

Arber W (1991) Elements in microbial evolution. J Mol Evol 33:4–12

Arber W (2006) The evolutionary strategy of DNA acquisition as a possible reason for a universal genetic code. Hist Philos Life Sci 28:525–532

Arber W (2012) Genetic variation and molecular darwinism. In: Meyers RA (ed) Systems biology: advances in molecular biology and medicine. Wiley, Weinheim, pp 145–168

Arber W (2014) Horizontal gene transfer among bacteria and its role in biological evolution. Life 4:217–224

Avery OT, MacLeod CM, McCarty M (1944) Studies on the chemical nature of the substance inducing transformation of pneumococcal types. J Exp Med 79:137–158

Blaser M, Bork P, Fraser C, Knight R, Wang J (2013) The microbiome explored: recent insights and future challenges. Nat Rev Microbiol 11:213–217

Goodenough U, Levine RP (1974) Genetics, Chap 9. Holt, Reinhart and Winston, New York

Hayes W (1964) The genetics of bacteria and their viruses: studies in basic genetics and molecular biology. Wiley, New York

Iida S, Hiestand Nauer R (1987) Role of the central dinucleotide at the crossover sites for the selection of quasi sites in DNA inversion mediated by the site-specific Cin recombinase of phage P1. Mol Gen Genet 208:464–468

Kühnlein U, Linn S, Arber W (1969) Host specificity of DNA produced by *Escherichia coli,* XI. In vitro modification of phage fd replicative form. Proc Natl Acad Sci USA 63:556–562

Lederberg J (1947) Gene recombination and linked segregation in *Escherichia coli.* Genetics 32:505–525

Sengstag C, Arber W (1983) IS2 insertion is a major cause of spontaneous mutagenesis of the bacteriophage P1: non-random distribution of target sites. EMBO J 2:67–71

Shapiro JA (1983) Mobile genetic elements. Academic Press, New York

Watson JD, Crick FHC (1953a) Molecular structure of nucleic acids. Nature 171:737–738

Watson JD, Crick FHC (1953b) The structure of DNA. Cold Spring Harb Symp Quant Biol 18:123–131

Zinder ND, Lederberg J (1952) Genetic exchange in *Salmonella.* J Bacteriol 64:679–699

Chapter 6
The Evolution of Gene Regulatory Mechanisms in Bacteria

Charles J. Dorman, Niamh Ní Bhriain, and Matthew J. Dorman

1 Introduction

Although genes can be expressed in a cell-free state, their regulation allows their expression to be bent to meet a need greater than just that of the gene itself—the need of the cell that houses them. Once cellular life emerged, gene regulation was obliged to evolve in concert with cellular evolution. The cell represents a community of molecules, and by acting in harmony in time and space in response to cues from the interior and exterior of the cell, the members of this community are more likely to thrive than those existing in one where there is no governance. An important principle that seems to underlie the need for efficient regulation concerns the costs to competitive fitness that accrue when a bacterium expresses gene products inappropriately (Price et al. 2016). Horizontal transmission of genes is a fact of bacterial life and one that adds richness and complexity to regulatory evolution. The need for regulatory harmony means that newly arrived genes, such as those in viruses or other horizontally acquired genetic elements, such as plasmids, must undergo regulatory integration in addition to physical integration (Dorman 2009; Syvanen 2012). We will consider below how the activities of the newcomer genes can be harmonized with those of the core genome so the new gene-cell combination can thrive.

What is it that is being 'regulated' when one considers gene regulatory processes? It is useful to recall that gene expression in bacteria involves the sequential processes of transcription initiation, elongation and termination and that all three stages can be placed under regulatory control. In the case of genes encoding proteins, the

C. J. Dorman (✉) · N. N. Bhriain
Department of Microbiology, Moyne Institute of Preventive Medicine, Dublin 2, Ireland
e-mail: cjdorman@tcd.ie

M. J. Dorman
Wellcome Sanger Institute, Wellcome Genome Campus, Hinxton, Cambridgeshire, UK

P. H. Rampelotto (ed.), *Molecular Mechanisms of Microbial Evolution*, Grand Challenges in Biology and Biotechnology,
https://doi.org/10.1007/978-3-319-69078-0_6

posttranscriptional processes of translation initiation, elongation and termination are also candidates for regulation, as are the many posttranslational modifications that proteins may experience, such as phosphorylation, methylation, compartmentalization (including secretion) and proteolysis (Kudva et al. 2013; Lan and Tu 2016; Schneewind and Missiakas 2014; Trentini et al. 2016). A further opportunity to control gene expression arises at the level of RNA stability, and this can be affected by interactions between mRNA and small non-coding RNAs, many of which rely on protein chaperones to facilitate RNA-RNA interaction (Desnoyers et al. 2009; Massé et al. 2003; Rajkowitsch and Schroeder 2007). In addition, opportunities for control of expression arise in the cases of prokaryotic messenger RNA molecules that can be spliced (this is not exclusively a eukaryotic process) (Doose et al. 2013; Martínez-Rodríguez et al. 2014) and in the case of bacterial proteins that undergo splicing (Novikova et al. 2014). The existence of these molecular maturation events indicates evolutionary links with their splicing-dependent eukaryotic counterparts.

The regulation of gene expression involves controlling, inter alia, the activities of RNA polymerase and ribosomes, large molecular machines that are responsible for transcription and translation, respectively (Ban et al. 2000; Feng et al. 2016; Mekler et al. 2000). The means by which these controls are exerted have been studied intensively for many years, and a wealth of molecular detail is available. It is evident that a multitude of regulatory mechanisms is employed, especially in controlling RNA polymerase, pointing to the importance of achieving control in response to a plethora of signals. These are unlikely to have evolved in a linear fashion, and they are probably capable of yet further evolution. This regulatory evolvability is indicative of the enormous adaptability of bacteria both at the level of single cells and as populations.

This chapter will survey briefly the major processes that regulate gene expression in bacteria. These processes will be considered in the context of bacterial nucleoid structure because gene regulatory processes frequently involve nucleoid architectural elements. In this way the reader can be encouraged to consider the co-evolution of gene control programmes and genome structure. Finally, the influence of horizontal gene transfer on the evolution of gene regulatory circuits and genome composition will be discussed.

2 The Emergence and Evolution of RNA Polymerase

The multisubunit RNA polymerase found in bacteria is closely related to those of the archaea and eukaryotes, suggesting that it has evolved from the polymerase of the last universal common ancestor (LUCA) (Booth et al. 2016; Koonin 2003). This close relationship exists at the levels of subunit amino acid sequence, subunit structure and function and the mechanism by which the complex carries out the different stages of transcription (Werner and Grohmann 2011; Zenkin 2014).

It has been suggested that the functional ancestor of modern multisubunit RNA polymerases was a ribozyme that existed in the putative 'RNA world' phase of cellular evolution and that the forerunner of the β-subunit and β′-subunit was a homodimeric β2 RNA-binding protein that had no catalytic activity (Iyer et al. 2003). This homodimer may first have become involved in RNA synthesis by performing the role of an RNA-binding chaperone on behalf of the ribozyme at the time that protein synthesis evolved and later evolved into a heterodimer with chemically distinct activities associated with its subunits. At some stage in evolution, the active site for RNA synthesis shifted from the ribozyme to the protein component, allowing the now redundant RNA element to be discarded. The next important evolutionary step in this speculative process (note that no contemporary RNA polymerizing ribozyme has been discovered) was the transition from an RNA template to a DNA template, leading finally to the emergence of a DNA-dependent RNA polymerase in LUCA (Werner and Grohmann 2011).

Bacterial RNA polymerase is responsible for all of the transcription in the bacterium, in contrast to eukaryotes where the tasks of transcribing mRNA, ribosomal RNA and small RNA are divided among the three different forms of the polymerase. The multisubunit structure of bacterial RNA polymerase distinguishes it from polymerases encoded by bacteriophage such as T7 in which all of the functions are found in a single polypeptide (Steitz 2009). The supply of subunits for RNA polymerase provides a mechanism for regulating its activity. This applies especially to the sigma factor, the subunit that is responsible for promoter recognition (Fig. 6.1). Sigma factor supply is influenced by sequestration by anti-sigma factors (Helmann 1999). The activity of RNA polymerase is also influenced by factors that do not bind to DNA, such as the alarmone guanosine tetraphosphate which, together with the DksA protein, reports accumulations of uncharged tRNAs and directs transcription away from genes involved in expression of translation machinery (Haugen et al. 2008).

2.1 Alternative Sigma Factors of RNA Polymerase

The model bacterium *Escherichia coli* has seven sigma factors, each of which can direct RNA polymerase to a different class of transcription promoter (Gourse et al. 2006; Gruber and Gross 2003; Nyström 2004). These differences are made manifest by way of promoter nucleotide sequence and the spacing between conserved nucleotide sequence motifs to which the sigma factor binds. The primary sigma factor is sigma-70 (also known as RpoD), and this protein is responsible for the initiation of transcription of the majority of the genes in the bacterium. These include genes that are expressed under very specific circumstances and genes that are expressed for long periods of every cell cycle. An example of the former is the *lac* operon that is expressed only when the bacterium detects lactose in its immediate environment, and examples of the latter include the genes that encode ribosomal RNA that is required to generate the ribosomes that are needed to build and maintain the cell through protein synthesis.

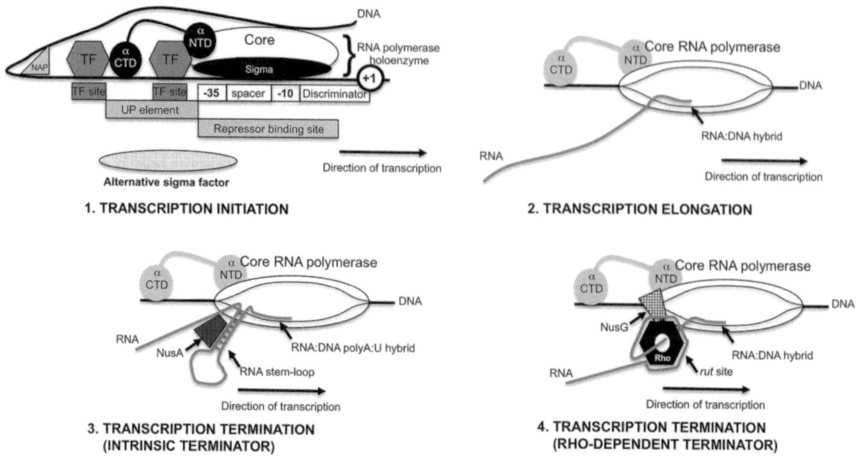

Fig. 6.1 Opportunities for the evolution of regulatory inputs at the major steps in transcription. (1) The key components of a transcription initiation complex are shown. The coordinates of the promoter elements (−10 and −35) that contact the sigma factor of the RNA polymerase holoenzyme are numbered with reference to the transcription start site (+1). For promoters that are recognized by the RpoD sigma factor (sigma-70), a DNA spacer that is typically 17 base pairs in length separates the −10 and −35 promoter 'boxes'. A G + C-rich discriminator sequence is found in the DNA of promoters of genes and operons that express stable RNA (rRNA, tRNA) and other components of the translational machinery of the cell. The A + T-rich UP element located immediately upstream of the −35 of some promoters acts to recruit RNA polymerase through interaction with the carboxyl terminal domain (CTD) of the alpha subunits of RNA polymerase. The same subunit is frequently contacted by transcription factors (TFs), proteins that bind to specific sequences in the DNA upstream of the promoter. TF-binding sites can be found in different locations and with different orientations, depending on the promoter. In addition to the CTD of the alpha subunit, TFs can also make contact with RNA polymerase at the amino terminal domain (NTD) of the alpha subunit and the sigma factor. DNA upstream of the promoter can fold back to contact RNA polymerase, guided by local DNA topology or by DNA-binding and DNA-bending proteins such as nucleoid-associated proteins (NAPs). Alternative sigma factors compete with RpoD for access to RNA polymerase core enzyme. If they succeed, they will redirect polymerase to those promoters whose architecture conforms to their requirements for binding. (2) During transcription elongation, there are opportunities for the imposition of control through pausing of the polymerase, possibly leading to premature termination. (3) Termination of transcription by intrinsic terminators involves the NusA-assisted formation of a stem-loop RNA structure in the transcript immediately adjacent to a weak RNA-DNA hybrid made up of A:U base pairs. The stem loop stalls the polymerase, and base pairing fails in the weak RNA-DNA hybrid, allowing the RNA to separate and causing transcription to terminate. (4) Rho-dependent transcription termination involves the binding to the transcript of the hexameric, ATP-dependent Rho helicase. The binding is guided by the C-rich *rut* site in RNA and does not require a secondary structure such as stem loop. Assisted by the NusG protein, Rho detaches the transcript from RNA polymerase, terminating transcription

In *E. coli*, the alternative sigma factors (and their signals) are sigma-19/FecI (ferric citrate uptake), sigma-24/RpoE (envelope stress), sigma-28 (flagella biosynthesis), sigma-38/RpoS (stress and stationary phase), sigma-32/RpoH (heat shock), and sigma-54/RpoN (nitrogen starvation) (Gruber and Gross 2003; Nyström 2004).

Alternative sigma factors must compete with sigma-70 for access to the core RNA polymerase. This competition takes place in every cell in the population in response to the appropriate sigma factor-specific environmental signal, and it is unlikely that a truly uniform outcome will be achieved across this population. The result is physiological diversity in a genetically homogeneous population. The alternative sigma factor may not be present under all growth conditions, so its expression must also be controlled so that it accumulates when it is required. The stress and stationary phase sigma factor sigma-38 (RpoS) is expressed throughout the growth cycle but is rapidly degraded by proteolysis. This process of degradation is arrested when the bacterium ceases to grow, allowing sigma-38 to accumulate and to complete with sigma-70 (Hengge 2009).

Adding genes for alternative sigma factors to a bacterial genome is a powerful way to enable the evolution of novel regulatory and physiological options. To have a physiological impact, the new sigma factor must be able to find transcription promoters that have the requisite sets of appropriately spaced DNA sequence motifs. Sigma-38 is closely related to sigma-70, suggesting that the genes encoding these homologous proteins are likely to share a common ancestor. The promoter sequences that are bound by sigma-38 are related to those recognized by sigma-70, yet they are sufficiently different to allow genuine reprogramming of transcription to occur when the population of sigma-38 proteins expands in the cell. In this way one can discern a sigma-38 regulon of genes, and many of these genes are involved in stress survival (Schellhorn 2014). Why was it advantageous for *E. coli* (and other bacteria) to evolve sigma-38 rather than relying on sigma-70 to transcribe stress survival genes? Firstly, not all stress survival genes depend on sigma-38 for transcription. Secondly, it is possible that sigma-38 utilizes a feature of the genome in non-growing bacteria that is antagonistic to efficient transcription initiation by sigma-70: i.e. a relaxed DNA template (Bordes et al. 2003). A relaxed DNA template is not an impediment to sigma-38, and DNA in stationary phase bacteria is relaxed compared with that in rapidly growing cells (Conter et al. 1997). This observation raises an important, and frequently overlooked, feature of gene regulation: the importance of DNA shape as a modulator of the regulatory process. We shall return to this topic presently (Sects. 5 and 6).

3 Transcription and Its Regulation

The three principal stages of transcription are initiation, elongation and termination. All three have the potential to be regulated, and it has been proposed that elongation was originally the main focus for control (Werner and Grohmann 2011). Controlling transcription during the elongation phase is attractive because the process is subject to discontinuities, allowing it to be paused or slowed by *cis*-acting features in the DNA template or the transcript, possibly aided by *trans*-acting factors such as small RNAs or proteins. In primitive transcription systems, how did the transcription process start prior to the elongation phase? The most straightforward model envisages non-specific transcription initiation, probably from DNA sequences that are A + T-rich. Such DNA sequences have a greater tendency to melt than G + C-rich ones, and melting of the DNA duplex is a prerequisite for transcription (Ross and Gourse 2009). It is interesting to recall that the promoters of most genes in *E. coli* and many other modern bacteria are A + T-rich (Pedersen et al. 2000).

The emergence of sigma factors and their preference for binding to DNA with specific sequences seem to mark a move away from spontaneous transcription initiation to a more orderly and predictable process. However, there is a great deal of sequence variety in the promoters of model bacteria indicting that modern RNA polymerase and its sigma factors have a high degree of tolerance when determining where to initiate transcription. The widespread phenomenon of pervasive transcription supports the low-fidelity of the relationship between RNA polymerase and promoters (Singh et al. 2014).

3.1 Promoter Architecture and Transcription Regulation

In modern bacteria there is much emphasis on regulating the transcription initiation step (Browning and Busby 2016). This raises questions about the roles of other factors that cooperate with the primary promoter elements to make a transcription initiation site. It is useful to remember that the sigma factor is not the only component of RNA polymerase to contact the DNA template: the alpha subunits, of which there are two copies, also do this (Fig. 6.1). The UP element is A + T-rich and located immediately upstream of the −35 box of some sigma-70 promoters (Ross et al. 1993) (Fig. 6.1). DNA even further upstream can loop across the promoter to make further contacts (sometimes called 'backside contacts') with RNA polymerase (Muskhelishvili and Travers 2003; Zhang and Schleif 1998) (Fig. 6.1). In combination, this group of protein-DNA contacts may help to distinguish a 'real' promoter from a spurious one.

From a regulatory perspective, there is a difficulty with relying on contacts between RNA polymerase and promoter-associated DNA. This concerns the need to vary the nature or the strength of the contacts in order to influence differentially transcription initiation. Because the DNA elements contacting RNA polymerase are always present, the interactions are likely to be subject to relatively minor variation. The promoter can evolve to be stronger or weaker through mutations that produce DNA sequence changes altering the nature of the contacts, but this does not provide a physiologically responsive regulatory mechanism. In contrast, changes to DNA structure that involve reversible chemical modification (e.g. methylation of bases) or topological changes can have both local and global regulatory effects that have a physiological connection.

DNA methylation is an important component of bacterial 'immunology', a process that allows the microbe to distinguish between its own DNA and DNA that is foreign. The foreign, unmethylated DNA is targeted for destruction by restriction endonucleases. The chemical modification of DNA also plays a role in controlling gene expression by influencing the efficiency with which DNA-binding proteins, including RNA polymerase, interact with DNA (Hernday et al. 2003; Waldron et al. 2002). The production of hemi-methylated DNA by semi-conservative DNA synthesis provides a mechanism by which methylation-sensitive genetic switches governing gene expression can be reset. Hemi-methylated DNA is chemically distinct from fully methylated DNA, and the former persists for a number of minutes following synthesis. Simple genetic switches in which transcriptional regulatory proteins compete with DNA methylases for access to the same sites in the genome are likely to evolve quite easily and to be retained when they prove to be advantageous under selective pressure (Hernday et al. 2003; Waldron et al. 2002).

3.2 *The Termination of Transcription Elongation*

Transcription termination is subject to regulation and occurs at sites in genes that contain either 'intrinsic terminators' or Rho-dependent terminators. Intrinsic transcription terminators consist of a G + C-rich inverted repeat followed by an oligo-T DNA sequence that gives rise to a stable stem-loop structure followed by a run of Us in RNA (Fig. 6.1). The stretch of T bases promotes pausing of the transcription elongation complex, and the instability of the A:U DNA:RNA hybrid causes the RNA polymerase core enzyme to backtrack, bringing the elongation complex to a halt. The NusA protein, a component of the transcription elongation complex, intervenes to stimulate the RNA-RNA base-pairing that forms the stem of the stem-loop structure; NusA can also prolong the period of transcription elongation complex pausing (Nudler and Gottesman 2002).

The Rho factor binds preferentially to pyrimidine-rich (especially cytidine-rich) RNAs by virtue of steric limitations within the surface cleft of the Rho monomer that interacts initially with RNA (Burgess and Richardson 2001; Skordalakes and Berger 2003) (Fig. 6.1). Mature Rho is hexameric, and the gaps between its six RNA-binding clefts determine the topography of the RNA that makes up the Rho utilization site (*rut*) (Grylak-Mielnicka et al. 2016). Unlike intrinsic terminators, Rho-dependent ones are free from secondary structures. Once Rho has bound RNA, it can thread it through the central pore of the hexameric complex and then translocate along the RNA using an ATP-dependent mechanism until it encounters a paused RNA polymerase (Richardson 1982) where it catalyses the release of the transcript from the stalled transcription elongation complex (Ephstein et al. 2010). Rho must compete with ribosomes for access to RNA, introducing the potential for a regulatory connection in which translating ribosomes and Rho compete for access to *rut* sites. If coupling between transcription and translation breaks down, Rho has an opportunity to terminate transcription (Cardinale et al. 2008). Work with *E. coli* suggests that Rho has an important role in silencing the transcription in the horizontally acquired portion of the genome. When a reduced genome derivative of *E. coli* lacking foreign DNA was examined, it was discovered that the transcription elongation factors NusA and NusG were no longer essential for the survival of the cell and that sensitivity to the Rho-inhibiting antibiotic bicyclomycin was reduced (Cardinale et al. 2008). These observations implicate not only Rho but also NusA and NusG in controlling preferentially the transcription of imported genes. It is interesting to consider these observations in the context of findings that the nucleoid-associated protein H-NS contributes both to the efficiency of Rho-dependent transcription termination and to the silencing of horizontally acquired genes (Dorman 2007; Kotlajich et al. 2015; Saxena and Gowrishankar 2011).

Rho plays an important role in suppressing the formation of R-loops during transcription (Harinarayanan and Gowrishankar 2003; Leela et al. 2013). These loops consist of an RNA-DNA hybrid with a displaced DNA strand and arise due to backtracking of RNA polymerase in G + C-rich DNA templates with elevated levels of negative supercoiling (Drolet et al. 2003) (Sect. 5). An important source of hyper-negative supercoiling is the tracking activities of RNA and DNA polymerases (Fig. 6.2). Topological changes to DNA link the control of gene expression to DNA replication and nucleoid architecture (Dorman 2013; Sobetzko et al. 2012). Transcription and replication are processes (or 'transactions') that have consequences for the topology of the DNA duplex. The duplex is unwound as the polymerase tracks along or the DNA is spooled through the polymerase, and the resulting overwinding of the DNA double helix ahead of the polymerase is matched by underwinding behind (Fig. 6.2). These disturbances to DNA topology are corrected by topoisomerases. The connections between topoisomerase activities, DNA topology and transcription will be considered in Sect. 5.

In this context it is interesting to note that *rho-15* mutants of *E. coli* have reduced DNA negative supercoiling (Fassler et al. 1986). If left unchecked, R-loop formation stimulates hyper-recombination leading to loss of genome stability (Nudler 2012; Wimberly et al. 2013). R-loops also contribute to potentially lethal double-stand breaks in DNA during collisions between transcription and DNA replication (Dutta et al. 2011; Gan et al. 2011). Rho inhibits R-loop formation principally by preventing backtracking by RNA polymerase (Nudler 2012).

4 Regulation with RNA

Historically, studies of bacterial gene regulation have focused on the control of RNA synthesis by proteins that bind to DNA (Brock 1990). If an RNA world preceded that in which DNA is the principal carrier of genetic information (Bowman et al. 2015; Cech 2009; Gilbert 1986), then the evolution of gene regulation has probably involved stages where RNA-based control was dominant among gene regulatory mechanisms (Ahmad et al. 2016). Work with modern bacteria shows that RNA-based control mechanisms are not simply vestigial but are central to the operation of gene expression control (Wagner and Romby 2015).

RNA that is cleaved can acquire a new chemical stability, allowing molecules with distinct half-lives to be produced from a common precursor transcript (Nilsson and Uhlin 1991). Intramolecular folding also influences the stability of an RNA

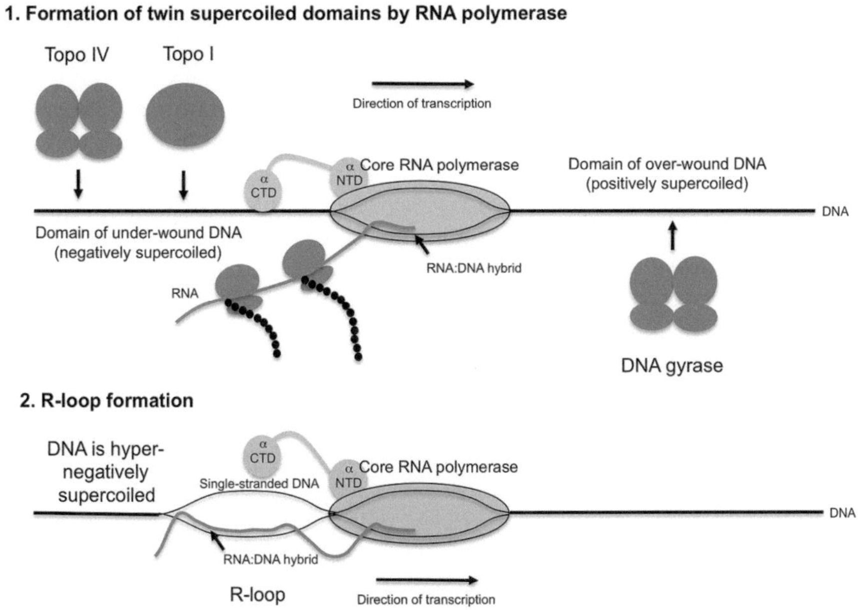

Fig. 6.2 The process of transcription has consequences for the topology of the DNA template and can lead to R-loop formation. (1) As the DNA template spools through RNA polymerase during transcription, the DNA duplex ahead of the moving transcription complex becomes overwound (or positively supercoiled), while the DNA behind is underwound, or negatively supercoiled. This situation arises because RNA polymerase, its growing RNA product and the associated ribosomes and nascent polypeptides cannot rotate around the DNA duplex quickly enough to relieve the topological tension and the DNA is itself unable to rotate. The solution is provided by the topoisomerases that eliminate the local domains of positive or negative supercoiling. DNA gyrase processes positive supercoils using a mechanism that is identical to the one that introduces negative supercoils to relaxed DNA. DNA topoisomerase IV (Topo IV) can relax negative supercoils using an ATP-dependent, double-stranded DNA breakage and strand passage mechanism. DNA topoisomerase I (Topo I) removes negative supercoils by a single-strand break-and-swivel mechanism. These topoisomerases are thought to accompany RNA (and DNA) polymerase as it translocates along the DNA duplex. The local disturbances to DNA topology that are caused by polymerase movement affect the activity of nearby promoters. This provides a mechanism of promoter coupling in which one gene can influence the expression of its neighbour without the involvement of a conventional gene regulatory protein (Wu et al. 1995). (2) When RNA polymerase traverses a region of G + C-rich DNA that becomes hyper-negatively supercoiled in its wake, stalling of the polymerase may allow the RNA to base pair with its DNA template, leaving the other DNA strand as a single-stranded bubble. If the R-loop is not resolved, it can led to DNA damage, including double-stranded breaks that may be lethal to the cell

molecule (Naville and Gautheret 2009). Folding can occur spontaneously as the molecule is created through a base-pairing process that is intrinsic to the nucleotide sequence of the RNA (Ma et al. 1994). Alternatively RNA folding can be guided by another molecule, usually a protein or another RNA, or by a protein and a regulatory RNA working in combination (Gottesman 2004). Intramolecular or intermolecular annealing and strand displacement are central to RNA-mediated gene regulation and to processes such as translation (Deighan et al. 2000; Rajkowitsch and Schroeder 2007). RNA-RNA interactions involving base pairing can enhance RNA lability or stability, depending on the nature of the participants and the interaction(s) between them (Barquist and Vogel 2015; Wagner and Simons 1994). RNA interacting with itself *in cis* provides the basis for gene expression control via transcription attenuation where alternative folding of the RNA reveals or sequesters translation signals within the message leading to alternative outputs (Fig. 6.3) (Naville and Gautheret 2009). In so-called riboswitches, the folding of the RNA is guided by the binding of a low-molecular-mass signal molecule (Fürtig et al. 2015). Other attenuators sense the presence of a specific, uncharged tRNA (Gutierrez-Preciado et al. 2009) or a particular ribosomal protein (Nomura et al. 1980). Still other attenuators exploit a short leader peptide whose translation rate forms the basis of the molecular signal (Yanofsky et al. 1981), while yet further examples involve a role for RNA-binding proteins such as cold shock proteins (Bae et al. 2000).

The expression of the stress and stationary phase sigma factor, RpoS, is regulated by the small RNAs DsrA, ArcZ and RprA via an Hfq-mediated RNA-RNA hybridization mechanism in which DsrA prevents *rpoS* mRNA from adopting a conformation that sequesters its translation initiation signals (Battesti et al. 2011). During exponential growth phase, the sRNAs are expressed at very low level and *rpoS* mRNA adopts the folded conformation that inhibits translation. When the bacterium stops growing, the associated build-up of the regulatory RNAs allows base-pairing with the 5′ end of *rpoS* mRNA, revealing the ribosome-binding site to ribosomes and allowing translation to begin (Majdalani et al. 1998, 2002; Mandin and Gottesman 2010). In addition, these sRNAs control Rho-dependent transcription termination within the 5′ untranslated region of *rpoS*; this appears to be a global function of sRNAs that operates throughout the genome at genes that have long untranslated leaders (Sedlyarova et al. 2016).

5 Transcription and Variable DNA Topology

DNA shape is described using the topological terms linking number, twist and writhe (Bauer et al. 1980; Boles et al. 1990; Vinograd et al. 1965). The linking number reports the number of times the two antiparallel strands of DNA cross one another in the duplex. The twist describes the number of complete rotations of one strand around the helical axis, and the writhe counts the number of times the helical axis winds around itself. A covalently closed, circular DNA molecule represents a closed topological system that can be described by its characteristic linking number, twist and writhe. The linking number can change if one or both the DNA strands are broken, the helix is rotated with or against the sense of the helix, and the broken strands are then resealed. Underwinding the DNA results in a reduction in the linking number, while overwinding the helix increases the value of this parameter. The resulting torsional stress changes the twist and writhe of the DNA. In the case of writhe, this is equivalent to the supercoiling of the helical axis about itself. The

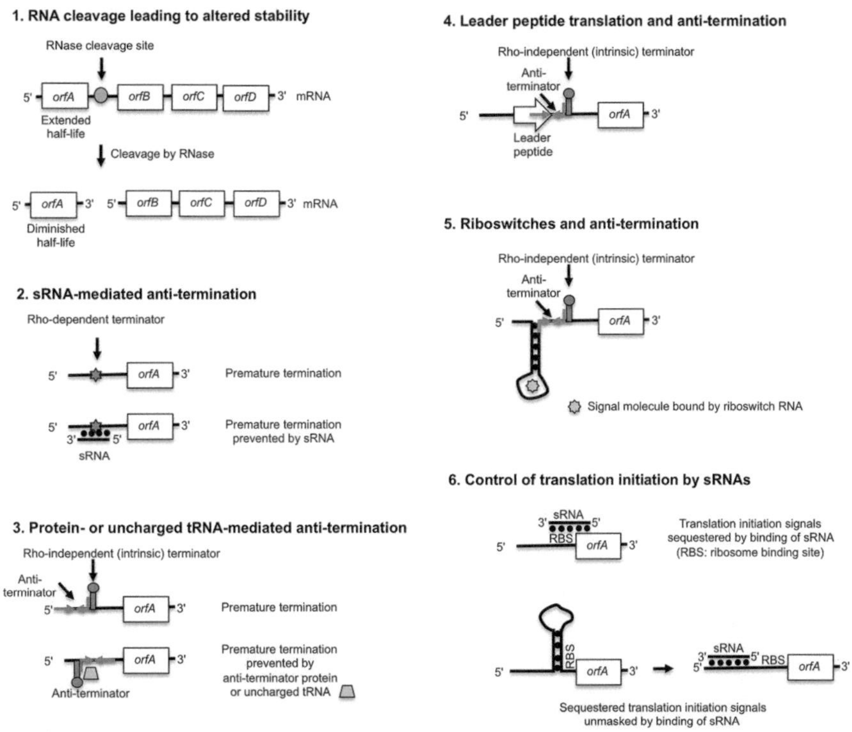

Fig. 6.3 Gene regulation using RNA. (1) The stability of an RNA molecule can be altered by RNase-mediated cleavage. Here, *orfA* lies at the beginning of an operon of four cistrons and has a relatively long half-life because RNA degradation in the 3′-to-5′ direction will target *orfA* last. Following cleavage of the *orfA* message from the remainder of the mRNA, it becomes exposed immediately to degradation and so has a shorter half-life. This will reduce the level of expression of the *orfA* gene product relative to that of the same open reading frame within the intact 4-cistron message. (2) A Rho-dependent terminator within the long untranslated leader of an open reading frame is disabled when a small regulatory RNA (sRNA) binds. The sRNA may act by impeding the movement of the Rho factor along the transcript toward RNA polymerase. (3) An intrinsic terminator and an anti-terminator overlap in the transcript where they can form mutually exclusive secondary structures. The binding of a protein (such as a ribosomal protein) or a suitable, uncharged tRNA can balance the terminator-anti-terminator competition in favour of the anti-terminator. (4) The translation of a short leader peptide influences the terminator-anti-terminator competition. If the leader peptide is expressed, the terminator forms, blocking the expression of *orfA*. (5) In the riboswitch the transcript forms a receptor that binds a signalling molecule that is relevant to the physiological role of *orfA*. When the signal is bound, the secondary structure of the riboswitch is stabilized in favour of the formation of the terminator. (6) sRNA molecules can influence the translation of *orfA* by sequestering its translation initiation signals (the ribosome-binding site, RBS, and the initiation codon of the reading frame). Alternatively, they may act by preventing the transcript from adopting spontaneously a secondary structure in which the translation initiation signals become sequestered

process of transcription causes simultaneously under- and overwinding of the template DNA, as outlined in the previous section. The topoisomerases that reset the linking number of the DNA in the zones affected by polymerase-mediated supercoiling changes can also influence the topological state of the genome.

Both local and global DNA supercoiling have the potential to influence transcription at several levels (Travers and Muskhelishvili 2005). The presentation of the binding site motifs to the sigma factor of RNA polymerase is affected by the spacing between the motifs and the superhelicity of the DNA. For sigma-70 promoters, an optimal spacing is 17 base pairs in length, but departures from this optimum are common (Fig. 6.1). Bringing the promoter motifs back into register for efficient recognition and binding by sigma-70 can be achieved by a local alteration to DNA twist (Ahmed et al. 2016). Torsional stress, arising from changes to twist in the DNA duplex, has the potential to drive the transition from a closed transcription complex to an open one. In the open complex, hydrogen bonds between bases are broken, and RNA polymerase can move from its transcription initiation phase to transcript elongation (Fig. 6.1). The topological state of the DNA template can also influence the processes of transcript elongation and termination (Ma and Wang 2014).

The process of transcription creates local changes to DNA topology. The DNA ahead of the moving transcription machinery becomes overwound, while the DNA behind becomes underwound (Fig. 6.2). If unchecked, these topological transitions have the potential to jam the polymerase, ending transcript elongation (Liu and Wang 1987). DNA topoisomerases relax the positive (or overwound) and negative (underwound) segments of DNA, facilitating the movement of RNA polymerase (Wu et al. 1988, 1995). Topoisomerases are essential components of the cell because they have the ability to maintain the superhelical density of the DNA within limits that are permissive for the essential processes of transcription and DNA replication. They are also important for the arrangement of the DNA in forms that facilitate its storage in the cell. For this reason, the architecture of the bacterial nucleoid and the processes that underpin gene expression are inextricably linked. It is perhaps unsurprising to discover that the nucleoid-associated proteins that help to organize genomic DNA also contribute to the control of transcription and translation (Dillon and Dorman 2010).

The DNA in bacterial cells is not maintained at a constant superhelical density throughout the growth cycle. Nor is the chromosome maintained at a constant superhelical density around its circumference: in the model bacterium *E. coli*, the segment of the chromosome that is known as the Ter macrodomain appears to adopt a superhelical density that is distinct from that of the rest of the chromosome (Lal et al. 2016; Sobetzko et al. 2012). This A + T-rich region of the chromosome has been reported as more relaxed and as more negatively supercoiled than the remainder of the molecule, depending on the stage of growth at which measurements are made. It has also been suggested that there may be a gradient of superhelical density extending along each replichore from *oriC* (the origin of chromosome replication) to the Ter macrodomain (Lal et al. 2016; Sobetzko et al. 2012). This suggestion is based on the finding that DNA gyrase, the type II topoisomerase that introduces negative supercoiling into DNA, has a spectrum of binding preferences that moves from high density near *oriC* to low density within Ter (Sobetzko et al. 2012). Genes closest to *oriC* are replicated first during the chromosomal DNA synthesis phase of the bacterial cell cycle (Cooper and Helmstetter, 1968). In rapidly growing populations, new rounds of replication are initiated before those already underway come to completion (Wang and Levin, 2009). This creates multiple copies of those genes lying closest to the origin. It has been suggested that this multicopy effect, working in concert with favourable levels of DNA supercoiling, ensures high levels of expression of genes that contribute to processes that are critical to support rapid growth (ATP synthesis, DNA synthesis, translation and transcription (Sobetzko et al. 2012). Recent work with *Salmonella enterica* serovar Typhimurium (*S.* Typhimurium) has shown that the *oriC*-proximity effect is tempered by genome architectural elements that are associated with specific macrodomains in the chromosome (Cameron et al. 2017).

DNA gyrase uses ATP as an energy source for the introduction of negative supercoils and is inhibited by ADP. It has been suggested that the ratio of [ATP]/[ADP] is critical for gyrase activity and that fluctuations in this ratio have a generalized effect on DNA topology throughout the genome (Hsieh et al. 1991; van Workum et al. 1996). Environmental shocks that alter metabolic flux have the potential to influence gyrase activity through changes in the relative levels of ATP and ADP in the cell (Hsieh et al. 1991). Examples of stresses that have been observed to correlate with changes to DNA supercoiling include osmotic, anaerobic and acid shock (Colgan et al. 2018; Dorman et al. 1988; Higgins et al. 1988; Hsieh et al. 1991; Karem and Foster 1993; Ní Bhriain et al. 1989; O'Byrne et al. 1992; Quinn et al. 2014). In principle, this link between environmental change, metabolic flux, gyrase activity, DNA topology and transcription provides a basis for a simple mechanism to adjust cellular gene expression at a cross-genomic level (Dorman and Dorman 2016). Changing the topology of the genetic material has the potential to alter the activity of each promoter in the genome, subject to the structural sensitivity of each promoter to the superhelical transition.

6 Nucleoid-Associated Proteins: An Evolving Picture

The nucleoid-associated proteins (NAPs) are abundant polypeptides that bind to a multitude of sites in the genome, altering its shape (Dillon and Dorman 2010). Some NAPs have strict DNA sequence requirements for their binding sites, while others have a much more promiscuous relationship with the chromosome. Promiscuity in binding site preferences may arise when the protein relies on the shape of the DNA rather than its base sequence as a guide to binding site selection. Sequence-dependent (direct readout) and DNA structure-dependent (indirect readout) mechanisms of binding are not mutually exclusive, and many DNA-binding proteins employ both (Dorman and Dorman 2017; Lawson et al. 2004; Rohs et al. 2009; Slattery et al. 2014). One can envision an evolutionary process in which a protein that relies exclusively on an indirect readout DNA-binding mechanism acquires some base sequence specificity and then evolves to use a binding mechanism that is entirely founded on direct readout of the base sequence. In this way, the protein may become restricted to binding fewer sites in the genome, making it more specialized in its interactions with DNA. These interactions may be architectural or have a gene regulatory purpose or they may facilitate both kinds of task.

Few NAPs bind signal molecules that control their DNA-binding activities (Dillon and Dorman 2010). Several are under growth phase or environmental control at the level of their expression, linking their appearance and activity in the cell to physiology. Their effects on gene expression can be positive or negative, with some binding to RNA as well as DNA. For example, the paralogous NAPs H-NS and StpA have both DNA- and RNA-binding activities (Park et al. 2010; Rajkowitsch and Schroeder 2007) and so does the ubiquitous NAP HU (Balandina et al. 2001). This may indicate a role for NAPs as regulatory and structural proteins in an RNA world prior to the emergence of DNA as the carrier of genetic information. NAPs may also provide an evolutionary reservoir from which transcription factors can emerge. For example, the NAP known as the factor for inversion stimulation (FIS) is closely related to the DNA-binding module of the NtrC protein, a DNA-binding protein that regulates transcription of genes involved in responding to nitrogen limitation (Klose et al. 1994). The activity of NtrC is controlled by covalent modification. It is phosphorylated in a signal receiver domain by the cytoplasmic sensor kinase NtrB, a protein that is autophosphorylated in response to changes in combined nitrogen levels in the bacterium (Weiss et al. 2002). In the model bacterium *E. coli* and its relatives, the *fis* gene lies in an operon with *dusB*, a gene involved in tRNA modification. Bioinformatic studies have detected examples of DusB-FIS fused proteins where the FIS component provides a DNA-binding motif (Morett and Bork 1998). These observations suggest that NAPs such as FIS can become components of larger DNA-binding proteins with more specific tasks in the cell. FIS itself is multifunctional and can influence transcription (positively or negatively), DNA replication, recombination and transposition. It serves as a signal molecule for the growth cycle by being present in large quantities in the early period of exponential growth when it stimulates the expression of genes that encode elements of the translational machinery of the cell (Nilsson et al. 1990; Ross et al. 1990; Lazarus and Travers 1993).

Some transcription factors seem to lie at the interface with NAPs. Two prominent examples are the catabolite repressor protein CRP and the leucine-responsive regulatory protein (LRP). CRP was one of the first global transcription factors to be characterized in *E. coli*. It regulates transcription positively or negatively, depending on the location of its binding site within the promoter region of the target gene (Browning and Busby 2016). It binds the second messenger cAMP, making CRP proficient for DNA binding. CRP has a well-characterized consensus DNA sequence for binding, and its physical relationships with DNA and RNA polymerase have been elucidated in great detail (Lloyd et al. 2002). Surprisingly, the cell appears to contain CRP in excess in relation to the number of matches to its binding site consensus sequence in the genome. This has led to speculation that CRP may play a structural role in the nucleoid in addition to its well-studied functions as a regulator of transcription (Grainger et al. 2005). It seems as though the boundaries of classification between NAPs and transcription factors and between architectural and regulatory functions are both blurred and porous.

Despite being a conventional transcription factor in many cases where its role has been examined, LRP has a weak requirement for a specific DNA sequence for binding and seems to prefer A + T-rich DNA (Beloin et al. 2000; Tapias et al. 2000; Peterson et al. 2007). LRP can wrap DNA, changing its shape with consequences for the architecture of the nucleoid and for gene expression (Peterson and Reich 2010). In addition to its role as a global regulator of transcription, LRP is involved in site-specific recombination systems, and it competes with DNA adenine methylase (Dam) to influence the performance of DNA methylation-dependent epigenetic switches such as those controlling expression of antigen 43 and Pap pili in *E. coli* (Hernday et al. 2003; Waldron et al. 2002).

7 An Integrated Evolutionary Proposal

Modern model bacteria such as *E. coli* possess layers of gene regulatory processes that differ in sophistication and specificity. Variable DNA topology plays a generalized role in affecting gene expression, and it also contributes to the organization of the bacterial nucleoid (Dorman et al. 2016) (Fig. 6.4, steps 1 to 6). Topoisomerases are essential enzymes that modify DNA topology by breaking DNA, driving or facilitating a change in linking number and rejoining the DNA. These enzymes are found even in the simplest microbial cells, such as *Mycoplasma* spp. The *Mycoplasma* example is an interesting one because the organism has few, if any, conventional transcription factors and has just one NAP, a protein related to HU, together with RNA polymerase and a set of topoisomerases (Fischer and Eisenberg 1997). This apparent regulatory simplicity is in keeping with the relatively constant nature of the environments in which these organisms live (Roh et al. 2013). Despite having such a small complement of components with the potential to regulate gene expression, experimental data have been obtained to show that transcriptional regulation is achieved in response to environmental stresses such as changes in osmotic pressure (Zhang and Baseman 2014) and heat shock (Musatovova et al. 2006). There is also a striking colinearity between the direction of transcription of the majority of the genes and the direction of DNA replication, with a strong correlation between the level of gene expression and proximity to the origin of chromosome replication (Herrmann and Reiner 1998). Such an arrangement is likely to reduce collisions between DNA and RNA polymerases and the creation of local DNA topological conflicts arising from convergent and divergent transcription. The *Mycoplasma* example shows that gene regulation is both necessary and achievable even with a minimal genome in an organism whose nutritional needs are met largely by its host (Citti and Blanchard 2013). One can envisage that these modest foundations can provide the basis for the evolution of a much more complex regulatory edifice (Dorman 2011).

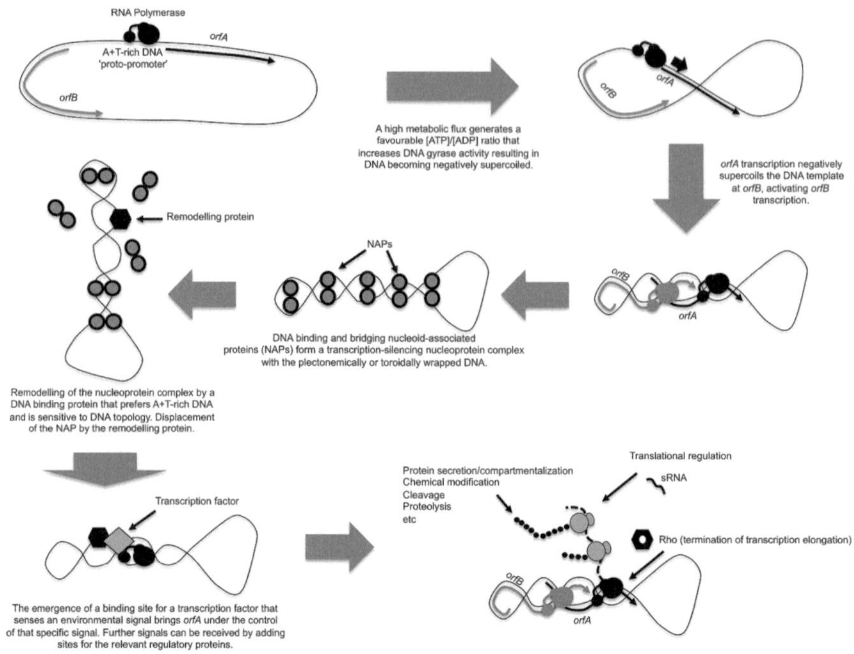

Fig. 6.4 The step-wise construction of a regulatory regime for a bacterial gene. The gene labelled *orfA* is part of a circular double-stranded DNA genome. It is located immediately upstream of and oriented divergently from another gene, *orfB* (1). The promoter of *orfA* has arisen from a patch of A + T-rich DNA that meets the minimum requirements of RNA polymerase for binding and transcription initiation. RNA polymerase can bind to this promoter in a relaxed DNA template, but the isomerization of the closed transcription complex to an open one requires a reduction in the linking number of the DNA circle. This is achieved by DNA gyrase in response to a favourable adjustment to the [ATP]/[ADP] ratio of the cell as a result of metabolic activity. Negative supercoiling of the DNA template allows RNA polymerase to initiate the transcription of *orfA*, and the local topological impact of the shift to transcription elongation at *orfA* includes the generation of a locally negatively supercoiled domain at the upstream gene *orfB* (2). This gene is now being transcribed, and the result is a further shift in the topology of the DNA template. The interwound strands of the DNA duplex are bound and cross-linked by a DNA-bridging nucleoid-associated protein, or NAP (3). H-NS is a NAP that has this type of activity (Dame et al. 2006). The resulting NAP-DNA nucleoprotein complex becomes transcriptionally silent as RNA polymerase reloading is blocked by the NAP. Fortuitously, a general DNA-binding protein with a preference for A + T-rich DNA is capable of remodelling the complex to readmit RNA polymerase (4). The competition between this remodeller and the NAP provides the basis for a crude genetic switch that governs transcription in the small genome. A more precise form of regulation arises when a DNA-binding protein that is under the control of a specific environmental signal is evolved and finds a good match to its preferred DNA target sequence in the *orfA* promoter (5). The promoter can gain targets for other transcription factors in the future, widening the range of influences to which it responds. It can also lose binding sites, freeing it from activities of the cognate transcription factors. The *orfA* gene will be co-regulated with other genes that share one or more of the regulatory sites found at the *orfA* promoter. This gene also has the potential to be controlled at levels beyond transcription initiation (6). For example, it may acquire sensitivity to the Rho factor, linking its DNA supercoiling response to transcription termination (and possibly R-loop formation). Rho will also connect the translation frequency of *orfA* to transcription termination. In common with many genes, *orfA* has the potential to be controlled by sRNA through a wide variety of mechanisms (Fig. 6.3), and its protein product can also acquire sensitivity to a range of posttranslational regulatory steps

7.1 Horizontal Gene Transfer and Gene Regulation

Bacteria benefit from horizontal gene transfer in building up a repertoire of new regulatory features together with new structural genes encoding useful cellular components or activities (Frost et al. 2005; Ochman et al. 2000; Thomas and Nielsen 2005). Regulatory integration of the newly arrived genes will be assisted if their promoter regions are already suitable (or can quickly become suitable) for control by the existing mechanisms in the recipient cell (Dorman 2009). Gene regulatory proteins with lax requirements for specific DNA sequences in their target genes may be particularly well placed to impose control on new genetic imports (Fig. 6.4, steps 3 to 5). Many of the global regulators that have been studied in *E. coli* and related model organisms rely wholly or partly on indirect readout mechanisms for DNA binding and have a preference for A + T-rich DNA (Dorman and Dorman 2017; Perez and Groisman 2009a). This leaves them well placed to recruit new genes to their existing regulons by binding to suitable sites in these newcomers. Subsequent changes to the sequence of the regulatory regions of new and established genes allow them to join and leave regulons, with selective pressure determining which regulatory connections are optimal for the survival of the bacterium.

Horizontal gene transfer is a major contributor to genome evolution (Dordet-Frisoni et al. 2014; Ochman et al. 2000; Syvanen 2012). It is estimated that the majority of the genes in *E. coli* were acquired in this way, including the majority of the regulatory genes (Price et al. 2008). It seems that horizontal acquisition of paralogous regulatory genes has been much more important than their appearance as a result of gene duplication (Price et al. 2008). It has been noted that genes that belong to the core genome (that portion which has not been acquired by lateral gene transfer) have less complicated regulatory mechanisms than genes that have arrived by horizontal transfer (Perez and Groisman 2009a). Experimental data that support this observation come from studies of the regulon of genes that is controlled by the PhoP/PhoQ two-component system in the pathogen *S.* Typhimurium. PhoQ is a cytoplasmic membrane-associated sensor kinase that relays information to the regulon about magnesium ion concentrations. PhoQ activates its DNA-binding partner PhoP by protein-protein phosphorylation, and the activated PhoP then binds and regulates its target genes. The regulatory inputs can be positive or negative depending on the locations of the binding sites (called PhoP boxes) with respect to the target gene promoters. In the case of core genome members (e.g. *mgtA*), the PhoP protein can operate independently of other DNA-binding proteins to achieve regulation; in the case of horizontally acquired genes, there are additional regulatory inputs (Perez et al. 2008).

The H-NS NAP silences the promoters of many genes that have been acquired by lateral transfer (Lucchini et al. 2006; Navarre et al. 2006; Oshima et al. 2006). PhoP is unable to activate the promoters of such genes when acting alone. Instead it must rely on the SlyA DNA-binding protein to modify the H-NS nucleoprotein complex in the vicinity of the promoter to allow PhoP-mediated transcription activation to proceed (Perez et al. 2008). SlyA is a LysR-like DNA-binding protein, and in common with other members of the group, it has very lax requirements for a specific DNA sequence to which to bind; the consensus binding site sequence for the LysR family is $TN_{11}A$, where N is any base (Oliver et al. 2016; Schell 1993; Sheehan and Dorman 1988). SlyA binds to A + T-rich DNA sequences using a winged helix-turn-helix motif (Haider et al. 2008; Wang et al. 2014). Several binding sites for SlyA are typically interspersed among and overlapping binding sites for H-NS (Lithgow et al. 2007; Perez et al. 2008). The mechanism by which SlyA overcomes H-NS silencing seems to involve a subtle remodelling of the complex rather than a simple displacement of H-NS (Stoebel et al. 2008) (Fig. 6.4, step 4). Importantly, H-NS also has the ability to remodel a SlyA-DNA complex, reimposing transcription silencing (Corbett et al. 2007; Lithgow et al. 2007). In contrast, the location and orientation of PhoP boxes seem to be determined with some precision, and the architecture of PhoP-regulated promoters determines the mechanism by which PhoP activates transcription. This can involve a dependency on the presence of the carboxyl terminal domain of the alpha subunit of RNA polymerase or it may be independent of this feature (Perez and Groisman 2009b). Comparisons of the PhoP regulons in *S.* Typhimurium and *Yersinia pestis* show that the highly conserved PhoP orthologues are used to control different, species-specific genes and that even orthologous target genes are controlled through distinct mechanisms (Perez and Groisman 2009b).

Similarly, the OmpR DNA-binding protein, which is 100% identical in *E. coli* and *S.* Typhimurium, controls widely different regulons in these two bacterial species (Quinn et al. 2014). OmpR is activated for DNA binding through phosphorylation by the sensor kinase EnvZ in response to osmotic and acid stress in *E. coli* and acid stress in *S.* Typhimurium (Chakraborty et al. 2015; Puente et al. 1991; Quinn et al. 2014; Stincone et al. 2011). It is an important regulator of horizontally acquired genes where it can function both as a conventional transcription factor to activate RNA polymerase and as a remodeller of H-NS-DNA transcription silencing complexes (Bang et al. 2002; Carroll et al. 2009). OmpR prefers A + T-rich DNA targets, and data from chromatin immunoprecipitation (ChIP) experiments performed in vivo (Quinn et al. 2014) have shown that OmpR senses the topological state of its DNA target: OmpR has a preference for DNA targets that are relaxed rather than negatively supercoiled, an observation that is supported by OmpR-binding studies performed in vitro with relaxed and supercoiled plasmids carrying one of its preferred binding sites—the ones from its own gene, *ompR* (Cameron and Dorman 2012).

The behaviour of OmpR exemplifies the importance of DNA topology as a determinant of regulatory inputs. Without modifying the nucleotide sequence of a promoter and without adding or subtracting any other regulatory factor (including chemical modifications to the bases), alterations to DNA shape can enhance or diminish a regulatory input. These alterations to DNA shape may be imposed by nearby transcription activity or the passage of a replication fork. They may also arise from the activities of topoisomerases. In the case of DNA gyrase, its sensitivity to the [ATP]/[ADP] ratio links DNA shape to metabolism and the synthesis of ATP (Hsieh et al. 1991; Snoep et al. 2002; van Workum et al. 1996). It is interesting to note that among the targets of OmpR in *S.* Typhimurium is the *mgtC* gene, whose product is an inhibitor of ATP synthase (Lee et al. 2013). These links suggest that a series of connections exist between ATP synthesis, DNA gyrase activity and OmpR-binding preferences that facilitate the operation of part of the global regulatory network of *S.* Typhimurium (Colgan et al. 2018; Dorman and Dorman 2016; Quinn et al. 2014). The expression of the genes within this network is also influenced by NAPs such as H-NS, FIS, HU and IHF (Cameron et al. 2011; Dillon et al. 2010; Mangan et al. 2006, 2011; Perez et al. 2008) and a long list of transcription factors (Fass and Groisman 2009; O'Byrne and Dorman 1994). In addition, there are multiple post-transcriptional regulatory inputs by small RNAs assisted by RNA chaperones (Hébrard et al. 2012; Kröger et al. 2012).

7.2 The Evolution of Regulatory Networks

Control networks arise from this multitude of gene-to-gene regulatory connections. Bioinformatic and experimental studies have suggested that the network map is 'written' into the four-dimensional geography of the nucleoid (Cameron et al. 2017; Dorman 2013; Lal et al. 2016; Sobetzko et al. 2012). This approach considers networks from a spatiotemporal point of view in which the origin and the terminus of chromosome replication represent cardinal points within the map and the temporal coordinates are represented by where an individual cell is in its cell cycle and where the population is in its growth cycle (Janga et al. 2009; Jeong et al. 2004). Regulatory connections are influenced by distances between regulatory genes and their targets in the folded genome within the nucleoid and not simply by gene positions along the circumference of the unfolded circular chromosome (Junier and Rivoire 2016; Junier et al. 2012; Képès 2004; Mathelier and Carbone 2010; Wright et al. 2007). They will also be influenced by the physical ability of regulatory molecules, RNA or protein, to traverse the spaces between the genes where they are born and their targets (Govindarajan and Amster-Choder 2016; Montero Llopis et al. 2010).

In this model, 'when' and 'where' an individual molecule is has importance for the survival of the cell. Variations in these spatiotemporal factors create variety among the genetically identical members of a bacterial population. Under selective pressure, this variety increases the probability that the population will include at least some individuals who are physiologically optimized for survival. The more dynamic the environment, the more important the creation of this variety may become. Stereotypical responses to environmental change, where every cell in the population perceives the same signal and responds in unison with the other cells, are certainly very important. However, because the time required for the detection of a stress and the mounting of a response may be too long to ensure survival, the creation of 'pre-prepared' individuals through exploitation of regulatory noise may be very beneficial. Persisters represent a medically relevant example of such individuals. These bacteria have entered a dormant state that renders them insensitive to antibiotics that are capable of killing all of their genetically identical but metabolically active siblings (Verstraeten et al. 2016).

A limited number of experimental studies have addressed the question of the importance of regulatory gene position on cellular fitness. The results suggest that gene location is important, especially when the gene encodes a regulator with global influence (Brambilla and Sclavi 2015; Bryant et al. 2014; Fitzgerald et al. 2015; Gerganova et al. 2015). The importance of global regulators to microbial evolution is also supported by the results of long-term experimental evolution studies where increases in fitness over time correlate with changes to genes encoding NAPs, topoisomerases and globally acting transcription factors (Crozat et al. 2005, 2010, 2011). Other work has shown that disruptions to regulatory connections are well tolerated, indicating that the networks in the model organism *E. coli* are robust and can withstand considerable rewiring without compromising the survival of the bacterium (Isalan et al. 2008). This suggests that the regulatory networks of modern bacteria possess considerable potential for further evolution.

8 Conclusion

Bacteria present a special case in evolutionary studies because horizontal gene transfer is central to their natural history. This is as true of their regulatory processes as it is of their structural features. Synthetic biology may involve attempting to rerun, at least partially, some of the genome-building experiments that have already been carried out by nature. In planning these attempts, we would do well to examine the regulatory wiring diagrams of as wide a range of microbes as possible rather than trying to infer too much from a narrow group of model organisms. Bacteria have evolved to colonize every environment on earth, and much of that evolution has involved gene regulatory processes. In conducting surveys of regulation, it is probably wise to widen the investigation beyond conventional transcription factors. Regulatory RNA is currently receiving a great deal of attention, but there is still much to learn about the contributions of regulatory peptides, DNA topology and the influence of genome architecture on what is possible and what is not possible in evolving and organizing an effective programme of gene expression that meets the needs of the individual bacterium and that of the wider microbial population.

Acknowledgements We thank AP Dorman for useful comments on the manuscript. This work was supported by Science Foundation Ireland Principal Investigator Award 13/IA/1875 to CJD. MJD is supported by the Wellcome Trust (grant 206194).

References

Ahmad M, Xue Y, Lee SK et al (2016) RNA topoisomerase is prevalent in all domains of life and associates with polyribosomes in animals. Nucleic Acids Res 44(13):6335–6349

Ahmed W, Menon S, Karthik PV et al (2016) Autoregulation of topoisomerase I expression by supercoiling sensitive transcription. Nucleic Acids Res 44(4):1541–1552

Bae W, Xia B, Inouye M et al (2000) *Escherichia coli* CspA-family chaperones are transcription antiterminators. Proc Natl Acad Sci USA 97(14):7784–7789

Balandina A, Claret L, Hengge-Aronis R et al (2001) The *Escherichia coli* histone-like protein HU regulates *rpoS* translation. Mol Microbiol 39(4):1069–1079

Ban N, Nissen P, Hansen J et al (2000) The complete atomic structure of the large ribosomal subunit at 2.4 A resolution. Science 289(5481):905–920

Bang IS, Audia JP, Park YK et al (2002) Autoinduction of the *ompR* response regulator by acid shock and control of the *Salmonella enterica* acid tolerance response. Mol Microbiol 44(5): 1235–1250

Barquist L, Vogel J (2015) Accelerating discovery and functional analysis of small RNAs with new technologies. Annu Rev Genet 49:367–394

Battesti A, Majdalani N, Gottesman S (2011) The RpoS-mediated general stress response in *Escherichia coli*. Annu Rev Microbiol 65:189–213

Bauer WR, Crick FHC, White JH (1980) Supercoiled DNA. Sci Am 243(1):100–113

Beloin C, Exley R, Mahe A et al (2000) Characterisation of LrpC DNA-binding properties and regulation of *Bacillus subtilis lrpC* gene expression. J Bacteriol 182(16):414–4424

Boles TC, White JH, Cozzarelli NR (1990) Structure of plectonemically supercoiled DNA. J Mol Biol 213(4):931–951

Booth A, Mariscal C, Doolittle WF (2016) The modern synthesis in the light of microbial genomics. Annu Rev Microbiol 70:279–297

Bordes P, Conter A, Morales V, Bouvier J, Kolb A, Gutierrez C (2003) DNA supercoiling contributes to disconnect signaS accumulation from sigmaS-dependent transcription in *Escherichia coli*. Mol Microbiol 48(2):561–571

Bowman JC, Hud NV, Williams LD (2015) The ribosome challenge to the RNA world. J Mol Evol 80(3):143–161

Brambilla E, Sclavi B (2015) Gene regulation by H-NS as a function of growth conditions depends on chromosomal position in *Escherichia coli*. G3 (Bethesda) 5(4):605–614

Brock TD (1990) The emergence of bacterial genetics. Cold Spring Harbor Press, Cold Spring Harbor

Browning DF, Busby SJ (2016) Local and global regulation of transcription initiation in bacteria. Nat Rev Microbiol 14(10):638–650

Bryant JA, Sellars LE, Busby SJ et al (2014) Chromosome position effects on gene expression in *Escherichia coli* K-12. Nucleic Acids Res 42(18):11383–11392

Burgess BR, Richardson JP (2001) RNA passes through the hole of the protein hexamer in the complex with the *Escherichia coli* Rho factor. J Biol Chem 276(6):4182–4189

Cameron AD, Dorman CJ (2012) A fundamental regulatory mechanism operating through OmpR and DNA topology controls expression of *Salmonella* pathogenicity islands SPI-1 and SPI-2. PLoS Genet 8(3):e1002615

Cameron AD, Stoebel DM, Dorman CJ (2011) DNA supercoiling is differentially regulated by environmental factors and FIS in *Escherichia coli* and *Salmonella enterica*. Mol Microbiol 80(1):85–101

Cameron AD, Dillon SC, Kröger C, Beran L, Dorman CJ (2017) Broad scale redistribution of mRNA abundance and transcriptional machinery in response to growth rate in *Salmonella enterica* serovar Typhimurium. Microb Genom 3(10):e000127

Cardinale CJ, Washburn RS, Tadigotia VR et al (2008) Termination factor Rho and its cofactors NusA and NusG silence foreign DNA in *E. coli*. Science 320(5878):935–938

Carroll RK, Liao X, Morgan LK, Cicirelli EM, Li Y, Sheng W, Feng X, Kenney LJ (2009) Structural and functional analysis of the C-terminal DNA binding domain of the *Salmonella typhimurium* SPI-2 response regulator SsrB. J Biol Chem 284(18):12008–12019

Cech TR (2009) Crawling out of the RNA world. Cell 136(4):599–602

Chakraborty S, Mizusaki H, Kenney LJ (2015) A FRET-based DNA biosensor tracks OmpR-dependent acidification of *Salmonella* during macrophage infection. PLoS Biol 13(4):e1002116

Citti C, Blanchard A (2013) Mycoplasmas and their host: emerging and re-emerging minimal pathogens. Trends Microbiol 21(4):196–203

Colgan AM, Quinn HJ, Kary SC, Mitchenall LA, Maxwell A, Cameron ADS, Dorman CJ (2018) Negative supercoiling of DNA by gyrase is inhibited in serovar Typhimurium during adaptation to acid stress. Mol Microbiol 107(6):734–746

Conter A, Menchon C, Gutierrez C (1997) Role of DNA supercoiling and *rpoS* sigma factor in the osmotic and growth phase-dependent induction of the gene *osmE* of *Escherichia coli* K12. J Mol Biol 273(1):75–83

Cooper S, Helmstetter CE (1968) Chromosome replication and the division cycle of *Escherichia coli* B/r. J Mol Biol 31(3):519–540

Corbett D, Bennett HJ, Askar H et al (2007) SlyA and H-NS regulate transcription of the *Escherichia coli* K5 capsule gene cluster, and expression of *slyA* in *Escherichia coli* is temperature-dependent, positively autoregulated, and independent of H-NS. J Biol Chem 282(46):33326–33335

Crozat E, Philippe N, Lenski RE et al (2005) Long-term experimental evolution in *Escherichia coli*. XII. DNA topology as a key target of selection. Genetics 169(2):523–532

Crozat E, Winkworth C, Gaffé J et al (2010) Parallel genetic and phenotypic evolution of DNA superhelicity in experimental populations of *Escherichia coli*. Mol Biol Evol 27(9):2113–2128

Crozat E, Hindré T, Kühn L et al (2011) Altered regulation of the OmpF porin by Fis in *Escherichia coli* during an evolution experiment and between B and K-12 strains. J Bacteriol 193(2): 429–440

Dame RT, Noom MC, Wuite GJL (2006) Bacterial chromatin organization by H-NS protein unravelled using dual DNA manipulation. Nature 444(7117):387–390

Deighan P, Free A, Dorman CJ (2000) A role for the *Escherichia coli* H-NS-like protein StpA in OmpF porin expression through modulation of *micF* RNA stability. Mol Microbiol 38(1): 126–139

Desnoyers G, Morissette A, Prévost K, Massé E (2009) Small RNA-induced differential degradation of the polycistronic mRNA *iscRSUA*. EMBO J 28(11):1551–1561

Dillon SC, Dorman CJ (2010) Bacterial nucleoid-associated proteins, nucleoid structure and gene expression. Nat Rev Microbiol 8(3):185–195

Dillon SC, Cameron AD, Hokamp K et al (2010) Genome-wide analysis of the H-NS and Sfh regulatory networks in *Salmonella typhimurium* identifies a plasmid-encoded transcription silencing mechanism. Mol Microbiol 76(5):1250–1265

Doose G, Alexis M, Kirsch R et al (2013) Mapping the RNA-Seq trash bin: unusual transcripts in prokaryotic transcriptome sequencing data. RNA Biol 10(7):1204–1210

Dordet-Frisoni E, Sagne E, Baranowski E, Breton M, Nouvel LX, Blanchard A, Marenda MS, Tardy F, Pascal S-P, Citti C (2014) Chromosomal transfers in Mycoplasmas: when minimal genomes go mobile. MBio 5(6):e01958-14

Dorman CJ (2007) H-NS, the genome sentinel. Nat Rev Microbiol 5(2):157–161

Dorman CJ (2009) Regulatory integration of horizontally-transferred genes in bacteria. Front Biosci (Landmark Ed) 14:4103–4112

Dorman CJ (2011) Regulation of transcription by DNA supercoiling in *Mycoplasma genitalium*: global control in the smallest known self-replicating genome. Mol Microbiol 81(2):302–304

Dorman CJ (2013) Genome architecture and global gene regulation in bacteria: making progress towards a unified model? Nat Rev Microbiol 11(5):349–355

Dorman CJ, Dorman MJ (2016) DNA supercoiling is a fundamental regulatory principle in the control of gene expression. Biophys Rev 8(3):209–220

Dorman CJ, Dorman MJ (2017) Control of virulence gene transcription by indirect readout in *Vibrio cholerae* and *Salmonella enterica* serovar Typhimurium. Environ Microbiol 19(10): 3834–3845. https://doi.org/10.1111/1462-2920

Dorman CJ, Barr GC, Ní Bhriain N et al (1988) DNA supercoiling and the anaerobic and growth phase regulation of *tonB* gene expression. J Bacteriol 170(6):2816–2826

Dorman CJ, Colgan A, Dorman MJ (2016) Bacterial pathogen gene regulation: a DNA-structure-centred view of a protein-dominated domain. Clin Sci (Lond) 130(14):1165–1177

Drolet M, Broccoli S, Rallu F et al (2003) The problem of hypernegative supercoiling and R-loop formation in transcription. Front Biosci 8:d210–d221

Dutta D, Shatalin K, Ephstein V et al (2011) Linking RNA polymerase backtracking to genome instability in *E. coli*. Cell 146(4):533–543

Ephstein V, Dutta D, Wade J et al (2010) An allosteric mechanism of Rho-dependent transcription termination. Nature 463(7278):245–249

Fass E, Groisman EA (2009) Control of *Salmonella* pathogenicity island-2 gene expression. Curr Opin Microbiol 12(2):199–204

Fassler JS, Arnold GF, Tessman I (1986) Reduced superhelicity of plasmid DNA produced by the *rho-15* mutation in *Escherichia coli*. Mol Gen Genet 204(3):424–429

Feng Y, Zhang Y, Ebright RH (2016) Structural basis of transcription activation. Science 352(6291):1330–1333

Fischer D, Eisenberg D (1997) Assigning folds to the proteins encoded by the genome of *Mycoplasma genitalium*. Proc Natl Acad Sci USA 94(22):11929–11934

Fitzgerald S, Dillon SC, Chao TC et al (2015) Re-engineering cellular physiology by rewiring high-level global regulatory genes. Sci Rep 5:17653. https://doi.org/10.1038/srep17653

Frost LS, Leplae R, Summers AO, Toussaint A (2005) Mobile genetic elements: the agents of open source evolution. Nat Rev Microbiol 3:722–732

Fürtig B, Nozinovic S, Reining A et al (2015) Multiple conformational states of riboswitches fine-tune gene regulation. Curr Opin Struct Biol 30:112–124

Gan W, Guan Z, Liu J et al (2011) R-loop-mediated genomic instability is caused by impairment of replication fork progression. Genes Dev 25(19):2041–2056

Gerganova V, Berger M, Zaldastanishvili E et al (2015) Chromosomal position shift of a regulatory gene alters the bacterial phenotype. Nucleic Acids Res 43(17):8215–8226

Gilbert W (1986) Origin of life: the RNA world. Nature 319:618

Gottesman S (2004) The small RNA regulators of *Escherichia coli*: roles and mechanisms. Annu Rev Microbiol 58:303–328

Gourse RL, Ross W, Rutherford ST (2006) General pathway for turning on promoters transcribed by RNA polymerases containing alternative sigma factors. J Bacteriol 188(13):4589–4591

Govindarajan S, Amster-Choder O (2016) Where are things inside a bacterial cell? Curr Opin Microbiol 33:83–90

Grainger DC, Hurd D, Harrison M et al (2005) Studies of the distribution of *Escherichia coli* cAMP-receptor protein and RNA polymerase along the *E. coli* chromosome. Proc Natl Acad Sci USA 102(49):17693–17698

Gruber TM, Gross CA (2003) Multiple sigma subunits and the partitioning of bacterial transcription space. Annu Rev Microbiol 57:441–466

Grylak-Mielnicka A, Bidnenko V, Bardowski J et al (2016) Transcription termination factor Rho: a hub linking diverse physiological processes in bacteria. Microbiology 162(3):433–447

Gutierrez-Preciado A, Henkin TM, Grundy FJ et al (2009) Biochemical features and functional implications of the RNA-based T-box regulatory mechanism. Microbiol Mol Biol Rev 73(1): 36–61

Haider F, Lithgow JK, Stapleton MR et al (2008) DNA recognition by the *Salmonella enterica* serovar Typhimurium transcription factor SlyA. Int Microbiol 11(4):245–250

Harinarayanan R, Gowrishankar J (2003) Host factor titration by chromosomal R-loops as a mechanism for runaway plasmid replication in transcription termination-defective mutants of *Escherichia coli*. J Mol Biol 332(1):31–46

Haugen SP, Ross W, Gourse RL (2008) Advances in bacterial promoter recognition and its control by factors that do not bind DNA. Nat Rev Microbiol 6(7):507–519

Hébrard M, Kröger C, Srikumar S et al (2012) sRNAs and the virulence of *Salmonella enterica* serovar Typhimurium. RNA Biol 9(4):437–445

Helmann JD (1999) Anti-sigma factors. Curr Opin Microbiol 2:135–141

Hengge R (2009) Proteolysis of sigmaS (RpoS) and the general stress response in *Escherichia coli*. Res Microbiol 160(9):667–676

Hernday AD, Braaten BA, Low DA (2003) The mechanism by which DNA adenine methylase and PapI activate the pap epigenetic switch. Mol Cell 12(4):947–957

Herrmann R, Reiner B (1998) *Mycoplasma pneumonia* and *Mycoplasma genitalium*: a comparison of two closely related bacterial species. Curr Opin Microbiol 1:572–579

Higgins CF, Dorman CJ, Stirling DA et al (1988) A physiological role for DNA supercoiling in the osmotic regulation of gene expression in *S. typhimurium* and *E. coli*. Cell 52(4):569–584

Hsieh LS, Rouvière-Yaniv J, Drlica K (1991) Bacterial DNA supercoiling and [ATP]/[ADP] ratio: changes associated with salt shock. J Bacteriol 173(12):3914–3917

Isalan M, Lemerle C, Michalodimitrakis K et al (2008) Evolvability and hierarchy in rewired bacterial gene networks. Nature 452(7189):840–845

Iyer LM, Koonin EV, Aravind L (2003) Evolutionary connection between the catalytic subunits of DNA-dependent RNA polymerases and eukaryotic RNA-dependent RNA polymerases and the origin of RNA polymerases. BMC Struct Biol 3:1

Janga SC, Salgado H, Martinez-Antonio A (2009) Transcriptional regulation shapes the organization of genes on bacterial chromosomes. Nucleic Acids Res 37:3680–3688

Jeong KS, Ahn J, Khodursky AB (2004) Spatial patterns of transcription activity in the chromosome of *Escherichia coli*. Genome Biol 5:R86

Junier I, Rivoire O (2016) Conserved units of co-expression in bacterial genomes: an evolutionary insight into transcriptional regulation. PLoS One 11:e0155740

Junier I, Hérison J, Képès F (2012) Genomic organization of evolutionarily correlated genes in bacteria: limits and strategies. J Mol Biol 419(5):369–386

Karem K, Foster JW (1993) The influence of DNA topology on the environmental regulation of a pH-regulated locus in *Salmonella typhimurium*. Mol Microbiol 10(1):75–86

Képès F (2004) Periodic transcriptional organization of the *E. coli* chromosome. J Mol Biol 340(5): 957–964

Klose KE, North AK, Stedman KM et al (1994) The major dimerization determinants of the nitrogen regulatory protein NtrC from enteric bacteria lie in its carboxy-terminal domain. J Mol Biol 241(2):233–245

Koonin EV (2003) Comparative genomics, minimal gene-sets and the last universal common ancestor. Nat Rev Microbiol 1(2):127–136

Kotlajich MV, Hron DR, Boudreau BA et al (2015) Bridged filaments of histone-like nucleoid structuring protein pause RNA polymerase and aid termination in bacteria. elife 4:e04970

Kröger C, Dillon SC, Cameron AD et al (2012) The transcriptional landscape and small RNAs of *Salmonella enterica* serovar Typhimurium. Proc Natl Acad Sci USA 109(20):E1277–E1286

Kudva R, Denks K, Kuhn P et al (2013) Protein translocation across the inner membrane of Gram-negative bacteria: the Sec and Tat dependent protein transport pathways. Res Microbiol 164(6): 505–534

Lal A, Dhar A, Trostel A et al (2016) Genome scale patterns of supercoiling in a bacterial chromosome. Nat Commun 7:11055

Lan G, Tu Y (2016) Information processing in bacteria: memory, computation, and statistical physics: a key issues review. Rep Prog Phys 79(5):052601

Lawson CL, Swigon D, Murakami KS (2004) Catabolite activator protein: DNA binding and transcription activation. Curr Opin Struct Biol 14:1):10–1):20

Lazarus LR, Travers AA (1993) The *Escherichia coli* FIS protein is not required for the activation of *tyrT* transcription on entry into exponential growth. EMBO J 12(6):2483–2494

Lee EJ, Pontes MH, Groisman EA (2013) A bacterial virulence protein promotes pathogenicity by inhibiting the bacterium's own F_1F_o ATP synthase. Cell 154(1):146–156

Leela JK, Syeda AH, Anupama K et al (2013) Rho-dependent transcription termination is essential to prevent excessive genome-wide R-loops in *Escherichia coli*. Proc Natl Acad Sci USA 110(1): 258–263

Lithgow JK, Haider F, Roberts IS et al (2007) Alternate SlyA and H-NS nucleoprotein complexes control *hlyE* expression in *Escherichia coli* K-12. Mol Microbiol 66(3):685–698
Liu LF, Wang JC (1987) Supercoiling of the DNA template during transcription. Proc Natl Acad Sci USA 84(20):7024–7027
Lloyd GS, Niu W, Tebbutt J et al (2002) Requirement for two copies of RNA polymerase alpha subunit C-terminal domain for synergistic transcription activation at complex bacterial promoters. Genes Dev 16(19):2557–2565
Lucchini S, Rowley G, Goldberg MD et al (2006) H-NS mediates the silencing of laterally acquired genes in bacteria. PLoS Pathog 2(8):e81
Ma J, Wang M (2014) Interplay between DNA supercoiling and transcription elongation. Transcription 5(3):e28636
Ma CK, Kolesnikow T, Rayner JC et al (1994) Control of translation by mRNA secondary structure: the importance of the kinetics of structure formation. Mol Microbiol 14(5):1033–1047
Majdalani N, Cunning C, Sledjeski D et al (1998) DsrA RNA regulates translation of RpoS message by an anti-sense mechanism, independent of its action as an antisilencer of transcription. Proc Natl Acad Sci USA 95(21):12462–12467
Majdalani N, Hernandez D, Gottesman S (2002) Regulation and mode of action of the second small RNA activator of RpoS translation, RprA. Mol Microbiol 46(3):813–826
Mandin P, Gottesman S (2010) Integrating anaerobic/aerobic sensing and the general stress response through the ArcZ small RNA. EMBO J 29(18):3094–3107
Mangan MW, Lucchini S, Danino V et al (2006) The integration host factor (IHF) integrates stationary-phase and virulence gene expression in *Salmonella enterica* serovar Typhimurium. Mol Microbiol 59(6):1831–1847
Mangan MW, Lucchini S, Ó Croinin T et al (2011) Nucleoid-associated protein HU controls three regulons that coordinate virulence, response to stress and general physiology in *Salmonella enterica* serovar Typhimurium. Microbiology 157(4):1075–1087
Martínez-Rodríguez L, García-Rodríguez FM, Molina-Sánchez MD et al (2014) Insights into the strategies used by related group II introns to adapt successfully for the colonisation of a bacterial genome. RNA Biol 11(8):1061–1071
Massé E, Escorcia FE, Gottesman S (2003) Coupled degradation of a small regulatory RNA and its mRNA targets in *Escherichia coli*. Genes Dev 17(19):2374–2383
Mathelier A, Carbone A (2010) Chromosomal periodicity and positional networks of genes in *Escherichia coli*. Mol Syst Biol 6:366
Mekler V, Kortkhonjia E, Mukhopadhyay J et al (2000) Structural organization of bacterial RNA polymerase holoenzyme and the RNA polymerase-promoter open complex. Cell 108(5): 599–614
Montero Llopis P, Jackson AF, Sliusarenko O et al (2010) Spatial organization of the flow of genetic information in bacteria. Nature 466(7302):77–81
Morett E, Bork P (1998) Evolution of new protein function: recombinational enhancer Fis originated by horizontal gene transfer from the transcriptional regulator NtrC. FEBS Lett 433(1–2): 108–112
Musatovova O, Dhandayuthapani S, Baseman JB (2006) Transcriptional heat shock response in the smallest known self-replicating cell, Mycoplasma genitalium. J Bacteriol 188(8):2845–2855
Muskhelishvili G, Travers A (2003) Transcription factor as a topological homeostat. Front Biosci 8: d279–d285
Navarre WW, Porwollik S, Wang Y et al (2006) Selective silencing of foreign DNA with low GC content by the H-NS protein in *Salmonella*. Science 313(5784):236–238
Naville M, Gautheret D (2009) Transcription attenuation in bacteria: theme and variations. Brief Funct Genomic Proteomic 8(6):482–492
Ní Bhriain N, Dorman CJ, Higgins CF (1989) An overlap between osmotic and anaerobic stress responses: a potential role for DNA supercoiling in the coordinate regulation of gene expression. Mol Microbiol 3(7):933–942

Nilsson P, Uhlin BE (1991) Differential decay of a polycistronic *Escherichia coli* transcript is initiated by RNaseE-dependent endonucleolytic processing. Mol Microbiol 5(7):1791–1799
Nilsson L, Vanet A, Vijgenboom E et al (1990) The role of FIS in trans activation of stable RNA operons of *E. coli*. EMBO J 9(3):727–734
Nomura M, Yates JL, Dean D et al (1980) Feedback regulation of ribosomal protein gene expression in *Escherichia coli*: structural homology of ribosomal RNA and ribosomal protein mRNA. Proc Natl Acad Sci USA 77(12):7084–7088
Novikova O, Topilina N, Belfort M (2014) Enigmatic distribution, evolution, and function of inteins. J Biol Chem 289(21):14490–14497
Nudler E (2012) RNA polymerase backtracking in gene regulation and genome instability. Cell 149(7):1438–1445
Nudler E, Gottesman ME (2002) Transcription termination and anti-termination in *E. coli*. Genes Cells 7(8):755–768
Nyström T (2004) Growth versus maintenance: a trade-off dictated by RNA polymerase availability and sigma factor competition? Mol Microbiol 54(4):855–862
O'Byrne CP, Dorman CJ (1994) The *spv* virulence operon of *Salmonella typhimurium* LT2 is regulated negatively by the cyclic AMP (cAMP)-cAMP receptor protein system. J Bacteriol 176(3):905–912
O'Byrne CP, Ní Bhriain N, Dorman CJ (1992) The DNA supercoiling-sensitive expression of the *Salmonella typhimurium his* operon requires the his attenuator and is modulated by anaerobiosis and by osmolarity. Mol Microbiol 6(17):2467–2476
Ochman H, Lawrence JG, Groisman EA (2000) Lateral gene transfer and the nature of bacterial innovation. Nature 405(6784):299–304
Oliver P, Peralta-Gil M, Tabche ML et al (2016) Molecular and structural considerations of TF-DNA binding for the generation of biologically meaningful and accurate phylogenetic footprinting analysis: the LysR-type transcriptional regulator family as a study model. BMC Genomics 17:686. https://doi.org/10.1186/s12864-016-3025-3
Oshima T, Ishikawa S, Kurokawa K et al (2006) *Escherichia coli* histone-like protein H-NS preferentially binds to horizontally acquired DNA in association with RNA polymerase. DNA Res 13(4):141–153
Park HS, Ostberg Y, Johansson J et al (2010) Novel role for a bacterial nucleoid protein in translation of mRNAs with suboptimal ribosome-binding sites. Genes Dev 24(13):1345–1350
Pedersen AG, Jensen LJ, Brunak S et al (2000) A DNA structural atlas for *Escherichia coli*. J Mol Biol 299(4):907–930
Perez JC, Groisman EA (2009a) Evolution of transcriptional regulatory circuits in bacteria. Cell 138(2):233–244
Perez JC, Groisman EA (2009b) Transcription factor function and promoter architecture govern the evolution of bacterial regulons. Proc Natl Acad Sci USA 106(11):4319–4324
Perez JC, Latifi T, Groisman EA (2008) Overcoming H-NS-mediated transcriptional silencing of horizontally acquired genes by the PhoP and SlyA proteins in *Salmonella enterica*. J Biol Chem 283(16):10773–10783
Peterson SN, Reich NO (2010) LRP: a nucleoid-associated protein with gene regulatory properties. In: Dame RT, Dorman CJ (eds) Bacterial chromatin. Springer, Dordrecht, pp 353–364
Peterson SN, Dahlquist FD, Reich NO (2007) The role of high affinity non-specific DNA binding by Lrp in transcriptional regulation and DNA organization. J Mol Biol 369(5):1307–1317
Price MN, Dehal PS, Arkin AP (2008) Horizontal gene transfer and the evolution of transcriptional regulation in *Escherichia coli*. Genome Biol 9:R4
Price MN, Wetmore KM, Deutschbauer AM et al (2016) A comparison of the costs and benefits of bacterial gene expression. PLoS One 11(10):e0164314
Puente JL, Verdugo-Rodríguez A, Calva E (1991) Expression of *Salmonella typhi* and *Escherichia coli* OmpC is influenced differently by medium osmolarity; dependence on *Escherichia coli* OmpR. Mol Microbiol 5(5):1205–1210

Quinn HJ, Cameron AD, Dorman CJ (2014) Bacterial regulon evolution: distinct responses and roles for the identical OmpR proteins of *Salmonella typhimurium* and *Escherichia coli* in the acid stress response. PLoS Genet 10(3):e1004215

Rajkowitsch L, Schroeder R (2007) Dissecting RNA chaperone activity. RNA 13(12):2053–2060

Richardson JP (1982) Activation of Rho protein ATPase requires simultaneous interaction at two kinds of nucleic acid binding sites. J Biol Chem 257(10):5760–5766

Roh K, Safaei FR, Hespanha JP, Proulx SR (2013) Evolution of transcription networks in response to temporal fluctuations. Evolution 67(4):1091–1104

Rohs R, West SM, Sosinsky A, Liu P, Mann RS, Honig B (2009) The role of DNA shape in protein-DNA recognition. Nature 461(7268):1248–1253

Ross W, Gourse RL (2009) Analysis of RNA polymerase-promoter complex formation. Methods 47(1):13–24

Ross W, Thompson JF, Newlands JT et al (1990) *E. coli* FIS protein activates ribosomal RNA transcription in vitro and in vivo. EMBO J 9(11):3733–3742

Ross W, Gosink KK, Salomon J et al (1993) A third recognition element in bacterial promoters: DNA binding by the alpha subunit of RNA polymerase. Science 262(5138):1407–1413

Saxena S, Gowrishankar J (2011) Compromised factor-dependent transcription termination in a *nusA* mutant of *Escherichia coli*: spectrum of termination efficiencies generated by perturbations of Rho, NusG, and H-NS family proteins. J Bacteriol 193(15):3842–3850

Schell MA (1993) Molecular biology of the LysR family of transcriptional regulators. Annu Rev Microbiol 47:597–626

Schellhorn HE (2014) Elucidating the function of the RpoS regulon. Future Microbiol 9(4): 497–507

Schneewind O, Missiakas D (2014) Sec-secretion and sortase-mediated anchoring of proteins in Gram-positive bacteria. Biochim Biophys Acta 1843(8):1687–1697

Sedlyarova N, Shamovsky I, Bharati BK et al (2016) sRNA-mediated control of transcription termination in *E. coli*. Cell 167(1):111–121

Sheehan BJ, Dorman CJ (1988) In vivo analysis of the interactions of the LysR-like regulator SpvR with the operator sequences of the *spvA* and *spvR* virulence genes of *Salmonella typhimurium*. Mol Microbiol 30(1):91–105

Singh SS, Singh N, Bonocora RP et al (2014) Widespread suppression of intragenic transcription initiation by H-NS. Genes Dev 28(3):214–219

Skordalakes E, Berger JM (2003) Structure of the Rho transcription terminator: mechanism of mRNA recognition and helicase loading. Cell 114(1):135–146

Slattery M, Zhou T, Yang L, Dantas Machado AC, Gordân R, Rohs R (2014) Absence of a simple code: how transcription factors read the genome. Trends Biochem Sci 39(9):381–399

Snoep JL, van der Weijden CC, Andersen HW et al (2002) DNA supercoiling in *Escherichia coli* is under tight and subtle homeostatic control, involving gene-expression and metabolic regulation of both topoisomerase I and DNA gyrase. Eur J Biochem 269(6):1662–1669

Sobetzko P, Travers A, Muskhelishvili G (2012) Gene order and chromosome dynamics coordinate spatiotemporal gene expression during the bacterial growth cycle. Proc Natl Acad Sci USA 109(2):E42–E50

Steitz TA (2009) The structural changes of T7 RNA polymerase from transcription initiation to elongation. Curr Opin Struct Biol 19(6):683–690

Stincone A, Daudi N, Rahman AS et al (2011) A systems biology approach sheds new light on *Escherichia coli* acid resistance. Nucleic Acids Res 39(17):7512–7528

Stoebel DM, Free A, Dorman CJ (2008) Anti-silencing: overcoming H-NS-mediated repression of transcription in Gram-negative enteric bacteria. Microbiology 154(9):2533–2545

Syvanen M (2012) Evolutionary implications of horizontal gene transfer. Annu Rev Genet 46: 341–358

Tapias A, Lopez G, Ayora S (2000) *Bacillus subtilis* LrpC is a sequence-independent DNA-binding and DNA-bending protein which bridges DNA. Nucleic Acids Res 28(2):552–559

Thomas CM, Nielsen KM (2005) Mechanisms of, and barriers to, horizontal gene transfer between bacteria. Nat Rev Microbiol 3(9):711–721

Travers A, Muskhelishvili G (2005) DNA supercoiling – a global transcriptional regulator for enterobacterial growth? Nat Rev Microbiol 3(2):157–169

Trentini DB, Suskiewicz MJ, Heuck A et al (2016) Arginine phosphorylation marks proteins for degradation by a Clp protease. Nature 539(7627):48–53

van Workum M, van Dooren SJ, Oldenburg N et al (1996) DNA supercoiling depends on the phosphorylation potential in *Escherichia coli*. Mol Microbiol 20(2):351–360

Verstraeten N, Knapen W, Fauvart M et al (2016) A historical perspective on bacterial persistence. Methods Mol Biol 1333:3–13

Vinograd J, Lebowitz J, Radloff R et al (1965) The twisted circular form of polyoma viral DNA. Proc Natl Acad Sci USA 53(5):1104–1111

Wagner EG, Romby P (2015) Small RNAs in bacteria and archaea: who they are, what they do, and how they do it. Adv Genet 90:133–208

Wagner EG, Simons RW (1994) Antisense RNA control in bacteria, phages, and plasmids. Annu Rev Microbiol 48:713–742

Waldron DE, Owen P, Dorman CJ (2002) Competitive interaction of the OxyR DNA-binding protein and the Dam methylase at the antigen 43 gene regulatory region in *Escherichia coli*. Mol Microbiol 44(2):509–520

Wang JD, Levin PA (2009) Metabolism, cell growth and the bacterial cell cycle. Nat Rev Microbiol 7(11):822–827

Wang D, Guo C, Gu L et al (2014) Comparative study of the *marR* genes within the family *Enterobacteriaceae*. J Microbiol 52(6):452–459

Weiss V, Kramer G, Dünnebier T et al (2002) Mechanism of regulation of the bifunctional histidine kinase NtrB in *Escherichia coli*. J Mol Microbiol Biotechnol 4(3):229–233

Werner F, Grohmann D (2011) Evolution of multisubunit RNA polymerases in the three domains of life. Nat Rev Microbiol 9(2):85–98

Wimberly H, Shee C, Thornton PC et al (2013) R-loops and nicks initiate DNA breakage and genome instability in non-growing *Escherichia coli*. Nat Commun 4:2115

Wright MA, Kharchenko P, Church GM, Segrè D (2007) Chromosomal periodicity of evolutionarily conserved gene pairs. Proc Natl Acad Sci USA 104(25):10559–10564

Wu HY, Shyy SH, Wang JC et al (1988) Transcription generates positively and negatively supercoiled domains in the template. Cell 53(3):433–440

Wu HY, Tan J, Fang M (1995) Long-range interaction between two promoters: activation of the *leu-500* promoter by a distant upstream promoter. Cell 82(3):445–451

Yanofsky C, Platt T, Crawford IP et al (1981) The complete nucleotide sequence of the tryptophan operon of *Escherichia coli*. Nucleic Acids Res 9(24):6647–6668

Zenkin N (2014) Ancient RNA stems that terminate transcription. RNA Biol 11(4):295–297

Zhang W, Baseman JB (2014) Functional characterization of osmotically inducible protein C (MG_427) from *Mycoplasma genitalium*. J Bacteriol 196(5):1012–1019

Zhang X, Schleif R (1998) Catabolite gene activator protein mutations affecting activity of the *araBAD* promoter. J Bacteriol 180(2):195–200

Chapter 7
Conservation of Two-Component Signal Transduction Systems in *E. coli*, *Salmonella*, and Across 100,000 Bacteria of Various Bacterial Phyla

Trudy M. Wassenaar, Visanu Wanchai, Duah Alkam, Intawat Nookaew, and David W. Ussery

1 Introduction

The environments in which bacteria live and thrive are diverse and can be extreme, but they are hardly ever constant. As single-cell organisms, bacteria need to adapt to their environment by responding to relevant external signals, which they do via a process that translates a variety of signals to a conserved chemical messaging system inside the cell. Most bacteria are well-equipped to respond to external changes by fine-tuning their gene expression and protein production. This adaptability is conveyed primarily by two-component signal transduction systems (2CSTS), which sense environmental cues and translate this information into changes in gene expression. The consequence of this is a fast, tailor-made adaptation to external conditions (Hoch 2000). The fact that 2CSTS are present in most bacterial species indicates how important their activity is for life; not only do most bacterial genomes contain 2CSTS, but they also usually contain multiple different variants.

The number of different 2CSTS bacteria possess can be expected to depend on the diversity of environmental changes to which their cells must respond. Species capable of surviving more diverse conditions are able to use more variable metabolic

---

Electronic supplementary material The online version of this chapter (https://doi.org/10.1007/978-3-319-69078-0_7) contains supplementary material, which is available to authorized users.

T. M. Wassenaar
Molecular Microbiology and Genomics Consultants, Zotzenheim, Germany

V. Wanchai · D. Alkam · I. Nookaew · D. W. Ussery (✉)
Department of BioMedical Informatics, University of Arkansas for Medical Sciences, Little Rock, AR, USA
e-mail: DWUssery@uams.edu

P. H. Rampelotto (ed.), *Molecular Mechanisms of Microbial Evolution*, Grand Challenges in Biology and Biotechnology,
https://doi.org/10.1007/978-3-319-69078-0_7

pathways, which are often reflected by larger genomes (Land et al. 2015). Thus, it can be expected that such organisms contain a higher number of different 2CSTS compared to species that live in relatively stable conditions. It is further presumed that all members of a given species will be able to survive more or less the same set of diverse conditions, which would predict that the nature of 2CSTS present is conserved between members of the same species. These hypotheses were tested by data mining 101,836 publicly available sequenced bacterial genomes.

The two components of a bacterial signal transduction system are a sensor histidine kinase (HK) and a response regulator (RR). The HK detects a signal from the outside of the cell by binding a signaling molecule, which results in activation of the RR. Some HKs respond to temperature, such as DesK of *Bacillus subtilis* (Abriata et al. 2017). Upon activation, the latter typically binds a DNA sequence to specifically alter the expression of one or several genes. Most HKs span the bacterial cell membrane (the inner membrane of Gram-negative species or the single membrane of Gram-positives) and contain an extracellular sensor domain that is highly variable, to allow the detection of diverse signals. Activation of the sensor domain by binding of their extracellular signal molecule, which in most cases can only take place when the HK is present as a homodimer, leads to a conformational change in the intracellular domain. Some HKs are always present in the dimeric form, whereas others only dimerize when the signaling molecule binds, and a few do not require dimerization, for example, EL346 from *Erythrobacter litoralis* (Rivera-Cancel et al. 2014). After signal binding, the conformational change results in activation of the catalytic domain responsible for kinase activity (Capra and Laub 2012; Attwood 2013; Bhate et al. 2015). The activated catalytic domain transfers a phosphoryl group from ATP to a conserved histidine residue present in the dimerization domain (historically described as the DHp domain). Thus, both ATPase and kinase activity are essential characteristics of an HK. The phosphoryl group is then transferred to an aspartate residue of the RR, which leads to its activation. Most activated RRs function as transcription factors and directly bind to DNA to modify gene expression (Bourret 2010).

Variations to this general setup are plentiful (see for example a review by Galperin 2010). Some HKs are not membrane-bound but reside in the cytosol, being activated by molecules that can enter the cell. The HK and RR activity can be combined into one protein; for instance, in HKs involved in phosphorelay, the phosphoryl group is passed on to an RR domain that is part of the same protein, whereby a domain with histidine phosphotransferase activity transfers it to the final RR (Varughese 2002; Hoch and Varughese 2001). In other instances, the phosphoryl group is transferred to a separate phosphotransferase before activating an RR.

The work presented here builds on earlier works, performed more than a decade ago, in which 167 bacterial genomes (Galperin 2005) were analyzed for HK content and 250 bacterial genomes for 2CSTS content (Kiil et al. 2005). With the increasing number of completed genome sequences, it has become possible to investigate the distribution of proteins within and across bacterial species more accurately. Such analyses may highlight the underappreciated strain diversity across certain species. However, with the wide functional variety in 2CSTS, which is reflected by high variation in their amino acid sequences, a simple comparison based on sequence homology does not suffice to identify all possible members of this functional group of proteins. It has been demonstrated that the more bacterial genomes are compared by sequence similarity, the fewer genes are found to be conserved (Carbone 2006); eventually, this number decreases to zero (Lagesen et al. 2010). Moreover, the quality of publicly published genomes varies considerably. At the extreme, we discovered at least one submission of a "genome" sequence that actually contained a collection of plasmids without any chromosome. Bacterial "genomes" with a gene content nearing 20,000 genes (approximating the number of genes estimated to be present in a human genome) can also be found. Clearly, the data need to be curated before analysis, which on this large scale needs to be done in an automated and standardized manner.

Herein, we compared the presence of the 2CSTS across approximately 100,000 bacterial genomes, including 5000 *E. coli* and 5000 *Salmonella* genomes, by searching for Pfam domains that are conserved in all HKs and RRs. All analyzed genomes were translated into protein sequences after which they were subjected to a quality scoring procedure prior to analysis. Genomes not passing this quality score were removed. The quality score assessed the number of ambivalent or unknown nucleotides, the number of contigs in which the sequences were recorded, presence of essential genes by means of detection of their Pfam domain, and the number of tRNA and rRNA genes (Land et al. 2014).

To begin with, we compiled a list of all HKs and RRs that are present in the genome of the *E. coli* type strain DSM30083 and in the genome of *Salmonella enterica* sv. Typhimurium LT2, the type strain of that species. Since the genome of *E. coli* K-12 MG1655 has been used as a reference in many comparative studies, this was also included at this stage. From this starting point, we investigated the conservation of the Pfam domains within and across *E. coli* and *Salmonella* genomes that had passed the quality score, after which we extended the comparison across all the bacterial genomes available in GenBank at the time of analysis. We then zoomed in at those species with very few or very large numbers of 2CSTS systems.

2 Selection of Genomes, Quality Scoring, and Protein Identification

2.1 Selection and Quality Scoring of the Investigated Genome Sequences

Two genome sequences were used as a reference for *E. coli*: strain K-12 MG1655 (GenBank accession number U00096.3) and the *E. coli* type strain, DSM 30083 (AGSE00000000.1). For *Salmonella* sp., we used the genome of the *S. enterica* LT2 type strain (AE006468.2), a representative of *S. enterica* serovar Typhimurium as a reference. In addition, 6270 *E. coli* and 7089 *Salmonella* genomes were downloaded from GenBank on July 11, 2017. For the comparison across a large collection of bacterial genomes, 101,836 genomes were extracted from GenBank.

All genomes had to pass a quality score before being included in the analysis. For this, we used a quality score threshold of $\geq$0.8, as calculated by a previously described method (Land et al. 2014). After application of the cutoff for low-quality score genomes, the size window for *E. coli* genomes that passed the score was between 3.98 and 5.90 MB and for *Salmonella* between 4.4 MB and 5.2 MB. This corresponds with the number of proteins ranging from 3621 to 6905 in *E. coli* and from 4013 to 6351 in *Salmonella*. The 101,836 bacterial genomes that were available at the time of analysis were subjected to the same quality score procedure.

2.2 *Identification of 2CSTS Proteins*

Since gene-calling can vary considerably between published genomes, we standardized this by applying Prodigal (Hyatt et al. 2010) to all genomes. This identified all protein-coding genes across all genomes in a standardized manner. The Pfam domains in these proteins were identified using HMMER (Johnson et al. 2010) to scan across the 16,712 hidden Markov models described in the current version of the Pfam database (Finn et al. 2016). Next, histidine kinases were identified based on the presence of the Pfam domain HATPase_c (PF02518) in addition to a kinase domain, as identified by one of several possible Pfam domains (PF07730, PF06580, or PF14689). Response regulator proteins were identified on the basis of presence of the Pfam domain Response_reg (PF00072). We validated the list of retrieved potential HKs and RRs using the three reference genomes. It was found that although the method was inclusive (all known HK and RR proteins had been retrieved) for HKs, there were a few additional false positives, including DNA gyrase B, topoisomerase IV, and HSP90-like proteins. Scripts were introduced to correct for this, after which the HK and RR proteins from all other genomes were retrieved.

3 Two-Component Signal Transduction Systems in *E. coli* and *Salmonella*

3.1 *Two-Component Signal Transduction Systems in Reference Genomes*

Beginning with the three reference genomes, all translated histidine kinase and response regulator genes were identified from the complete genomes of *E. coli* K-12 strain MG1655, from the *E. coli* type strain DSM 30083, and from the *Salmonella* type strain *S. enterica* sv. Typhimurium LT2. Since the searches were carried out with translated gene sequences, we use the term "protein" to report the results. The results are summarized in Table 7.1. Of the 35 different HKs shown in the table, 29 and 31 are found in *E. coli* K-12 and DSM 30083, respectively, while *S.* Typhimurium LT2 contains 31 HK proteins. Twenty-six of the 35 HK proteins are conserved in all three genomes, as indicated by the shaded cells in Table 7.1. An extended table, showing accession numbers for these proteins, alternative names, and the individual protein domains detected in the translated protein sequences, is available as Supplementary Table S1.

The identified HK proteins vary in length from 363 (BasS) to 654 (CheA) amino acids (aa), while six proteins possess a combination of HK and RR domains, with a length between 778 and 1197 aa. Since cytosolic HKs do not contain transmembrane domains, it is expected that the shorter proteins are cytosolic, as has been described for 349 aa-long GlnL (Sanders et al. 1992).

The corresponding RR of each listed HK is presented on the same row of Table 7.1. With the exception of three cases (*arcA/arcB*, *barA/yvrU*, and *narP/narQ*), the HK and corresponding RR gene are present in the same locus, though there are several RR genes without a corresponding HK. As a consequence, all three reference genomes contain slightly more RRs than HKs. Twenty-eight of the 40 RRs are conserved in all three reference genomes.

The RR proteins vary in length between 196 aa (UhpA) and 461 aa (AtoC). Several cases were identified where the signal transduction is extended beyond a one-to-one protein interaction; for instance, histidine kinase ArcB is annotated to phosphorylate two response regulators (ArcA and RssB, both included in the Table), while two proteins (histidine kinase RscC and phosphotransferase RscD) can both activate the response regulator RscB (Baxter and Jones 2015; Mika and Hengge 2005). Note that RscD is not included in Table 7.1 since it is not a histidine kinase. Some of the translated genes listed in Table 7.1 code for proteins that have functions related to flagellar motility, chemotaxis, or sigma factor protein stability (Baxter and Jones 2015; Rogov et al. 2006). Though these processes may not be generally regarded as "classical" signal responses, they aid to the repertoire of the bacterial cell to respond to environmental triggers. The specific function of the 2CSTS listed in Table 7.1, in terms of the signals to which their HK responds, or the gene expression regulated by the corresponding RR, was not further investigated. The analysis was not extended to distinguish phosphorelay proteins from classic HKs.

Although *E. coli* strain K-12 has been extremely well characterized, its genome is smaller than most other *E. coli* genomes; thus, it can be questioned if it is suitable as a reference in comparative studies. Here we demonstrate that the K-12 strain lacks two 2CSTS in comparison with the species type strain DSM 30083 and also lacks a further putative response regulator, though it contains the gene coding for FimZ which the type strain does not. It is plausible that strain K-12 has lost a number of genes over the years since it was first isolated. The strain originated from a diphtheria patient in 1922, as described in an historic overview of this strain and its derivatives (Bachmann 1972). The author cited a book chapter by J. Lederberg as the source of the information regarding the origin of K-12 (Lederberg 1951). Originally, the strain contained phage lambda and the F plasmid. In 1944, the strain was treated with

Table 7.1 Histidine kinase and response regulator proteins identified in three reference genomes by computational analysis

Histidine kinases					Response regulators					
Protein name	Protein length (aa)[a]	K-12	30,083	LT2	Protein name	Protein length (aa)[a]	Comment	K-12	30,083	LT2
ArcB	**778**	+	+	+	**ArcA**	**238**	**ArcB: phosphorelay. *ArcB* and *arcA* not in 1 locus**	+	+	+
AtoS	608	+	+	–	AtoC	461		+	+	–
BaeS	**467**	+	+	+	**BaeR**	**240**		+	+	+
BarA	**918**	+	+	+	**UvrY**	**218**	**BarA: phosphorelay. *barA* and *uvrY* not in 1 locus**	+	+	+
BasS	**363**	+	+	+	**BasR**	**222**		+	+	+
CheA	**654**	+	+	+	**CheB**	**349**		+	+	+
CitA	**552**	+	+	+	**CitB**	**226**		+	+	+
CpxA	**457**	+	+	+	**CpxR**	**232**		+	+	+
CreC	**474**	+	+	+	**CreB**	**229**		+	+	+
CusS	480	+	+	–	CusR	227		+	+	–
DcuS	**543**	+	+	+	**DcuR**	**239**		+	+	+
DpiB	539	–	–	+	DpiA	228		–	–	+
EnvZ	**450**	+	+	+	**OmpR**	**239**		+	+	+
EvgS	1197	+	+	–	EvgA	204	EvgS: phosphorelay	+	+	–
GlnL	**349**	+	+	+	**GlnG**	**469**		+	+	+
KdpD	**894**	+	+	+	**KdpE**	**225**		+	+	+
NarQ	**566**	+	+	+	**NarP**	**215**	***NarQ* and *narP* not in 1 locus**	+	+	+
NarX	**598**	+	+	+	**NarL**	**216**		+	+	+
PgtB	669	–	+	+	PgtA	415		–	+	+
PhoQ	**486**	+	+	+	**PhoP**	**223**		+	+	+
phoR	**431**	+	+	+	**PhoB**	**229**		+	+	+
QseC	**449**	+	+	+	**QseB**	**219**		+	+	+

RcsC	**949**	+	+	+	**RcsB**	**216**	**RcsC: phosphorelay**	+	+	+
RstB	**433**	+	+	+	**RstA**	**239**		+	+	+
SsrA	920	–	–	+	SsrB	212		–	–	+
TctE	471	–	–	+	TctD	224		–	–	+
TorS	**914**	+	+[b]	+	**TorR**	**230**		+	+	+
TtrS	592	–	–	+	TtrR	206		–	–	+
UhpB	**500**	+	+	+	**UhpA**	**196**		+	+	+
YedV	**452**	+	+	+	**YedW**	**223**		+	+	+
YehU	**561**	+	+	+	**YehT**	**239**		+	+	+
yfhK	**475**	+	+	+	**YfhG**	**444**		+	+	+
YpdA	**565**	+	+	+	**YpdB**	**245**		+	+	+
ZraS	**458**	+	+	+	**ZraR**	**441**		+	+	+
Putative	606	–	+	–	Putative	452		–	+	–
					RssB	**337**	**RssB responds to ArcB**	+	+	+
					CheY	**129**		+	+	+
					fimZ	**210**		+	–	+
					putative	351		–	+	–
					putative	475		–	–	+
Total nr. of HK proteins		29	31	31	Total nr. of RR proteins			32	34	35

Cells with bold emphasis represent proteins conserved in all three genomes
[a]When present protein length is deduced from the *E. coli* MG1665 K-12 genome
[b]Contains a frameshift

X-rays to induce autotrophies (Gray and Tatum 1944). Blattner and colleagues state in their publication presenting the genome sequence of strain K-12 MG1665 (Blattner et al. 1997) that the sequenced variant "has been maintained as a laboratory strain with minimal genetic manipulation, having only been cured of the temperate bacteriophage lambda and F plasmid by ultraviolet light and acridine orange." Presumably, the "MG" in its name stands for "minimal genetic manipulation." But in fact, as Bachmann described (whose paper is cited by Blattner *c.s.*) the ancestor K-12 strain was treated by X-ray twice. The first X-ray treatment resulted in strain "58F$^+$, *bio-1*," and the second in strain "58-161F$^+$, *bio-1*, *metB1*," from which the spontaneous mutant W6 (F$^+$, *bio-1*, *metB1*, *rel-1*) was isolated. Treatment with UV light then resulted in strain W1665 that had lost its λ^- phage, and this was treated with acridine orange to produce W1665F$^-$ (*metB1*, *rel-1*, λ^-). It was this W1665F$^-$ strain that was renamed "MG1665" when it was sequenced, nearly a half-century later (Blattner et al. 1997). Blattner and colleagues specified that their strain contains "a frameshift mutation at the end of *rph*, causing low expression of the downstream gene *pyrE* and, in turn, a pyrimidine starvation phenotype. In addition, a mutation in *ilvG* disrupts one of the isoleucine-valine biosynthesis pathways in all K-12 isolates. Finally, almost all K-12 derivatives, including MG1655, carry the *rfb-50* mutation, where an IS5 insertion results in the absence of O-antigen synthesis in the lipopoly-saccharide" (Blattner et al. 1997). But it is obvious that the ancestor of the sequenced strain has undergone severe mutagenic treatments that can hardly be described as "minimal genetic manipulation." Whether the absence of two HK and two RR proteins, compared to the type strain DSM 30083, had resulted from the in vitro treatments of its ancestor, or belonged to the natural strain variation within the species, can no longer be investigated. However, by comparison of a large number of *E. coli* genomes, the overall conservation of 2CSTS proteins in this species could be assessed. Before we performed this, the architecture of the identified HK and RR proteins is briefly summarized.

3.2 HK and RR Subdomain Architecture

We reviewed the Pfam domains present in all 2CSTS proteins translated from the genes as shown in Table 7.1. A generalized protein structure of a canonical HK is schematically represented in Fig. 7.1. The position of the catalytic ATP-binding domain HATPase_c (PF02518), which we used as a telltale domain to identify HKs, is indicated in the drawing. The crucial histidine that is phosphorylated is present in the dimerization DHp part of the protein, located in an alpha-helix. In the majority of proteins identified, the kinase Pfam domain present was HisKA (PF00512), while alternative histidine kinase domains (PF07730, PF06580, or PF14689) were found eight times (Table S1). A number of other Pfam domains were identified in the HK proteins, for instance HAMP (PF00672), which is thought to be involved in the conformational change that activates the kinase (Aravind and Ponting 1999). However, in 60% of the proteins, no HAMP domain was found (Table S1), suggesting

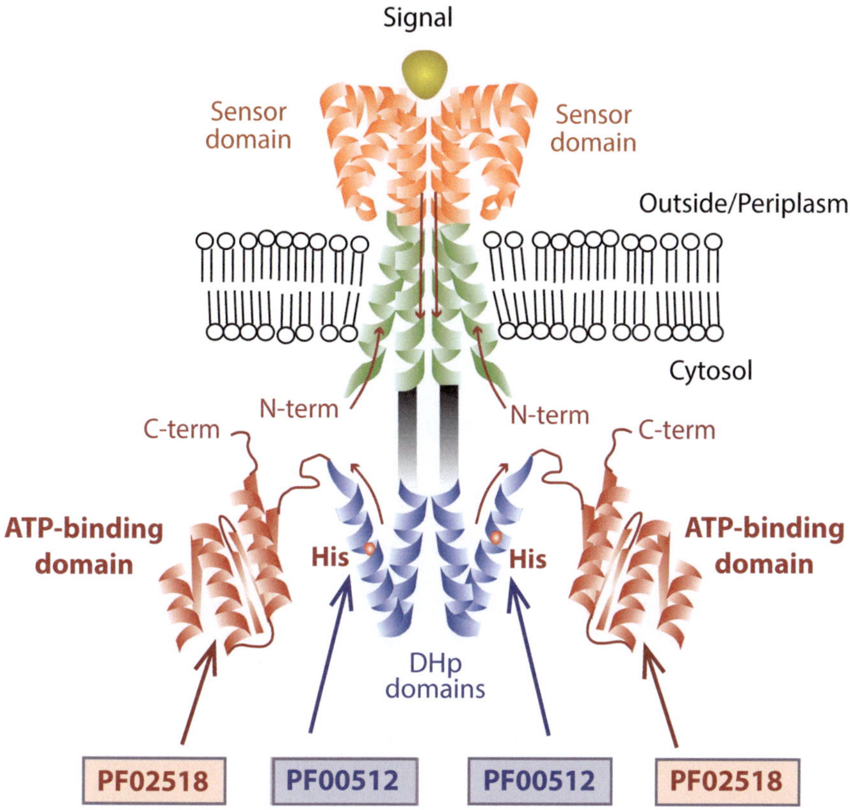

Fig. 7.1 Schematic structure of canonical histidine kinase in its dimeric form embedded in a membrane. The position of the absolutely conserved ATPase_c Pfam domain PF02518, as part of the ATP-binding functional domain, is indicated in red. The strongly conserved kinase Pfam domain PF00512 is located in the functional DHp domain (shown in blue)

that other, less well-characterized domains may be responsible for the conformational change. The PAS domain, responsible for signal detection (Ponting and Aravind 1997), was found in only six cases. A variety of other domains were found in single or a few proteins, which are all listed in Table S1. The automated process had missed a few domains due to incomplete coverage of the Pfam database. For instance, the ATPase domain in YehU and the Hpt domain in EvgS were not automatically identified, but since these domains are specified in the GenBank files of the proteins, they were manually added to Table S1.

The RR proteins of the two reference *E. coli* and the type strain *Salmonella* genomes invariably contained the receiver domain Response_reg (PF00072). Six other types of domains were recognized in RR proteins, of which Trans_reg_C (PF00486) was most common (found 15 times). This domain is responsible for DNA binding and was first described in OmpR (Martínez-Hackert and Stock 1997). Alternatively, the LuxR-like helix-turn-helix DNA-binding domain can be present

(GerE PF00196, found eight times). This domain was first characterized in GerE of *Bacillus* (Ducros et al. 2001). In six RR proteins (AtoC, PgtA, GlnG, YfhG, and ZraR; see Table S1 for alternative names), the presence of the domain Sigma54_activat (PF00158) suggests a role in more general cellular responses, as activation of this alternative sigma factor would have downstream effects on expression of a number of genes and operons (Studholme and Dixon 2003).

The Response_reg domain was also present in six phosphorelay hybrid proteins (ArcB, BarA, EvgS, RcsD, TorS, and SsrA), but only in four of these the Hpt domain (PF01627) was identified that is presumably responsible to pass the phosphoryl group on to the final RR (Varughese 2002).

3.3 Quality Score for Over 5000 E. coli *and Over 5000* S. enterica *Genomes*

As described previously (Cook and Ussery 2013; Land et al. 2014), genome quality scores can be used to filter out poor-quality genomes. A genome quality score was calculated by combining four different scores with values between 0 (minimum) and 1 (maximum) that assessed (a) sequence quality (punishing higher numbers of incomplete chromosomal contigs and ambiguous nucleotides), (b) coverage of rRNA genes (requiring the presence of at least one full-length copy of the rRNA genes coding for 5S, 16S, and 23S for a maximum score), (c) coverage of tRNA genes (requiring the presence of all tRNA genes coding for the 20 amino acids for maximum score), and (d) coverage of 102 conserved genes as assessed by their functional domains. These four scores were combined and averaged to give a genome quality score for each of the bacterial genomes. The obtained quality score of the 6270 *E. coli* and 7437 *Salmonella* genomes downloaded from GenBank is shown by differently colored and sized dots in Fig. 7.2, sorted for increasing score values on the *x*-axis and plotted with the number of proteins predicted from prodigal for each genome on the *y*-axis. As can be seen from the figure, the range of the latter is far greater than one would expect for *E. coli* genomes, ranging from close to zero proteins to more than 14,000 proteins for a single *E. coli* genome. Such extreme findings were the result of incorrectly submitted or severely under- or over-annotated sequences, and, as indicated by their gray-colored dots, none of these entries passed the quality score cutoff of 0.8. When the genomes with low-quality scores were ignored, the dataset was reduced to the ranges shown in the red boxes in Fig. 7.2, with 5243 *E. coli* and 7041 *Salmonella* genomes that passed the test. These were further used to analyze conservation of HK and RR proteins within and across these species.

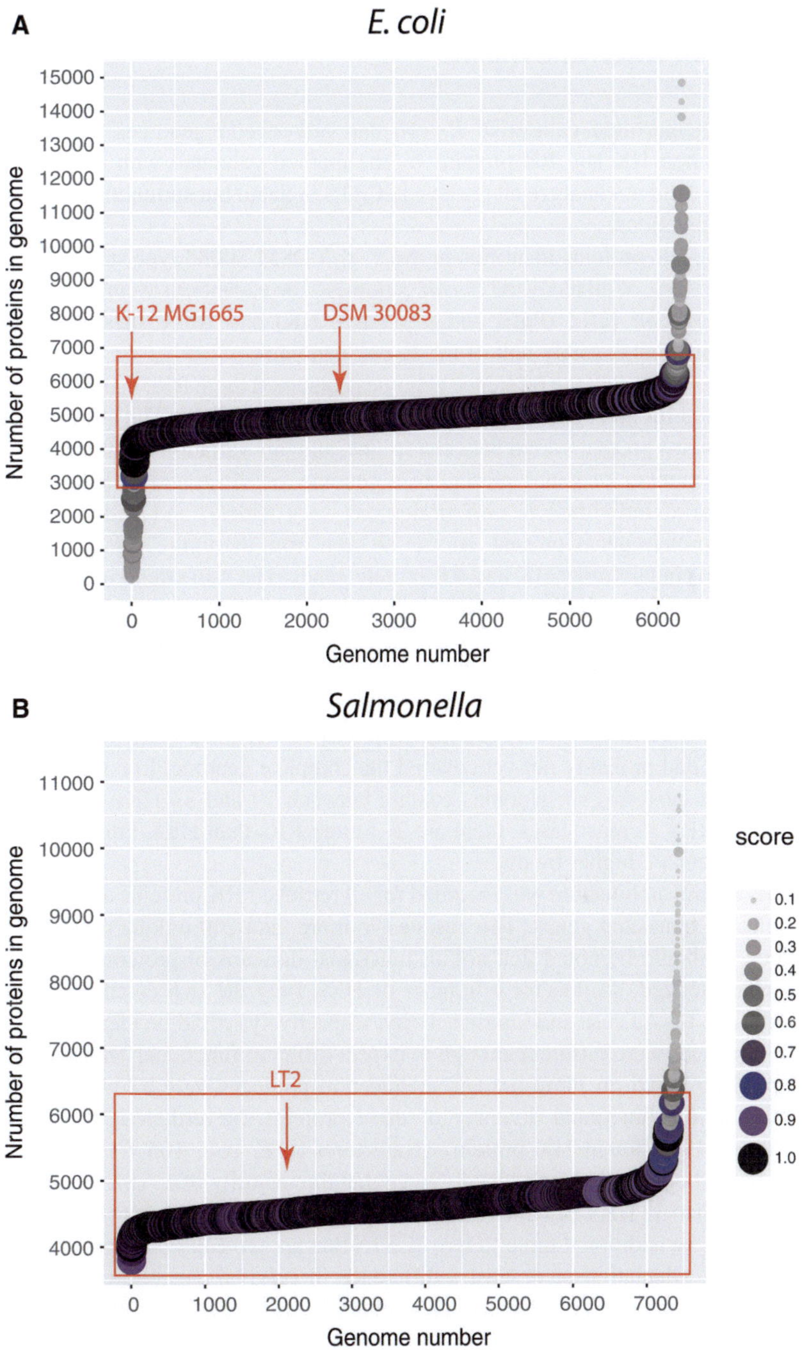

Fig. 7.2 Quality scores of 6270 *E. coli* (**a**) and 7437 *S. enterica* (**b**) genomes. The red squares include the genomes selected that passed the quality score. The positions of the three reference genes are indicated by arrows

3.4 *Total Number of HK and RR Proteins Across* E. coli *and* S. enterica *Genomes*

The quality-controlled dataset of *E. coli* and *Salmonella* genomes was used to retrieve all their HK and RR proteins by the method validated with the reference genomes. We did not analyze all retrieved 2CSTS in detail but summarize here some general observations.

Of the 31 HK proteins identified in the *E. coli* DSM 30083 type strain, 28 were conserved in 95% of all analyzed *E. coli* genomes. The exceptions were the two HKs missing in *E. coli* K-12 (PgtB and a putative HK) and AtoS: these were only conserved in approximately half of the *E. coli* genomes.

The observation that the reference genomes contained more RRs than HKs turned out to be valid for most genomes of both *E. coli* and *Salmonella*. The number of RR and HK proteins present varied between strains, and this variation weakly correlated to the total number of genes present in a genome, as illustrated in Fig. 7.3. The genome size of individual *E. coli* strains can vary widely, and with it the total number of protein genes present. As can be seen from the figure, a larger genome also tends to contain more HK and RR protein genes. The same trend is seen for the *Salmonella* genomes, though the variation in total number of genes is lower for this genus than is observed for the genomes that all belong to the single-species *E. coli*. The *E. coli* genome with 49 reported RRs was checked manually, which revealed 16 of these were duplicated in different contigs; this probably means the real number of RR proteins is closer to 41. A genome with only 32 RR and 26 HK was still a draft version that had probably not yet covered the complete genome. In conclusion, we observe that most *E. coli* genomes contain between 29 and 33 HKs and between 31 and 35 RRs; in most cases, there are 2–3 more RRs than HKs. For *Salmonella*, these numbers are higher by one.

The domain architecture was recorded for all retrieved HK proteins deduced from the selected translated genes. This resulted in more than 200 unique protein architectures. It should be noted that not all functional domains of proteins are equally well characterized, so that for a number of HKs, only the two essential domains (responsible for ATPase and kinase activity, respectively) could be identified, while their sequences were different enough to expect different functions. Thus, grouping HK proteins based on Pfam domain architecture produces only a lowest possible estimate of the functional diversity of these proteins. Nevertheless, this already identified 119 different HK protein architectures in *E. coli*, 105 for *Salmonella*, and 148 for both of these combined. Given that the majority of *E. coli* strains don't have more than 33 HKs, of which 28 are conserved, it follows that there exists a wide variation in the additional three to five 2CSTS that are typically present. This could reflect niche adaptation or at least a difference in survival capabilities for a wide range of environmental conditions under which *E. coli* is known to be found (Van Elsas et al. 2011). Thus, for *E. coli* it can be concluded that histidine kinase proteins, and their accompanying response regulators, fall into two categories: a core set of 28 HKs and 31 RRs that are found in nearly all genomes and a much larger set (of at

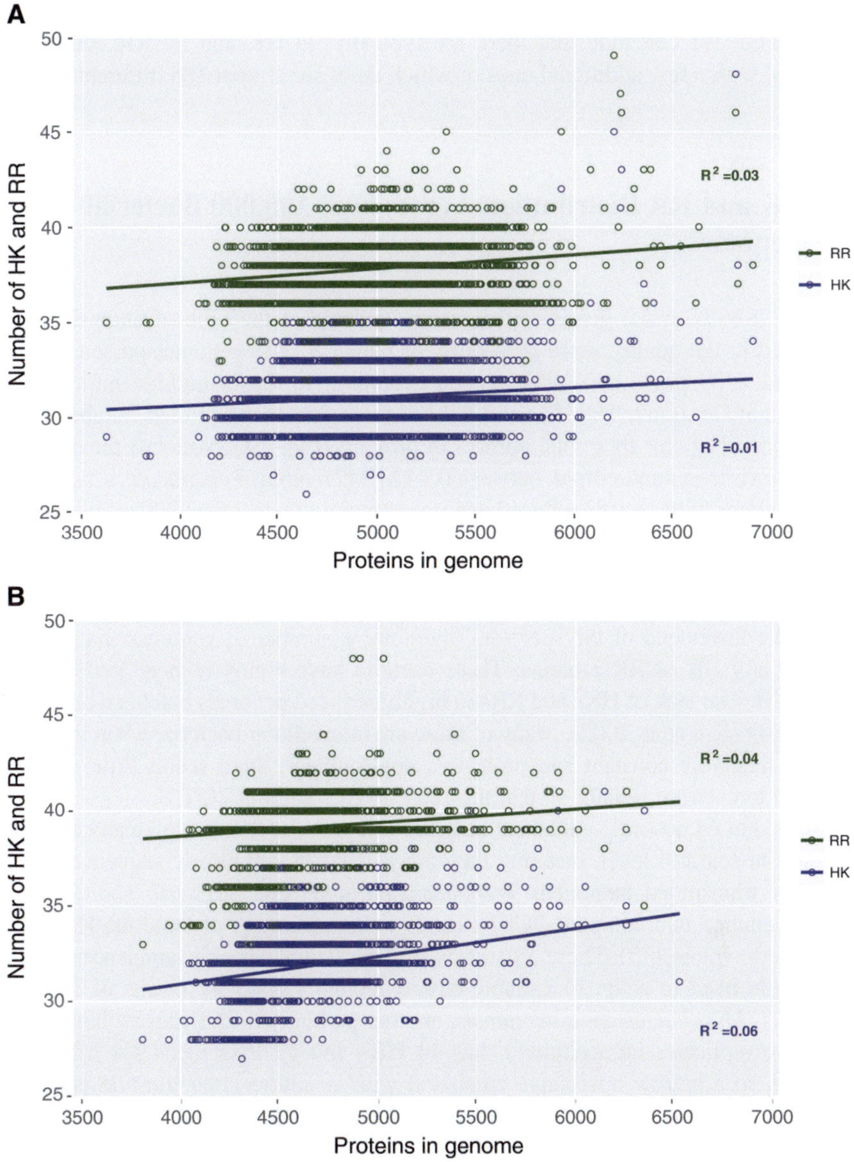

Fig. 7.3 Relationship between the number of histidine kinase (blue) and response regulator genes (green) with total number of protein genes in *E. coli* (**a**) and *Salmonella* (**b**) genomes. The lines represent the linear regression of the data with values for R^2 as indicated. The two panels were plotted using the same scale to enable direct comparison

least 119 HKs) that are part of the variable gene pool, of which a few may be present in any given strain. This finding reflects the strain-to-strain variation of this species. Strain *E. coli* K-12 contains only one extra 2CSTS compared to the core 2CSTS proteins, which is at the lower end of the spectrum but not exceptional. For *Salmonella*, we conclude that there are typically 29 HK and 32 RR conserved proteins, with a few additional ones of which there are at least 105 different kinds.

4 HK and RR Distribution Across over 100,000 Bacterial Genomes

The 101,836 bacterial genomes that were available at the time of analysis were subjected to the quality score procedure, of which 87,248 genomes passed. These represented 33 phyla and 1942 genera, according to the taxonomic information recorded at GenBank. The HKs and RRs of these genomes and their number were again plotted against their total number of proteins (Fig. 7.4). Note that the genome size now varies a hundredfold, between 0.14 MB (*Tremblaya phenacola*, a symbiont of mealybugs with a vastly reduced genome containing fewer than 200 proteins) and 16 MB (*Minicystis rosea*, a soil bacterium resembling *Myxobacteria*, with more than 14,000 proteins). The number of HK proteins varies from 0 to over 300 and the number of RR from 0 to nearly 250, with one outlier of 308 (Fig. 7.4b).

At the lower end of the spectrum, there are a number of genomes that do not contain any HK or RR proteins. These bacteria have highly reduced genomes of <1.0 MB. The lack of HKs and RRs in highly reduced genomes had been observed previously (Kiil et al. 2005). Many of these are intracellular bacteria; when living in such a relatively constant and protective environment, there seems little need to respond to external signals, so that they can survive without 2CSTS.

A total of 64 genomes contained only one HK and one RR, though many of these are still at scaffold level, meaning they are as yet not completely sequenced. The sexually transmitted pathogens *Treponema pallidum* and *Neisseria ghonorrhoea* only contained two complete 2CSTS (in *T. pallidum* in one of these the HK gene contained a frameshift). These organisms are strictly host-specific, and it seems they have little need to adapt to variable environmental signals by means of 2CSTS. However, *Mycobacterium tuberculosis*, another pathogen with a narrow host range that also replicates intracellularly, has 14 HKs and 12 RRs in its 4.4 MB-large genome, so a narrow host range combined with an intracellular lifestyle does not automatically imply there is no need for 2CSTS. Note that *M. tuberculosis* spreads via droplet contamination and can colonize a variety of tissues. This may explain its need for more flexible niche adaptation.

As expected, with increasing genome size and the total number of genes present, the number of HK and RR proteins increases as well. This trend has been reported before, for instance, when 167 bacterial genomes were compared for their HK content, which correlated with genome size (Galperin 2005, RRs were not included in that study) or in an analysis of 250 genomes that was published in the same year (Kiil et al. 2005). Our present findings, based on a much larger dataset, indicates that, as genome size increases, the number of HK and RR proteins goes up, and, as a

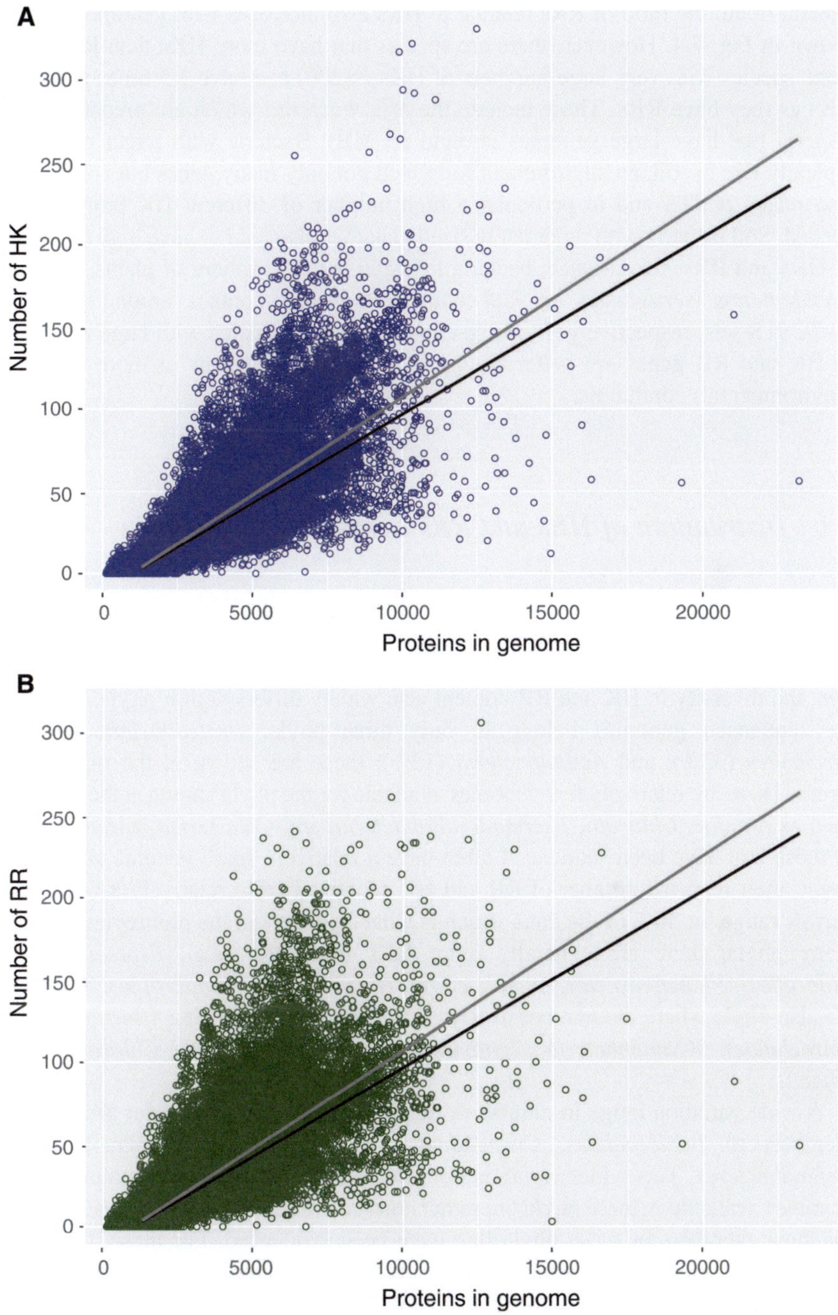

Fig. 7.4 Relationship between number of histidine kinase (**a**, blue) and response regulator genes (**b**, green) with total number of genes in close to 100,000 bacterial genomes. The linear regression lines for HKs (gray, $R^2 = 0.57$) and RRs (black, $R^2 = 0.55$) are shown in both panels for comparison

general trend, the ratio of RRs relative to HKs also increases with genome size, as shown in Fig. 7.4. However, there are species that have more HKs than RRs, and some species with very large numbers of HKs (>200) can have 1.5 times as many HKs as they have RRs. These include the *Myxobacteria*, which are predatory soil bacteria that have large genomes (around 13 MB). Bacteria with larger genomes typically live in soil, an environment for which not only many genes but apparently also many 2CSTS and in particular a high number of different HK proteins are needed. Soil bacteria with moderately sized genomes also tend to have high numbers of HKs and RRs; for instance, bacteria living in the rhizosphere of plants, such as *Pseudomonas aeruginosa* or *Rhizobium miluonense* contain around 70 and 50 2CSTS sets, respectively. It can be expected that organisms with large numbers of HK and RR genes are better equipped to survive and grow at more variable (environmental) conditions.

4.1 Distribution of HKs and RRs Across Bacterial Phyla

The reported HK and RR proteins were further analyzed per bacterial phylum. The results are shown in Fig. 7.5. Since most phyla contain a highly diverse collection of families, genera, and species, with huge variation in genome size and niche adaptation, the diversity in HK and RR content also widely differs within phyla. Most of the sequenced genomes belong to only three phyla: *Proteobacteria* (45%), *Firmicutes* (32%), and *Actinobacteria* (12%); these are shown at the top of the figure. There are relatively few genomes available for the phyla shown at the bottom, such as *Aquifae*, *Chlorobi*, *Thermodesulfobacteria*, and *Caldiserica*, and members of those that have been sequenced often have a relatively small genome size. As a result, their reported number of RR and HK proteins is also relatively low, with a narrow range for 50% of the data, which is what the boxes in the plot represent. For many phyla, there are generally more HKs than RRs (e.g., *Proteobacteria*, *Chloroflexi*, *Planctomycetes*, *Acidobacteria*, *Nitrospirae*, *Elusimicrobia*), but there are also phyla where the number of RRs is in strong excess to HKs (*Bacteroidetes*, *Spirochaetes*, *Cyanobacteria*, *Ignavibacteriae*, *Fibrobacteres*, *Chlorobi*, and others).

A wide variation range in number of 2CSTS proteins is reported for 50% of the members of *Bacteroidetes*, *Cyanobacteria*, *Spirochaetes*, *Lentisphaerae*, and *Deferribacteres*. This wider variation does not necessarily reflect large numbers of genomes available in these phyla but rather correlates to the variation in genome size for those members of these phyla that have been sequenced. For these, there are generally more RRs than HKs. A striking observation is the long whiskers and the large number of outliers for the three phyla containing almost 90% of the total genomes, the *Proteobacteria*, *Firmicutes*, and *Actinobacteria*. Some outliers have exceptionally high numbers of RR and HK proteins, compared to the majority of species in those phyla.

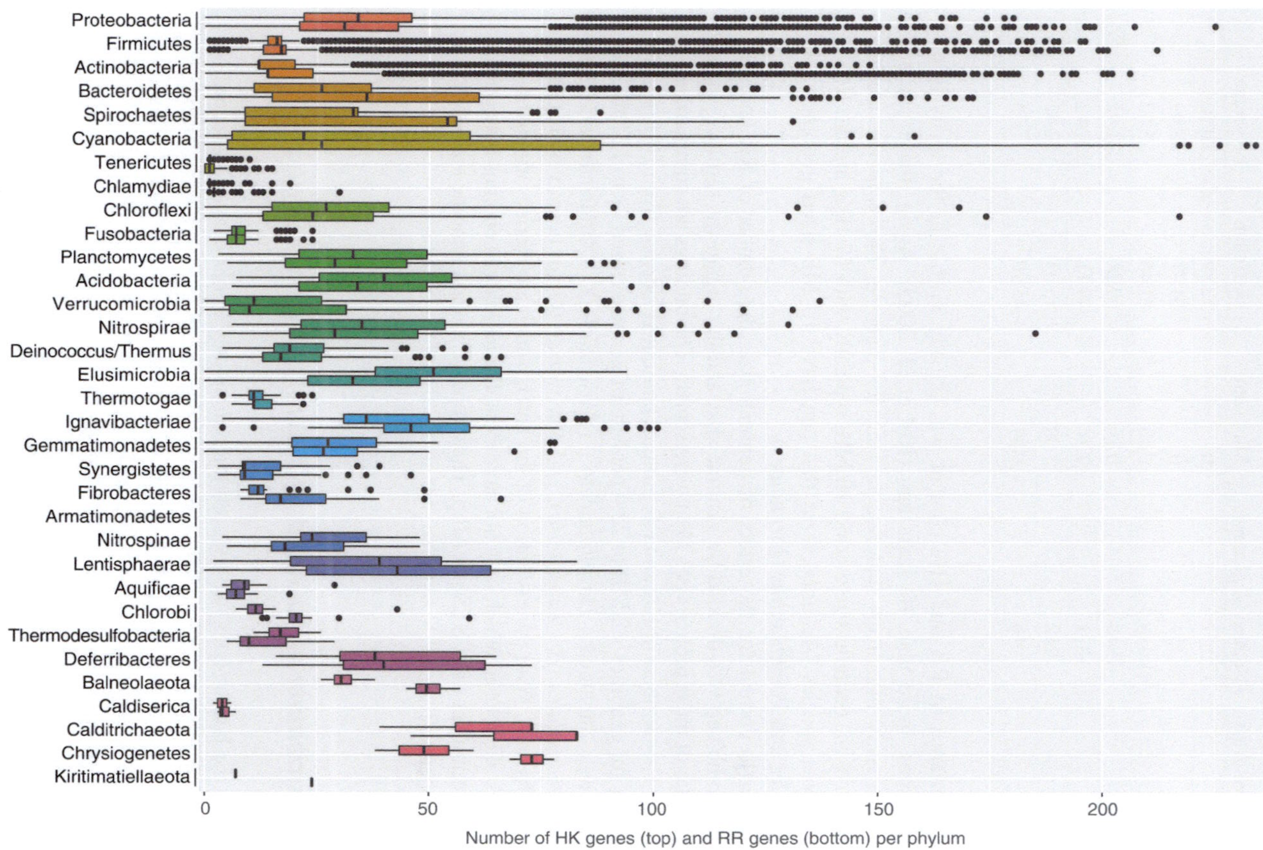

Fig. 7.5 Box-and-whisker plots of the number of histidine kinases and response regulator proteins across 33 bacterial phyla, represented by individual colors. For each phylum, the numbers of HKs are given above those of RRs. The phyla are ordered for the number of sequenced members from top (highest numbers) to bottom (lowest numbers)

Table 7.2 Number of histidine kinase and response regulator proteins (combined) related to genome size for all analyzed bacterial genomes

		Number of HKs and RRs combined per Mb		
Genome size (Mb)	Number of genomes	Minimum	Maximum	Average
<2 Mb	12,147	0	72.99	10.10
<4 Mb	38,754	0	84.77	13.09
<6 Mb	41,416	0.18	81.46	15.02
<8 Mb	7848	0.15	69.24	21.93
<10 Mb	1373	7.16	49.11	21.42
<12 Mb	252	7.09	45.63	22.42
>12 Mb	46	0.64	45.90	17.76

The variance in RR and HK content was already visible in the analysis of 250 genomes, more than 10 years ago (Kiil et al. 2005). Again, these outliers represent mostly organisms that live in soil, for instance, *Burkholderia* species (*Proteobacteria*, genome size around 10 MB, typically with 67 HKs and 80 RRs), *Streptomyces* species (*Actinobacteria* that can have 10 MB-sized genomes), or *Clostridiales* and *Paenibacillus* (*Firmicutes*).

In order to correct for the effect of genome size, the number of RR and HK proteins combined was divided by genome size, and the organisms resulting in the highest ratio of these proteins per Mbp were scored. The findings are summarized in Table 7.2. The numbers of HK and RR proteins combined increase from 10.1 per Mb on average for genomes <2 MB to 22.4 per Mb for genomes sized up to 12 Mb. The top scorers included *Desulfobacula* (*Clostridiales*), *Sulfuricurvum* and *Desulfuromonadales* (*Proteobacteria*), and *Nitrospirae* (Gram negatives that are a separate phylum). With the exception of the latter, which are marine organisms, all others are soil bacteria, which may not have exceptionally large genomes but contain the highest number of 2CSTS per Mbp. Thus, relative to their genome size, they are best equipped to respond to external signals by signal transduction.

4.2 Distribution of HKs and RRs Across Bacterial Genera

Finally, in the last analysis, we reduced the redundancy in our dataset by recording the average number of HK and RR proteins per sequenced species, also recording the range of minimum and maximum values. This nonredundant dataset was used to compare the average number of HK and RR proteins per bacterial genus. The distribution of numbers of HK and RR proteins, grouped in increments of 5, is shown in Fig. 7.6.

The general trend reported above, that genomes in general tend to have more RR than HK, is also visible in this analysis, mostly for species containing over 60 HK proteins. The most frequently encountered numbers in this nonredundant dataset is

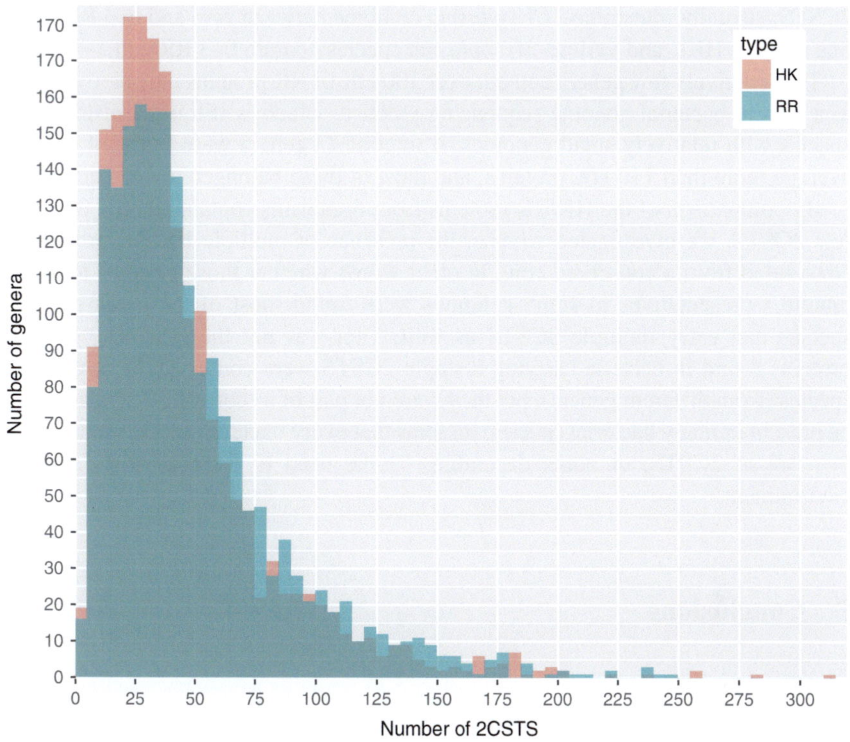

Fig. 7.6 Distribution of average number of HK (green) and RR (blue) proteins per bacterial genus, ordered for a total number of 2CSTS, in a nonredundant dataset. The average data are presented for in blocks of five

the presence of 20–35 HK and 20–40 RR proteins, observed in over 500 bacterial genera.

Only one genus could be identified that consistently lacked HKs or RRs, though its number of species sequenced to date is still limited: none of the eight genomes of *Blattabacterium* species had a recognizable 2CSTS. These are obligate insect endosymbionts belonging to the *Bacteroidetes*.

Although in total five genera completely lacked RRs, most of these did contain species with a few HK proteins (note that this cannot be seen in Fig. 7.6, where the first column represents all genomes combined containing between zero and five HK or RR proteins). Even of the 20 sequenced *Buchnera* genomes (like *C. ruddii* these *Gammaproteobacteria* live as obligate arthropod endosymbionts with highly reduced genomes), some contained a single HK. Other insect endosymbionts, that more typically belong to the *Tenericutes* (e.g., *Entomoplasma or Spiroplasma*), also completely lack RR proteins but may contain between zero and one or between zero and two HK proteins, respectively. Likewise, the *Tenericutes Mesoplasma* (insect endosymbionts and plant pathogens) also survive without RR but may contain between zero and two HKs. Other *Tenericutes* also contain none or very few

2CSTS: sexually transmitted *Ureaplasma* combine between zero and two RRs with one to three HKs, and various *Mycoplasma* species contain 0–3 RR and 1–4 HKs.

That relatively few genera contain over 100 HK or RR proteins may partly be due to a bias in bacterial genome sequencing practice, which is still often restricted to species with relatively small genomes. A total of 73 genera were recorded with on average more than 110 HK proteins, and most of these belonged to soil bacteria.

By comparison of the HK/RR proteins from organisms containing only one, or a few 2CSTS, we investigated whether the presence of multiple HK/RR proteins is constituted from a conserved core, in other words whether those proteins found as unique representatives in some genomes were found most of the times in other species that carry multiple 2CSTS as well. That was not the case. The HK/RR proteins found in bacteria carrying a single 2CSTS were conserved in some other species (from different genera), but they were not part of a "core 2CSTS" that would be present in many bacterial species. It seems that every bacterial species contains its own set of 2CSTS, or none, depending on the need to respond to its external environment.

5 Conclusions

Here we have shown the distribution of 2CSTS proteins across bacterial genomes. When present, the number of these proteins weakly correlates with genome size. For *E. coli*, there are approximately 29–33 HK and a slightly higher number of RRs, with higher numbers for larger genomes. These comprise mostly of species-conserved 2CSTS proteins, but a few variable proteins are present in every genome, and their nature can vary considerably. When all bacterial genomes were compared, it was confirmed that endosymbionts may lack 2CSTS, while species with large genomes, typically inhabitants of soil and/or those that are well equipped to survive variable conditions, often have high numbers of 2CSTS. Even soil bacteria with moderately sized genomes have more of these proteins per million bp than members living in other environments.

Our data provide a swift and accurate method to compare protein content of thousands of genomes by using the Pfam domains. The method demonstrated here serves as a valid way to compare genomes based on protein functions and is more reliable to predict conserved gene function than comparisons that depend on sequence alignments. Here we show that the number of 2CSTS that occur in bacterial genomes correlates with the environment in which the bacteria live and that larger genomes require more of these signal transduction systems, as does a life in soil. This is expected, as the larger the genome, the more complex the bacteria are and the more capable they must be of adapting to diverse environments. There is no core containing particular HK and RR proteins that are conserved in most bacterial species; instead, each species contains those 2CSTS that best matches their needs, which ranges from no need at all to high numbers of different two-component signal transduction systems.

Acknowledgments This research was funded in part by the College of Medicine and the Department of Biomedical Informatics at UAMS, the Helen Adams & Arkansas Research Alliance Endowment, and by discretionary funding from the Joint Institute of Computational Sciences (JICS), and Oak Ridge National Laboratory sponsored laboratory director's research and development project 7899 and by US DOE Office of Biological and Environmental Research, Genomic Science Program. Oak Ridge National Laboratory is managed by UT-Battelle, LLC, for the US Department of Energy under Contract no. DEAC05-00OR22725.

References

Abriata LA, Albanesi D, Dal Peraro M, de Mendoza D (2017) Signal sensing and transduction by histidine kinases as unveiled through studies on a temperature sensor. Acc Chem Res 50:1359–1366

Aravind L, Ponting CP (1999) The cytoplasmic helical linker domain of receptor histidine kinase and methyl-accepting proteins is common to many prokaryotic signalling proteins. FEMS Microbiol Lett 176:111–116

Attwood PV (2013) Histidine kinases from bacteria to humans. Biochem Soc Trans 41:1023–1028

Bachmann BJ (1972) Pedigrees of some mutant strains of *Escherichia coli* K-12. Bacteriol Rev 36:525–557

Baxter MA, Jones BD (2015) Two-component regulators control *hilA* expression by controlling *fimZ* and *hilE* expression within *Salmonella enterica* serovar Typhimurium. Infect Immun 83:978–985

Bhate MP, Molnar KS, Goulian M et al (2015) Signal transduction in histidine kinases: insights from new structures. Structure 23:981–994

Blattner FR, Plunkett G 3rd, Bloch CA et al (1997) The complete genome sequence of *Escherichia coli* K-12. Science 277:1453–1462

Bourret RB (2010) Receiver domain structure and function in response regulator proteins. Curr Opin Microbiol 13:142–149

Capra EJ, Laub MT (2012) Evolution of two-component signal transduction systems. Annu Rev Microbiol 66:325–347

Carbone A (2006) Computational prediction of genomic functional cores specific to different microbes. J Mol Evol 63:733–746

Cook H, Ussery DW (2013) Sigma factors in a thousand *E. coli* genomes. Environ Microbiol 15:3121–3129

Ducros VM, Lewis RJ, Verma CS et al (2001) Crystal structure of GerE, the ultimate transcriptional regulator of spore formation in *Bacillus subtilis*. J Mol Biol 306:759–771

Finn RD, Coggill P, Eberhardt RY et al (2016) The Pfam protein families database: towards a more sustainable future. Nucleic Acids Res 44(D1):D279–D285

Galperin MY (2005) A census of membrane-bound and intracellular signal transduction proteins in bacteria: bacterial IQ, extroverts and introverts. BMC Microbiol 14(5):35

Galperin MY (2010) Diversity of structure and function of response regulator output domains. Curr Opin Microbiol 13:150–159

Gray CH, Tatum EL (1944) X-ray induced growth factor requirements in bacteria. Proc Natl Acad Sci USA 30:404–410

Hoch JA (2000) Two-component and phosphorelay signal transduction. Curr Opin Microbiol 3:165–170

Hoch JA, Varughese KI (2001) Keeping signals straight in phosphorelay signal transduction. J Bacteriol 183:4941–4949

Hyatt D, Chen GL, Locascio PF et al (2010) Prodigal: prokaryotic gene recognition and translation initiation site identification. BMC Bioinf 11:119

Johnson LS, Eddy SR, Portugaly E (2010) Hidden Markov model speed heuristic and iterative HMM search procedure. BMC Bioinf 11:4319

Kiil K, Ferchaud JB, David C et al (2005) Genome update: distribution of two-component transduction systems in 250 bacterial genomes. Microbiology 151:3447–3452

Lagesen K, Ussery DW, Wassenaar TM (2010) Genome update: the 1000th genome – a cautionary tale. Microbiology 156(Pt 3):603–608

Land ML, Hyatt D, Jun SR et al (2014) Quality scores for 32,000 genomes. Stand Genomic Sci 9:20

Land M, Hauser L, Jun SR et al (2015) Insights from 20 years of bacterial genome sequencing. Funct Integr Genomics 15:141–161

Lederberg J (1951) Genetic studies with bacteria. In: Dunn LC (ed) Genetics in the 20th century. Macmillan, New York, pp 263–289

Martínez-Hackert E, Stock AM (1997) The DNA-binding domain of OmpR: crystal structures of a winged helix transcription factor. Structure 5:109–124

Mika F, Hengge R (2005) A two-component phosphotransfer network involving ArcB, ArcA, and RssB coordinates synthesis and proteolysis of sigmaS (RpoS) in *E. coli*. Genes Dev 19:2770–2781

Ponting CP, Aravind L (1997) PAS: a multifunctional domain family comes to light. Curr Biol 7: R674–R677

Rivera-Cancel G, Ko WH, Tomchick DR, Correa F, Gardner KH (2014) Full-length structure of a monomeric histidine kinase reveals basis for sensory regulation. Proc Natl Acad Sci USA 111:17839–17844

Rogov VV, Rogova NY, Bernhard F (2006) A new structural domain in the *Escherichia coli* RcsC hybrid sensor kinase connects histidine kinase and phosphoreceiver domains. J Mol Biol 364:68–79

Sanders DA, Gillece-Castro BL, Burlingame AL et al (1992) Phosphorylation site of NtrC, a protein phosphatase whose covalent intermediate activates transcription. J Bacteriol 174:5117–5122

Studholme DJ, Dixon R (2003) Domain architectures of sigma54-dependent transcriptional activators. J Bacteriol 185:1757–1767

Van Elsas JD, Semenov AV, Costa R, Trevors JT (2011) Survival of *Escherichia coli* in the environment: fundamental and public health aspects. ISME J 5:173–183

Varughese KI (2002) Molecular recognition of bacterial phosphorelay proteins. Curr Opin Microbiol 5:142–148

Chapter 8
Effects of Spatial Structure and Reduced Growth Rates on Evolution in Bacterial Populations

Michael T. France, Ben J. Ridenhour, and Larry J. Forney

1 Introduction

The biological diversity among and within microbial species is important because it is the pool from which solutions to novel challenges are selected. The diversity of species present in many microbial communities and populations far exceeds that found in their macro-organism counterparts (Dykhuizen 1998; Whitman et al. 1998), with a single gram of soil estimated to contain as many as 50,000 bacterial species (Schloss and Handelsman 2006; Roesch et al. 2007). While some of this diversity can be explained by functional differences between the constituent species, much of it exists within ecotypes, which are lineages of genetically and ecologically distinct strains within a named species (Kopac et al. 2014; Shapiro and Polz 2014; Cohan 2016). These ecotypes coexist in small spaces seemingly in defiance of Gause's law of competitive exclusion (Gause 1934). Thus, there is a need to understand evolutionary processes as they play out in naturally occurring microbial habitats so that we can better comprehend the emergence and maintenance of genetic diversity in the microbial world.

Evolution is defined as changes in the frequency of genotypes in a population over time (Barton et al. 2007). The processes that drive evolution can be divided into four fundamental forces: mutation, selection, gene flow, and genetic drift. Decades of prior research on microbial evolution has characterized the action of these evolutionary forces in well-mixed laboratory populations (Kawecki et al. 2012).

M. T. France
Institute for Bioinformatics and Evolutionary Studies, University of Idaho, Moscow, ID, USA

B. J. Ridenhour · L. J. Forney (✉)
Institute for Bioinformatics and Evolutionary Studies, University of Idaho, Moscow, ID, USA

Department of Biological Sciences, University of Idaho, Moscow, ID, USA
e-mail: lforney@uidaho.edu

P. H. Rampelotto (ed.), *Molecular Mechanisms of Microbial Evolution*, Grand Challenges in Biology and Biotechnology,
https://doi.org/10.1007/978-3-319-69078-0_8

Yet the vast majority of microbial populations found outside the laboratory differ dramatically in at least two key ways. First, they typically reside in structured environments such as soils, sediments, and biofilms (Costerton et al. 1978; Whitman et al. 1998). Biofilms in particular can be found in nature (Costerton et al. 1978; Besemer et al. 2009; Burmolle et al. 2012), the built environment (Rogers et al. 1994; Wang et al. 2013), and even on/within the human body (Palestrant et al. 2004; Macfarlane and Dillon 2007; Verstraelen and Swidsinski 2013). Second, the majority of environmental populations are growing markedly slower than their laboratory counterparts. The growth of microbial populations is limited by the abundance of routinely scarce essential nutrients leading to generation times for populations in soils and the oceans frequently on the order of days and weeks (Jannasch 1969; Harris and Paul 1994; Rousk and Bååth 2011; Kirchman 2016). If we want to understand how diversity is created and maintained in natural microbial populations, we must endeavor to characterize the effects of spatial structure and reduced growth rates on microbial evolution.

While there are many examples of microbial populations that either exhibit spatial structure or are growing slowly, in this chapter we chose to focus our discussion on populations of bacteria growing within biofilms. Biofilms are complex assemblages of microbial cells that are enclosed in a self-made extracellular matrix (Flemming and Wingender 2010). They are often attached to surfaces but can also be found as free floating pellicles or globules of cells. This choice was made for several reasons: the first of which is that populations within biofilms both exhibit inherent spatial structure (Nadell et al. 2016) and often contain large subpopulations that are growing slowly (Sternberg et al. 1999; Stewart et al. 2016). However, they are also fairly ubiquitous in the natural world and, among other things, play critical roles in human disease (Parsek and Singh 2003; James et al. 2007; Hall-Stoodley and Stoodley 2009), biofouling of industrial equipment (Brooks and Flint 2008; Flemming 2011), and the cycling of nutrients in the environment (Paul et al. 1991; Battin et al. 2003; Meckenstock et al. 2015). Despite their prevalence and importance, surprisingly few studies have explored evolution within biofilm populations (Adams and Rosenzweig 2014; Steenackers et al. 2016), although this subject has gained more attention recently (Poltak and Cooper 2011; Traverse et al. 2013; Ellis et al. 2015). This is perhaps because biofilms are inherently more complex than their planktonic counterparts (Stoodley et al. 2002; Stewart and Franklin 2008; Flemming et al. 2016), which complicates both the technical execution of experiments and the analysis and interpretation of the resulting data. In the following chapter, we review how growth within a biofilm influences microbial evolution. We break our discussion down into three overarching topics: spatial structure, the effects of growing slowly, and horizontal gene transfer (HGT) in biofilms. While the discussion within focuses on biofilms, our observations may apply equally well to other, non-biofilm populations that are either spatially structured or growing slowly.

1.1 Biofilms and the Evolution of Antibiotic Resistance

The evolution of antibiotic resistance is a notable example of why it is imperative that we advance our understanding of how evolutionary processes operate in biofilms populations. Biofilms have been shown to play a critical role in the pathogenesis of many bacterial species (Lynch and Robertson 2008; Percival et al. 2012; Mulcahy et al. 2014). Estimates from the Centers for Disease Control and Prevention indicate that over 65% of all bacterial infections are caused by pathogens residing with biofilms (Klevens et al. 2007). Prominent examples of biofilm infections include those caused by *Pseudomonas aeruginosa* in the lungs (Mulcahy et al. 2014), *Acinetobacter baumannii* in wounds (Qi et al. 2016), and *Staphylococcus aureus* on indwelling devices (Otto 2008). Infections caused by pathogens residing within biofilms have been repeatedly shown to be extremely difficult to eradicate, regardless of the method used, including antibiotics (Høiby et al. 2010), disinfectants (Kumar and Anand 1998), and mechanical removal (Simões et al. 2005). Furthermore, residing in biofilms has been shown to provide protection against the human immune system (Jensen et al. 2010; Domenech et al. 2013). Consequently these infections are associated with higher rates of morbidity and mortality (Lynch and Robertson 2008; Høiby et al. 2011) and are more likely to progress from an acute to a chronic state (Wolcott et al. 2010). We speculate that because biofilm infections are both pervasive and persistent, they may be a hot spot for the evolution of antibiotic resistance.

Perhaps the most alarming characteristic of populations within biofilms is their inherent recalcitrance to antibiotic treatment. Initial investigations demonstrated that this trait was not heritable, suggesting that it did not result from genetic change. This led to the proposition of several phenotypic mechanisms which have been proposed to explain the characteristic recalcitrance of populations within biofilms. Early studies suggested the biofilm matrix might slow down the diffusion of antibiotics, providing for zones of subinhibitory antibiotic concentrations (Hoyle et al. 1992; Stewart 1996). While others focused on the wide range of physiological states present in biofilm populations, some of which are more tolerant to certain antibiotics (Xu et al. 1998; Sternberg et al. 1999). For example, β-lactam antibiotics that inhibit bacterial cell wall synthesis may not be effective against nongrowing or slowly growing cells found in the interior of a biofilm. More recent studies have shown that biofilm populations harbor antibiotic-tolerant individuals that emerge from the differential expression of toxin-antitoxin system genes (e.g., *hipAB*) (Lewis 2007, 2010; Maisonneuve and Gerdes 2014). The tolerant phenotype exhibited by these individuals allows them to persist, but not grow, during antibiotic treatment. This phenotype is readily reversed (Amato et al. 2014; Schumacher et al. 2015) and is not the result of heritable genetic mutations that confer antibiotic resistance (Olsen 2015; Brauner et al. 2016). While these phenotypic mechanisms certainly play a role in the recalcitrance of biofilm populations to antibiotic treatment, they do not preclude a significant contribution of genetic change. In fact, by allowing portions of the population to survive during treatment, tolerance mechanisms may actually facilitate the evolution of antibiotic resistance. This was demonstrated in a recent study that showed antibiotic-tolerant *Escherichia coli* subpopulations acquired mutations that provided resistance to ampicillin (Levin-Reisman et al. 2017). More work is needed to detail the potential interplay between phenotypic tolerance and the evolution of heritable antibiotic resistance in biofilm infections.

There is also a growing body of research that suggests that the accumulation of genetic diversity in biofilm populations may also directly contribute to their recalcitrance (Boles and Singh 2008; Eastman et al. 2011; Tyerman et al. 2013). This was perhaps missed by the initial investigations into biofilm recalcitrance because these studies were largely conducted by exposing relatively young biofilms to antibiotics for relatively short periods of time and then examining survivors (Williams et al. 1997). By not providing enough time for evolution to enact change, this experimental design may have precluded any observation of the evolution of heritable antibiotic resistance. This is especially important given the often- prolonged nature of biofilm infections. In their 2004 paper, Boles et al. demonstrated that *Pseudomonas aeruginosa* biofilm populations accumulate genetic diversity and that this diversity allows populations to better resist environmental stress. They connected these observations to the "insurance hypothesis" from ecological theory, which states that genetic diversity fortifies populations against future challenges (Yachi and Loreau 1999). Other studies have shown that the diversity accumulated in biofilm populations includes spontaneous genetic mutants that are resistant to antibiotics (Boles and Singh 2008; Driffield et al. 2008; Ponciano et al. 2009; Tyerman et al. 2013). Likewise, our own unpublished data from studies on *Escherichia coli* biofilms suggests this observation may be true for a wide array of antibiotics (Fig. 8.1). Mutants resistant to kanamycin, chloramphenicol, cycloserine, and tetracycline spontaneously increased in frequency by more than two orders of magnitude over the course of 15 days of growth in the absence of antibiotics, as compared to the frequency in the inoculum. Future efforts to study microbial evolution in biofilms are likely to reveal the mechanisms behind the accumulation of these antibiotic-resistant variants.

2 Effect of Spatial Structure on Microbial Evolution

Populations of bacteria housed within biofilms exhibit inherent spatial structure (Nadell et al. 2016). The extracellular matrix that encloses a biofilm provides the population with a three-dimensional configuration and restricts the movement of individual cells (Flemming and Wingender 2010). Studies on the evolution and ecology of macroorganisms have demonstrated the profound effects of spatial structure on adaptation and diversity (Tillman 1994; Pannell and Fields 2014). The key findings from these studies are the following: (1) spatial structure promotes diversity; (2) interactions among and between species occur within localized niches; (3) organisms adapt at a local scale; and (4) forces such as migration, gene flow, and genetic drift can either promote or retard local adaptation, depending on specific

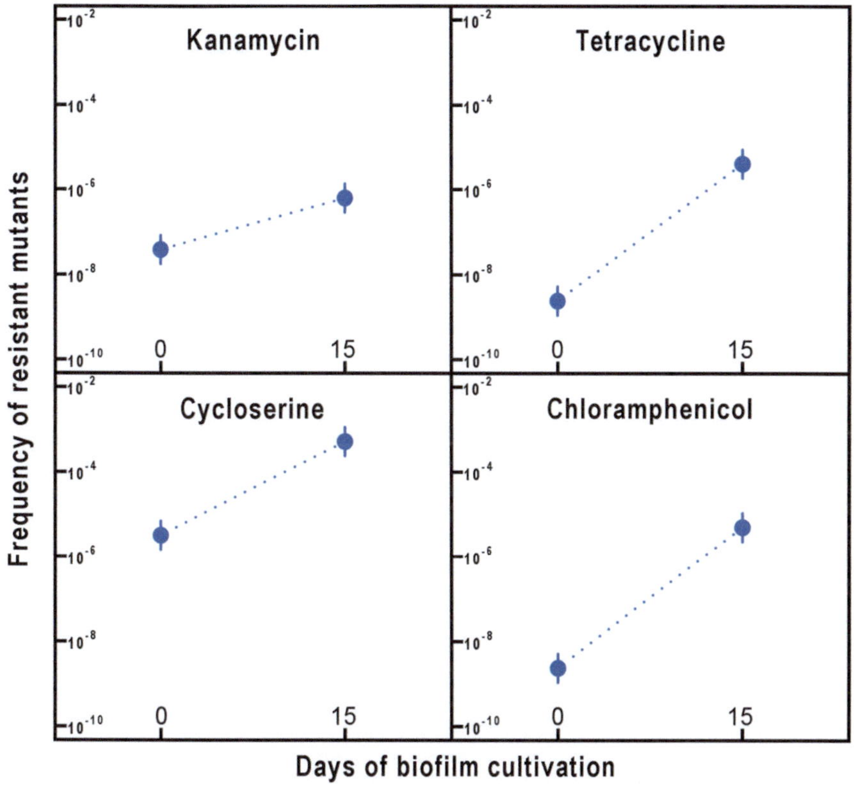

Fig. 8.1 Increase in the frequency of antibiotic-resistant mutants in a single biofilm population of *Escherichia coli* MG1655 during 15 days of antibiotic-free cultivation. Biofilms were grown in flow cells as described in Ponciano et al. (2009). Resistance was determined by plating on selective media containing one of the following antibiotic concentrations (kanamycin 25 μg/mL, tetracycline 10 μg/mL, cycloserine 40 μM, chloramphenicol 25 μg/mL)

circumstances. The limited information available on the evolution of microbial populations in spatially structured environments suggests that many of these findings will apply equally well to populations of microorganisms.

2.1 Spatial Structure and the Maintenance of Diversity

Previous studies have shown that spatial structure has a profound impact on the evolution of microbial populations (Rainey and Travisano 1998; Perfeito et al. 2008; Kryazhimskiy et al. 2012; Nahum et al. 2015). In structured populations the position of individuals within the spatial landscape matters because competitive interactions do not occur on a global scale like they do in well-mixed populations (Kerr et al. 2002; Kim et al. 2014). Instead, individual cells only compete for resources with a

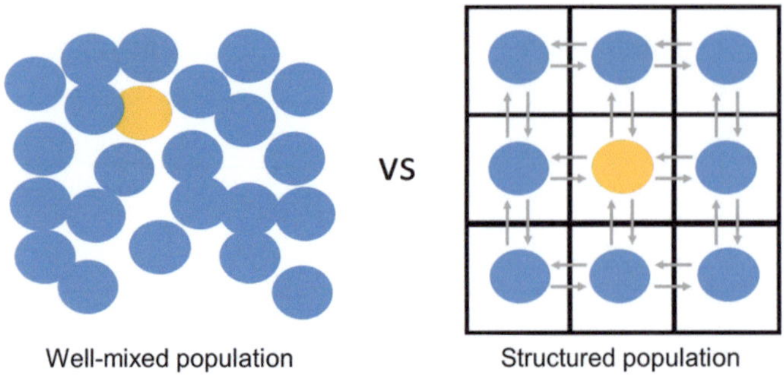

Fig. 8.2 Simplified example of competition in well-mixed versus structured populations. In well-mixed populations, competition occurs at a global scale, and the focal (yellow) individual must compete against the entire population. While in structured population, competition is limited to a local scale, and the focal individual (yellow) only competes against the subset of the population that is physically near it

small subset of the population that is in close physical proximity (Fig. 8.2). This means that the relative fitness of a given genotype is dependent not on the average fitness of the entire population, but rather just a subset of cells that are in the immediate neighborhood. Thus, spatial structure helps maintain diversity in microbial populations by reducing the scale of competitive interactions. This was elegantly demonstrated by Kerr et al. (2002) who showed that when toxin-producing, toxin-resistant, and toxin-susceptible strains of *Escherichia coli* compete on the surface of an agar plate (a spatially structured environment), then all three strains persisted. In contrast, when environmental spatial structure was removed by growing the populations in well-mixed cultures, then only a single genotype—the toxin-resistant strain—persisted (Kerr et al. 2002). The fate of only three genotypes was tracked in this study. Yet mutational processes in naturally occurring biofilm populations that have very large population sizes are likely to create a plethora of genotypes, many of which might coexist due to spatial structure. We postulate that spatial structure is likely to play a critical role in the maintenance of this diversity through time. This is consistent with the results from several studies that have already shown that biofilms tend to accumulate extraordinary genetic diversity over relatively short periods of time (Fig. 8.3; Boles et al. 2004; Ponciano et al. 2009; Tyerman et al. 2013).

2.2 Spatial Structure and Clonal Interference

Previous studies have shown that spatial structure can increase the amount of time required for beneficial mutations to fix (Whitlock 2003; Habets et al. 2007; Perfeito et al. 2008). Instead of rapidly sweeping through the entire population all at once, beneficial mutations in biofilms must sweep through the physical space as well,

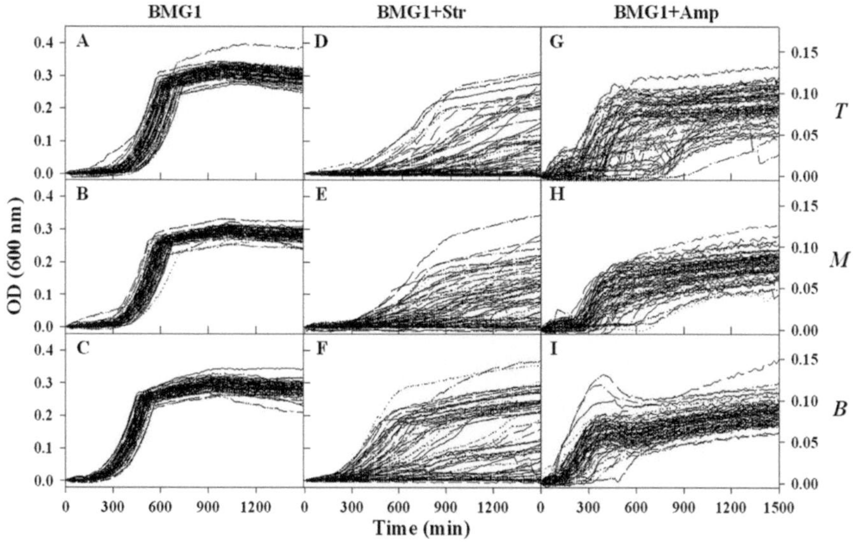

Fig. 8.3 Diversity in growth kinetics displayed by 192 clones isolated from a single antibiotic naïve *Escherichia coli* k12 MG1655 biofilm. Sixty-four clones were isolated from the top (2250 to 2400 μm; **a**, **d**, and **g**), middle (1050 to 1200 μm; **b**, **e**, and **h**), and bottom (0 to 100 μm; **c**, **f**, and **i**) of a biofilm. The growth of each clone was tested in three different media, two of which contain sublethal antibiotic concentrations: BMG1 (panels **a**, **b**, and **c**), BMG1 + streptomycin (5 μg/mL; panels **d**, **e**, and **f**), and BMG1 + ampicillin (5 μg/mL; panels **g**, **h**, and **i**). OD, optical density. Figure reproduced from: Ponciano, J. M., La, H., Joyce, P., Forney, L. J. (2009) Evolution of diversity in spatially structured *Escherichia coli* populations. Appl. Environ. Microbiol. 75(19): 6047–6054

thereby slowing the overall increase in biofilm population fitness (Nahum et al. 2015). This was clearly shown in the work of Perfeito et al. (2008). They demonstrated that the increase in fitness after 275 generations of evolution was significantly lower in *Escherichia coli* populations that were grown on the surface of an agar medium as compared to populations grown in a liquid medium (Perfeito et al. 2008). This stands in contrast to the idealized process of evolution in asexual populations wherein adaptation is driven by a series of beneficial mutations that occur and sweep to high frequency in the population one at a time (Elena et al. 1996), with each new mutation building on the fitness gains provided by prior mutations. In reality, the events seem to play out differently in large populations of microorganisms, and there is increasing evidence that multiple beneficial mutations arise simultaneously and compete against each other (Lang et al. 2013). This effect, known as clonal interference (Sniegowski and Gerrish 2010), has been theorized to slow the rate of adaption by creating "wasted genetic potential" (Muller 1932). We posit that spatial structure is likely to enhance and amplify the effects of clonal interference in biofilm populations. This can be attributed to the increased amount of time required for a beneficial mutation to sweep through a biofilm population, which provides an

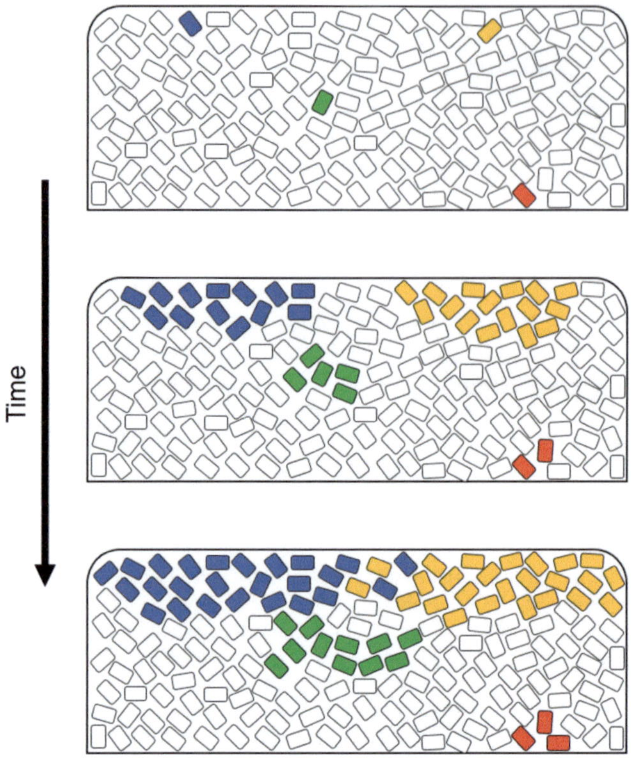

Fig. 8.4 Spatial structure causes the beneficial mutations colored in blue, yellow, red, and green to sweep locally through the biofilm. In a well-mixed population, the beneficial mutation with the highest fitness would rapidly sweep through the population, thereby collapsing genetic diversity

opportunity for new beneficial mutations to emerge. As a consequence, there are likely countless beneficial mutations present in a biofilm at any given time (Fig. 8.4).

While the clonal interference that results from evolution in structured environments is generally predicted to slow adaptation in the short term, it may ultimately allow members of the population to attain higher fitness values in the long run. This "tortoise and hare" effect (Nahum et al. 2015) is best visualized using Sewall Wright's fitness landscape, which relates genotypic space to fitness (Wright 1932). When epistasis is rare and the fitness effects of mutations are largely additive, the landscape is a smooth surface (Fig. 8.5a). Conversely, when epistasis is common, the landscape becomes "rugged" with sharp transitions in fitness over few mutational steps (Fig. 8.5b). In the absence of spatial structure, adaptation is expected to drive the entire population up the nearest fitness peak (Fig. 8.5a; Gillespie 1983). In contrast, spatial structure effectively fragments populations into metapopulations that simultaneously adapt along multiple paths in the fitness landscape (Wakeley 1998; Cherry and Wakeley 2003). If the fitness landscape is smooth, no benefit is derived from increased exploration as the metapopulations are all expected to climb

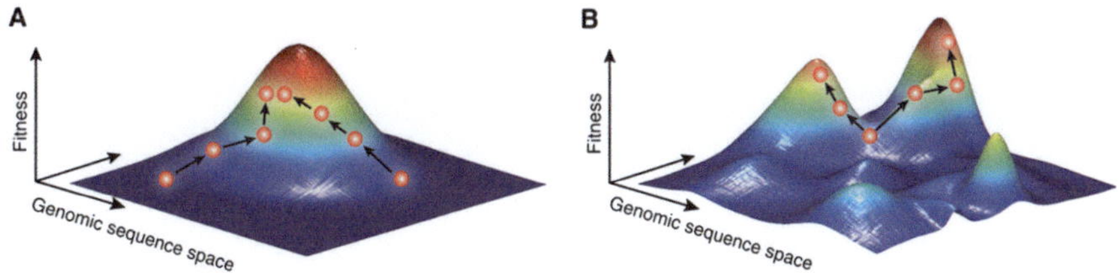

Fig. 8.5 Example of smooth (**a**) and rugged (**b**) fitness landscapes. For a smooth landscape with a single fitness peak (**a**), sampling more of the landscape is not beneficial because all paths lead to the same location. Whereas, for a rugged landscape with multiple peaks (**b**), spatial structure allows a population to simultaneously climb multiple fitness peaks, increasing the likelihood that the population reaches higher peaks. Figure reproduced from: Conrad. T.M., Lewis, N.E., Palsson, B. Ø. (2011) Microbial laboratory evolution in the era of genome-scale science. Molecular Systems Biology 7:509

the same fitness peak. However, if the fitness landscape is rugged, the largely independent exploration of the landscape can prevent the population from becoming stuck at local fitness optima and instead allow certain metapopulations to reach fitness peaks that might otherwise be unattainable (Fig. 8.5b; Burch and Chao 2000; Rozen et al. 2008; Nahum et al. 2015).

2.3 *Emergence of Antagonistic Pleiotropy in Biofilms*

Spatial structure creates environmental heterogeneity in biofilm populations (Xu et al. 1998; Stewart and Franklin 2008; Stewart et al. 2016). Instead of being uniformly distributed as they are in well-mixed populations, the availability of nutrients and terminal electron acceptors in biofilms is altered by reaction-diffusion processes (Erban and Chapman 2007). As molecules diffuse into the biofilm structure, they are consumed by bacterial cells creating environmental gradients wherein resource concentrations are highest at the interface with the bulk media (exterior) and lowest in interior regions of biofilms (Fig. 8.6; Stewart 1998). This heterogeneity is further compounded by the complex three-dimensional structure of biofilms and the fact that biofilms are often comprised of multiple species with distinctive physiologies and a patchy distribution (Tolker-Nielsen and Molin 2000; Webster et al. 2006; Elias and Banin 2012; Wierzchos et al. 2015). This environmental heterogeneity drives adaptation in response to selective pressures that are localized and specific to individual microenvironments (Kraemer and Boynton 2017). For example, mutations that improve aerobic metabolism may be selected for in the upper reaches of a biofilm where oxygen is readily available but selected against in the lower, anaerobic parts of a biofilm (de Beer et al. 1994; Stewart and Franklin 2008).

Not surprisingly, mutations that are beneficial in one environment might be detrimental in another. Such instances of antagonistic pleiotropy have been shown

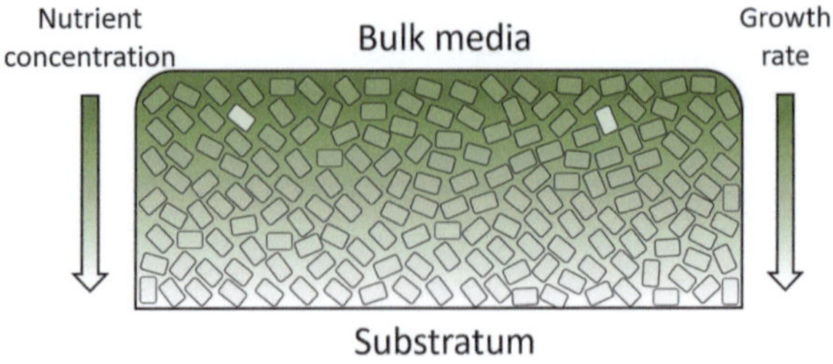

Fig. 8.6 Simple schematic of the distribution of nutrient concentrations and growth rates in a theoretical biofilm population

to be relatively common in nature (Anderson et al. 2013; Kassen 2014; Schenk et al. 2015; Ferenci 2016). Should antagonistic pleiotropy be common in biofilms, the complex array of microenvironments would be expected to fragment the population into multiple subpopulations (metapopulations). Antagonistic pleiotropy would drive these metapopulations to diverge and evolve toward specialization in their particular environment (Futuyma and Moreno 1988; Devictor et al. 2010). The effect of local adaption to specific microenvironments is also likely to be amplified by the spatial structure of biofilm populations, which can prevent "entire biofilm" selective sweeps during reasonable time periods. Indeed, population-wide selective sweeps in biofilms may only result from strong directional selective pressures like that provided by the application of antibiotics.

3 Effect of Growth Rate on Mutation and Selection

Variation in cellular growth rates is an important consequence of the complex array of environments found in biofilms (Sternberg et al. 1999; Stewart and Franklin 2008; Stewart et al. 2016). Those cells located at or near the source of incoming nutrients are likely to exhibit higher growth rates than those located in more nutrient-depleted zones. Large swaths of the interior of biofilms may not be growing (dividing) at all and exist in a quiescent state. Although it has been chronically overlooked, we suggest that a microbial population's growth rate is likely to strongly influence its evolution. While mutations can occur independent of chromosomal replication and cellular division, selection requires some combination of cellular growth and death to affect evolutionary changes in microbial populations. The relative influence of these two processes, mutation and selection, may therefore vary with the rate at which the population is growing.

3.1 Replication-Dependent and Replication-Independent Mutational Processes

Mutation, unlike selection, can occur independent of cell growth and death. The mutations that underpin evolutionary change in biofilm populations fall into two

Table 8.1 Processes that cause mutations independent of genome duplication

Mutational mechanism	References
Repair of DNA damage by:	
UV radiation	Bridges (1992), Truglio et al. (2006)
Gamma radiation	Wijker et al. (1998), Rodgers and Mcvey (2016)
Cytosine deamination	Duncan (1980), Frederico et al. (1993)
Depurination	Schaaper and Loeb (1980), Suzuki et al. (1994)
DNA oxidation	Imlay (2003), Bjelland (2003)
Nonenzymatic methylation	Rydberg and Lindahl (1982), Mazin et al. (1985)
Movement of mobile elements	Kidwell and Lisch (2001), Frost et al. (2005)
Recombination	Anderson and Roth (1981), Bull et al. (2001)
Slipped-strand mispairing	Levinson and Gutman (1987), Torres-cruz and Van Der Woude (2003)

broad categories: *replication-dependent* and *replication-independent* mutations. The former largely consist of point mutations introduced during the replication of bacterial chromosomes in actively growing cells, while the latter arise through several different mechanisms (Table 8.1), including error-prone repair of damaged DNA (Goodman 2002; Rodgers and Mcvey 2016), recombination (Anderson and Roth 1981; Bull et al. 2001), movement of mobile genetic elements (Frost et al. 2005; Foster 2007), and slipped-strand mispairing (Levinson and Gutman 1987; Torres-cruz and Van Der Woude 2003). These replication-independent mutations occur in all cells regardless of their growth rate, probably even in those that are quiescent (Bull et al. 2001; Kivisaar 2003, 2010). This phenomenon can be seen in Fig. 8.7, which demonstrates a gradual linear increase in rifampicin-resistant mutants in stationary-phase populations of *E. coli*.

The failure to adequately consider replication-independent mutations is reflected in the fact that mutation rates are expressed in terms of "mutations per generation" which by logical extension implies that if there are no generations (i.e., cells are not replicating their chromosome and dividing), then the mutation rate is undefined by virtue of dividing by zero. Additionally, the vast majority of experiments done to measure mutation rates are done in studies in which mutation rates are determined using exponentially growing cells (Rosche and Foster 2000; Foster 2006). This could be misleading since microbes in many natural habitats grow slowly or episodically and exponential growth is uncommon (Debellis et al. 1998; Dixon and Turley 2001; Rousk and Bååth 2011; Kirchman 2016). Harris and Paul (1994) estimated the generation time of bacteria in agricultural soils to be 160 days and 107 days in grassland soils. Other investigators have estimated the generation times of bacteria in pelagic marine environments to be 8–9 days (Carlucci and Williams 1978), while Lomstein et al. (2012) have projected biomass turnover times of hundreds to thousands of years in deep sub-seafloor sediments.

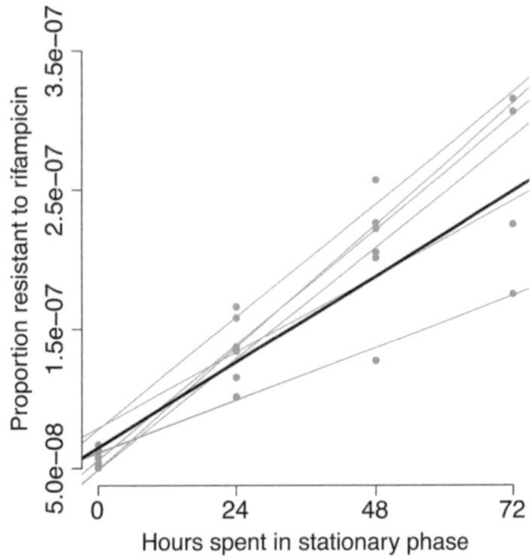

Fig. 8.7 Increase in the frequency of rifampicin-resistant mutants in seven planktonic stationary-phase cultures of *E. coli*. Gray points are estimates of the frequency from each individual culture at each time point. Gray and black lines resulted from a simple linear regression of the mutant frequency versus the time spent in stationary phase for the individual cultures (gray) and the overall average of all cultures (black)

Given the low bacterial growth rates in most habitats, it could be that replication-independent mutations are the principal means by which genetic diversification occurs. Thus, in studies of evolution in biofilms, it may be important to consider both replication-dependent and replication-independent mutations (Eastman et al. 2011) because growth rates may vary depending on the location of cells in the matrix. Lastly, because of the potentially important role of mutation in slowly growing or quiescent cells, we suggest that mutation rates be expressed as mutations per unit time instead of mutations per generation.

3.2 Strength of Selection When Growing Slowly

On the other hand, cell growth or death must occur for selection to affect changes in allele frequencies in the population. This implies that the rate at which a population is growing (or dying) has a direct influence on the rate at which selection can alter allele frequencies. Consider the following simplified example: if a beneficial mutation requires 250 generations to reach fixation, then this would require 25 days in a population that has a growth rate of 10 generations of per day. In contrast, fixation would require 1750 days in a population that experiences only one generation per week. When applied to biofilms, this implies that the times required for beneficial mutations to sweep may well depend on their location in the matrix, with fixation occurring more quickly near the interface with the bulk media and much more slowly in deeper regions near the substratum. Moreover, as previously discussed, many of the molecular mechanisms that cause mutations are not bound to chromosomal replication or cellular division. Perhaps mutation rates might be best expressed as a continuous time process rather than instead of being indexed to the production of offspring—an event that is required for natural selection to operate. By extension of this reasoning, the effect of selection is undetectable in populations that are in stasis (i.e., cells are neither replicating nor dying).

The connection between growth rate and selection in bacterial populations can be made clear by examining mathematical models that relate mutation and selection in well-mixed populations. If we ignore gene flow and genetic drift, a deleterious mutant accumulates at a rate defined in terms of mutation, which provides new variants, and selection, which purges lower-fitness mutants from the population. Traditional models (Eq. 8.1) define the change in mutant abundance (Δx) as being approximately equal to the mutation rate μ minus the selection coefficient s multiplied by the current proportion of mutants x (Fisher 1928), i.e.:

$$\begin{aligned} \Delta x &\approx \mu - sx \\ x_t &= (1 - s)x_{t-1} + \mu. \end{aligned} \tag{8.1}$$

The selection coefficient s in this model is defined as the as covariance of x and its fitness (measured at time t). If we assume the time step between $t - 1$ and t is 1 (note that the units here are not specified; the usual choice would be one generation), we can break the mutation rate up into two portions: (1) k is the portion of the time step during which chromosomal replication is not occurring, and (2) $(1 - k)$ is the proportion of the time step where chromosomal replication occurs. Now assume that replication-dependent and replication-independent mutations occur at rates η and ν, respectively. Then:

$$\begin{aligned} \mu &\equiv \nu k + \eta(1 - k) \\ \therefore x_t &= (1 - s)x_{t-1} + \nu k + \eta(1 - k). \end{aligned} \tag{8.2}$$

Equation (8.2) highlights the balance between the two types of mutation. If k is large (as it might be in slow-growing populations), then ν has increased importance in evolution. Conversely, if most of the time the bacteria are undergoing chromosomal replication (i.e., k is small), then η has increased importance. Note that, because the selection coefficient is estimated at unit length, the change due to natural selection is of fixed size in comparison.

This analysis suggests that it is important to consider the possible effect of growth rate on selection when thinking about evolution in biofilm populations. This matter might be important for at least two reasons. First, there is large variation in growth rates throughout a biofilm (Wentland et al. 1996; Sternberg et al. 1999), so it is likely that the relative contribution of replication-dependent and replication-independent mutations varies through the population (Eastman et al. 2011). Specifically, cells near the bulk fluid interface are likely to experience faster growth rates and may therefore incur more growth-dependent mutations (Sternberg et al. 1999), while nutrient-starved cells may incur a higher proportion of replication-independent mutations (Bull et al. 2001; Saumaa et al. 2002; Kivisaar 2003). Second, cells in the basal layers of biofilms experience conditions that severely limit or preclude growth (shown in Fig. 8.6). The influence of selection on these slowly growing or quiescent populations is likely minimal ($s \sim 0$). Yet mutations are expected to accumulate unabatedly through replication-independent processes.

4 Horizontal Gene Transfer in Biofilms

Genetic diversity can also be created by the introduction of genetic elements into cells (Thomas and Nielsen 2005). This process is termed horizontal gene transfer (HGT), and it stands in contrast to the vertical inheritance of genes through common descent (Thomas and Nielsen 2005). HGT is an important component of microbial evolution because it allows for the rapid dissemination of genetic information. Comparative genomics studies have shown that the genomes of most microorganisms include considerable amounts of genetic material have been acquired through HGT and not by common descent (Ochman et al. 2000). Considerable attention has been given to HGT because it has played a critical role in the acquisition of virulence determinants and the spread of antibiotic resistance (Davies and Davies 2010). Below we describe how residing within a biofilm might influence HGT. We focus our attention on conjugation and transformation because relatively few studies have examined transduction in biofilms.

Conjugation is probably the HGT process most studied in biofilms. Many have argued that conjugation rates are likely to be elevated in densely populated biofilms because cell-cell contact is required for plasmid transfer, and immobilization in the extracellular matrix may facilitate successful mating by decreasing the chances that the donor and recipient conjugation machinery is not disrupted by sheer forces (Hausner and Wuertz 1999). On the other hand, spatial structure may slow and limit plasmid spread in biofilms through the same mechanisms that slow selective sweeps (Fig. 8.8). Correspondingly, some studies have indicated that plasmid transfer rates are elevated in biofilms (Hausner and Wuertz 1999; Kajiura et al. 2006; Van Meervenne et al. 2014), while others have suggested the exact opposite (Christensen et al. 1996, 1998). Arguments against elevated transfer rates in biofilms

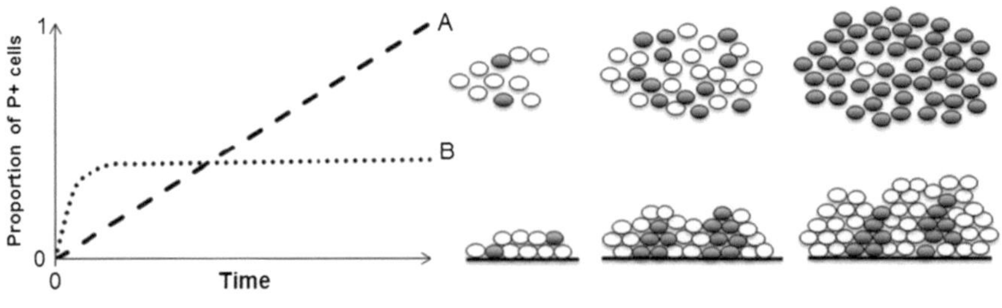

Fig. 8.8 Theoretical spread of a self-transmissible plasmid in a well-mixed (**a**) and structured population (**b**). This is demonstrated schematically in the drawing on the right. Plasmid-free bacteria (white) and plasmid-containing bacteria (dark gray) in well-mixed (top) and structured (bottom) populations. Figure is reproduced from: Stalder, T., Top, E. (2016) Plasmid transfer in biofilms: a perspective on limitations and opportunities. NPJ Biofilms and Microbiomes 2: 16022

are based on either the need for cell-cell contact that may limit plasmid transfer to nearest neighbors in the biofilm matrix (Tolker-Nielsen and Molin 2000; Seoane et al. 2011) or the depletion of energy sources (nutrients) in the deeper regions of a biofilm (Fox et al. 2008). Furthermore, Król et al. (2011) demonstrated that conjugal transfer of IncP-1 in *E. coli* biofilms was dependent on oxygen levels and population densities (Król et al. 2011). This is was supported by the findings of Stalder and Top (2016) who observed limited plasmid transfer and spread in mixed-species biofilms of *E. coli* and *P. putida* (Fig. 8.9; Stalder and Top 2016). Additional studies are needed to define the conditions under which conjugation is favored or limited in biofilm populations.

Transformation occurs when extracellular DNA (eDNA) is taken up by another bacterium and integrated into the chromosome. This process may be a prominent mechanism of horizontal gene transfer in microbial biofilms because eDNA has been shown to be a common component of the extracellular matrix (Molin and Tolker-Nielsen 2003; Chiang and Tolker-Nielsen 2010). Extracellular DNA is thought to

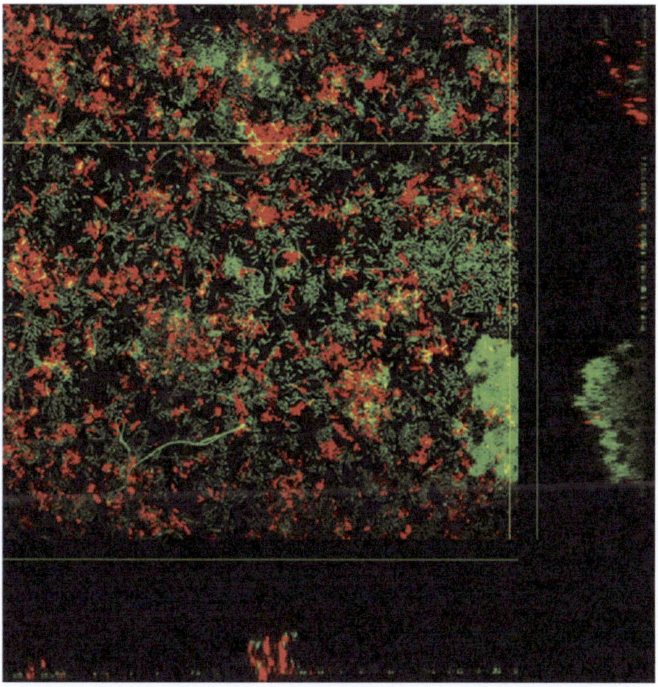

Fig. 8.9 Confocal laser scanning microscopy photographs of plasmid transfer in a dual-species biofilm. Plasmid donor cells (red) are *E. coli* K12 MG1655 carrying plasmid pB10 marked with *dsRed*, and recipient cells (green) are *P. putida* KT2244 marked with *gfp*. Transfer of the dsRed-marked plasmid from *E. coli* into the gfp-marked P. putida generates the yellow/orange transconjugants seen in the image. Figure reproduced from: Stalder, T., Top, E. (2016) Plasmid transfer in biofilms: a perspective on limitations and opportunities. NPJ Biofilms and Microbiomes 2: 16022

play a structural role in biofilms but may also be taken up by the bacteria embedded in the matrix (Whitchurch et al. 2002; Hobley et al. 2015). Furthermore, studies have identified transformation in biofilms formed by *Streptococcus mutants* (Li et al. 2001), *Acinetobacter* spp. (Hendrickx et al. 2003; Merod and Wuertz 2014), *Gonococcus* sp. (Kouzel et al. 2015), and multispecies oral biofilms (Hannan et al. 2010). The study on *Gonococcus* biofilms by Nadzeya et al. further demonstrated that transformation rates were higher in early stage biofilms than in planktonic populations. However, they also showed that transformed rates were diminished in the biofilm populations after 24 h, a result that was also observed in the studies on *Acinetobacter* sp. and *Streptococcus mutants* (Li et al. 2001; Hendrickx et al. 2003). The mechanism responsible for this trend has not been identified, although Nadzeya et al. suggest that it could result from either oxygen deprivation or the increased matrix density of older biofilms (Kouzel et al. 2015). While it is still not clear whether transformation rates are generally elevated or reduced in biofilms, we speculate that this may be an important and underappreciated means of HGT in biofilms. We base our speculation on the abundance of eDNA in the matrix and the fact that many environmental biofilms are complex multispecies conglomerates.

5 Conclusions

Despite the fact that few studies have been done on the evolution of populations that reside in biofilms, we have synthesized the available findings along with principles of macroecology and evolutionary theory to speculate about the factors that influence adaptive radiation and evolution in biofilms. In our view, four fundamental characteristics of biofilms must be taken into account. The first is that the extracellular matrix restricts the movement of individual bacteria within biofilms, limiting the scale of competitive interactions. Second, reaction-diffusion processes within biofilms create microenvironments that exert a spectrum of selective pressures. Third, mutagenic processes that are independent of chromosome replication and cell division undoubtedly occur, which leads to genetic diversification of cells, even those that are growing slowly or not at all. Finally, the growth rates of certain metapopulations within the biofilm are greatly reduced, which can protract selective sweeps thus allowing clonal interference and the persistence of genetic variants. These characteristics lead to the emergence and maintenance of genetic diversity within biofilms and create metapopulations that, by chance, may survive better if environmental conditions change. Our observations should apply equally well to other spatially structured, non-biofilm environments such as soils and sediments.

It has been proposed that adaptive evolution in bacterial (haploid) populations is driven by strong selection and weak mutation (SSWM; Orr 2002; Joyce et al. 2008). In this scenario, there is a finite number of rare mutations that can only be beneficial or deleterious. Selection is strong enough that the entire population is essentially composed of only one genotype and adaption occurs through the stepwise fixation of single mutations. In many ways, this scenario captures our understanding of how evolution operates in microbial populations residing in homogeneous, unstructured environments. Here, we propose that evolution in biofilms and other spatially structured environments is more akin to weak selection and strong mutation; a conceptual model that we refer to as "anti-SWWM." Perhaps the ideas put forward here can be used in the future to guide the development of new models of bacterial evolution in spatially structured environments and a better understanding of the extraordinary diversity found in the microbial world.

References

Adams J, Rosenzweig F (2014) Experimental microbial evolution: history and conceptual underpinnings. Genomics 104:393–398

Amato SM, Fazen CH, Henry TC et al (2014) The role of metabolism in bacterial persistence. Front Microbiol 5:1–9

Anderson P, Roth J (1981) Spontaneous tandem genetic duplications in *Salmonella typhimurium* arise by unequal recombination between rRNA (rrn) cistrons. Proc Natl Acad Sci USA 78:3113–3117

Anderson JT, Lee C, Rushworth CA et al (2013) Genetic trade-offs and conditional neutrality contribute to local adaptation. Mol Ecol 22:699–708

Barton NH, Briggs DEG, Eisen JA et al (2007) Evolution, 1st edn. Cold Spring Harbor Laboratory Press, New York

Battin TJ, Kaplan LA, Denis Newbold J, Hansen CME (2003) Contributions of microbial biofilms to ecosystem processes in stream mesocosms. Nature 426:439–442

Besemer K, Singer G, Hödl I, Battin TJ (2009) Bacterial community composition of stream biofilms in spatially variable-flow environments. Appl Environ Microbiol 75:7189–7195

Bjelland S (2003) Mutagenicity, toxicity and repair of DNA base damage induced by oxidation. Mutat Res 531:37–80

Boles BR, Singh PK (2008) Endogenous oxidative stress produces diversity and adaptability in biofilm communities. Proc Natl Acad Sci USA 105:12503–12508

Boles BR, Thoendel M, Singh PK (2004) Self-generated diversity produces "insurance effects" in biofilm communities. Proc Natl Acad Sci USA 101:16630–16635

Brauner A, Fridman O, Gefen O, Balaban NQ (2016) Distinguishing between resistance, tolerance and persistence to antibiotic treatment. Nat Rev Microbiol 14:320–330

Bridges B (1992) Mutagenesis after exposure of bacteria to ultraviolet-light and delayed photoreversal. Mol Gen Genet 233:331–336

Brooks JD, Flint SH (2008) Biofilms in the food industry: problems and potential solutions. Int J Food Sci Technol 43:2163–2176

Bull HJ, Lombardo MJ, Rosenberg SM (2001) Stationary-phase mutation in the bacterial chromosome: recombination protein and DNA polymerase IV dependence. Proc Natl Acad Sci USA 98:8334–8341

Burch CL, Chao L (2000) Evolvability of an RNA virus is determined by its mutational neighbourhood. Nature 406:625–628

Burmolle M, Kjoller A, Sorensen SJ (2012) An invisible workforce: biofilms in the soil. In: Lear G, Lewis GD (eds) Microbial biofilm: current research and applications. Caister Academic Press, Norfolk, pp 61–71

Carlucci AF, Williams PM (1978) Simulated in situ growth rates of pelagic marine bacteria. Naturwissenschaften 65:541–542

Cherry JL, Wakeley J (2003) A diffusion approximation for selection and drift in a subdivided population. Genetics 163:421–428

Chiang W, Tolker-Nielsen T (2010) Extracellular DNA as a matrix component in microbial biofilms. In: Kikuchi Y, Rykova E (eds) Extracellular nucleic acids. Springer, Berlin, pp 1–14

Christensen BB, Sternberg C, Molin S (1996) Bacterial plasmid conjugation on semi-solid surfaces monitored with the green fluorescent protein (GFP) from *Aequorea victoria* as a marker. Gene 173:59–65

Christensen BB, Sternberg C, Andersen JB et al (1998) Establishment of new genetic trait in a microbial biofilm community. Appl Environ Microbiol 64:2247–2255

Cohan FM (2016) Bacterial speciation: genetic sweeps in bacterial species. Curr Biol 26: R112–R115

Costerton JW, Geesey GG, Cheng K-J (1978) How bacteria stick. Sci Am 238:86–95

Davies J, Davies D (2010) Origins and evolution of antibiotic resistance. Microbiol Mol Biol Rev 74:417–433

de Beer D, Stoodley P, Roe F, Lewandowski Z (1994) Effects of biofilm structures on oxygen distribution and mass-transport. Biotechnol Bioeng 43:1131–1138

Debellis T, Kernaghan G, Bradley R, Widden P (1998) Growth rate of bacterial communities in soils at varying pH: a comparison of the thymidine and leucine incorporation techniques. Microb Ecol 36:316–327

Devictor V, Clavel J, Julliard R et al (2010) Defining and measuring ecological specialization. J Appl Ecol 47:15–25

Dixon JL, Turley CM (2001) Measuring bacterial production in deep-sea sediments using 3H-thymidine incorporation: ecological significance. Microb Ecol 42:549–561

Domenech M, Ramos-Sevillano E, García E et al (2013) Biofilm formation avoids complement immunity and phagocytosis of *Streptococcus pneumoniae*. Infect Immun 81:2606–2615

Driffield K, Miller K, Bostock JM et al (2008) Increased mutability of *Pseudomonas aeruginosa* in biofilms. J Antimicrob Chemother 61(5):1053–1056

Duncan B (1980) Mutagenic deamination of cytosine residues in DNA. Nature 287:560–561

Dykhuizen DE (1998) Santa Rosalia revisited: why are there so many species of bacteria? Antonie van Leeuwenhoek 73:25–33

Eastman JM, Harmon LJ, La HJ et al (2011) The onion model, a simple neutral model for the evolution of diversity in bacterial biofilms. J Evol Biol 24:2496–2504

Elena SF, Cooper VS, Lenski RE (1996) Punctuated evolution caused by selection of rare beneficial mutations. Science 272:1802–1804

Elias S, Banin E (2012) Multi-species biofilms: living with friendly neighbors. FEMS Microbiol Rev 36(5):990–1004. https://doi.org/10.1111/j.1574-6976.2012.00325.x

Ellis CN, Traverse CC, Mayo-smith L et al (2015) Character displacement and the evolution of niche complementarity in a model biofilm community. Evolution (NY) 69:283–293

Erban R, Chapman SJ (2007) Reactive boundary conditions for stochastic simulations of reaction-diffusion processes. Phys Biol 4:16–28

Ferenci T (2016) Trade-off mechanisms shaping the diversity of bacteria. Trends Microbiol 24: 209–223

Fisher RRA (1928) The possible modification of the response of the wild type to recurrent mutations. Am Nat 62:115–116

Flemming H-C (2011) Microbial biofouling: unsolved problems, insufficient approaches, and possible solutions. In: Flemming H-C, Wingender J, Szewzyk U (eds) Biofilm highlights. Springer, Berlin, pp 81–109

Flemming HC, Wingender J (2010) The biofilm matrix. Nat Rev Microbiol 8:623–633

Flemming H-C, Wingender J, Szewzyk U et al (2016) Biofilms: an emergent form of bacterial life. Nat Rev Microbiol 14:563–575

Foster PL (2006) Methods for determining spontaneous mutation rates. Methods Enzymol 409: 195–213

Foster PL (2007) Stress-induced mutagenesis in bacteria. Crit Rev Biochem Mol Biol 42:373–397

Fox RE, Zhong X, Krone SM, Top EM (2008) Spatial structure and nutrients promote invasion of IncP-1 plasmids in bacterial populations. ISME J 2:1024–1039

Frederico LA, Kunkel TA, Shaw BR (1993) Cytosine deamination in mismatched base pairs. Biochemistry 32:6523–6530
Frost LS, Leplae R, Summers AO, Toussaint A (2005) Mobile genetic elements: the agents of open source evolution. Nat Rev Microbiol 3:722–732
Futuyma DJ, Moreno G (1988) The evolution of ecological specialization. Annu Rev Ecol Syst 19: 207–233
Gause GF (1934) The struggle for existence. Williams & Wilkins, Baltimore, MD
Gillespie JH (1983) Some properties of finite populations experiencing strong selection and weak mutation. Am Nat 121:691–708
Goodman MF (2002) Error-prone repair DNA polymerases in prokaryotes and eukaryotes. Annu Rev Biochem 71:17–50
Habets MGJL, Czárán T, Hoekstra RF, de Visser JAGM (2007) Spatial structure inhibits the rate of invasion of beneficial mutations in asexual populations. Proc R Soc B Biol Sci 274:2139–2143
Hall-Stoodley L, Stoodley P (2009) Evolving concepts in biofilm infections. Cell Microbiol 11: 1034–1043
Hannan S, Ready D, Jasni AS et al (2010) Transfer of antibiotic resistance by transformation with eDNA within oral biofilms. FEMS Immunol Med Microbiol 59:345–349
Harris D, Paul EA (1994) Measurement of bacterial growth rates in soil. Appl Soil Ecol 1:277–290
Hausner M, Wuertz S (1999) High rates of conjugation in bacterial biofilms as determined by quantitative in situ analysis. Appl Environ Microbiol 65:3710–3713
Hendrickx L, Hausner M, Wuertz S (2003) Natural genetic transformation in monoculture *Acinetobacter* sp. Appl Environ Microbiol 69:1721–1727
Hobley L, Harkins C, MacPhee CE, Stanley-Wall NR (2015) Giving structure to the biofilm matrix: an overview of individual strategies and emerging common themes. FEMS Microbiol Rev 39: 649–669
Høiby N, Bjarnsholt T, Givskov M et al (2010) Antibiotic resistance of bacterial biofilms. Int J Antimicrob Agents 35:322–332
Høiby N, Ciofu O, Johansen HK et al (2011) The clinical impact of bacterial biofilms. Int J Oral Sci 3:55–65
Hoyle BD, Alcantara J, Costerton JW (1992) *Pseudomonas aeruginosa* biofilm as a diffusion barrier to piperacillin. Antimicrob Agents Chemother 36:2054–2056
Imlay JA (2003) Pathways of oxidative damage. Annu Rev Microbiol 57:395–418
James GA, Swogger E, Wolcott R et al (2007) Biofilms in chronic wounds. Wound Repair Regen 16:37–44
Jannasch HW (1969) Estimations of bacterial growth rates in natural waters. J Bacteriol 99:156–160
Jensen PØ, Givskov M, Bjarnsholt T, Moser C (2010) The immune system vs. *Pseudomonas aeruginosa* biofilms. FEMS Immunol Med Microbiol 59:292–305
Joyce P, Rokyta DR, Beisel CJ, Orr HA (2008) A general extreme value theory model for the adaptation of DNA sequences under strong selection and weak mutation. Genetics 180: 1627–1643
Kajiura T, Wada H, Ito K et al (2006) Conjugative plasmid transfer in the biofilm formed by *Enterococcus faecalis*. J Heal Sci 52:358–367
Kassen R (2014) Experimental evolution and the nature of biodiversity. Roberts, Greenwood Village, CO
Kawecki TJ, Lenski RE, Ebert D et al (2012) Experimental evolution. Trends Ecol Evol 27: 547–560
Kerr B, Riley MA, Feldman MW, Bohannan BJM (2002) Local dispersal promotes biodiversity in a real-life game of rock-paper-scissors. Nature 418:171–174
Kidwell M, Lisch D (2001) Perspective: transposable elements parasitic DNA and genome evolution. Int J Org Evol 55:1–24
Kim W, Racimo F, Schluter J et al (2014) Importance of positioning for microbial evolution. Proc Natl Acad Sci USA 111:E1639–E1647
Kirchman DL (2016) Growth rates of microbes in the oceans. Annu Rev Mar Sci 8:285–309

Kivisaar M (2003) Minireview: Stationary phase mutagenesis: mechanisms that accelerate adaptation of microbial populations under environmental stress. Environ Microbiol 5:814–827
Kivisaar M (2010) Mechanisms of stationary-phase mutagenesis in bacteria: mutational processes in pseudomonads. FEMS Microbiol Lett 312:1–14
Klevens RM, Edwards JR, Richards CL Jr et al (2007) Estimating health care-associated infections and deaths in U.S. hospitals, 2002. Public Heal Rep 122:160–166
Kopac S, Wang Z, Wiedenbeck J et al (2014) Genomic heterogeneity and ecological speciation within one subspecies of *Bacillus subtilis*. Appl Environ Microbiol 80:4842–4853
Kouzel N, Oldewurtel ER, Maier B (2015) Gene transfer efficiency in Gonococcal biofilms: role of biofilm age, architecture, and pilin antigenic variation. J Bacteriol 197:2422–2431
Kraemer S, Boynton P (2017) Evidence for microbial local adaptation in nature. Mol Ecol 26: 1860–1876
Król JE, Nguyen HD, Rogers LM et al (2011) Increased transfer of a multidrug resistance plasmid in *Escherichia coli* biofilms at the air-liquid interface. Appl Environ Microbiol 77:5079–5088
Kryazhimskiy S, Rice DP, Desai MM (2012) Population subdivision and adaptation in asexual populations of *Saccharomyces cerevisiae*. Evolution (NY) 66:1931–1941
Kumar CG, Anand SK (1998) Significance of microbial biofilms in food industry: a review. Int J Food Microbiol 42:9–27
Lang GI, Rice DP, Hickman MJ et al (2013) Pervasive genetic hitchhiking and clonal interference in forty evolving yeast populations. Nature 500:571–574
Levin-Reisman I, Ronin I, Gefen O et al (2017) Antibiotic tolerance facilitates the evolution of resistance. Science 355:826–830
Levinson G, Gutman GA (1987) Slipped-strand mispairing: a major mechanism for DNA sequence evolution. Mol Biol Evol 4:203–221
Lewis K (2007) Persister cells, dormancy and infectious disease. Nat Rev Microbiol 5:48–56
Lewis K (2010) Persister cells. Annu Rev Microbiol 64:357–372
Li Y, Lau PCY, Lee JH et al (2001) Natural genetic transformation of *Streptococcus mutans* growing in biofilms. J Bacteriol 183:897–908
Lomstein B, Langerhuus A, D'Hondt S, Jorgensen B, Spivack A (2012) Endospore abundance, microbial growth, and necromass turnover in deep sub-seafloor sediment. Nature 484 (7392):101–104
Lynch AS, Robertson GT (2008) Bacterial and fungal biofilm infections. Annu Rev Med 59: 415–428
Macfarlane S, Dillon JF (2007) Microbial biofilms in the human gastrointestinal tract. J Appl Microbiol 102:1187–1196
Maisonneuve E, Gerdes K (2014) Molecular mechanisms underlying bacterial persisters. Cell 157: 539–548
Mazin A, Gimadutdinov O, Turkin S et al (1985) Non-enzymatic DNA methylation by S-adenosylmethionine results in the formation of minor thymine residues and 5-methylcytosine from cytosine. Mol Biol 19:903–914
Meckenstock RU, Elsner M, Griebler C et al (2015) Biodegradation: updating the concepts of control for microbial cleanup in contaminated aquifers. Environ Sci Technol 49:7073–7081
Merod RT, Wuertz S (2014) Extracellular polymeric substance architecture influences natural genetic transformation of *Acinetobacter baylyi* in biofilms. Appl Environ Microbiol 80:7752–7757
Molin S, Tolker-Nielsen T (2003) Gene transfer occurs with enhanced efficiency in biofilms and induces enhanced stabilisation of the biofilm structure. Curr Opin Biotechnol 14:255–261
Mulcahy LR, Isabella VM, Lewis K (2014) *Pseudomonas aeruginosa* biofilms in disease. Microb Ecol 68:1–12
Muller CJ (1932) Some genetic aspects of sex. Am Nat 66:118–138
Nadell CD, Drescher K, Foster KR (2016) Spatial structure, cooperation and competition in biofilms. Nat Rev Microbiol 14:589–600

Nahum JR, Godfrey-Smith P, Harding BN et al (2015) A tortoise–hare pattern seen in adapting structured and unstructured populations suggests a rugged fitness landscape in bacteria. Proc Natl Acad Sci USA 112:201410631

Ochman H, Lawrence J, Groisman E (2000) Lateral gene transfer and the nature of bacterial innovation. Nature 405:299–304

Olsen I (2015) Biofilm-specific antibiotic tolerance and resistance. Eur J Clin Microbiol Infect Dis 34:877–886

Orr HA (2002) The population genetics of adaptation: the adaptation of DNA sequences. Evolution 56:1317–1330

Otto M (2008) Staphylococcal biofilms. Curr Top Microbiol Immunol 322:207–228

Palestrant D, Holzknecht ZE, Collins BH et al (2004) Microbial biofilms in the gut: visualization by electron microscopy and by acridine orange staining. Ultrastruct Pathol 28:23–27

Pannell JR, Fields PD (2014) Evolution in subdivided plant populations: concepts, recent advances and future directions. New Phytol 201:417–432

Parsek MR, Singh PK (2003) Bacterial biofilms: an emerging link to disease pathogenesis. Annu Rev Microbiol 57:677–701

Paul BJ, Duthie HC, Taylor WD (1991) Nutrient cycling by biofilms in running waters of differing nutrient status. J North Am Benthol Soc 10:31–41

Percival SL, Hill KE, Williams DW et al (2012) A review of the scientific evidence for biofilms in wounds. Wound Repair Regen 20:647–657

Perfeito L, Pereira MI, Campos PRA, Gordo I (2008) The effect of spatial structure on adaptation in *Escherichia coli*. Biol Lett 4:57–59

Poltak SR, Cooper VS (2011) Ecological succession in long-term experimentally evolved biofilms produces synergistic communities. ISME J 5:369–378

Ponciano JM, La HJ, Joyce P, Forney LJ (2009) Evolution of diversity in spatially structured *Escherichia coli* populations. Appl Environ Microbiol 75:6047–6054

Qi L, Li H, Zhang C et al (2016) Relationship between antibiotic resistance, biofilm formation, and biofilm-specific resistance in *Acinetobacter baumannii*. Front Microbiol 7:1–10

Rainey PB, Travisano M (1998) Adaptive radiation in a heterogeneous environment. Nature 394: 69–72

Rodgers K, Mcvey M (2016) Error-prone repair of DNA double-strand breaks. J Cell Physiol 231: 15–24

Roesch L, Fulthorpe R, Riva A et al (2007) Pyrosequencing enumerates and contrasts soil microbial diversity. ISME J 1:283–290

Rogers J, Dowsett AB, Dennis PJ et al (1994) Influence of temperature and plumbing material selection on biofilm formation and growth of *Legionella pneumophila* in a model potable water system containing complex microbial flora. Appl Environ Microbiol 60:1585–1592

Rosche WA, Foster PL (2000) Determining mutation rates in bacterial populations. Methods 20: 4–17

Rousk J, Bååth E (2011) Growth of saprotrophic fungi and bacteria in soil. FEMS Microbiol Ecol 78:17–30

Rozen DE, Habets MGJL, Handel A, de Visser JAGM (2008) Heterogeneous adaptive trajectories of small populations on complex fitness landscapes. PLoS One 3:14–17

Rydberg B, Lindahl T (1982) Nonenzymatic methylation of DNA by the intracellular methyl group donor S-adenosyl-L-methionine is a potentially mutagenic reaction. EMBO J 1:211–216

Saumaa S, Tover A, Kasak L, Kivisaar M (2002) Different spectra of stationary-phase mutations in early-arising versus late-arising mutants of *Pseudomonas putida*: involvement of the DNA repair enzyme MutY and the stationary-phase sigma factor RpoS. J Bacteriol 184:6957–6965

Schaaper R, Loeb L (1980) Depurination causes mutations in SOS-induced cells. PNAS 78: 1773–1777

Schenk MF, Witte S, Salverda MLM et al (2015) Role of pleiotropy during adaptation of TEM-1 β-lactamase to two novel antibiotics. Evol Appl 8:248–260

Schloss PD, Handelsman J (2006) Toward a census of bacteria in soil. PLoS Comput Biol 2: 0786–0793

Schumacher MA, Balani P, Min J et al (2015) HipBA-promoter structures reveal the basis of heritable multidrug tolerance. Nature 524:59–64

Seoane J, Yankelevich T, Dechesne A et al (2011) An individual-based approach to explain plasmid invasion in bacterial populations. FEMS Microbiol Ecol 75:17–27

Shapiro BJ, Polz MF (2014) Ordering microbial diversity into ecologically and genetically cohesive units. Trends Microbiol 22:235–247

Simões M, Pereira MO, Vieira MJ (2005) Effect of mechanical stress on biofilms challenged by different chemicals. Water Res 39:5142–5152

Sniegowski PD, Gerrish PJ (2010) Beneficial mutations and the dynamics of adaptation in asexual populations. Proc R Soc B Biol Sci 365:1255–1263

Stalder T, Top EM (2016) Plasmid transfer in biofilms: a perspective on limitations and opportunities. NPJ Biofilms Microbiomes 2:16022

Steenackers HP, Parijs I, Foster KR, Vanderleyden J (2016) Experimental evolution in biofilm populations. FEMS Microbiol Rev 40:373–397

Sternberg C, Christensen BB, Johansen T et al (1999) Distribution of bacterial growth activity in flow-chamber biofilms. Appl Environ Microbiol 65:4108–4117

Stewart PS (1996) Theoretical aspects of antibiotic diffusion into microbial biofilms. Antimicrob Agents Chemother 40:2517–2522

Stewart PS (1998) A review of experimental measurements of effective diffusive permeabilities and effective diffusion coefficients in biofilms. Biotechnol Bioeng 59:261–272

Stewart PS, Franklin MJ (2008) Physiological heterogeneity in biofilms. Nat Rev Microbiol 6: 199–210

Stewart PS, Zhang T, Xu R et al (2016) Reaction–diffusion theory explains hypoxia and heterogeneous growth within microbial biofilms associated with chronic infections. NPJ Biofilms Microbiomes 2:16012

Stoodley P, Sauer K, Davies DG, Costerton JW (2002) Biofilms as complex differentiated communities. Annu Rev Microbiol 56:187–209

Suzuki T, Ohsumi S, Makino K (1994) Mechanistic studies on depurination and apurinic site chain breakage in oligodeoxyribonucleotides. Nucleic Acids Res 22:4997–5003

Thomas CM, Nielsen KM (2005) Mechanisms of, and barriers to, horizontal gene transfer between bacteria. Nat Rev Microbiol 3:711–721

Tillman D (1994) Competition and biodiversity in spatially structured habitats. Ecology 75:2–16

Tolker-Nielsen T, Molin S (2000) Spatial organization of microbial biofilm communities. Microb Ecol 40:75–84

Torres-cruz J, Van Der Woude MW (2003) Slipped-strand mispairing can function as a phase variation mechanism in *Escherichia coli*. J Bacteriol 185:6990–6994

Traverse CC, Mayo-smith LM, Poltak SR, Cooper VS (2013) Tangled bank of experimentally evolved *Burkholderia* biofilms reflects selection during chronic infections. Proc Natl Acad Sci USA 110(3):E250–E259

Truglio JJ, Croteau DL, van Houten B, Kisker C (2006) Prokaryotic nucleotide excision repair: the UvrABC system. Chem Rev 106:233–252

Tyerman JG, Ponciano JM, Joyce P et al (2013) The evolution of antibiotic susceptibility and resistance during the formation of *Escherichia coli* biofilms in the absence of antibiotics. BMC Evol Biol 13:1–7

Van Meervenne E, De Weirdt R, Van Coillie E et al (2014) Biofilm models for the food industry: hot spots for plasmid transfer? Pathog Dis 70:332–338

Verstraelen H, Swidsinski A (2013) The biofilm in bacterial vaginosis: implications for epidemiology, diagnosis and treatment. Curr Opin Infect Dis 26:86–89

Wakeley J (1998) Segregating sites in Wright's island model. Theor Popul Biol 53:166–174

Wang H, Ding S, Wang G et al (2013) In situ characterization and analysis of *Salmonella* biofilm formation under meat processing environments using a combined microscopic and spectroscopic approach. Int J Food Microbiol 167:293–302

Webster NS, Negri AP, Base S (2006) Site-specific variation in Antarctic marine biofilms established on artificial surfaces. Environ Microbiol 8:1177–1190

Wentland EJ, Stewart PS, Huang CT, McFeters GA (1996) Spatial variations in growth rate within *Klebsiella pneumoniae* colonies and biofilm. Biotechnol Prog 12:316–321

Whitchurch CB, Tolker-nielsen T, Ragas PC et al (2002) Extracellular DNA required for bacterial biofilm formation. Science 295:59–60

Whitlock MC (2003) Fixation probability and time in subdivided populations. Genetics 164: 767–779

Whitman WB, Coleman DC, Wiebe WJ et al (1998) Prokaryotes: the unseen majority. Proc Natl Acad Sci USA 95:6578–6583

Wierzchos J, Vincent WF, Quesada A (2015) Microstructure and cyanobacterial composition of microbial mats from the High Arctic. Biodivers Conserv 24(4):841–863

Wijker CA, Wientjes NM, Lafleur MVM (1998) Mutation spectrum in the lacI gene, induced by gamma radiation in aqueous solution under oxic conditions. Mutat Res 403:137–147

Williams I, Venables WA, Lloyd D et al (1997) The effects of adherence to silicone surfaces on antibiotic susceptibility in *Staphylococcus aureus*. Microbiology 143:2407–2413

Wolcott RD, Rhoads DD, Bennett ME, et al (2010) Chronic wounds and the medical biofilm paradigm. J Wound Care 19:45–46, 48–50, 52–53

Wright S (1932) The roles of mutation, inbreeding, crossbreeding and selection in evolution. In: Proceedings of the sixth international congress on genetics. Brooklyn Botanic Garden, Brooklyn, NY, pp 356–366

Xu KD, Stewart PS, Xia F et al (1998) Spatial physiological heterogeneity in *Pseudomonas aeruginosa* biofilm is determined by oxygen availability. Appl Environ Microbiol 64:4035–4039

Yachi S, Loreau M (1999) Biodiversity and ecosystem productivity in a fluctuating environment: the insurance hypothesis. Proc Natl Acad Sci USA 96:1463–1468

Chapter 9
Integrons as Adaptive Devices

José Antonio Escudero, Céline Loot, and Didier Mazel

1 Integrons

Integrons are genetic platforms that allow bacteria to evolve rapidly through the acquisition, stockpiling, excision, and reordering of coding sequences (CDS) found in genetic structures named cassettes. Cassettes generally contain promoterless genes that act as *plug and play* functions that are exaptable from their genetic context and interchangeable among integrons. Therefore, integrons confer enhanced genetic plasticity to their hosts by providing access to the virtually infinite variety of functions encoded in these elements. They serve as a long-term *memory* of adaptive traits that, through its fine intertwining with bacterial physiology, act when bacteria need to adapt (Escudero et al. 2015; Hocquet et al. 2012; Baharoglu et al. 2010).

As we will see below, integrons were first discovered embedded in mobile genetic platforms, like transposons and conjugative plasmids, and circulating among clinical

J. A. Escudero
Institut Pasteur, Unité de Plasticité du Génome Bactérien, Département Génomes et Génétique, Paris, France

CNRS, UMR3525, Paris, France

Molecular Basis of Adaptation, Departamento de Sanidad Animal, Facultad de Veterinaria, Universidad Complutense de Madrid, Madrid, Spain

Centro de Vigilancia Sanitaria Veterinaria (VISAVET), Universidad Complutense de Madrid, Madrid, Spain

C. Loot · D. Mazel (✉)
Institut Pasteur, Unité de Plasticité du Génome Bactérien, Département Génomes et Génétique, Paris, France

CNRS, UMR3525, Paris, France
e-mail: didier.mazel@pasteur.fr

P. H. Rampelotto (ed.), *Molecular Mechanisms of Microbial Evolution*, Grand Challenges in Biology and Biotechnology,
https://doi.org/10.1007/978-3-319-69078-0_9

isolates. Yet the discovery of integrons in the chromosomes of many bacterial species revealed the origin and the real history of these structures.

1.1 Sedentary Chromosomal Integrons

Integrons are ancient structures that have influenced bacterial evolution for eons (Rowe-Magnus et al. 2001). They have been found in approximately 9–17% of the genomes available in databases (Cambray et al. 2010; Cury et al. 2016; Boucher et al. 2007), although this figure is subject to biases in genome abundance and depends on the date of the analysis. Integrons are most prevalent among γ-proteobacteria but have been found in all Gram-negative bacterial phyla except among α-proteobacteria (Cury et al. 2016). These integrons are commonly referred to as chromosomal integrons (CIs), to distinguish them from plasmid-borne integrons, but these terms have lately become ambiguous, since mobile integrons can sometimes be found on chromosomes as a consequence of subsequent mobilization events (Cury et al. 2016). Therefore, *sedentary* chromosomal integrons (SCIs) seems a better term to refer to integrons that share an evolutionary history with the chromosome they are encoded in (Escudero et al. 2015). Phylogenetic studies of SCIs show that the branching of integrases is largely coherent with the organismal phylogeny, proving the ancestry and stability of integrons (Rowe-Magnus et al. 2001; Nemergut et al. 2008; Cambray and Mazel 2008). Three large groups can be distinguished in the phylogeny of integrases: (1) the soil-freshwater proteobacteria group, (2) the marine γ-proteobacteria group, and (3) the inverted integrase group (Cambray et al. 2010; Boucher et al. 2007). The first two groups form ecologically relevant taxa (Mazel 2006), suggesting, together with some phylogenetic incongruences in the branching of integrases, that horizontal transfer of chromosomal integrons occurs among bacteria sharing the same environment over long evolutionary periods. Interestingly, integron loss has also been reported. Although it could be anecdotal, the only report of integron loss occurred in the inverted integrase group, suggesting that these platforms might have specific properties that change their evolutionary dynamics (Wu et al. 2013).

Sedentary chromosomal integrons can be large, representing a substantial percentage of a species' genome. The best examples of large integrons are the so-called superintegrons (SI) found in species of the *Vibrio* genus. Some superintegrons have been shown to harbor more than 200 cassettes. Of them, the best studied superintegron is that of *V. cholerae* (Mazel et al. 1998), the causative agent of cholera disease. Located in the secondary chromosome, this SI harbors 172 cassettes and represents 3% of the bacterium's genome. The function of the majority of its cassettes remains cryptic, with only some exceptions that have been characterized mainly through their role in pathogenicity (Ogawa and Takeda 1993; Barker et al. 1994). This lack of knowledge applies to almost all SCI cassettes from other species. Indeed, among the cassettes found in superintegrons of *Vibrio* species, the majority are of unknown function (Boucher et al. 2007), and those whose function can be inferred are homologous to proteins related to a wide variety of functions. This highlights the genomic and phenotypic plasticity that integrons confer to their hosts.

Interestingly, the attachment site in cassettes (*attC* sites) of a given SCI often shows a high degree of sequence identity. This is interpreted as reflecting a relationship between the recombination sites and the host (Rowe-Magnus et al. 2001, 2003). Such sequence conservation is not true for all SCIs (Boucher et al. 2007), yet its existence is relevant to our understanding of the hitherto unexplored phenomenon of *de novo* generation of cassettes, as discussed below (see Sect. 5).

1.2 Mobile Integrons

During the 1950s, clinical isolates of *Shigella flexneri* showing resistance against streptomycin, tetracycline, and chloramphenicol became prevalent in Japan (Mitsuhashi et al. 1961). This unexpected and alarming phenotype triggered the investigations that yielded the discovery of integrons in the late 1980s (Stokes and Hall 1989; Martinez and de la Cruz 1988). Indeed, integrons were first discovered encoded in plasmids of clinical isolates showing multidrug resistance phenotypes. After decades of study, it is now broadly accepted that these *mobile* integrons (MIs) are not the native form of integrons. Instead they are the result of the association of SCIs with transposable elements and conjugative plasmids that have served as vehicles for their spread among clinically relevant bacteria (Cambray et al. 2010; Boucher et al. 2007; Mazel 2006; Gillings 2014). Selection of such mobilization events has likely occurred through intense antibiotic pressure in humans, animals, and the environment (Gonzalez-Zorn and Escudero 2012). Indeed, MIs played a major role in the early rise of multidrug resistance among clinically relevant bacteria in the 1960s (Mitsuhashi et al. 1961; Nakaya et al. 1960; Liebert et al. 1999). Since then, and according to the sequence of their integrase, five different mobile integrons have been described. Of these, classes 1 to 3 are the most clinically relevant, while classes 4 and 5 were identified in a subset of the SXT elements in *Vibrio cholerae* (Hochhut et al. 2001) and on the pRSV1 plasmid of *Aliivibrio salmonicida* (GenBank AK277063) (Sorum et al. 1992) where they were responsible for the trimethoprim resistance phenotype. Integron classes 1 and 3 (Arakawa et al. 1995) are found associated with Tn*402* (Collis et al. 2002; Xu et al. 2007), while class 2 integrons are almost exclusively linked to Tn*7* derivatives (Ramirez et al. 2010). Of all these, the class 1 integron is the most prevalent and most relevant from a medical perspective since it is detected in 22–59% of Gram-negative clinical isolates (Labbate et al. 2009). It is also the only class for which evidence of activity *in persona* is available (Hocquet et al. 2012). Hence, this class has become the archetype of mobile integron and represents today the main experimental model of integrons. Classes 2 and 3 have also been historically involved in the spread of multiresistance, yet to a lesser extent due to their lower prevalence. The success of any determinant to be stably transferred horizontally can be divided in four parts (the four Ps): *penetration*, *promiscuity*, *plasticity*, and *persistence* (Gonzalez-Zorn and

Escudero 2012; Baquero et al. 2011). We are ignorant of the reasons for the differences in prevalence for each integron class among clinical isolates. It could be the result of differences in the penetrance, persistence, and promiscuity of the mobilizing platform to which they are associated for their intrinsic rates of transposition and conjugation or the fitness cost that platforms impose on the host. It could also be the result of intrinsic adaptive traits of each integron class, such as (1) differences in the fitness cost they impose on their host that could affect their persistence in a new environment after mobilization, (2) distinct rates of activity of the integrases, or (3) differences in the range of *attC* sites recognized. Indeed, IntI1 has a broader substrate specificity when compared to other integrases, such as IntIPstQ from *Pseudomonas stutzeri* and VchIntIA from *V. cholerae* (Biskri et al. 2005; Holmes et al. 2003a). It is possible that a greater plasticity is the reason for the success of class 1 integrons, since they explore a larger proportion of functions encoded in cassettes. Concerning the influence of integron classes in persistence, it could be argued that MIs are *artificially* stabilized by high antibiotic pressure. As we will explain later, the integrated vision we now have of integrons makes us believe that their stability is likely assured through subtle connections with the host physiology that allows them to provide highly adaptive traits at critical moments while representing a very low fitness cost at any other time (Baharoglu et al. 2010; Starikova et al. 2012; Lacotte et al. 2017).

Compared to the size of some SCI's arrays, the cargo of mobile integrons is small, with the longest array described bearing only eight cassettes (Naas et al. 2001). Yet, their dissemination capabilities grant them access to the virtually endless content of SCIs arrays from environmental bacteria. Indeed, the codon usage, GC content, and *attC* sequences of cassettes found in MIs suggest that they can gather these elements from a variety of genetic backgrounds, from where they take them back to the clinical setting, where antibiotic pressure is high. It is therefore not surprising that despite the myriad of unknown functions found in the arrays of SCIs, the functions of MI cassettes are, almost exclusively, related to resistance. Dozens of different cassettes conferring resistance against almost all families of clinically relevant antibiotics and antiseptics have been characterized in MIs (Mazel 2006; Partridge et al. 2009; Fluit and Schmitz 2004). In anthropocentric environments, where antibiotics are prevalent, MIs confer such an adaptive advantage that they have become commonplace among Gram-negative clinical isolates. They have occasionally been identified in some Gram-positive bacteria too (Martin et al. 1990; Nandi et al. 2004; Nesvera et al. 1998; Shi et al. 2006), but these findings remain controversial. Altogether, it is not surprising that once mobilized, an adaptive device such as the integron served bacteria to thrive in the rapidly changing environment shaped by humans.

2 Anatomy of the Integron

The typical structure of an integron includes a stable platform and a variable cassette array. The platform contains three main components: first the *intI* gene that encodes the integrase governing integration and excision of cassettes; second, the *att*achment site in the *i*ntegron (the *attI* site), where cassettes are integrated in the array. *attI* sites are generally found upstream of the *intI* gene, with the only notable exception of integrons from the *Treponema* genus, where they are located downstream; and third, the P_C promoter that is located within the *intI* gene or between *intI* and the *attI* site and that is oriented toward the integration point to drive the expression of cassettes (Jove et al. 2010; Levesque et al. 1994; Collis and Hall 1995) (Fig. 9.1a). Cassettes are circular non-replicative elements (Collis and Hall 1992a) that generally contain a promoterless gene and a second recombination site: the *attC* site. These genes become functional once integrated into the platform where the P_C promoter drives their expression. Consecutive integration events give rise to an array of cassettes, of which only the most recently acquired, those close enough to the Pc promoter, are

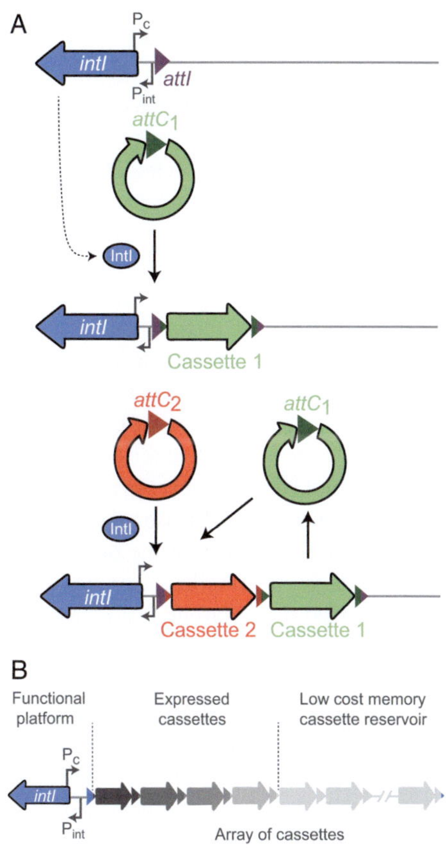

Fig. 9.1 Organization of integrons. (**a**) *Insertion and excision of cassettes.* Functional platform composed of the *intI* gene encoding integrase, both the cassette promoter (P_C) and the integrase promoter (Pint) as well as the primary *attI* recombination site (blue triangle) are shown. Cassette insertion (*attC* x *attI*) and excision (*attC* x *attC)* catalyzed by the IntI integrase are represented. Hybrid *attI* and *attC* sites are indicated. Arrowheads indicate the direction of transcription of the open reading frame. (**b**) *Expression of cassettes.* Small arrows represent cassette genes within the array. Their expression level is reflected by the color intensity of each arrow. Only the first cassettes of the array are expressed, and the subsequent ones can be seen as a low cost cassette reservoir

expressed (Fig. 9.1b). Cassettes that are too far from the P_C promoter to be expressed can be randomly excised and reintegrated back in first position where their expression becomes maximal again. Hence, the cassette array of an integron represents a low-cost memory of valuable functions for the cell. Its content is variable and reflects a nonlinear history of adaptive events (Fig. 9.1b).

Along this chapter, we will explain the many structural and functional peculiarities that make of integrons unique recombination systems.

2.1 The Integrase

2.1.1 Structural Features of the Integrase

Integrases govern the recombination reactions of the integron. Phylogenetic analyses have revealed that they belong to the family of Y-recombinases (Nunes-Duby et al. 1998) where they group together in a specific clade, close to the chromosome dimer resolution proteins Xer (Boyd et al. 2009). Integrases possess the typical 3D fold of Y-recombinases, as well as other structural features, such as patches I to III, boxes I and II, and the conserved RKHRHY residues in the catalytic site (Grindley et al. 2006). But recombination in integrons is qualitatively different to that performed by the rest of the members of this family. Indeed, Y-recombinases recognize specific sequences within double-strand (ds) DNA molecules and recombine them in an archetypical process involving the sequential transfer of two strands with the transient formation of a Holliday junction (HJ) intermediate. Instead, integrases can recognize either a sequence on ds-DNA (the *attI* site) or a very specific secondary structure formed by a single strand (ss) of DNA without almost no sequence conservation (the *attC* site). The latter substrate imposes such constrains to the reaction that a new recombination pathway is put in place to avoid resolution through a second strand transfer. Furthermore, these recombinases can deliver different recombination pathways depending on the set of substrates that they process (Escudero et al. 2016). The structural basis for such a novel activity in integrases is the presence of a 19-amino-acid-long domain within their catalytic core, which has not been found yet in any other protein and is hence considered a defining element of these recombinases. This domain contains an α-helix named I2, which is essential for the activity of the protein (Messier and Roy 2001). Its main role is to accommodate and stabilize a unique feature from *attC* sites, a set of extrahelical bases (EHBs) that protrude from the site in its active structured form, allowing for the specific recognition of the hairpin structure and its correct processing through a dedicated resolution pathway (see below) (MacDonald et al. 2006). The presence of the I2 domain and of EHBs suggests a coevolutionary process between integrases and *attC* sites driving the innovation of the system to what it is today. The study of a variety of integrase mutants has allowed confirming the role of the catalytic residues as well as that of the I2 domain (Messier and Roy 2001; MacDonald et al. 2006; Demarre et al. 2007; Gravel et al. 1998a; Johansson et al. 2009). A surprising aspect of integron recombination is that *attI* sites do not have EHBs, highlighting the fact that these proteins can recognize and bind to DNA molecules that are structurally very different. There is an evident lack of knowledge about the structural basis of this dual site recognition. It seems likely that the differential recognition of these sites depends on certain structural aspects of the synaptic complex, such as the strength and flexibility of the interactions between protein monomers. These parameters are likely finely balanced, since we have observed a trade-off in the recognition of one or the other site (Demarre et al. 2007).

2.1.2 Expression of the Integrase

A key feature of the adaptive value of integrons is that they are tightly intertwined with bacterial physiology to provide adaptation on demand. It is known that genetic elements that promote variability can have a negative impact in bacterial fitness if their activity is not firmly controlled. Accordingly, Lacotte and collaborators have recently shown that class 1 integrases entail a low, yet detectable, fitness cost that is dependent on the level of expression of the *intI1* gene as well as on the catalytic activity of the protein (Lacotte et al. 2017). In the case of integrons, the cell controls the expression of the integrase through the SOS response, a widespread regulatory network that detects DNA damage and repairs or bypasses lesions (Erill et al. 2007). The promoter regions of integrases possess a LexA-binding motif overlapping the -10 box of the P_{int} promoter (Cambray et al. 2011). LexA is the master regulator of the SOS response. When bound to a LexA box in a promoter, it represses gene expression by allosteric interference with the RNA polymerase. Abnormal amounts of ss-DNA trigger the polymerization of RecA nucleofilaments that ultimately induce the autocatalysis of LexA, releasing the repression and triggering the expression of the set of genes that compose the SOS response. In the case of integrons, the SOS response has a variable effect on the induction of integrase expression, with a 4.5- and a 37-fold induction for the class 1 and *V. cholerae* integrases, respectively (Guerin et al. 2009). Several biological processes can increase the amount of ss-DNA in the cell and trigger the SOS response, of which DNA damage and horizontal gene transfer are likely the most relevant ones. Hence, by linking the activity of the integron to an alarm signal, recombination events are limited to the moments when there is a need to evolve and adapt, when cassette acquisition or reshuffling can have a dramatic impact on cell survival. This is also supported by the interplay between the SOS response and the stringent response (a genetic network induced upon starvation) in the induction of integrase expression in biofilms (Strugeon et al. 2016).

The link through the SOS response between stress and resistance cassettes makes the selective pressure driving the capture and spread of antimicrobial resistance genes easy to understand. Especially so in the case of genes conferring resistance to antibiotics that cause DNA damage and trigger the SOS response, such as quinolones. A less clear matter was the mechanism leading to the high prevalence in cassettes of genes that confer resistance against antibiotics that do not induce the

SOS response in *E. coli,* like aminoglycosides. This conundrum led us to the discovery that these antibiotics do induce the SOS response (even at subinhibitory concentrations) in species other than *E. coli*, such as *V. cholerae*, *Photorhabdus luminescens*, and *Klebsiella pneumoniae* (Baharoglu and Mazel 2011; Baharoglu et al. 2013; Gutierrez et al. 2013). The underlying mechanism is the generation of reactive oxygen species (ROS) intermediates that cause DNA damage through the oxidizing of nucleotides, ultimately triggering the SOS response. A subtler consequence of belonging to the SOS regulon is that ss-DNA becomes central to the integron since it is at the same time the alarm signal and the substrate of integrases. Indeed, a major source of ss-DNA in the cell is horizontal gene transfer through conjugation and natural transformation (Baharoglu et al. 2010, 2012). Thus, the connection to the SOS response enables integrons to screen incoming DNA for new cassettes to capture.

The constant repression of the integrase in other situations in which stress is low and the host is well adapted, (i.e.: when the configuration of the array is optimal) prevents the movement and reordering of cassettes and precludes random recombination events at secondary sites that could be deleterious for the cell (Starikova et al. 2012; Harms et al. 2013). Good proof of the latter is the recent finding that the expression of the integrase gene from class 2 integrons is not regulated by the SOS response (Jové et al. 2017), explaining the long-known presence of a premature stop codon in the *intI2*-coding gene. This is likely the reflection of the deleterious effect that integrases have on host genomes if not well repressed. It seems probable that class 2 integrases lost the LexA box or that they are under the control of an unknown regulatory network in their native host. SOS-mediated regulation of the integron activity allowed understanding the puzzling observation that integron content is stable under laboratory conditions, yet it is the most variable part of the genome among field isolates of *V. cholerae* (Rowe-Magnus et al. 1999; Chowdhury et al. 2010; Feng et al. 2008; Labbate et al. 2007). Interestingly, the integrase coding gene of *V. cholerae* (*intIA*) is under the control of a second regulatory network through the cAMP receptor protein (CRP). CRP is the master regulator of the carbon catabolite repression response, adapting the metabolism of the cell to the available carbon sources. A CRP binding box is present between P_{int} and P_C in the promoter region of *intIA*. It regulates its expression independently of the SOS response. This connects the superintegron to environmental conditions. It is noteworthy that CRP is also directly and indirectly linked to the uptake of ss-DNA, via the regulators of natural competence HapR and TfoX (Baharoglu et al. 2012). Finally, low-level induction of the expression of *intIA* has also been observed at high temperature (42 °C) (Krin et al. 2014).

Altogether, the regulation of integrase expression highlights the intimate connection between integrons, the host's physiology, and the environment, providing enhanced adaptation capabilities when necessary.

2.2 *The Cassettes*

2.2.1 Structural Features of Cassettes

Integron cassettes (ICs) constitute the mobile and variable part of integrons. They seem ubiquitous, since they have been recovered from every environment investigated, including soil, riverine sediment, seawater, biofilms, plant surfaces, and even eukaryotes' symbionts (Elsaied et al. 2007, 2011; Stokes et al. 2001; Koenig et al. 2008; Gillings et al. 2005, 2008, 2009; Holmes et al. 2003b). Cassettes are generally composed of a single promoterless CDS and the *attC* recombination site recognized by the integrase. The size of ICs is therefore relatively small, generally falling in the range between 500 and 1000 bps (Loot et al. 2017). ICs can be found integrated within the integron array, as a linear ds molecule or, when excised, as a free, non-replicative mobile circular element.

Despite being mobile and readily interchangeable, several differences can be found between cassettes in MIs and SCIs. For instance, the size of the array of cassettes found in sedentary chromosomal integrons, its *cargo*, is highly variable, ranging from 0 to up to the 217 cassettes found in the *Vibrio vulnificus* integron. This represents a prodigious reservoir of exchangeable genes in the environment. On the other hand, the cargo of MIs is typically smaller, with the longest array reported being of only eight cassettes (Naas et al. 2001).

2.2.2 Functions of Cassettes

As we have previously mentioned, there are functional differences between cassettes in mobile and sedentary integrons. MIs dedicate their cargo almost exclusively to antimicrobial resistance genes. More than 130 ICs comprising resistance determinants against almost all antibiotic families have been found in class 1 integrons. These include genes conferring resistance to trimethoprim, β-lactams, chloramphenicol, all aminoglycosides, streptothricin, rifampin, fosfomycin, quinolones, macrolides, and antiseptics (Partridge et al. 2009; MacDonald et al. 2006). Sedentary chromosomal integrons, instead, contain mainly cassettes of unknown function. Among those found in *Vibrio* species, only 20% of genes code for proteins with homologs of known functions. These were related to a variety of processes such as metabolism, information storage, or cellular processes. Thirteen percent of the genes in these cassettes have homologs of unknown functions, and the remainder 66% coded for proteins with no homologs in the databases (Boucher et al. 2007). Additionally, some cassettes have functions mediating interactions with the external environment. Some examples are cassettes involved in adhesion and biofilm formation, protection from grazers, or bacterial aggregation. This is also supported by the presence in some genes of signatures of multiple transmembrane domains or signal peptide regions; the relation between cassette arrays and pathovars in *Xanthomonas* species; and the presence of cassettes in *Vibrio rotiferianus* DAT722 involved in host surface polysaccharide modifications suggesting a role in bacteriophage resistance (Rowe-Magnus et al. 2003; Koenig et al. 2008; Gillings et al. 2005; Rapa and Labbate 2013).

A distinct type of cassettes found in chromosomal integrons is that encoding toxin-antitoxin (TA) systems (Rowe-Magnus et al. 2003; Gerdes et al. 2005; Yamaguchi et al. 2011). TAs are addiction systems that stabilize DNA segments through a process called *post-segregational killing*. Briefly, TAs encode a stable toxin and its labile neutralizing antitoxin so that if the genes encoding the TA system are lost, the rapid decay of the antitoxin enables the toxin to kill the cell (Van Melderen and Saavedra De Bast 2009). TA systems might play a role in stabilizing long cassette arrays (Rowe-Magnus et al. 2003; Guerout et al. 2013). However, TAs have also been shown to mediate phage resistance (Sberro et al. 2013), and a dual role for these elements cannot be ruled out. Seventeen cassettes carrying type II TAs have been found in the superintegron of *V. cholerae* N16961(Guerout et al. 2013; Iqbal et al. 2015). A puzzling observation is that these cassettes seem to be intrinsically distinct and streamlined not to interfere with the array. For instance, contrarily to the majority of cassettes, TA systems do contain promoters that enable their functioning independently of their position in the array. Furthermore, nine of them are integrated in opposite orientation within the array, likely to avoid an influence of their promoter in the expression of cassettes downstream. These TA systems show a remarkable orthogonality, with every toxin being specific of its cognate antitoxin. Indeed, no cross talk between systems has been found, even between those belonging to the same family (Iqbal et al. 2015).

2.2.3 Expression of Cassettes

Promoterless cassettes are rendered functional upon insertion in the integron platform, where their expression is ensured by the P_C promoter located within the integron platform, either within the coding region of the *intI* gene or between the gene and the *attI* site (Fig. 9.1). As we will see below, *attC* sites govern the directionality of the integration reaction to ensure the correct orientation relative to the P_C promoter of the genes it encodes (Nivina et al. 2016). The experimental model of cassette expression is the class 1 integron, where the P_C promoter is located within the *intI1* coding sequence. In a few cases, a second promoter (P_C2) can also be found in *attI1*. To date, a variety of P_C variants of different strengths for both P_C and P_C2 promoters have been described: 13 for P_C and 3 for P_C2 (Jove et al. 2010; Collis and Hall 1995). Five P_C–P_C2 combinations, showing distinct levels of promoter strength, are highly prevalent among clinical isolates. The diversity of strength of these promoters produces differential phenotypes for identical arrays of cassettes.

The extensive work of Ploy's lab has led to a deeper understanding of the complexity of the system, revealing a transcriptional interference phenomenon between the opposing P_{int} and P_C promoters in class 1 integrons. Indeed, strong P_C variants show lower levels of P_{int} activity, even in conditions where P_{int} is de-repressed (SOS response). This could be advantageous, acting as a repression system for the integrase in the case of mobile integrons moving to bacterial species lacking LexA, such as *Acinetobacter baumannii* (Couvé-Deacon, pers. communication), where the unregulated expression of the integrase is costly (Starikova et al. 2012; Harms et al. 2013). When present, the P_C2 promoter disrupts the sequence of the LexA box and therefore abolishes SOS regulation of the integrase. P_C2 does not interfere transcriptionally with P_{int}, because both promoters are close enough for their transcription starts (+1) not to overlap (Guerin et al. 2011). Transcriptional interference between P_C and P_{int} means a trade-off between the expression of the adaptive traits encoded in cassettes and the expression of the integrase, which allows to capture them. The analysis of P_C variants in clinical and environmental *E. coli* isolates shows a higher prevalence of weak P_C variants suggesting that plasticity in the content of the array is more important than high expression levels (Vinue et al. 2011; Moura et al. 2012). In class 1 integrons, where the P_C is located within the coding region of the integrase, P_C variants modify the IntI1 sequence. Interestingly, the amino acid substitutions associated with the presence of strong promoters entail a decrease in the excision (but not in the integration) activity of the integrase, regardless of the transcriptional interference between promoters (Jove et al. 2010).

Note also that, exceptionally, some cassettes harbor their own promoter so that their expression is assured regardless of their position in the array. Some examples are the *cmlA1* chloramphenicol resistance gene (Bissonnette et al. 1991; Stokes and Hall 1991), the *ere*(A) cassette encoding an erythromycin resistance gene (Biskri and Mazel 2003), the quinolone resistance *qnrVC1* genes found in the class 1 integron (da Fonseca and Vicente 2012), and, as mentioned earlier, the toxin-antitoxin (TA) gene pairs found in the *V. cholerae* SI (Rowe-Magnus et al. 2003; Guerout et al. 2013; Szekeres et al. 2007).

The distance from a given cassette to the P_C promoter affects its level of expression. This produces a gradient of expression from the first gene in the array that decreases gradually until a point from which cassettes are silent. In long arrays, a large portion of the cargo is too far from the P_C promoter to be expressed. Therefore, these cassettes are carried at the minimum possible cost, the cost of replication. Note that, although silent, these cassettes remain available for the cell if needed (see cassette reshuffling). The gradient of expression depends on the strength of the promoter, the position of the cassette within the array, and the nature of the inserted cassettes (Collis and Hall 1995). The latter can influence the phenotype produced by a cassette through two posttranscriptional mechanisms. On one hand, the translation rate of genes in cassettes is going to be deeply influenced by whether these genes possess a ribosome binding site (RBS) to trigger the assembly of ribosomes or not (Shultzaberger et al. 2001). Indeed, some genes in integron cassettes are devoid of this motif. In these cases, translation can be initiated at an upstream RBS site and proceed to the next gene as if it was an operon. In class I integrons, the *attI1* site encodes a small ORF (*orf11*) containing a functional RBS that is present in all transcripts generated from P_C and drives a significant part of the translation of cassettes devoid of an RBS (Hanau-Bercot et al. 2002). On the other hand, the presence of a translated ORF increases the translation rates of the gene downstream by favoring the destabilization of *attC* sites on the mRNA transcript and facilitating ribosome progression (Cambray et al. 2010; Jacquier et al. 2009). This likely influences the *ruggedness* of the expression gradient of arrays (Fig. 9.2).

2.3 The *attI* Site

The *attI* site of the integron is the sequence into which cassettes are integrated. It is the most *canonical* of both sites since it is recognized for its sequence and recombined as a ds-DNA molecule. *attI* sites are composed of two integrase binding regions termed L and R that form the *core* site. Occasionally, some *attI* sites can present accessory sequences whose function is yet to be fully understood. In the cassette integration reaction (*attI* x *attC*), the crossover in *attI* takes place in the 5′-GTT-3′ triplet located in the R box of the core site. More precisely, the cleavage takes place between the A and the C on the complementary strand, the bottom strand (bs) (Fig. 9.3a). *attI* sites are difficult to detect because sites from different integrons diverge significantly, paralleling the evolutionary pattern observed for integrases (Rowe-Magnus et al. 2001). Furthermore, the L binding domains are degenerate with respect to R and the central regions differ greatly between *attI* sites. In vitro cross recombination assays between non-cognate *attI*/IntI partners have proven that integrases recognize preferentially their cognate site, suggesting the coevolution of both partners. Nevertheless, in some cases, cross talk between two systems can take place albeit less efficiently (Biskri et al. 2005). In the case of IntI1/*attI1*, from the class 1 integron, four monomers bind cooperatively to the site: two in the R and L boxes of the core site, as expected, and two within imperfect direct repeats dubbed DR1 and DR2, located upstream of the core site (Gravel et al. 1998b) (Fig. 9.3a). The role of these repeats is not fully understood. They are dispensable for *attI1* recombination, yet they seem to enhance the activity of IntI1 on the site. A hypothesis about the function of these repeats is that they could serve to keep integrase monomers in the vicinity of the core site (Gravel et al. 1998b). Yet the difference on the impact they have on the reaction depending on the partner substrate (an *attC* or an *attI1* site) (Partridge et al. 2000) could mean that they act as a topological filter. Accessory sequences are not a common feature of *attI* sites with only 12% of known *attI* sites (Escudero et al. 2015; Nield et al. 2001) showing putative additional binding domains.

As we have mentioned before, only the bs of the *attI* site is reactive in the *attC* x *attI* reaction. This is a consequence of the ss nature of *attC* sites that need the second strand transfer to be abolished. As we will see below, we have recently shown that in the symmetrical *attI* x *attI* reaction, both strands of *attI* sites are reactive (Escudero et al. 2016). In these cases, the cleavage on the top strand (ts) occurs in the border of the L box, with the crossover located between both adenines in the 5′-AAC-3′ triplet.

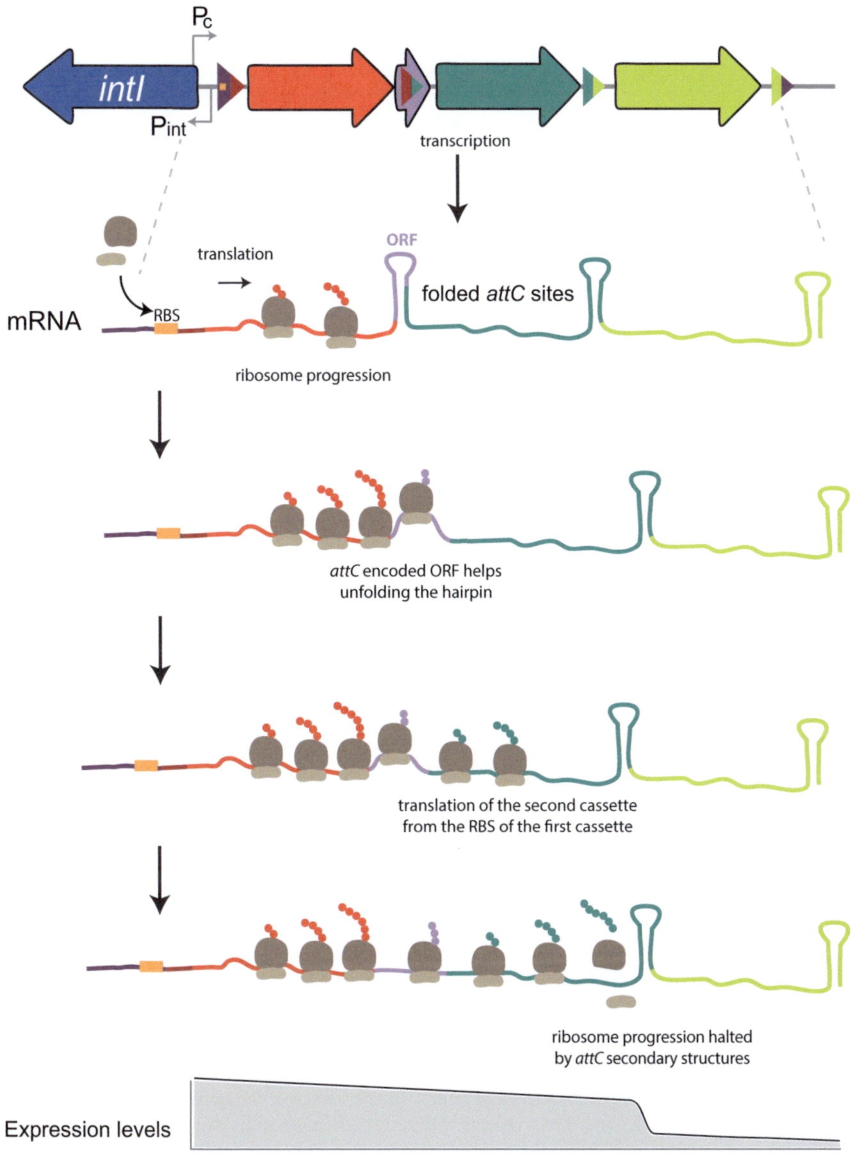

Fig. 9.2 Diagram of the influence of *attC* sites in cassette translation. Genes are depicted as arrows and recombination sites as triangles. RBS, ribosome binding site. P_C and P_{int} are the promoters of the cassette array and the integrase, respectively. Transcription is not influenced by the presence of *attC* sites, but the structuring of these sites in the mRNA molecule can impede ribosome progression. Coding of an ORF within the site (purple arrow between the first and second cassettes) allows ribosomes to unfold the hairpin structure of *attC* sites and increases the translation levels of the gene downstream

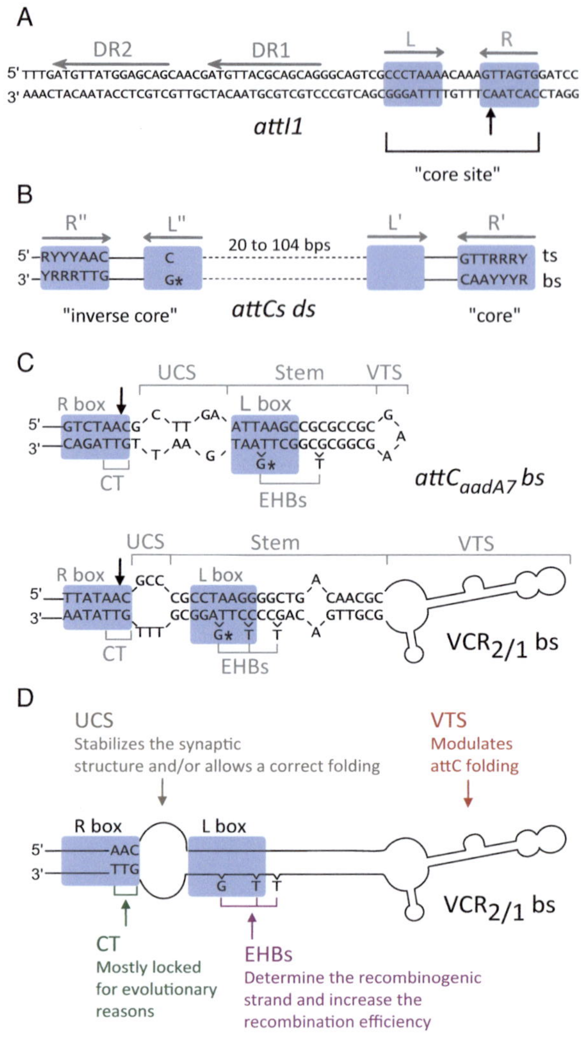

Fig. 9.3 Integron recombination sites. The putative IntI1 binding domains are marked with blue boxes. The black arrows show the cleavage points. (**a**) *Sequence of the double-stranded attI1 site.* Inverted repeats (R and L) and direct repeats (DR1 and DR2) are indicated with gray arrows. (**b**) *Schematic representation of double-stranded (ds) attC sites.* Inverted repeats (R″, L″, L′, and R′) are indicated with gray arrows. The dotted lines represent the variable central part. The conserved nucleotides are indicated. Asterisks (*) show the conserved G nucleotides, which generate extrahelical bases (EHB) in the folded *attC* site bottom strand (bs). The top strand (ts) and bottom strand (bs) are marked. (**c**) *Proposed secondary structures of the* $attC_{aadA7}$ *and* $VCR_{2/1}$ *bottom strands.* Structure of the *attC* sites of the *aadA7* cassette and of the second cassette in the SI were determined by the UNAFOLD online interface at the Institut Pasteur. Structural features of *attC* sites, namely, the unpaired central spacer (UCS), the extrahelical bases (EHBs), the stem, and the variable terminal structure (VTS), are indicated. Asterisks (*) show the conserved G extrahelical base. The conserved triplet (CT) is indicated. Primary sequences of the *attC* sites are shown (except for the VTS of the $VCR_{2/1}$ site). (**d**) *Schematic representation of structural features of the* $VCR_{2/1}$ *site and their roles.* The structural features and their roles are indicated

2.4 The *attC* Site

2.4.1 *Structural Features of* attC *Sites*

attC sites represent an essential part of integron cassettes since they grant their mobility. They are unique recombination sites that differ significantly from canonical sites of Y-recombinases. Indeed, many of the features that make the integron a distinct recombination system derive from the structural peculiarities of *attC* sites.

attC sites present an abnormal organization on ds-DNA, with two regions of inverted homology, R″–L″ and L′–R′. They are separated by a central region that is highly variable in size and sequence. The total lengths of *attC* sites can hence range from 57 to 141 bp (Stokes et al. 1997) (Fig. 9.3b). An even more striking feature of these sites is the almost complete lack of sequence conservation. The explanation for such an abnormal arrangement is that *attC* sites are not recognized for their sequence on ds-DNA. Instead, they all display a palindromic organization that can form hairpins through intra-strand DNA pairing (Fig. 9.3c) (Hall et al. 1991). Upon folding, the bottom strand of *attC* sites shows a secondary structure that reconstitutes an almost canonical core site (the result of the pairing of R″–L″ and R′–L′ to form the R and L boxes). This specific structure, with some other features that appear upon folding and that we will describe briefly below, is the substrate recognized (Francia et al. 1999; Johansson et al. 2004) and recombined by the integrase (Bouvier et al. 2005). Therefore, contrarily to what happens with canonical Y-recombinase sites, the genetic information allowing proper recombination of *attC* sites is not limited to the primary sequence of the site. DNA folding is a well-regulated process in living cells. Therefore, *attC* site folding permits the inclusion of a new layer of information and regulation in *attC* sites that connects recombination in integrons to the host's physiology.

A comparison of *attC* sites shows that sequence conservation is restricted to two inverted triplets, 5′-AAC-3′ and 5′-GTT-3′, located in the R″ and R′ boxes, respectively (Fig. 9.3b; see ts). As we have seen before, this triplet is also present in *attI* sites, and it is the region where the single strand exchange takes place. In the *attI* x *attC* reaction, the crossover point is located between the C and A nucleotides in both sites. Sequence conservation in *attC* sites can be extended to the inverted repeat sequences of 7 bps designated as *inverse core* and *core* (Fig. 9.3b), consisting, respectively, of a 5′-RYYYAAC-3′ (R, purine; Y, pyrimidine) at the 5′end of the ds-*attC* site, and the complementary 5′-GTTRRRY-3′ localized at the 3′extremity (Fig. 9.3b). Recombination of *attC* sites is deeply influenced by three structural features that emerge from the folding of the bs (MacDonald et al. 2006; Bouvier et al. 2009).

First, a set of single bases located on the R″–L″ arm, which do not have complementary nucleotides on the R′–L′ arm, protrude from the structured site as extrahelical bases (EHBs) (Fig. 9.3c). Depending on the *attC* site, two or three EHBs can be present (Bouvier et al. 2009). These bases play three roles of the utmost importance in the recombination reaction:

1. They determine the strand that will be recombined from a ds molecule containing an *attC* site (the bottom strand). This is of extreme importance for the correct functioning of the integron, since the recombination of the ts leads to the insertion of the cassette in opposite orientation to the P_C promoter. This cassette would therefore remain silent. Strand recognition is not trivial, since similar hairpins can be formed from either the top or the bottom strand of a ds-*attC* site. In both cases, a core site is reconstituted, and a 5′-AAC-3′ triplet is available for the crossover. Yet the bs is approximately 10^3 times more recombinogenic than the ts (Bouvier et al. 2005). The location of EHBs relative to the 5′-AAC-3′ triplet is not a mirror image between the bottom and the top strand (Nivina et al. 2016). This allows differentiating both strands and maintaining the correct functionality of the integron by favoring bs recombination.
2. EHBs also stabilize the synaptic complex formed by two DNA molecules and four monomers of the integrase. This stabilization takes place through contacts between EHB and protein monomers across the synapse.
3. The docking of an EHB within the integrase monomer bound to the L″–L′ box produces a conformational change that pulls apart the tyrosine residue of that monomer and avoids the second cut in the *attC* site and the abortive second strand exchange (see *attI* x *attC* reaction below) (MacDonald et al. 2006; Bouvier et al. 2009).

The second structural feature is the *u*ncoupled *c*entral *s*pacer (UCS) between the L and R boxes. It arises from the imperfect annealing of the region between R″ and L″ to that between L′ and R′. The UCS is essential to achieve high-level bs recombination, likely through a dual role in active site exclusion and in hindering the reverse reaction after the first strand exchange.

The third feature that arises through the folding of *attC* sites is the variable terminal structure (VTS). It corresponds to the end of the stem and the loop of the hairpin and arises through the pairing of the region between R″–L″ and R′–L′ (Fig. 9.3c) (Bouvier et al. 2009). VTSs can vary in length from a three nucleotide loop as in $attC_{aadA7}$ to a complex branched structure in large sites like VCRs (*Vibrio cholerae* repeats, the *attC* sites of the superintegron (Mazel et al. 1998)). The VTS has a strong impact in recombination through the modulation of *attC* folding (Loot et al. 2010). Indeed, the length and sequence of VTSs determine the tendency and strength of intra-strand pairing and hence the propensity of ds-*attC* sites to form the recombinogenic ss-*attC* hairpin structure. The VTS has a distinct impact on *attC* site recombination depending on the context. On one hand, it has a poor influence in the structuring of *attC* sites that are found in a ss form, as it is the case of sites contained in DNA fragments undergoing natural transformation or conjugal transfer. On the other hand, its impact is critical when the bs has to extrude from a symmetric and paired ds-DNA molecule to form a hairpin on each strand. This phenomenon, called cruciform formation, is strongly burdened by long VTSs with highly unfavorable folding energy (Loot et al. 2010). The UCS and the VTS also contribute to the bs recombination bias that ensures the correct orientation of inserted cassettes. This is likely due to a skew in base composition of the unpaired regions. It has been observed that purines as well as non-Watson-Crick G/T pairings are enriched in the bs of *attC* sites (Nivina et al. 2016). This skew likely serves to render bs folding more stable than ts, although this might not be a universal feature since VCRs show the opposite skew (Loot et al. 2017).

This atypical sequence-independent recognition of ss-*attC* sites readily explains how cassettes containing different sites can be efficiently recombined by the same integrase as well as how the proper orientation with respect to the P_C promoter is ensured through the recombination of the bs.

Despite the lack of sequence conservation among *attC* sites, and the fact that integrases can recombine very different sites, the sites of long SCIs generally show a high degree of identity (>80% in the case of VCRs in *V. cholerae*). This seems to indicate that there is a connection between the *attC* site sequence and the bacterial species harboring the array that is not necessarily related to the correct functioning of the site (see also *cassette genesis*, (Rowe-Magnus et al. 2001; Cambray et al. 2010; Mazel 2006)). On the contrary, the *attC* sites of cassettes in MIs show a large diversity in length and sequence, as well as an inconsistent codon usage in the ORFs encoded. This suggests that mobile integrons can explore and capture the cassette content of sedentary chromosomal integrons from a variety of bacterial species and genetic backgrounds. Integron cassettes contained in MIs are likely representatives of a specific SCI in the environment (Rowe-Magnus et al. 2001).

2.4.2 attC *Site Folding*

Secondary structures are common in DNA and they can have biological functions. Nevertheless, they represent a risk for cell viability if they are stable enough to interfere with replication or translation machineries (Collins et al. 1982). This imposes constraints on the size and energy of the inverted repeats and has fostered the development of host factors regulating the formation of secondary structures. Indeed, bacteria have evolved mechanisms to mitigate the deleterious effects of secondary structures, managing to find a subtle balance between the benefit provided by encoding biological functions in these structures and the risk of an excess of them (Bikard et al. 2010a). Good examples of such host factors are proteins like SbcCD, which destroy stable palindromes, and the single-stranded DNA-binding protein (SSB) that unfolds hairpins in ss-DNA without sequence specificity. Therefore, the ss structure of *attC* sites connects integrons to their hosts through physiological processes that regulate the natural formation of secondary structures in the chromosome, ultimately allowing the cell to control integron recombination.

SSB plays an important role in DNA replication and repair and in homologous recombination. It migrates along ss-DNA, preventing premature annealing while stabilizing single-stranded DNA, protecting it from exonucleases and removing secondary structures (Meyer and Laine 1990; Roy et al. 2009). It was initially thought that SSB has a paradoxical effect on integron recombination. On one hand, it promotes the viability of *attC* sites within long cassette arrays by flattening them so that they do not interfere with the replication machinery. On the other hand, this effect interferes with integron recombination by impeding the structuring of *attC* sites. We have recently shown that the integrase can overcome the effect of SSB on *attC* site folding, efficiently stabilizing them at the moment of their extrusion. Integrases and SSB interact in specific ways so that SSB does not interfere with integron recombination, but rather helps maintain the integrity of *attC* sites in the absence of the integrase (Grieb et al. 2017).

attC site folding is controlled differently depending on whether the site folds from a ss-DNA molecule (e.g., during horizontal gene transfer) or when it extrudes from a ds-DNA molecule to form a cruciform.

From ss-DNA

Four common cellular processes involve the presence of ss-DNA in prokaryotes: replication, transcription, DNA repair, and horizontal gene transfer (HGT). Of these, replication and HGT have been linked to the function of the integron. During replication, the synthesis of the lagging strand is discontinuous, producing stretches of up to 2kb of ss-DNA: the Okazaki fragments. Structuring of *attC* sites contained in the template strand between fragments is hence favored (Trinh and Sinden 1991), but the impact it has on recombination depends on the orientation of the integron relative to the replication fork, i.e., whether the lagging strand template carries the bottom or the top strand of the *attC* site. Indeed, when it carries the *attC* bs, recombination occurs at higher rates than when it carries the ts (Loot et al. 2010, 2017). Also, the defined length between Okazaki fragments favors the recombination of *attC* sites that are not further apart than 2 kb, resulting in the excision of a limited number of cassettes (1 or 2). Notably, SCIs are generally oriented in such a way the *attC* sites bs are carried on the leading strand template, limiting the excision of cassettes. This is particularly true for the SCIs of the *Vibrio* genus for which we never found the inverse orientation (Loot et al. 2017).

Horizontal gene transfer mechanisms, namely, conjugation, natural transformation, and transduction, involve mainly the entry of a single strand of DNA into the cell. This strongly favors the folding of *attC* sites encoded in it, limiting the influence of intrinsic features of the site, such as long or complex VTSs. This has been clearly demonstrated in the case of plasmid conjugation (Bouvier et al. 2009). Furthermore, conjugation and natural competence simultaneously induce the expression of the integrase through the triggering of the SOS (Baharoglu et al. 2010, 2012; Loot et al. 2010). Altogether, this allows the integron to recruit incoming cassettes.

From ds-DNA

Formation of cruciforms in ds-DNA, which occurs by intra-strand hybridization of inverted repeats, can lead to *attC* site extrusion and folding, a process that relies heavily on DNA topology and superhelicity. Cruciforms are rarely formed at significant rates under normal supercoiling conditions, since they involve a deep structural disruption and reorganization of base pairing, and very stable cruciforms are strongly selected against for their interference with the bacterial replication machinery (Pearson et al. 1996). However, common biological processes such as transcription and replication (Liu and Wang 1987) increase locally and transiently the levels of superhelicity, allowing for the formation of cruciforms (Baharoglu et al. 2010). Superhelicity can also vary depending on growth phase and growth conditions or through the action of certain stimuli such as the presence of antibiotics (Balke and Gralla 1987; Jaworski et al. 1991; Ferrandiz et al. 2010) or the induction of the SOS response (Majchrzak et al. 2006). Crucially for mobile integrons, bacterial species can present different levels of superhelicity, so that their recombination dynamics might be altered depending on the host strain (Champion and Higgins 2007). Cruciform extrusion of *attC* sites from ds-DNA and the influence of superhelicity have been proven using mutants of topoisomerase I and gyrase, the enzymes in charge of maintaining topology and supercoiling levels in *E. coli* (Loot et al. 2010).

Despite the evident usefulness of integrons, the elegance of how they provide enhanced evolvability to bacteria is best appreciated through the subtle and complex coupling to the physiology and needs of its host. In this section, we have given a brief overview of such connection, of which Fig. 9.4 provides a snapshot scheme that should help the reader.

3 Recombination Reactions Tell a Story of Evolutionary Innovation

Proteins of the Y-recombinase family deliver recombination reactions between two DNA molecules in a well-defined set of steps. The first one is the formation of the synaptic complex, composed of the two DNA molecules to be recombined and four monomers of the recombinase. Then one strand from each substrate is cleaved by opposing protein monomers and transferred to the other molecule, forming an H-like structure called the Holliday junction (HJ). This HJ is then resolved through the cleavage and exchange of the remaining set of strands (Grindley et al. 2006). This recombination pathway is conserved in all known Y-recombinases with the only exception of the integron. This is mainly due to the single-stranded nature of *attC* sites that imposes changes in the process and the mechanistics of the recombination reactions.

In the next few paragraphs, we will give a brief yet comprehensive view of the recombination reactions emphasizing the innovative aspects of this unique process. A detailed explanation on the mechanistic aspects of these reactions can be found in (Escudero et al. 2016). Integrases can deliver three reactions using *attI* and *attC* sites: (1) the *attI* x *attC* reaction that serves to integrate cassettes into the integron platform (Fig. 9.5), (2) the *attC* x *attC* reaction that excises cassettes from the array, and (3) the *attI* x *attI* reaction (Fig. 9.6) (the ancestral reaction), a rather cryptic one,

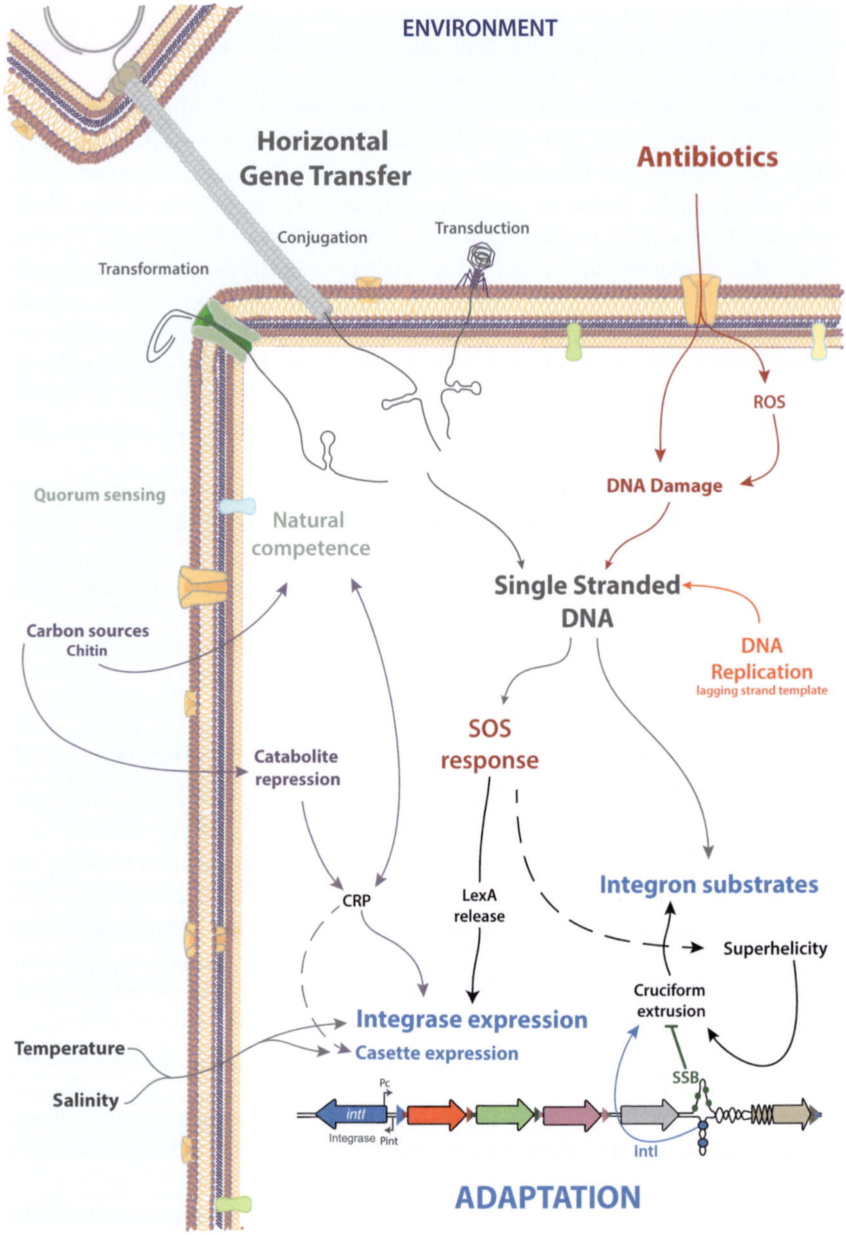

Fig. 9.4 Intimate connection between the integron and cell physiology (modified from (Escudero et al., 2015)). Snapshot representation of the links between integrons' activity and bacterial physiology is shown. The main triggering signal for integrase expression is the bacterial SOS response. A detailed explanation is available in the section entitled: *Integrons are intimately connected to bacterial physiology*

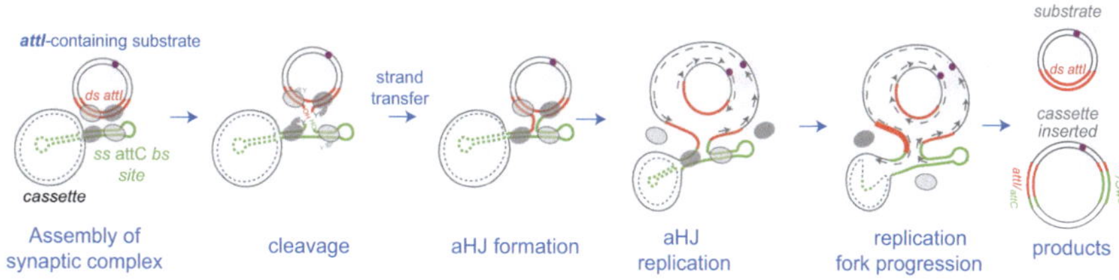

Fig. 9.5 Replicative resolution of integron cassette insertion. Recombination between a double-stranded *attI* site (bold red lines) and a single-stranded bottom *attC* site (bold green lines) terminating a cassette is shown. The top strand of the *attC* site is represented as a dotted line as we do not exactly know the nature of the cassettes (ss or ds). The synaptic complex comprises two DNA duplexes bound by four integrase protomers. The two activated protomers are represented by dark gray shade. One strand from each duplex is cleaved and transferred to form an atypical Holliday junction (aHJ). The non-abortive resolution implies a replication step. The origin of replication is represented by a purple circle and the newly synthesized leading and lagging strands by dashed gray lines. Both products are represented: the initial substrate resulting from the top strand replication and the molecule containing the inserted cassette resulting from the bottom strand replication. Hybrid *attC* and *attI* sites are indicated

which can have biological consequences in the case of multicopy mobile integrons. The combination of excision and integration reactions leads to the shuffling of integron cassettes within the array.

3.1 Innovative Reactions

3.1.1 *Cassette Integration: The* attI *x* attC *Reaction*

The *attI* x *attC* reaction is the most efficient of the reactions catalyzed by the integrase to ensure that cassette insertion takes place preferentially at the *attI* site rather than at an arbitrary *attC* site within the array(Collis et al. 1993, 2001) (a phenomenon that does happen, albeit at very low frequencies (Baharoglu et al. 2010)). This ensures that newly acquired promoterless cassettes are expressed from the P_C promoter within the integron and therefore directly tested for their adaptive value. Subsequent insertions lead to the formation of an array of cassettes of variable length. New acquisition events relocate previous cassettes at a higher distance from the Pc promoter, attenuating their expression. It is likely that in long arrays, such as those of superintegrons, a majority of cassettes are silent. Arrays act as an adaptive *memory*, encoding functions that have been valuable for the cell in the past and which can be recalled on demand (see Sect. 4). The higher efficiency of the *attI* x *attC* reaction is likely achieved by the distinct structure of both sites. Indeed, since bacterial DNA is generally in a ds form, *attI* sites are almost constantly in their recombinogenic state (ds-DNA). Instead, under these circumstances, *attC* sites in the array are not generally available to serve as integration sites. This difference in substrate availability is also at the basis of the tendency of integrons to accumulate

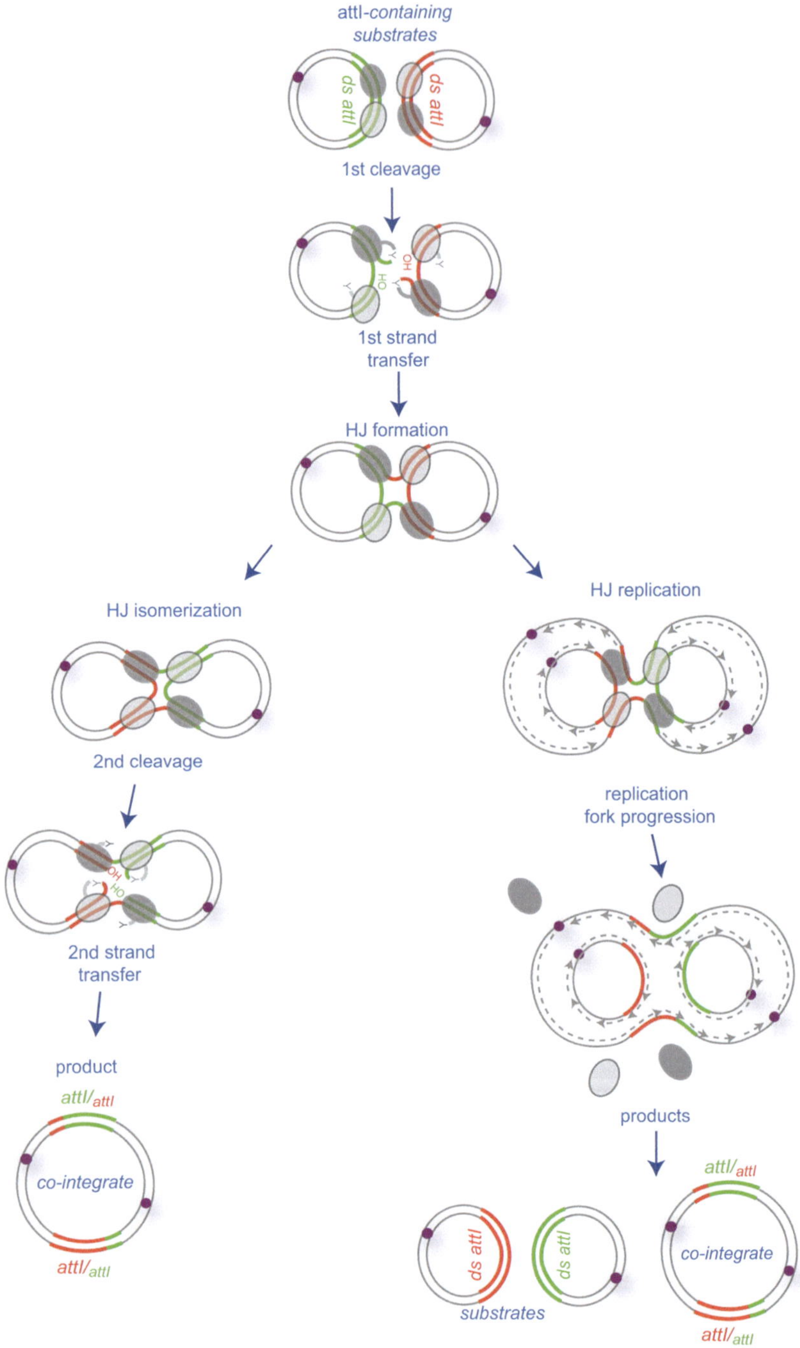

Fig. 9.6 Two resolution pathways proposed for *attI* x *attI* recombination. Recombination between two double-stranded *attI* sites (bold green and red lines) is shown. The first proposed pathway is

cassettes instead of losing them. Indeed, being the *attI* x *attC* reaction so efficient, one could expect integrases to excise the cassette in first position very frequently, until the integron becomes empty. The lack of expression of the integrase in the absence of SOS induction (see below) likely contributes to the stability of the array once an adaptive conformation has been found. Yet the SOS response can be triggered for many reasons that are not related to the integron, and cassette loss can take place. The distinct recombinogenic form of both sites makes highly improbable that a ds-*attI* site and a ss-*attC* site are found close enough within the same molecule, to produce the excision of the first cassette. The duality in the nature of the sites is at the basis of the necessary biases among reactions that ensures the correct functioning of the system.

The origin of the innovative nature of integron recombination comes from the fact that this reaction is mechanistically atypical, since it involves on one side a double-stranded substrate (*attI*) and on the other a hairpin-like single-stranded one (*attC*). The mechanistic problem of recombining ss substrates arises after the first strand exchange that leads to the formation of an asymmetric and therefore atypical HJ (aHJ). This aHJ cannot be resolved through the classical second strand exchange, because the second cut on the *attC* site would result in a lethal outcome: the linearization of the replicon containing the *attI* site (Fig. 9.5). As mentioned before, the ingenious mechanism that enables avoiding the second cleavage is the presence in the L box of *attC* sites of an extrahelical "T" that docks within the I2 domain of the integrase. This causes a conformational change in the integrase monomer that pulls apart the catalytic Y impeding the nucleophilic attack exclusively on that side of the synapse (MacDonald et al. 2006). The aHJ remains unresolved until the passage of the replisome produces the integration of the cassette in only one of both strands without the need for a second strand exchange of any kind (Fig. 9.5) (Loot et al. 2012). The involvement of a replicative event is biologically relevant since it makes cassette insertion a semiconservative process. This allows integrons to explore the adaptive value of incoming cassettes limiting the risk of acquiring maladaptive ones, since the non-recombined strand of the integron (the top strand) serves as a backup of the original conformation.

At the nucleotide level, the crossover point of the reaction is localized in the R boxes of both substrates between the C and AA of the bs. Consequently, six of the seven bases of the R box in the *attI* sites are exchanged at each integration event by

Fig. 9.6 (continued) similar to the classical site-specific recombination catalyzed by Y-recombinases. The synaptic complex comprises two DNA duplexes bound by four recombinase protomers. The first two activated protomers are represented by dark gray color. One strand from each duplex is cleaved and transferred to form a HJ. Isomerization of this junction alternates the catalytic activity between the two pairs of protomers (dark- and light-gray ovals) ensuring the second strand exchange and recombination product formation (co-integrate). The second pathway proposes a resolution of the HJ by replication. The origin of replication is represented by a purple circle and the newly synthesized leading and lagging strands by dashed gray lines. Products are represented: two initial substrates resulting from the top strand replication and co-integrate resulting from the bottom strand replication. Hybrid *attI* sites are indicated

the corresponding sequence of the incoming *attC* site resulting in a chimeric site. Given the lack of sequence conservation of *attC* sites (limited to 5′-RYYYAAC-3′), this highlights an unanswered question about integrons: how can integrases recognize *attI* sites for their sequence if this is continually changing? This is especially troubling since we know IntI1 binding to the *attI1* is a cooperative process that starts at the R box (Gravel et al. 1998b and unpublished results). The alteration of the recombination site in every recombination reaction without any biological consequences is a feature specific to integron recombination. Indeed, the sites in other recombination systems are often identical, and the moieties of each substrate are therefore unrecognizable in the recombined site.

3.1.2 *Cassette Excision: The* attC *x* attC *Reaction*

The recombination between *attC* sites located in the same array excises a covalently closed integron cassette (Collis and Hall 1992a, b). For this reaction to take place, two *attC* sites must be folded and available simultaneously, a phenomenon influenced by intrinsic characteristics of these sites (see the role of the VTS, in the *attC* section). Other external parameters play a role in the likelihood of cassette excision by influencing the frequency at which the bs can be found as a single strand (and hence favor its folding). Indeed, we have observed that excision reactions are more frequent when the bs of two *attC* sites are encoded in the lagging strand template, within a distance compatible with the length of Okazaki fragments(Loot et al. 2017). Integrases probably play a role in this phenomenon too, since they are known to capture *attC* sites at the beginning of the folding process and stabilize them to enable their complete structuring (Loot et al. 2014).

Again, a first exchange of strands in this reaction forms an atypical HJ that cannot be resolved through the classical resolution pathway. In fact, in this reaction, a second strand exchange is not abortive (since it does not linearize the replicon), but it only produces the exchange of stems between both *attC* sites. Although not formally demonstrated, it seems likely that this atypical Holliday junction is also resolved by a replicative (and hence semiconservative) event, leading to the excision of the cassette exclusively from the bs. Replication here can play an additional role, since the cassette is presumably separated from the ts and released from the integron as a consequence of the passage of the replication fork. In the case of recombination events taking place between sites located on an Okazaki fragment, the cassette is directly excised after recombination without the need for a replicative release.

Similarly to what happens with *attI* sites during recombination, the *attC* sites in the array are also chimeric, with the last 6 bps of any given *attC* belonging, originally, to the next site in the array (Fig. 9.1a). The 6 bps of the last *attC* in the array are in fact those of the original *attI* site (Fig. 9.1a and b). It is also remarkable that the imperfect complementarity found in such chimeric *attC* sites does not limit their propensity to fold efficiently nor their stability once folded.

3.2 The Ancestral Reaction: *attI* x *attI*.

Integrases can recombine two *attI* sites (Collis et al. 2001; Hansson et al. 1997), but this reaction is between 10^3 and 10^4 times less efficient than the *attI* x *attC* reaction. From a biological perspective, bearing in mind the chromosomal origin of integrons, it can be argued that this reaction has little adaptive meaning, since the only moment in which two *attI* sites can be recombined is during the transient cohabitation of replicated chromosomes before cell division. In this setting, the *attI* x *attI* reaction would produce an undesirable dimer between chromosomes that needs further resolution prior to segregation. Furthermore, since both integrons are identical, the reaction would not produce any relevant genotypic change. Hence, this reaction seems to be the vestige of the main activity of integrase ancestors, since it is structurally more similar to that of other Y-recombinases. We therefore suspected that the recombination of *attI* sites was in fact the *ancestral* reaction of the integrase. Interestingly, this reaction can in principle be resolved through a second strand exchange because it is symmetrical and it involves two double-stranded identical partners. Therefore, a second strand exchange is neither abortive nor impeded by EHBs. Nevertheless, this reaction can also follow the innovative resolution pathway of the two other reactions, with only one strand exchange and resolution through replication (Fig. 9.6, HJ replication). If the later was true, this would mean that, despite its appearance, the *attI* x *attI* reaction would not be the ancestral activity of the integrase because it would be qualitatively different from reactions delivered by other Y-recombinases. We have recently investigated the recombination pathway followed by the integrase when processing two symmetrical *attI1* sites (Escudero et al. 2016). Our results show that this reaction can indeed follow the classical resolution pathway and that it is therefore the *bona fide* ancestral reaction. In this reaction, both bottom and top strands of the *attI1* site are reactive. The crossover point within the R box of the bs is conserved (3′-CAA-5′), and the cleavage on the L box of the ts occurs between the adenines in the 5′-AAC-3′ triplet overlapping the spacer region and the L box (Fig. 9.3). In clear contrast to what is observed in reactions involving *attC* sites, we found that both strands of the *attI* site can start the reaction with similar frequencies (Escudero et al. 2016). It therefore follows that *attC* sites govern the directionality of the integration and excision reactions.

Despite the limited and likely deleterious impact that this reaction can have in sedentary chromosomal integrons, the clinical environment can provide a different outcome for this reaction with a more relevant biological meaning. In clinical isolates, integrons are plasmid-borne and have become prevalent to the point of redundancy (Roy Chowdhury et al. 2009; Gonzalez-Zorn et al. 2005). In these clones, *attI* sites can be constantly present in more than one copy (rather than transiently after replication). Since different integrons can have distinct arrays of cassettes under the control of Pc's of different strengths, the *attI* x *attI* reaction can exchange en bloc the arrays of different integrons. This would produce, in a single event, changes in the levels of expression of multiple antimicrobial resistance genes, potentially impacting the outcome of antibiotic therapy. As we mentioned previously, *attI* sites are recombinogenic during almost the whole cell cycle, being the brief passage of the replisome the only exception. Therefore, despite the low frequency of this reaction when tested in laboratory conditions, it is plausible that *attI* x *attI* reactions are more prevalent among MIs and have a more important impact in the evolution of antimicrobial resistance than we expect. This is especially the case if one considers that horizontal gene transfer is frequent among clinical isolates and that these events produce pulses of transient, low-level expression of the integrase (Baharoglu et al. 2010, 2012). Therefore, the impact of this reaction in the evolution of antibiotic resistance in the clinic is yet to be unveiled.

3.3 Biological Consequences of Innovation in Integrons

Evolutionary innovation is loosely defined as the acquisition of qualitatively new and adaptive traits that confer organisms with game-changing capabilities. Some of the examples that better convey the importance of innovation are the evolution of the eyes, flowers, or flight (Wagner 2011). Integrases represent molecular examples of evolutionary innovation because they have evolved to be capable of recognizing distinct substrates (ds vs. ss hairpins) in different ways (sequence vs. structure specific) and processing the reaction through different pathways (double strand exchange vs. replicative resolution). As we have briefly mentioned and we will see below, semiconservative recombination is indeed a game-changing ability for a recombinase. It is remarkable that integron integrases have conserved the activity of their canonical Y-recombinase ancestors despite the acquisition of a 20-residue-long domain within the catalytic core and millions of years of evolution toward a system specialized in structured ss-DNA recombination. Many examples of enzymes that break and rejoin ds-DNA (like Y-recombinases) or ss-DNA (like HUH endonucleases (Chandler et al. 2013)) can be found, but integrases are the first to show full dual activity on both types of sites, as well as a substrate-dependent switch in recombination pathways. As mentioned earlier, this duality has deep biological implications, since the innovative pathway is semiconservative and dependent on host machinery, while the ancestral is not. There is no obvious reason why integrons could not function exclusively through the classical recombination pathway on ds-DNA to deliver cassette integration and excision reactions. It is therefore tempting to speculate that the force driving the evolution of integrons toward a mixed ss-/ds-DNA system derives from the benefits of semiconservative recombination. The game-changing ability here is that by producing recombined and non-recombined offspring, integron recombination allows testing the adaptive value of incoming DNA while minimizing the deleterious effect of capturing maladaptive genes. This bet-hedging strategy is likely of high evolutionary value. Furthermore, it also represents a mechanism for gene duplication, as observed in the *Vibrio cholerae* superintegron (Escudero et al. 2015) and possibly in some mobile integrons (San Millan et al. 2015).

Once we assume that semiconservative recombination represents an evolutionary advantage that has driven the differentiation of integrons, what suddenly becomes striking is the preservation of the ancestral pathway after eons of evolution (Rowe-Magnus et al. 2001). Indeed, the duality in the recognition and binding of integron integrases to the sites is independent of the pathway, and the classical recombination pathway could have been lost through the specialization towards the innovative one, without affecting any of the known aspects of integron dynamics. Its preservation can hence mean that there has been an absence of strong negative selection (Tawfik 2014) or rather that this reaction still represents a biologically relevant function that we have not unveiled yet. The fact that top and bottom strand recombination in *attI* sites occurs at similar frequencies in the *attI* x *attI* reaction (Fig. 9.5b) suggests that ts cleavage is not a promiscuous activity of the integrase and tilts the balance toward the biological function hypothesis. The flexibility of the integrase is probably important for the integron, and it is possible that functions relying to some extent on this plasticity will be discovered. For instance, the second strand exchange resolution pathway opens new avenues for possible mechanisms of cassette genesis, a subject for which no reliable model is yet available (Escudero et al. 2015), as we will see below.

The evolutionary innovation process toward the recombination of folded ss-*attC* sites imposed strong mechanistic constraints to the recombination process. These ranged from fundamental changes in site recognition to the need for the development of a specific resolution pathway to deliver productive reactions when atypical Holliday junctions are involved. The solution to such constraints, and the physical support allowing innovation to take place, comes in the form of the additional I2 domain that recognizes the EHBs in *attC* sites. Docking of EHBs in the I2 domain permits the differentiation of ss and ds sites. It impedes the second cleavage on ss sites (but not on ds sites (Escudero et al. 2016)) and ultimately produces the replicative resolution of the aHJ. *attC* sites and the integrase I2 domain are a good example of coevolution. The appearance of both structures was likely among the first steps in the evolutionary transition that turned an ancestral recombination system into the hybrid ss-/ds-DNA recombination platform that integrons are today. Yet, the origin of both features raises a chicken-egg paradox because neither the I2 domain nor the *attC* site provides an evolutionary advantage in the absence of the other. Such a conundrum is solved with the discovery of functional intermediary forms in which at least one of the elements does not have a strong negative impact on fitness and is not lost through purifying selection. Integrases are examples of such an intermediary bifunctional state and are proof of a smooth evolutionary transition during functional innovation. We can now hypothesize that an I2-like domain was acquired at a given time by the ancestor of today's integron integrases, conserving to some degree its original activity on the initial site while permitting the recognition of new ss substrates. From this starting point, the optimization process leading to the origin of integrons is easily conceivable through natural selection.

The only other element known to share some similarities with integron recombination is the CTXϕ phage of *V. cholerae*. Indeed, the phage genome is integrated as a single strand and adopts a hairpin structure that is similar to that of *attC* sites.

Nevertheless, CTXϕ does not encode an integrase itself, but instead hijacks the host XerCD recombinases (Val et al. 2005) that lack the structural adaptations for ss-DNA recombination (I2-like domains). Accordingly, the phage site stem does not have EHBs docking on XerCD and simply mimics a double-stranded site. Still, some common features between cassette and CTXϕ recombination, together with the close phylogenetic relation between integrases and XerCD recombinases and the fact that both elements cohabit nowadays in some bacterial species (for instance, in *V. cholerae*), foster the speculation about a possible viral origin for integrons. Indeed, some other elements of the integron physiology, such as its connection to the bacterial SOS response, could also have a viral origin, since it is a strategy well spread among phages, including CTXϕ (Quinones et al. 2005). This remains purely speculative since there are no reports of XerCD proteins that possess an I2-like domain. Nevertheless, within the context of a viral origin of integrons, it is possible to conceive a situation in which the accidental acquisition of a similar domain would be advantageous for both the host and the phage and would therefore be stable. If this domain helped avoid the second strand exchange during ss-phage integration, it would protect the integrity of the chromosome while favoring correct phage integration events. We now know that the acquisition of the domain is not as destabilizing as previously assumed and that the recombinase could still conserve its chromosome-dimer resolution activity. In this situation, phages (or any other *i*ntegrative *m*obile element *e*xploiting *X*er (IMEX (Das et al. 2013))) would serve as a vehicle of new adaptive genes, just as they currently do (Midonet and Barre 2014; Waldor and Mekalanos 1996), representing an ancestor of integron cassettes.

4 Cassette Shuffling and Dynamics

The coupling of an excision and an integration reaction produces a change in the location of a cassette within the array. Crucially, this relocation comes with a difference in expression levels of the gene encoded within the cassette. Coupling the two reactions allows bacteria to render functional genes with adaptive traits that had been kept silent (see Sect. 2.2.3). If we stick to the metaphor in which integrons act as a *memory* of adaptive functions, the shuffling of cassettes would be the equivalent to *recalling memories*. Excised cassettes could be lost through degradation or segregation if not rapidly reinserted, a process favored by the high efficiency of the *attI* x *attC* reaction. Excision and reinsertion of cassettes can foster long-term evolution of gene functions through diversification, since it can lead to cassette duplication. This occurs if, after the passage of the replication fork, the excised cassette reintegrates back at the *attI* site that derives from the replication of the non-recombined ts (the backup version of the integron where the cassette is still present in its original location) (Gestal et al. 2005).

Cassettes are streamlined to be reshuffled within integrons, but their role in horizontal gene transfer remains unclear. Indeed, the replicative resolution pathway imposes that cassettes come in a circular, covalently closed conformation in order for the integration reaction not to be abortive. Given that natural competence involves linear DNA, it is unlikely that individual cassettes are viable intercellular gene vehicles *per se*. Nevertheless, linear fragments of DNA containing more than one *attC* site could be integrated through two simultaneous reactions to overcome this limitation.

As we have seen above, cassette excision rate and dynamics are highly regulated by replication and by the intrinsic properties of cassette recombination sites and differ between MIs and SCIs. Indeed, in MIs, efficient cassette recombination is favored and timed to conditions when generation of diversity upon which selection can act ensures a rapid response to environmental stresses (evolvability). Moreover, MIs are essentially carried by conjugative plasmids, in which the folding and recombination of *attC* sites is favored during ss-DNA transfer (Fig. 9.7).

In contrast, for SCIs, cassette excisions are less frequent, limiting cassette loss and favoring the maintenance of large cassette arrays and vertical transmission (genetic capacitance). Moreover, in SCIs, large sections of cassette arrays are maintained due to the presence of TA systems in some cassettes (Fig. 9.7) (Jove et al. 2010; Iqbal et al. 2015). The most recombinogenic cassettes in SCIs would be more likely mobilized in MIs and further selected because of their higher capacity to disseminate the associated adaptive functions (Fig. 9.7).

Despite a clear working model, the dynamics of cassette reshuffling are not clear. Some articles provide evidence that the simplest way to locate a cassette closer to the P_C promoter is to lose the ones in between rather than to excise it and reintegrate it in *attI*. Unfortunately, the experimental setup that yields such results is one in which integrons are located in multicopy plasmids, where it is impossible to differentiate *bona fide* loss (cassette excision not followed by reintegration) from the recombination between *attC* sites in integrons borne by different copies of the plasmid (Barraud and Ploy 2015). The only experimental setup in which cassette movement was assessed using a monocopy integron (the superintegron) proved the movement of a single cassette without the loss of intermediary ones, supporting the excision/integration model (Baharoglu and Mazel 2011). It could be the case that cassette reorganization occurs through different mechanisms in MIs and SCIs and that integron properties shape a trade-off between evolvability and genetic capacitance.

5 Genesis of Integron Cassettes

One of the major questions yet to find an answer in the field of integrons is the mechanism of de novo creation of cassettes. Indeed, we now have a valuable knowledge on *attC* sites, their structure, folding and intertwining with host processes, but we do not have a working model on how a gene can be associated to an *attC* site for the first time. The only theory available nowadays is one postulating an RNA origin (Leon and Roy 2009), based on the general absence of promoters and the paucity of pseudogenes or noncoding sequences in cassettes. The agent in charge of cassette genesis in this model is the group IIC-*attC* intron, a complex protein with

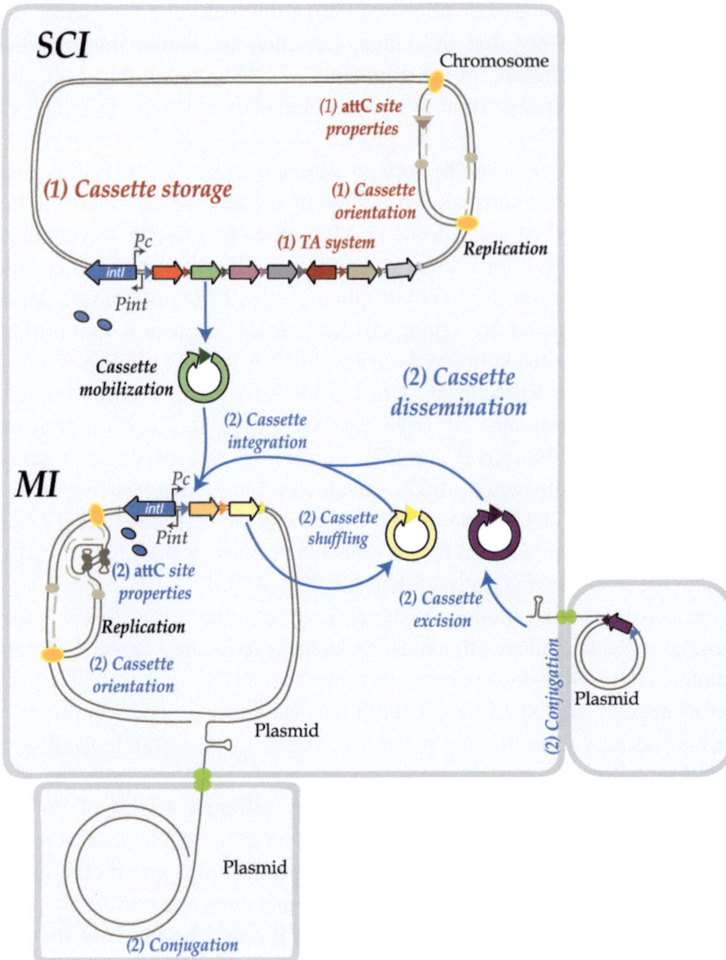

Fig. 9.7 Difference between integrons in their genetic capacitance. Cassette storage in sedentary chromosomal integrons (SCIs) (red fonts) is favored by a low rate of cassette excision depending on the *attC* site properties, the orientation of the integron relative to the replication fork (cassette orientation) and the presence of TA (toxin-antitoxin) systems. Cassette dissemination in mobile integrons (MIs) (blue fonts) is favored by a high rate of cassette excision and shuffling, depending on *attC* site properties, cassette orientation, and conjugation. P_{int} and P_C promoters are represented by black arrows. The *intI1* integrase gene is represented by a blue arrow. Cassettes are represented by colour arrows followed by triangles (*attC* sites). Replication forks are shown by yellow ovals

RNA splicing and retrotranscriptase activities that shows affinity for palindromic sequences including *attC* sites (Quiroga and Centron 2009). The mechanism is not simple: presumably, two identical group IIC-*attC* introns would integrate at a solitary *attC* site and at the transcriptional terminator of the gene of interest. Homologous recombination would fuse both molecules forming a gene-intron-

attC intermediate. Once this intermediate is transcribed into RNA, the intron would then splice from it and retrotranscribe the resulting gene-*attC* mRNA molecule, forming a DNA-encoded cassette that could be then incorporated to the integron array by the integrase. This model of cassette genesis is fundamentally based on the fact that such introns have seldom been found within integrons, but direct evidence supporting the mechanism is lacking (Leon and Roy 2009; Quiroga and Centron 2009). Although the rationale for the RNA origin of cassettes is interesting, some facts are difficult to be accommodated in this model. For instance, it does not explain the creation of cassettes that do encode promoters, such as toxin-antitoxin systems, suggesting that one additional mechanism is necessary to produce this type of cassettes. Also, and bearing in mind the subtleties of the intertwining between integrons and the host machinery, it seems improbable that a major function of integrons relies on an independent entity that is found in a very low percentage of integrons. Indeed, if group IIC-*attC* introns were cassette generators, it seems likely that after millions of years of coevolution, they would have become a domesticated element encoded in the stable part of the integron platform, especially in the case of SIs in which *attC* sites show high levels of identity, suggesting that the cassette generator is stably linked to the host. Unfortunately, the creation of cassettes remains a subject of the utmost importance for which an uncontested model remains to be postulated.

6 Biotechnological Applications of Integrons

The ability to reshuffle long arrays of cassettes, together with the large population size of bacterial cultures, makes integrons extremely powerful combinatorial tools. The possibility of harnessing such recombination power for biosynthetic purposes led to the concept of the synthetic integron: an in vivo genetic shuffling device in which the elements to be reshuffled are designed by the researcher (Bikard et al. 2010b). The synthetic integron is of particular interest in the streamlining of metabolic pathways in which higher yields of a metabolite can be achieved by improving the order of the genetic elements encoding the proteins of the metabolic route. A step further in the engineering of synthetic integrons is the rational design of synthetic *attC* sites taking advantage of the lack of sequence conservation of these elements. This allows to embed sites into functional genetic elements with minimal disturbances of the genetic code. *attC* sites can be engineered, for example, to be encoded within a protein sequence or a promoter. The synthetic integron is a novel tool that can lead to interesting applications in bioengineering.

Some other applications of the integron have already been implemented, although their scope is limited to the integron field. For instance, integrons have been engineered to explore the cassette content of different environments. This tool allows for the recovery of cassettes from genomic libraries, regardless of their sequence or genetic context (Rowe-Magnus 2009). This is particularly useful to access integron cassettes from unculturable organisms. Given the large number of cassettes encoding genes of unknown function among environmental bacteria, it is likely that such tools can lead to the discovery of new, currently uncharacterized protein families with possible novel functions (Koenig et al. 2008, 2009; Sureshan et al. 2013).

Last, a cloning technique based on the integron has also been developed. It uses natural transformation to deliver synthetic cassettes into mobile or chromosomal integrons (Gestal et al. 2011). This method can be used with environmental bacteria, yielding large numbers of stable recombinants without a vector and in the absence of selection. Notably, the lack of antibiotic resistance genes for vector selection makes the technique particularly safe and appealing.

7 Integrons as Markers of Anthropogenic Activity

Class 1 integrons have become extremely prevalent among clinical isolates, due to the adaptive advantage that multiresistance represents in anthropomorphized environments. They are readily detectable from metagenomic samples of a variety of human-related environments, and they are therefore considered as proxies of human activity (Moura et al. 2012; Gillings et al. 2015). In accordance with the integrated view on integrons provided along this chapter, that helps limit their impact on the host's fitness, integrons have recently been shown to persist in the environment in the absence of selective pressure (Chamosa et al. 2017). Yet, there might be a drawback to studies assessing anthropogenic pollution through the presence of the *intI1* gene. As we have already seen, plasmid-encoded mobile integrons are not the native form of these integrons but result from the association of sedentary chromosomal integrons to transposons. Indeed, class I integrons have been found in their native *sedentary* form in environmental bacteria of the *Imtechium*, *Hydrogenophaga*, and *Aquabacterium* genera, i.e., encoded in the chromosome, containing genes of functions other than antibiotic resistance and lacking a transposon in the surroundings (Gillings et al. 2008). It is therefore possible that PCR-based approaches in which the *intI1* gene is amplified from environmental samples yield positive results in the absence of human activity.

8 Open Questions in the Field

Almost three decades of research on integrons have yielded a full working model and a complex understanding of these structures. From the uniqueness of the recombination process to the subtle blending with bacterial physiology through host regulatory networks, integrons have proven to be elegant genetic solutions to the delicate balance between risk and gain in evolution. They have also taken our understanding of bacterial evolvability to a completely different level, giving support to Louis Pasteur's intuition that "*Messieurs, ce sont les microbes qui auront le dernier mot*" (Gentlemen, it is the microbes who will have the last word). Many questions about integrons remain unanswered. We have already mentioned them in their specific sections, but they are worth being repeated here to put together the current challenges in the field. These questions hold fascinating secrets of bacterial ecology, physiology, and evolution awaiting to be revealed.

From a mechanistic point of view, the main question remains *how* and *who* creates cassettes de novo, i.e., how a given gene is exapted from the chromosome to become a mobile, modular, and accessory part of the genome. What are the genetic cues recognized in these genes, and how can this process be so precise that cassettes rarely contain pseudogenes? What protein is capable of producing cassettes, and is it an unrelated element in the genome or a gene encoded in the integron platform of only a subset of integrons? The answer to all these questions might reveal a dedicated system for gene mobilization and is therefore relevant for all fields of biology.

A second relevant mechanistic question is how integrases bind to *attI* sites. It is indeed a sequence-specific process since they recognize their cognate *attI* very efficiently, but they do not recognize, or they do at very low frequency, the *attI* sites from other integron classes. The R box is generally the first one to be bound by an IntI monomer, and this produces a cooperative binding of other monomers to the L box (and to the direct repeats of the class 1 *attI*). Yet the sequence of the R box changes after every cassette integration event, except for the 5′-CAA-3′, and this triplet is common to all integron sites, so it cannot help distinguish cognate from non-cognate sites. So, altogether, it is unclear how *attI* site recognition occurs, but this process is relevant to understand the plasticity of Y-recombinases that has influenced their evolution.

From a physiologic standpoint, the most relevant open issue in the field is the cryptic functions encoded in cassettes. The vast amount of novel functions and protein families that are encoded in environmental integrons holds a prodigious biotechnological potential. Given the presumed adaptive value of cassettes, they will also shed light on relevant yet unknown aspects of the ecology and lifestyle of bacteria. Hence, the importance of the functions that will be found in cassettes is as difficult to predict as it is to overestimate.

From an evolutionary perspective, many aspects of integrons remain unexplored. Yet some relevant questions are starting to find answers. As we have already seen, the low fitness cost of class 1 integrons—and the role that the tight control of integrase expression plays in keeping it low—have given the experimental basis to understand many field observations. These range from the great evolutionary success of integrons to the presence of premature stop codons in unrepressed class 2 integrases or in integrons found in species lacking LexA. Still, the precise behavior of integrons under selective pressure, their adaptive value, the differential cost of cassettes, the impact of cassette reshuffling and of co-selection, and the role played by cassettes that bear a promoter are aspects that need to be studied in greater depth. A good understanding of the evolutionary dynamics of integrons is crucial to design approaches to limit their spread and impact in public health.

9 Online Resources

Several databases compile the information available on a variety of aspects of integrons, such as integrase and cassette DNA sequences. Among them, the Integrall database has listed more than 8500 cassettes (Moura et al. 2009, http://integrall.bio.ua.pt/), RAC is focused on annotations of cassettes related to antibiotic resistance (Tsafnat et al. 2011), and ACID has contains a collection of 5622 integron cassettes (Joss et al. 2009). Annotation systems allowing cassette identification have also been developed, such as XXR (https://galaxy.pasteur.fr motif tools tab Rowe-Magnus et al. 2003). Last, a new tool allowing to identify integrons or parts of integrons in bacterial genomes, Integron finder, has recently been developed (Cury et al. 2016, https://galaxy.pasteur.fr).

Acknowledgments This work was supported by the Institut Pasteur, the Centre National de la Recherche Scientifique (CNRS-UMR3525), the European Union Seventh Framework Program (FP7-HEALTH- 2011-single-stage), the "Evolution and Transfer of Antibiotic Resistance" (EvoTAR) project, the Fondation pour la Recherche Medicale (FRM) project DBF20160635736 and the French Government's Investissement d'Avenir program Laboratoire d'Excellence "Integrative Biology of Emerging Infectious Diseases" (ANR-10-LABX-62-IBEID), and the French National Research Agency (ANR-12-BLAN-DynamINT). JAE is supported by the Marie Curie Intra-European Fellowship for Career Development (FP-7-PEOPLE-2011-IEF, ICADIGE) and by the 2016-T1/BIO-1105 project of the *Atracción de Talento* Program of the *Comunidad de Madrid*.

References

Arakawa Y et al (1995) A novel integron-like element carrying the metallo-beta-lactamase gene *blaIMP*. Antimicrob Agents Chemother 39:1612–1615

Baharoglu Z, Mazel D (2011) *Vibrio cholerae* triggers SOS and mutagenesis in response to a wide range of antibiotics: a route towards multiresistance. Antimicrob Agents Chemother 55:2438–2441. https://doi.org/10.1128/AAC.01549-10

Baharoglu Z, Bikard D, Mazel D (2010) Conjugative DNA transfer induces the bacterial SOS response and promotes antibiotic resistance development through integron activation. PLoS Genet 6:e1001165. https://doi.org/10.1371/journal.pgen.1001165

Baharoglu Z, Krin E, Mazel D (2012) Connecting environment and genome plasticity in the characterization of transformation-induced SOS regulation and carbon catabolite control of the Vibrio cholerae integron integrase. J Bacteriol 194:1659–1667. https://doi.org/10.1128/JB.05982-11

Baharoglu Z, Krin E, Mazel D (2013) RpoS plays a central role in the SOS induction by sub-lethal aminoglycoside concentrations in *Vibrio cholerae*. PLoS Genet 9:e1003421. https://doi.org/10.1371/journal.pgen.1003421

Balke VL, Gralla JD (1987) Changes in the linking number of supercoiled DNA accompany growth transitions in *Escherichia coli*. J Bacteriol 169:4499–4506

Baquero F, Coque TM, de la Cruz F (2011) Ecology and evolution as targets: the need for novel eco-evo drugs and strategies to fight antibiotic resistance. Antimicrob Agents Chemother 55:3649–3660. https://doi.org/10.1128/AAC.00013-11

Barker A, Clark CA, Manning PA (1994) Identification of VCR, a repeated sequence associated with a locus encoding a hemagglutinin in *Vibrio cholerae* O1. J Bacteriol 176:5450–5458

Barraud O, Ploy MC (2015) Diversity of Class 1 integron gene cassette rearrangements selected under antibiotic pressure. J Bacteriol 197:2171–2178. https://doi.org/10.1128/JB.02455-14

Bikard D, Loot C, Baharoglu Z, Mazel D (2010a) Folded DNA in action: hairpin formation and biological functions in prokaryotes. Microbiol Mol Biol Rev 74:570–588. https://doi.org/10.1128/MMBR.00026-10

Bikard D, Julie-Galau S, Cambray G, Mazel D (2010b) The synthetic integron: an in vivo genetic shuffling device. Nucleic Acids Res 38:e153. https://doi.org/10.1093/nar/gkq511

Biskri L, Mazel D (2003) Erythromycin esterase gene ere(A) is located in a functional gene cassette in an unusual class 2 integron. Antimicrob Agents Chemother 47:3326–3331

Biskri L, Bouvier M, Guerout AM, Boisnard S, Mazel D (2005) Comparative study of class 1 integron and *Vibrio cholerae* superintegron integrase activities. J Bacteriol 187:1740–1750

Bissonnette L, Champetier S, Buisson JP, Roy PH (1991) Characterization of the nonenzymatic chloramphenicol resistance (*cmlA*) gene of the In4 integron of Tn1696: similarity of the product to transmembrane transport proteins. J Bacteriol 173:4493–4502

Boucher Y, Labbate M, Koenig JE, Stokes HW (2007) Integrons: mobilizable platforms that promote genetic diversity in bacteria. Trends Microbiol 15:301–309

Bouvier M, Demarre G, Mazel D (2005) Integron cassette insertion: a recombination process involving a folded single strand substrate. EMBO J 24:4356–4367. https://doi.org/10.1038/sj.emboj.7600898

Bouvier M, Ducos-Galand M, Loot C, Bikard D, Mazel D (2009) Structural features of single-stranded integron cassette *attC* sites and their role in strand selection. PLoS Genet 5:e1000632. https://doi.org/10.1371/journal.pgen.1000632

Boyd EF, Almagro-Moreno S, Parent MA (2009) Genomic islands are dynamic, ancient integrative elements in bacterial evolution. Trends Microbiol 17:47–53. https://doi.org/10.1016/j.tim.2008.11.003

Cambray G, Mazel D (2008) Synonymous genes explore different evolutionary landscapes. PLoS Genet 4:e1000256. https://doi.org/10.1371/journal.pgen.1000256

Cambray G, Guerout AM, Mazel D (2010) Integrons. Annu Rev Genet 44:141–166. https://doi.org/10.1146/annurev-genet-102209-163504

Cambray G et al (2011) Prevalence of SOS-mediated control of integron integrase expression as an adaptive trait of chromosomal and mobile integrons. Mob DNA 2:6. https://doi.org/10.1186/1759-8753-2-6

Chamosa LS et al (2017) Lateral antimicrobial resistance genetic transfer is active in the open environment. Sci Rep 7:513. https://doi.org/10.1038/s41598-017-00600-2

Champion K, Higgins NP (2007) Growth rate toxicity phenotypes and homeostatic supercoil control differentiate *Escherichia coli* from *Salmonella enterica* serovar Typhimurium. J Bacteriol 189:5839–5849. https://doi.org/10.1128/JB.00083-07

Chandler M et al (2013) Breaking and joining single-stranded DNA: the HUH endonuclease superfamily. Nat Rev Microbiol 11:525–538. https://doi.org/10.1038/nrmicro3067

Chowdhury N et al (2010) Development of simple and rapid PCR-fingerprinting methods for *Vibrio cholerae* on the basis of genetic diversity of the superintegron. J Appl Microbiol 109(1):304–312. https://doi.org/10.1111/j.1365-2672.2009.04658.x

Collins J, Volckaert G, Nevers P (1982) Precise and nearly-precise excision of the symmetrical inverted repeats of Tn5; common features of recA-independent deletion events in *Escherichia coli*. Gene 19:139–146. https://doi.org/10.1016/0378-1119(82)90198-6

Collis CM, Hall RM (1992a) Gene cassettes from the insert region of integrons are excised as covalently closed circles. Mol Microbiol 6:2875–2885

Collis CM, Hall RM (1992b) Site-specific deletion and rearrangement of integron insert genes catalyzed by the integron DNA integrase. J Bacteriol 174:1574–1585

Collis CM, Hall RM (1995) Expression of antibiotic resistance genes in the integrated cassettes of integrons. Antimicrob Agents Chemother 39:155–162

Collis CM, Grammaticopoulos G, Briton J, Stokes HW, Hall RM (1993) Site-specific insertion of gene cassettes into integrons. Mol Microbiol 9:41–52

Collis CM, Recchia GD, Kim MJ, Stokes HW, Hall RM (2001) Efficiency of recombination reactions catalyzed by class 1 integron integrase IntI1. J Bacteriol 183:2535–2542

Collis CM, Kim MJ, Partridge SR, Stokes HW, Hall RM (2002) Characterization of the class 3 integron and the site-specific recombination system it determines. J Bacteriol 184:3017–3026

Cury J, Jove T, Touchon M, Neron B, Rocha EP (2016) Identification and analysis of integrons and cassette arrays in bacterial genomes. Nucleic Acids Res 44:4539–4550. https://doi.org/10.1093/nar/gkw319

da Fonseca EL, Vicente AC (2012) Functional characterization of a Cassette-specific promoter in the class 1 integron-associated qnrVC1 gene. Antimicrob Agents Chemother 56:3392–3394. https://doi.org/10.1128/AAC.00113-12

Das B, Martinez E, Midonet C, Barre FX (2013) Integrative mobile elements exploiting Xer recombination. Trends Microbiol 21:23–30. https://doi.org/10.1016/j.tim.2012.10.003

Demarre G, Frumerie C, Gopaul DN, Mazel D (2007) Identification of key structural determinants of the IntI1 integron integrase that influence *attC* x *attI1* recombination efficiency. Nucleic Acids Res 35:6475–6489. https://doi.org/10.1093/Nar/Gkm709

Elsaied H et al (2007) Novel and diverse integron integrase genes and integron-like gene cassettes are prevalent in deep-sea hydrothermal vents. Environ Microbiol 9:2298–2312. https://doi.org/10.1111/j.1462-2920.2007.01344.x

Elsaied H et al (2011) Marine integrons containing novel integrase genes, attachment sites, attI, and associated gene cassettes in polluted sediments from Suez and Tokyo Bays. ISME J 5:1162–1177. https://doi.org/10.1038/ismej.2010.208

Erill I, Campoy S, Barbe J (2007) Aeons of distress: an evolutionary perspective on the bacterial SOS response. FEMS Microbiol Rev 31:637–656

Escudero JA, Loot C, Nivina A, Mazel D (2015) The integron: adaptation on demand. Microbiol Spectr 3: MDNA3-0019-2014. https://doi.org/10.1128/microbiolspec.MDNA3-0019-2014

Escudero JA et al (2016) Unmasking the ancestral activity of integron integrases reveals a smooth evolutionary transition during functional innovation. Nat Commun 7:10937. https://doi.org/10.1038/ncomms10937

Feng L et al (2008) A recalibrated molecular clock and independent origins for the cholera pandemic clones. PLoS One 3:e4053. https://doi.org/10.1371/journal.pone.0004053

Ferrandiz MJ, Martin-Galiano AJ, Schvartzman JB, de la Campa AG (2010) The genome of *Streptococcus pneumoniae* is organized in topology-reacting gene clusters. Nucleic Acids Res 38:3570–3581. https://doi.org/10.1093/nar/gkq106

Fluit AC, Schmitz FJ (2004) Resistance integrons and super-integrons. Clin Microbiol Infect 10:272–288

Francia MV, Zabala JC, de la Cruz F, Garcia-Lobo JM (1999) The IntI1 integron integrase preferentially binds single-stranded DNA of the *attC* site. J Bacteriol 181:6844–6849

Gerdes K, Christensen SK, Lobner-Olesen A (2005) Prokaryotic toxin-antitoxin stress response loci. Nat Rev Microbiol 3:371–382

Gestal AM, Stokes HW, Partridge SR, Hall RM (2005) Recombination between the dfrA12-orfF-aadA2 cassette array and an aadA1 gene cassette creates a hybrid cassette, aadA8b. Antimicrob Agents Chemother 49:4771–4774. https://doi.org/10.1128/AAC.49.11.4771-4774.2005

Gestal AM, Liew EF, Coleman NV (2011) Natural transformation with synthetic gene cassettes: new tools for integron research and biotechnology. Microbiology 157:3349–3360. https://doi.org/10.1099/mic.0.051623-0

Gillings MR (2014) Integrons: past, present, and future. Microbiol Mol Biol Rev 78:257–277. https://doi.org/10.1128/MMBR.00056-13

Gillings MR, Holley MP, Stokes HW, Holmes AJ (2005) Integrons in Xanthomonas: a source of species genome diversity. Proc Natl Acad Sci USA 102:4419–4424

Gillings M et al (2008) The evolution of class 1 integrons and the rise of antibiotic resistance. J Bacteriol 190:5095–5100. https://doi.org/10.1128/JB.00152-08

Gillings MR, Holley MP, Stokes HW (2009) Evidence for dynamic exchange of qac gene cassettes between class 1 integrons and other integrons in freshwater biofilms. FEMS Microbiol Lett 296:282–288. https://doi.org/10.1111/j.1574-6968.2009.01646.x

Gillings MR et al (2015) Using the class 1 integron-integrase gene as a proxy for anthropogenic pollution. ISME J 9:1269–1279. https://doi.org/10.1038/ismej.2014.226

Gonzalez-Zorn B, Escudero JA (2012) Ecology of antimicrobial resistance: humans, animals, food and environment. Int Microbiol 15:101–109. https://doi.org/10.2436/20.1501.01.163

Gonzalez-Zorn B et al (2005) Genetic basis for dissemination of *armA*. J Antimicrob Chemother 56:583–585. https://doi.org/10.1093/jac/dki246

Gravel A, Messier N, Roy PH (1998a) Point mutations in the integron integrase IntI1 that affect recombination and/or substrate recognition. J Bacteriol 180:5437–5442

Gravel A, Fournier B, Roy PH (1998b) DNA complexes obtained with the integron integrase IntI1 at the *attI1* site. Nucleic Acids Res 26:4347–4355. https://doi.org/10.1093/Nar/26.19.4347

Grieb MS et al (2017) Dynamic stepwise opening of integron attC DNA hairpins by SSB prevents toxicity and ensures functionality. Nucleic Acids Res 45(18):10555–10563. https://doi.org/10.1093/nar/gkx670

Grindley ND, Whiteson KL, Rice PA (2006) Mechanisms of site-specific recombination. Annu Rev Biochem 75:567–605. https://doi.org/10.1146/annurev.biochem.73.011303.073908

Guerin E et al (2009) The SOS response controls integron recombination. Science 324:1034. https://doi.org/10.1126/science.1172914

Guerin E, Jove T, Tabesse A, Mazel D, Ploy MC (2011) High-level gene cassette transcription prevents integrase expression in class 1 integrons. J Bacteriol 193:5675–5682. https://doi.org/10.1128/JB.05246-11

Guerout AM et al (2013) Characterization of the phd-doc and ccd toxin-antitoxin cassettes from Vibrio superintegrons. J Bacteriol 195:2270–2283. https://doi.org/10.1128/JB.01389-12

Gutierrez A et al (2013) beta-Lactam antibiotics promote bacterial mutagenesis via an RpoS-mediated reduction in replication fidelity. Nat Commun 4:1610. https://doi.org/10.1038/ncomms2607

Hall RM, Brookes DE, Stokes HW (1991) Site-specific insertion of genes into integrons – role of the 59-base element and determination of the recombination cross-over point. Mol Microbiol 5:1941–1959. https://doi.org/10.1111/J.1365-2958.1991.Tb00817.X

Hanau-Bercot B, Podglajen I, Casin I, Collatz E (2002) An intrinsic control element for translational initiation in class 1 integrons. Mol Microbiol 44:119–130

Hansson K, Skold O, Sundstrom L (1997) Non-palindromic *attI* sites of integrons are capable of site-specific recombination with one another and with secondary targets. Mol Microbiol 26:441–453

Harms K, Starikova I, Johnsen PJ (2013) Costly Class-1 integrons and the domestication of the the functional integrase. Mob Genet Elements 3:e24774. https://doi.org/10.4161/mge.24774

Hochhut B et al (2001) Molecular analysis of antibiotic resistance gene clusters in *Vibrio cholerae* O139 and O1 SXT constins. Antimicrob Agents Chemother 45:2991–3000

Hocquet D et al (2012) Evidence for induction of integron-based antibiotic resistance by the SOS response in a clinical setting. PLoS Pathog 8:e1002778. https://doi.org/10.1371/journal.ppat.1002778

Holmes AJ et al (2003a) Recombination activity of a distinctive integron-gene cassette system associated with Pseudomonas stutzeri populations in soil. J Bacteriol 185:918–928

Holmes AJ et al (2003b) The gene cassette metagenome is a basic resource for bacterial genome evolution. Environ Microbiol 5:383–394

Iqbal N, Guerout AM, Krin E, Le Roux F, Mazel D (2015) Comprehensive functional analysis of the 18 *Vibrio cholerae* N16961 toxin-antitoxin systems substantiates their role in stabilizing the superintegron. J Bacteriol 197:2150–2159. https://doi.org/10.1128/JB.00108-15

Jacquier H, Zaoui C, Sanson-le Pors MJ, Mazel D, Bercot B (2009) Translation regulation of integrons gene cassette expression by the attC sites. Mol Microbiol 72:1475–1486. https://doi.org/10.1111/j.1365-2958.2009.06736.x

Jaworski A, Higgins NP, Wells RD, Zacharias W (1991) Topoisomerase mutants and physiological conditions control supercoiling and Z-DNA formation in vivo. J Biol Chem 266:2576–2581

Johansson C, Kamali-Moghaddam M, Sundstrom L (2004) Integron integrase binds to bulged hairpin DNA. Nucleic Acids Res 32:4033–4043

Johansson C, Boukharta L, Eriksson J, Aqvist J, Sundstrom L (2009) Mutagenesis and homology modeling of the Tn21 integron integrase IntI1. Biochemistry 48:1743–1753. https://doi.org/10.1021/bi8020235

Joss MJ et al (2009) ACID: annotation of cassette and integron data. BMC Bioinformatics 10:118. https://doi.org/10.1186/1471-2105-10-118

Jove T, Da Re S, Denis F, Mazel D, Ploy MC (2010) Inverse correlation between promoter strength and excision activity in class 1 integrons. PLoS Genet 6:e1000793. https://doi.org/10.1371/journal.pgen.1000793

Jové T, Da Re S, Tabesse A, Gassama-Sow A, Ploy MC (2017) Gene expression in class 2 integrons is SOS-independent and involves two Pc promoters. Front Microbiol 8:1499. https://doi.org/10.3389/fmicb.2017.01499

Koenig JE et al (2008) Integron-associated gene cassettes in Halifax Harbour: assessment of a mobile gene pool in marine sediments. Environ Microbiol 10:1024–1038. https://doi.org/10.1111/j.1462-2920.2007.01524.x

Koenig JE et al (2009) Integron gene cassettes and degradation of compounds associated with industrial waste: the case of the Sydney tar ponds. PLoS One 4:e5276. https://doi.org/10.1371/journal.pone.0005276

Krin E, Cambray G, Mazel D (2014) The superintegron integrase and the cassette promoters are co-regulated in *Vibrio cholerae*. PLoS One 9:e91194. https://doi.org/10.1371/journal.pone.0091194

Labbate M et al (2007) Use of chromosomal integron arrays as a phylogenetic typing system for *Vibrio cholerae* pandemic strains. Microbiology 153:1488–1498

Labbate M, Case RJ, Stokes HW (2009) The integron/gene cassette system: an active player in bacterial adaptation. Methods Mol Biol 532:103–125. https://doi.org/10.1007/978-1-60327-853-9_6

Lacotte Y, Ploy MC, Raherison S (2017) Class 1 integrons are low-cost structures in *Escherichia coli*. ISME J 11:1535–1544. https://doi.org/10.1038/ismej.2017.38

Leon G, Roy PH (2009) Potential role of group IIC-attC introns in integron cassette formation. J Bacteriol 191:6040–6051. https://doi.org/10.1128/JB.00674-09

Levesque C, Brassard S, Lapointe J, Roy PH (1994) Diversity and relative strength of tandem promoters for the antibiotic-resistance genes of several integrons. Gene 142:49–54

Liebert CA, Hall RM, Summers AO (1999) Transposon Tn21, flagship of the floating genome. Microbiol Mol Biol Rev 63:507–522

Liu LF, Wang JC (1987) Supercoiling of the DNA template during transcription. Proc Natl Acad Sci USA 84:7024–7027

Loot C, Bikard D, Rachlin A, Mazel D (2010) Cellular pathways controlling integron cassette site folding. EMBO J 29:2623–2634. https://doi.org/10.1038/emboj.2010.151

Loot C, Ducos-Galand M, Escudero JA, Bouvier M, Mazel D (2012) Replicative resolution of integron cassette insertion. Nucleic Acids Res 40:8361–8370. https://doi.org/10.1093/nar/gks620

Loot C et al (2014) The integron integrase efficiently prevents the melting effect of *Escherichia coli* single-stranded DNA-binding protein on folded attC sites. J Bacteriol 196:762–771. https://doi.org/10.1128/JB.01109-13

Loot C et al (2017) Differences in integron cassette excision dynamics shape a trade-off between evolvability and genetic capacitance. MBio 8:e02296-16. https://doi.org/10.1128/mBio.02296-16

MacDonald D, Demarre G, Bouvier M, Mazel D, Gopaul DN (2006) Structural basis for broad DNA-specificity in integron recombination. Nature 440:1157–1162. https://doi.org/10.1038/nature04643

Majchrzak M, Bowater RP, Staczek P, Parniewski P (2006) SOS repair and DNA supercoiling influence the genetic stability of DNA triplet repeats in *Escherichia coli*. J Mol Biol 364:612–624. https://doi.org/10.1016/j.jmb.2006.08.093

Martin C et al (1990) Transposition of an antibiotic resistance element in mycobacteria. Nature 345:739–743

Martinez E, de la Cruz F (1988) Transposon Tn21 encodes a RecA-independant site-specific integration system. Mol Gen Genet 211:320–325

Mazel D (2006) Integrons: agents of bacterial evolution. Nat Rev Microbiol 4:608–620. https://doi.org/10.1038/nrmicro1462

Mazel D, Dychinco B, Webb VA, Davies J (1998) A distinctive class of integron in the *Vibrio cholerae* genome. Science 280:605–608

Messier N, Roy PH (2001) Integron integrases possess a unique additional domain necessary for activity. J Bacteriol 183:6699–6706

Meyer RR, Laine PS (1990) The single-stranded DNA-binding protein of *Escherichia coli*. Microbiol Rev 54:342–380

Midonet C, Barre FX (2014) Xer site-specific recombination: promoting vertical and horizontal transmission of genetic information. Microbiol Spectr 2. https://doi.org/10.1128/microbiolspec.MDNA3-0056-2014

Mitsuhashi S, Harada K, Hashimoto H, Egawa R (1961) On the drug-resistance of enteric bacteria. Jpn J Exp Med 31:47–52

Moura A et al (2009) INTEGRALL: a database and search engine for integrons, integrases and gene cassettes. Bioinformatics 25:1096–1098. https://doi.org/10.1093/bioinformatics/btp105

Moura A, Jove T, Ploy MC, Henriques I, Correia A (2012) Diversity of gene cassette promoters in class 1 integrons from wastewater environments. Appl Environ Microbiol 78:5413–5416. https://doi.org/10.1128/AEM.00042-12

Naas T, Mikami Y, Imai T, Poirel L, Nordmann P (2001) Characterization of In53, a class 1 plasmid- and composite transposon-located integron of *Escherichia coli* which carries an unusual array of gene cassettes. J Bacteriol 183:235–249

Nakaya R, Nakamura A, Murata Y (1960) Resistance transfer agents in *Shigella*. Biochem Biophys Res Commun 3:654–659

Nandi S, Maurer JJ, Hofacre C, Summers AO (2004) Gram-positive bacteria are a major reservoir of Class 1 antibiotic resistance integrons in poultry litter. Proc Natl Acad Sci USA 101:7118–7122

Nemergut DR et al (2008) Insights and inferences about integron evolution from genomic data. BMC Genomics 9:261

Nesvera J, Hochmannova J, Patek M (1998) An integron of class 1 is present on the plasmid pCG4 from gram-positive bacterium *Corynebacterium glutamicum*. FEMS Microbiol Lett 169:391–395

Nield BS et al (2001) Recovery of new integron classes from environmental DNA. FEMS Microbiol Lett 195:59–65

Nivina A, Escudero JA, Vit C, Mazel D, Loot C (2016) Efficiency of integron cassette insertion in correct orientation is ensured by the interplay of the three unpaired features of attC recombination sites. Nucleic Acids Res 44:7792–7803. https://doi.org/10.1093/nar/gkw646

Nunes-Duby SE, Kwon HJ, Tirumalai RS, Ellenberger T, Landy A (1998) Similarities and differences among 105 members of the Int family of site-specific recombinases. Nucleic Acids Res 26:391–406. https://doi.org/10.1093/Nar/26.2.391

Ogawa A, Takeda T (1993) The gene encoding the heat-stable enterotoxin of *Vibrio cholerae* is flanked by 123-base pair direct repeats. Microbiol Immunol 37:607–616

Partridge SR et al (2000) Definition of the *attI1* site of class 1 integrons. Microbiology 146 (Pt 11):2855–2864

Partridge SR, Tsafnat G, Coiera E, Iredell JR (2009) Gene cassettes and cassette arrays in mobile resistance integrons. FEMS Microbiol Rev 33:757–784. https://doi.org/10.1111/j.1574-6976.2009.00175.x

Pearson CE, Zorbas H, Price GB, Zannis-Hadjopoulos M (1996) Inverted repeats, stem-loops, and cruciforms: significance for initiation of DNA replication. J Cell Biochem 63:1–22

Quinones M, Kimsey HH, Waldor MK (2005) LexA cleavage is required for CTX prophage induction. Mol Cell 17:291–300. https://doi.org/10.1016/j.molcel.2004.11.046

Quiroga C, Centron D (2009) Using genomic data to determine the diversity and distribution of target site motifs recognized by class C-attC group II introns. J Mol Evol 68:539–549. https://doi.org/10.1007/s00239-009-9228-3

Ramirez MS, Bello H, Gonzalez Rocha G, Marquez C, Centron D (2010) Tn7::In2-8 dispersion in multidrug resistant isolates of *Acinetobacter baumannii* from Chile. Rev Argent Microbiol 42:138–140. https://doi.org/10.1590/S0325-75412010000200015

Rapa RA, Labbate M (2013) The function of integron-associated gene cassettes in *Vibrio* species: the tip of the iceberg. Front Microbiol 4:385. https://doi.org/10.3389/fmicb.2013.00385

Rowe-Magnus DA (2009) Integrase-directed recovery of functional genes from genomic libraries. Nucleic Acids Res 37:e118. https://doi.org/10.1093/nar/gkp561

Rowe-Magnus DA, Guerout A-M, Mazel D (1999) Super-integrons. Res Microbiol 150:641–651

Rowe-Magnus DA et al (2001) The evolutionary history of chromosomal super-integrons provides an ancestry for multiresistant integrons. Proc Natl Acad Sci USA 98:652–657

Rowe-Magnus DA, Guerout AM, Biskri L, Bouige P, Mazel D (2003) Comparative analysis of superintegrons: engineering extensive genetic diversity in the vibrionaceae. Genome Res 13:428–442

Roy Chowdhury P et al (2009) Tn6060, a transposon from a genomic island in a *Pseudomonas aeruginosa* clinical isolate that includes two class 1 integrons. Antimicrob Agents Chemother 53:5294–5296. https://doi.org/10.1128/AAC.00687-09

Roy R, Kozlov AG, Lohman TM, Ha T (2009) SSB protein diffusion on single-stranded DNA stimulates RecA filament formation. Nature 461:1092–1097. https://doi.org/10.1038/Nature08442

San Millan A et al (2015) Sequencing of plasmids pAMBL1 and pAMBL2 from *Pseudomonas aeruginosa* reveals a blaVIM-1 amplification causing high-level carbapenem resistance. J Antimicrob Chemother 70(11):3000–3003. https://doi.org/10.1093/jac/dkv222

Sberro H et al (2013) Discovery of functional toxin/antitoxin systems in bacteria by shotgun cloning. Mol Cell 50:136–148. https://doi.org/10.1016/j.molcel.2013.02.002

Shi L et al (2006) Unnoticed spread of class 1 integrons in gram-positive clinical strains isolated in Guangzhou, China. Microbiol Immunol 50:463–467. https://doi.org/10.1111/j.1348-0421.2006.tb03815.x

Shultzaberger RK, Bucheimer RE, Rudd KE, Schneider TD (2001) Anatomy of *Escherichia coli* ribosome binding sites. J Mol Biol 313:215–228. https://doi.org/10.1006/jmbi.2001.5040

Sorum H, Roberts MC, Crosa JH (1992) Identification and cloning of a tetracycline resistance gene from the fish pathogen *Vibrio salmonicida*. Antimicrob Agents Chemother 36:611–615

Starikova I et al (2012) A trade-off between the fitness cost of functional integrases and long-term stability of integrons. PLoS Pathog 8:e1003043. https://doi.org/10.1371/journal.ppat.1003043

Stokes HW, Hall RM (1989) A novel family of potentially mobile DNA elements encoding site-specific gene-integration functions: integrons. Mol Microbiol 3:1669–1683

Stokes HW, Hall RM (1991) Sequence analysis of the inducible chloramphenicol resistance determinant in the Tn1696 integron suggests regulation by translational attenuation. Plasmid 26:10–19

Stokes HW, O'Gorman DB, Recchia GD, Parsekhian M, Hall RM (1997) Structure and function of 59-base element recombination sites associated with mobile gene cassettes. Mol Microbiol 26:731–745

Stokes HW et al (2001) Gene cassette PCR: sequence-independent recovery of entire genes from environmental DNA. Appl Environ Microbiol 67:5240–5246

Strugeon E, Tilloy V, Ploy MC, Da Re S (2016) The stringent response promotes antibiotic resistance dissemination by regulating integron integrase expression in biofilms. MBio 7: e00868-16. https://doi.org/10.1128/mBio.00868-16

Sureshan V et al (2013) Integron gene cassettes: a repository of novel protein folds with distinct interaction sites. PLoS One 8:e52934. https://doi.org/10.1371/journal.pone.0052934

Szekeres S, Dauti M, Wilde C, Mazel D, Rowe-Magnus DA (2007) Chromosomal toxin-antitoxin loci can diminish large-scale genome reductions in the absence of selection. Mol Microbiol 63:1588–1605

Tawfik DS (2014) Accuracy-rate tradeoffs: how do enzymes meet demands of selectivity and catalytic efficiency. Curr Opin Chem Biol 21:73–80. https://doi.org/10.1016/j.cbpa.2014.05.008

Trinh TQ, Sinden RR (1991) Preferential DNA secondary structure mutagenesis in the lagging strand of replication in *E. coli*. Nature 352:544–547

Tsafnat G, Copty J, Partridge SR (2011) RAC: repository of antibiotic resistance cassettes. Database (Oxford) 2011:bar054. https://doi.org/10.1093/database/bar054

Val ME et al (2005) The single-stranded genome of phage CTX is the form used for integration into the genome of *Vibrio cholerae*. Mol Cell 19:559–566. https://doi.org/10.1016/j.molcel.2005.07.002

Van Melderen L, Saavedra De Bast M (2009) Bacterial toxin-antitoxin systems: more than selfish entities? PLoS Genet 5:e1000437. https://doi.org/10.1371/journal.pgen.1000437

Vinue L, Jove T, Torres C, Ploy MC (2011) Diversity of class 1 integron gene cassette Pc promoter variants in clinical *Escherichia coli* strains and description of a new P2 promoter variant. Int J Antimicrob Agents 38:526–529. https://doi.org/10.1016/j.ijantimicag.2011.07.007

Wagner A (2011) The molecular origins of evolutionary innovations. Trends Genet 27:397–410. https://doi.org/10.1016/j.tig.2011.06.002

Waldor MK, Mekalanos JJ (1996) Lysogenic conversion by a filamentous phage encoding cholera toxin. Science 272:1910–1914

Wu YW, Doak TG, Ye Y (2013) The gain and loss of chromosomal integron systems in the Treponema species. BMC Evol Biol 13:16. https://doi.org/10.1186/1471-2148-13-16

Xu H, Davies J, Miao V (2007) Molecular characterization of class 3 integrons from *Delftia* spp. J Bacteriol 189(17):6276–6283

Yamaguchi Y, Park JH, Inouye M (2011) Toxin-antitoxin systems in bacteria and archaea. Annu Rev Genet 45:61–79. https://doi.org/10.1146/annurev-genet-110410-132412

Chapter 10
Experimental Evolution to Explore Adaptation of Terrestrial Bacteria to the Martian Environment

Wayne L. Nicholson

1 Introduction

The question of the possible existence of life beyond the confines of Earth has been a broad topic of philosophical debate for over two millennia (O'Leary 2008; Nicholson 2009). But only relatively recently, beginning in the late nineteenth century, has this question been addressed scientifically. Beginning in the 1970s, scientists began to fully appreciate the broad range of extreme physical conditions under which Earth life can grow, and it rapidly became apparent that single-celled microorganisms of the domains Bacteria and Archaea were capable of growth at the widest extremes of temperature, pressure, pH, and salinity (Horikoshi et al. 2011).

But is Earth unique among planets, or could microbial life actually exist elsewhere in our solar system or beyond? Until recently, the nearly complete lack of data about the environments of planets beyond Earth posed an insurmountable obstacle to answering this question. Major advances in understanding of the physical conditions prevailing on other planets within our solar system have come only since the last half of the twentieth century, with (1) orbiters, landers, and rovers providing detailed analyses of the surface environment of Mars and (2) orbiters around Jupiter and Saturn revealing the existence of subsurface oceans beneath some of their icy moons. In parallel, the discovery of the first planets orbiting other stars was reported, prompting scientists to search for clues of their possibly harboring life. These recent advances in planetary exploration have resulted in the emergence of the field of "habitability" to astrobiology, i.e., studies of the ability of planetary environments to host life [reviewed recently in Cockell et al. (2016), Heller and Armstrong (2014)].

W. L. Nicholson (✉)
Department of Microbiology and Cell Science, University of Florida, Merritt Island, FL, USA
e-mail: WLN@ufl.edu

P. H. Rampelotto (ed.), *Molecular Mechanisms of Microbial Evolution*, Grand Challenges in Biology and Biotechnology,
https://doi.org/10.1007/978-3-319-69078-0_10

What makes a planet habitable? Until we discover life elsewhere, "life as we know it" is Earth life. Thus the question immediately becomes constrained to "what makes a planet habitable to Earth life?". At present it is generally agreed upon among scientists that the requirements for Earth life are:

1. Liquid water (hence, temperature/pressure/solute conditions at which water can exist as a liquid)
2. A number of elements, some needed in bulk (mainly C, H, O, N, P, S), with lesser amounts of Na, Mg, Ca, Cl, K, and Fe and trace amounts of several other elements such as Cr, Co, Cu, F, I, Mn, Mo, Se, and Zn
3. A source of energy (supplied either by sunlight or redox chemistry)

And of course, life needs to be protected from, or be able to mitigate, the lethal effects of bombardment by ionizing or UV radiation or exposure to toxic chemicals.

A substantial amount of environmental data regarding Mars habitability has been gained from robotic exploration missions. For example, orbiters have observed large features indicative of past or present water on the surface of Mars, such as channels, ancient lake beds, and sedimentary features. Mars landers and rovers have also observed sedimentary features at close range and have measured such environmental parameters as air pressure and composition, wind speed, relative humidity, UV and ionizing radiation flux, and the chemical compositions of myriad rocks and soils. Taken together, the data suggest that present-day Mars may be habitable and that conditions in the distant past were even more so. No definitive evidence exists of past or present life on Mars, but future robotic missions are being planned to directly test for signs of life (so-called "biosignatures").

The potential habitability of icy moons such as Europa and Enceladus derives from orbital observations of the moons' wobble and the patterns of surface ice cracks and chaotic terrain, coupled with deep-penetrating radar, magnetic, and surface altimetry data. Taken together, these observations indicate the presence of global oceans of liquid water beneath the surface of both moons. In addition, the Cassini spacecraft directly observed jets of material emerging from the south pole of Enceladus and actually flew through and sampled this material with its onboard mass spectrometer—finding predominantly water, CO_2, N_2, methane, and traces of acetylene and propane (Waite et al. 2006). It is thus tantalizing to suggest that both Mars and icy moons are places with environments capable of supporting life as we know it.

In this chapter I will attempt to summarize recent attempts at understanding of how terrestrial life forms might be capable of growth in extraterrestrial environments, either due to their natural endowments or due to directed evolution experiments. Most of this discussion will focus on the environment of Mars, because of its proximity and similarity to Earth, its relative accessibility to exploration, and the relative wealth of environmental data we possess. This review updates and expands upon previous reviews, to which the reader is directed for further information (Nicholson et al. 2000, 2005, 2009; Nicholson 2009; Fajardo-Cavazos et al. 2007).

2 Evolutionary Trajectories of Earth and Mars

In addressing the question of whether or not a particular form of life can inhabit a particular extraterrestrial environment, first we need to know the physicochemical characteristics of the environment in question. Although current evidence suggests that the environments of ancient Earth and Mars were once rather similar [reviewed in Nicholson (2009)], present-day Mars has a very different environment from that of present-day Earth, summarized in Table 10.1.

The different evolutionary trajectories of Earth and Mars stem from the time of their formation by accretion in the early solar system. Both are terrestrial (rocky) planets, but Mars is only 1/10 as massive as Earth and its gravity is only 38% that of Earth's. Mars is ~1.5 times further from the Sun than Earth, receives only about 40% as much solar radiation, and takes 687 Earth days to orbit the Sun. Both Earth and Mars have roughly 24-hour days, and their axes of rotation are both inclined to the Sun, at 23.5° (Earth) and 25° (Mars); hence, both planets have four seasons. Both planets have moons, but Earth's moon is much more massive than Mars' tiny moons Phobos and Deimos (which are likely small asteroids captured by Mars' gravity). Earth's massive moon stabilizes its inclination to an approximately 40,000-year period of "wobble" between 22.1° and 24.5°, which has a stabilizing effect on our long-term climate. In the absence of a large stabilizing moon, Mars' wobble is much more pronounced, varying between ~15° and ~35° with a period of ~100,000 years. This extreme wobble results in the planet cycling between relatively warm, wet periods and relative cold, dry periods.

Due to the energy released during accretion, the terrestrial planets started their lives as near-molten balls, far too hot for liquid water to form. As bombardment slowed and the planets cooled, they developed dense atmospheres and crusts floating on liquid magma, and the crusts cooled to the point that oceans of liquid water were stable on the surface. Because Mars formed further from the Sun and was less massive than Earth, it cooled at a faster rate, likely becoming "habitable" before Earth did. Both planets originally had an internal dynamo generating a protective magnetic field. However, the faster rate of Mars core cooling led to faster core solidification and subsequent collapse of its protective magnetic field. Losing its magnetic protection from the charged particles of the solar wind, the lighter components of the Mars atmosphere were blown into space, leaving behind a thin, predominantly CO_2 atmosphere.

2.1 *Atmospheric Composition and Pressure (P)*

Earth contains a relative thick (~101.3 kPa), humid atmosphere. Nitrogen comprises the majority gas (~78%), with both oxygen (~21%) and carbon dioxide (~0.04%) to support abundant and diverse photosynthetic and aerobic life (Table 10.1). In stark contrast, the average atmospheric P on Mars (0.7 kPa) is only 1/150th that of Earth

Table 10.1 Comparison of the physical environments of Earth and Mars[a]

Parameter	Earth	Mars
Temperature:		
Air, global average	+15 °C	−55 °C
Air, measured extremes	−89 to +56 °C	−75 to −2 °C (MSL)
Ground, range at surface	−93 to +94 °C	−91 to +3 °C (MSL)
Ground (~10 m depth)	+4 to +27 °C	Approx. −93 °C
Ocean, range	−2 to +36 °C	n.a.
Air pressure at surface:		
Average	101.3 kPa (sea level)	~0.7 kPa
Range	~29 (Mt. Everest) to ~106 kPa (Dead Sea)	~0.1 to ~10 kPa
Atmospheric composition:	N_2: 78.1%	CO_2: 96.0%
	O_2: 20.9%	Ar: 1.93%
	Ar: 1%	N_2:1.89%
	CO_2: 0.04%	O_2: 0.145%
	H_2O vapor: (v), up to 100%	CO: 0.07% H_2O vapor: ~0.03% (v)
Magnetic field:	Global; ~65,000 nT	Patchy; ~1500 nT
Solar UV:		
Solar constant	1368 W/m^2	590 W/m^2
Spectrum at surface	≥290 nm (UV−B + A)	≥190 nm (UV−C + B + A)
Ionizing radiation at surface:		
SEP[b], episodic	Negligible	Spikes, ~0.3 mGy/day
GCR, continuous	Negligible	0.18–0.23 mGy/day
Organic compounds	Abundant and diverse	Chlorobenzene and chlorinated C_2–C_4 alkanes (ppb range)

[a]Modified from Schuerger (2004), Nicholson et al. (2005), Rummel et al. (2014), Millan et al. (2016)

[b]Abbreviations: *GCR* galactic cosmic radiation, *nT* nanoTesla, *ppb* parts per billion, *SEP* solar energetic particles, (*v*) varies

(Table 10.1). Furthermore, the composition of the martian atmosphere differs greatly from that of Earth; CO_2 makes up the majority (96%) of the atmosphere, followed by N_2 and Ar, each at ~2% of the total. Note that O_2 (0.145%) and water vapor (~0.03%) are only minor constituents of the atmosphere. This has several profound consequences for Mars habitability.

2.2 *Temperature (T)*

The surface T of a planet is determined in large part by its mass, proximity to its parent star, and the greenhouse effect supplied by its atmosphere. Present-day martian surface T's remain cold year-round, because (1) Mars orbits farther from the Sun and thus receives only about 40% as much solar radiation as does Earth and, (2) despite its high concentration of the greenhouse gas CO_2, the thin martian atmosphere provides a relatively weak atmospheric greenhouse effect.

Earth's global average T is ~15 °C but can vary dramatically depending on location, season, time of day, and type of material being measured (air, ground, water, ice). Recorded air T's on Earth vary over a span of nearly 145 °C, from a record low of −89 °C (Vostok Station, Antarctica) to a record high of +56 °C (Death Valley, USA). Mars is clearly much colder than the Earth; by comparison, the global average air T on Mars is estimated to be approx. −55 °C, and diurnal T's are estimated to vary from lows of approx. −153 °C at the poles to daytime highs of around −10 °C. In addition, brief excursions of up to approx. +20 °C at the summertime equator are estimated (Schuerger 2004; Rummel et al. 2014).

At the surface, ground T's can vary even more widely than air T's but rapidly stabilize with increasing depth; even so, the T in the martian near-subsurface is substantially lower than that of Earth (Table 10.1).

2.3 *UV and Ionizing Radiation*

The composition and density of the Mars atmosphere have important implications for the radiation environment at the surface. First, the low abundance of oxygen in the atmosphere precludes the development of a UV-absorbing stratospheric ozone layer as found on Earth. The bulk gas in the martian atmosphere is CO_2, which is transparent to UV down to ~190 nm (Table 10.1); thus, the surface is bathed in high levels of lethal UV-C radiation, with roughly 2–3 orders of magnitude higher biologically relevant doses than the Earth (Cockell et al. 2000). Ionizing radiation species (X-rays, γ-rays, protons, electrons, and high-energy nuclei of atomic number ≥ 2) originating from the Sun and from outside the solar system (galactic cosmic radiation) are either absorbed by Earth's atmosphere or deflected by Earth's strong global magnetic field (Table 10.1). By contrast, Mars' attenuated atmosphere and weak remnant magnetic field result in the ionizing radiation flux being significantly higher on the martian surface (Table 10.1).

2.4 *Liquid Water*

The boiling and freezing points of water are a function of T and P. Most of the Earth's surface is at a T–P regime conducive to the presence of abundant liquid water available to biological processes. However, on the surface of Mars, there exists only a small T–P window within which liquid water can exist stably; thus the vast majority of water on Mars is frozen either underground or in the polar caps, and the remainder is present mainly as vapor or ice crystals in the atmosphere.

Interestingly, there is abundant evidence from orbiters and rovers that liquid water flowed freely on the martian surface in the past, indicating that the ancient Mars environment may have been warmer with a higher-P atmosphere capable of supporting liquid water. How could water remain in the liquid state at the T–P conditions of present-day Mars?

2.5 *Salinity*

Mars orbiters, rovers, and landers have sampled various sites associated with ancient sedimentary features and have detected high concentrations of various salts, including chloride, sulfate, and perchlorate salts of Ca, Na, or Mg. In particular, perchlorate salts have been detected at high levels in soil and ground ice samples from various locations on the martian surface (Hecht et al. 2009; Glavin et al. 2013). Dissolved salts can dramatically lower the freezing point of water and slow its evaporation. High salinity is a double-edged sword however; high concentrations of solutes, such as salts or organic compounds (e.g., sugars), lower the water activity (a_w), effectively sequestering water away from cells and inhibiting microbial growth. At present, no microbes have been found which are able to grow at a_w less than 0.61 [reviewed in Rummel et al. (2014)].

2.6 *pH*

Knowledge about the pH of martian rocks and regoliths (soils) are by and large poorly constrained. Because direct measurement of pH relies on suspension of sample material in water, most pH estimates have been derived indirectly from dry chemical composition data, yielding qualitative measures such as "acidic" or "alkaline." In fact, the only direct measurement of martian soil pH was conducted at the Phoenix landing site, where a slightly alkaline pH of 7.7 ± 0.5 was measured (Hecht et al. 2009).

2.7 *Organic Compounds*

Early Mars probes failed to find concrete evidence for organic compounds down to the ppb range, and until relatively recently, it was thought that organic compounds were essentially absent on Mars. This finding was paradoxical, as it was widely presumed that organics were constantly being delivered to the surface of Mars in meteoritic dust from space, as it is on Earth. Organic detection experiments rely on heating soil samples in an oven, then separating and identifying the released volatile compounds using gas chromatography/mass spectrometry (GC/MS). Evidence has been mounting that during the heating step, strongly oxidizing compounds (particularly perchlorates) present in the martian soil samples were reacting with organics in the soil, breaking them down to organochlorine compounds and CO_2 (Sephton et al. 2014). Consistent with this notion, the Sample Analysis at Mars (SAM) instrument on the Mars Science Lab recently detected chlorobenzene and C_2–C_4 dichloroalkanes at 70–300 ppb in martian soil samples (Glavin et al. 2013; Freissinet et al. 2015). More recent sampling by MSL of material collected from the interiors of rocks resulted in detection of both aromatic and aliphatic organic compounds (Eigenbrode et al. 2018). Therefore, Mars soils and rocks do appear to contain organic compounds, but oxidants, mainly perchlorate, present in martian soils make their detection, identification, and quantification difficult.

2.8 *Transport of Life from Earth to Mars*

At first glance, it would seem that Mars and Earth are isolated by at least 50 million kilometers of raw space. How could microbes possibly traverse this void? Two mechanisms have been postulated by which Earth microbes could be transported to Mars: natural impacts and human spaceflight activities.

2.8.1 Natural Impacts

Currently scientists have found over 80 meteorites on Earth that originated from Mars (for details, the reader is referred to the Martian Meteorite Compendium maintained by NASA at https://curator.jsc.nasa.gov/antmet/mmc/). These meteorites were ejected into space by large impacts striking the martian surface, and it is reasoned that large impacts on Earth's surface could also transport Earth rocks bearing viable microbes to Mars. Thus it is postulated that Earth and Mars have been continuously exchanging life since the early days of the solar system [reviewed extensively in Mileikowsky et al. (2002), Nicholson (2009), Nicholson et al. (2000)].

2.8.2 Human Spaceflight Activities

Beginning in the early 1960s, humans have launched over 40 flyby, orbiter, lander, and rover missions toward Mars (for details, see http://mars.nasa.gov/programmissions/missions/log/). Several of these missions were unsuccessful and crashed onto the martian surface. Even a successful lander or rover mission leaves debris strewn across the martian surface, including parachutes, heat shields, back shells, etc. which are discarded during entry and landing operations. Because microorganisms are ubiquitous on Earth, it is virtually certain that they have contaminated flight hardware and been deposited on the surface or near-subsurface of Mars. Spacefaring nations have addressed this concern by international "Planetary Protection" agreements dictating the strict biological cleanliness of spacecraft destined for Mars [for recent extensive reviews, see Rummel et al. (2014), Rettberg et al. (2015)], but it should be stressed that current Planetary Protection protocols do not mandate *sterilization* of spacecraft, but the *reduction* of microbial bioburdens to below specified levels. A major unanswered question remains: is transfer of Earth microbes to Mars ecologically relevant? Just because an Earth microbe is deposited into the Mars environment does not necessarily mean that it can survive, let alone propagate, in that environment. This question has been addressed by the use of simulations of the martian environment as described in Sect. 3.

2.8.3 Candidate Microorganisms

A wide diversity of prokaryotes exist in various Earth niches, but which ones would be likely candidates for Earth-to-Mars transport? When one considers the physics of natural impacts, only microbes inside (1) well-consolidated rocks located (2) at or near the surface of (3) a ring-shaped *spallation zone* surrounding the impact site are most likely to be accelerated to escape velocity without shock and heating sufficient to sterilize the rock [discussed in Melosh (1984, 1989), Mileikowsky et al. (2000), Nicholson et al. (2000), Nicholson (2009)].

Considering Earth-to-Mars transport by human spaceflight activities, numerous bacteria have been identified by sampling Mars-bound spacecraft and their ultraclean Spacecraft Assembly Facilities (SAFs) (Venkateswaran et al. 2001; La Duc et al. 2003, 2004, 2007). Spore-formers, mainly *Bacillus* spp., comprise a major fraction of these isolates, and resistance of their spores to extreme disinfecting treatments (UV, ionizing radiation, hydrogen peroxide) has been documented to greatly exceed the resistance properties of common laboratory strains (Kempf et al. 2005; Link et al. 2004; Newcombe et al. 2005). SAFs are maintained in an ultraclean state by HEPA filtration, rigorous cleaning and disinfection, and strict protocols for contamination control. It has been suggested that these very conditions render SAFs excellent selective environments for the very hardiest microorganisms—paradoxically, the very contaminants most likely to survive an Earth-to-Mars transit (Link et al. 2004; Crawford 2005).

3 Growth of Terrestrial Microbes in Simulated Mars Environments

Considerable effort has been expended since the 1970s in testing the ability of Earth microbes to survive and grow in the martian environment, driven mostly by the need to mitigate the so-called forward contamination of Mars by Earth microorganisms [reviewed in Rummel et al. (2014), Rummel (2001), Nicholson et al. (2009)]. These experiments have been conducted in various chambers designed to simulate the environmental conditions on Mars. As our understanding of Mars has become more detailed and refined, such simulations have become increasingly more sophisticated and representative of the martian environment. With the exceptions of gravity and ionizing radiation, a myriad of modern Mars simulation chambers have been designed that can more or less faithfully replicate nearly all physical aspects of the martian environment such as T, atmospheric P and composition, and the solar fluence and spectrum at Mars through the UV-visible-near-IR range under atmospheric opacity conditions ranging from clear sky to global dust storms. In addition, our increasing knowledge about the composition and properties of martian soils has enabled the preparation of increasingly accurate Mars soil simulants (Schuerger et al. 2012). Below is a brief summary of our current knowledge regarding the ability of Earth microbes to survive and grow under simulated Mars conditions.

3.1 UV and Ionizing Radiation

On the surface of Mars, solar UV is by far the strongest environmental factor limiting bacterial survival [reviewed in Schuerger (2004); Nicholson et al. (2000, 2005)]. However, the results from experiments conducted in low Earth orbit or in Mars simulation chambers have indicated that viable microbes can be shielded effectively from UV by relatively thin layers of UV-opaque materials such as dust, regolith, irregularities in spacecraft surfaces, or even the upper layers of cells deposited in multiple layers (Horneck 1993; Schuerger et al. 2005, 2006). Thus, to avoid the rapidly lethal effects of solar UV, microbes would likely have to reside in subsurface environments (Rummel et al. 2014). Direct measurements of the ionizing radiation flux on Mars have recently been obtained using the Mars Science Lab's Radiation Assessment Detector (MSL RAD) instrument (Hassler et al. 2014). The combined GCR and SEP doses measured at the martian surface by MSL RAD were extrapolated to an annual dose of <0.3 Gy, much lower than the dose required to inactivate microbes by even one order of magnitude (ranging from ~200 to ~12,000 Gy). These observations led to the conclusion that with minimal shielding, UV and ionizing radiation would exert negligible lethal effects on the viability of microbes on Mars (Rummel et al. 2014).

3.2 The Mars Atmosphere

Many of the key physical constraints to the growth of terrestrial life on Mars are manifested in its atmosphere.

3.2.1 Atmospheric Composition (AC)

The martian atmosphere is CO_2-rich and oxygen-poor (Table 10.1). These factors per se are not inhibitory to microbes, as numerous prokaryotes are capable of growth under oxygen-limited or completely anaerobic conditions (Horikoshi et al. 2011). In particular, the low concentration of oxygen (~0.145%) is further exacerbated by low P. It has been calculated that the pO_2 in the "average" martian atmosphere (~700 Pa at −10 °C) is ~1 Pa, which corresponds to a dissolved O_2 concentration of ~3 nM; in comparison, the O_2 concentration on sea-level Earth (~101.3 kPa, +25 °C) is ~250 μM. At an O_2 concentration of 3 nM, *E. coli* cells were demonstrated to be capable of growth using only aerobic respiration (Stolper et al. 2010), suggesting that even aerobic microorganisms could utilize the scant O_2 present on Mars (Rummel et al. 2014).

3.2.2 Atmospheric Temperature and Pressure

As discussed above in Sect. 2, a key factor for Mars habitability is the presence of liquid water. However, water is largely in the form of either ice or vapor at martian surface conditions, where T and P hover around the triple point. What are the lowest T's and P's at which life can function? A recent exhaustive review of the literature (Rummel et al. 2014) revealed that the lowest temperature at which cell division has been reported to occur in the laboratory is −18 °C, by the yeast *Rhodotorula glutinis* (Collins and Buick 1989). Regarding P, early experiments used an approach in which growth of laboratory bacteria or bacteria isolated from Spacecraft Assembly Facilities was tested in simulation chambers under conditions where T, P, and AC were systematically altered one at a time; in other words, cells were cultivated at fixed T and AC under various P conditions. Results from these experiments indicated that the group of terrestrial microbes tested were unable to grow at P's lower than ~2.5 kPa (Schuerger and Nicholson 2006; Berry et al. 2010; Kral et al. 2011; Smith et al. 2009).

3.2.3 Improved Simulations of Temperature, Pressure, and Atmospheric Composition

Changing one parameter at a time in a simulation chamber raises the possibility of missing combined effects. For example, T and P exert opposite effects on

Table 10.2 Terrestrial bacteria capable of growth under simulated Mars atmospheric conditions[a]

Phylum	Genus	Species	Reference
Actinobacteria	*Rhodococcus*	*qingshengii* (1 isolate)	Schuerger and Nicholson (2016)
Firmicutes	*Bacillus*	3 unclassified isolates	Schuerger and Nicholson (2016)
Firmicutes	*Carnobacterium*	*alterfunditum* *divergens* *funditum* *gallinarum* *inhibens* subsp. *gilichinskyi* *inhibens* subsp. *inhibens* *maltaromaticum* *mobile* *pleistocenium* *viridans* 5 isolates tentatively classified as *viridans*	Nicholson et al. (2013)
Firmicutes	*Carnobacterium*	5 unclassified isolates	Schuerger and Nicholson (2016)
Firmicutes	*Clostridium*	2 unclassified isolates	Schuerger and Nicholson (2016)
Firmicutes	*Cryobacterium*	7 unclassified isolates	Schuerger and Nicholson (2016)
Firmicutes	*Exiguobacterium*	*sibiricum* (5 isolates)	Schuerger and Nicholson (2016)
Firmicutes	*Paenibacillus*	*antarcticus* (6 isolates) *macquariensis* (1 isolate) 9 unclassified isolates	Schuerger and Nicholson (2016)
Firmicutes	*Trichococcus*	*collinsii* (2 isolates) *pasteurii* (10 isolates) 1 unclassified isolate	Schuerger and Nicholson (2016)
Proteobacteria	*Serratia*	*ficaria* *fonticola* *grimesii* *liquefaciens* *plymuthica* *quinivorans*	Schuerger and Nicholson (2016), Schuerger et al. (2013)

[a]0 °C, 0.7 kPa, CO_2-dominated, O_2-limited atmosphere

membranes: membranes become more rigid as T decreases but become more fluid as P decreases. These effects may result in a microbe being unable to grow at low T or low P applied singly but capable of growth when both P and T are lowered simultaneously. Recently a more holistic approach was pursued in experiments testing the growth of microbes from so-called "Mars analog" environments, i.e., sites such as alpine sites, high deserts, permafrost soils, arctic and Antarctic sites, etc. In this approach, microbial growth was tested in a chamber simultaneously simulating Mars T, P, and AC conditions (0 °C, 0.7 kPa, CO_2-dominated, O_2-limited atmosphere). Using these conditions, we discovered a small subset of the total microbial populations originating from environmental samples capable of growth at 0.7 kPa, belonging to a variety of bacterial taxonomic groupings (Table 10.2)

(Nicholson et al. 2013; Schuerger et al. 2013; Schuerger and Nicholson 2016). Examination of Table 10.2 immediately reveals that *Firmicutes* comprise the preponderance of environmental bacteria capable of growth under simulated Mars atmosphere (66 out of 73 total isolates, consisting of 3 *Bacillus*, 20 *Carnobacterium*, 2 *Clostridium*, 7 *Cryobacterium*, 5 *Exiguobacterium*, 16 *Paenibacillus*, and 13 *Trichococcus* isolates). In contrast, only the genus *Serratia* from the phylum *Proteobacteria* was found to be capable of growth under simulated Mars atmospheric conditions; furthermore, two *Serratia* species (*S. marcescens* and *S. rubidaea*) could not grow under these conditions (Schuerger and Nicholson 2016).

3.3 Nutrients, Liquid Water, CO_2, and Perchlorate Salts

In the experiments described above, some bacteria were able to grow under simulated martian atmospheric conditions; but it must be noted that the samples were cultivated in Petri dishes containing complex rich agar media (LB, R2A, TSA, or TSY) and abundant liquid water. At 0 °C, the water in these plates did not turn to ice, because the presence of solutes and salts in the media depressed its freezing point; nor did the water evaporate rapidly, because the P–T conditions were below the liquid/vapor boundary, again partly due to dissolved organic solutes and salts. Clearly the Mars surface poses a distinctly different environment than that provided by a well-hydrated Petri dish.

3.3.1 Potential Nutrients, Including CO_2

Mars appears to contain essentially all of the bulk elements, macro- and micronutrients necessary for life (see Sect. 1). Carbon and oxygen are present in the atmosphere as CO_2 and CO, as carbonates in the soil, and as the parent compounds to the organics detected by the MSL SAM instrument (see Sect. 2). The martian atmosphere contains nitrogen gas, and nitrogen was recently detected in Mars soils heated in the SAM instrument, suggestive of nitrate in the soil samples (Stern et al. 2015). Sulfur is widespread on the martian surface in the form of sulfate-dominated mineral deposits [reviewed in Gaillard et al. (2013)]. Hydrogen is obviously present in water, but direct detection of hydrogen gas, postulated to be an essential component of the subsurface biosphere, is lacking. However, H_2 can be produced by geochemical metamorphism in which water interacts with certain igneous (ultramafic) rocks to produce the mineral serpentine and H_2. The detection of serpentine mineral deposits from orbit argues in favor of this process having occurred on Mars in the past (Ehlmann et al. 2010). Autotrophic microorganisms are capable of growth using purely inorganic compounds. A large class of photosynthetic microbes, the photoautotrophs, are not considered to be strong candidates for inhabitants of Mars, as their need for exposure to sunlight would necessarily expose them to the harshly biocidal UV environment (Cockell et al. 2000). However, a category of autotrophic microbes, the methanogenic archaea, has recently gained much attention concerning possible life on Mars. These organisms use CO_2 as a source of carbon, producing methane (CH_4) in the process. Indeed, the detection of trace amounts of methane in the martian atmosphere, both by remote sensing (Mumma et al. 2009) and by the MSL rover (Webster et al. 2015, 2018), has been taken by some as evidence of methanogenic life, although abiotic sources of methane are certainly possible.

3.3.2 Liquid Water and Perchlorate Salts

As discussed in Sect. 2 above, evidence to date indicates that perchlorate salts permeate the soils at various sampling locations on Mars. Chloride ($NaCl$, $MgCl_2$) and perchlorate [$NaClO_4$, $Mg(ClO_4)_2$, $Ca(ClO_4)_2$] salts can lower the freezing point of water to eutectic points of −33 °C to −74 °C (Stillman and Grimm 2011), and water containing these salts can further be supercooled to even lower temperatures (Toner et al. 2014). Furthermore, measurements of water activities in solutions of $NaClO_4$ from +25 °C down to −98 °C revealed that a_w actually increased with decreasing T, maintaining values of >0.6 even at temperatures as low as −73 °C (Toner and Catling 2016). Therefore, it appears that liquid water at the T–P conditions of the martian surface could be stabilized significantly by high concentrations of perchlorate salts. Although perchlorates are toxic to humans via inhibition of iodine uptake by the thyroid, at least 40 species of bacteria can utilize perchlorate as a redox compound, reducing it sequentially through chlorate and chlorite to chloride [reviewed in Nicholson et al. (2012)]; thus perchlorate could be exploited by putative martian microbes as a redox chemical. To test the possible toxicity of perchlorate to bacteria, a Mars soil simulant was produced based on the chemical composition of martian soil at the Phoenix landing site, where water was detected along with perchlorate (at ~1.5 wt.%) (Schuerger et al. 2012). An aqueous extract of this soil simulant was tested for its ability to inhibit spore germination or vegetative growth of *B. subtilis* and *B. pumilus*; both microbes were found to grow and germinate normally in the presence of the extract, indicating that levels of perchlorate found at the Phoenix landing site were not inhibitory to these two commonly used model spacecraft contaminants (Nicholson et al. 2012).

In summary, (1) Mars appears to contain all of the chemical ingredients for life, although perhaps not optimally available; (2) different groups of prokaryotes possess attributes enabling them to deal with different aspects of the martian environment; and (3) some terrestrial bacteria are able to grow under the physical constraints of martian T, P, and AC. However, to date, no single Earth bacterium has been isolated that exhibits all of the attributes necessary for growth under the myriad constraints posed simultaneously by the environments on Mars.

4 Experimental Evolution to Mars Environmental Extremes

Our understanding of the question of whether Earth life could grow on Mars has been greatly enhanced through the utilization of chambers simulating various aspects of the martian environment, particularly in identifying various physicochemical factors constraining Mars habitability. It is important to define the limits at which life can function but perhaps even more important to understand *why* some organisms can survive and proliferate under a set of environmental conditions (say, Mars) which would be lethal to other organisms. Increasingly, these studies have aimed toward understanding at the molecular level how microbes sense and respond to factors in the martian environment at the limits of their growth range. And as discussed below, experimental evolution has been used as a tool to use the constraints of the martian environment as selective forces to uncover mechanisms by which cells expand their growth ranges to inhabit new niches.

4.1 Low Temperature (T)

As mentioned above, Mars presents an extreme cold environment to life. Psychrophiles are cold-adapted microorganisms whose cardinal growth temperatures are minimum <0 °C, optimum ~15 °C, and maximum ~20 °C (Fig. 10.1). Terrestrial psychrophiles are required to withstand wide fluctuations in temperature (from −50 °C to +25 °C for Antarctic soil organisms), as well as desiccation, low nutrient levels, and high radiation impacts; in contrast, deep marine psychrophiles experience constant low T's (~0–3 °C) and high P's (up to ~100 MPa) [reviewed in Casanueva et al. (2010)]. Cellular effects of low T include freezing of intracellular water, increased rigidity of membranes, lowering of enzyme catalytic efficiency, and increase in reactive oxygen species (ROS) due to increased solubility of oxygen.

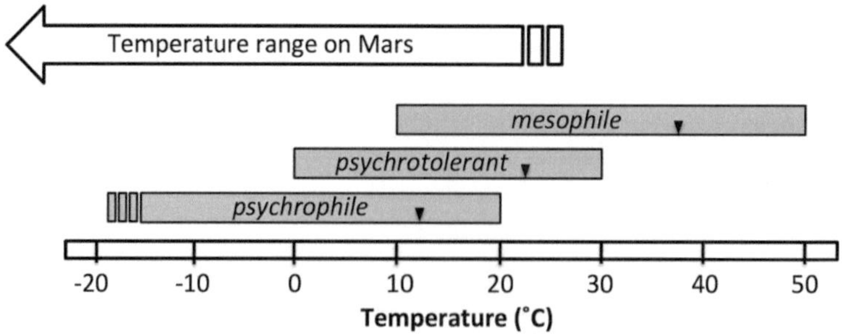

Fig. 10.1 Approximate temperature ranges for growth of psychrophilic, psychrotolerant, and mesophilic microorganisms (gray boxes). Downward arrowheads denote the approximate optimum growth T of each class. The temperature range on Mars is shown for comparison. See text for details

Accordingly, psychrophiles exhibit several adaptations to life in the cold. They import or produce compatible solutes, antifreeze proteins, and ice-binding proteins to prevent intracellular water from freezing. Their membranes typically contain large amounts of unsaturated fatty acids which result in increased fluidity at low T's. The enzymes of psychrophiles show numerous adaptations for efficient function in the cold, and several psychrophilic protein-refolding enzymes (i.e., chaperonins), nucleic acid-binding proteins, and helicases have been identified that improve DNA and RNA function in the cold [reviewed in Casanueva et al. (2010), D'Amico et al. (2006)].

In contrast, psychrotrophs and mesophiles demonstrate minimum growth T's of ~0 °C and ~10 °C, respectively, but it should be noted that their lower T ranges overlap considerably with the upper T range of psychrophiles (Fig. 10.1). A sudden lowering of T induces a "cold-shock" response in many mesophiles, involving upregulation of a group of cold-shock proteins (Csp's) including small nucleic acid-binding proteins and helicases which enable replication, transcription, and translation at low T. Interestingly, homologs of many of the cold-shock-inducible Csp proteins in mesophiles are also present and expressed constitutively in psychrophiles.

Although the literature lacks actual experiments describing the evolution of mesophiles to psychrophily, it seems reasonable to expect that mesophilic organisms could be induced to evolve enhanced growth ability at low T. For example, *E. coli* is a mesophile whose growth rate drops off dramatically at temperatures lower than ~20 °C, and growth essentially ceases at ~7 °C (Ferrer et al. 2003). Researchers noted that activity of the *E. coli* GroES/EL chaperonins, involved in refolding of denatured proteins, was inhibited at low T, leading them to propose that inhibition of *E. coli* growth at low T was due to cold sensitivity of its GroES/EL proteins. In support of this notion, expression of the cold-active chaperonins Cpn60 and Cpn10 from the psychrophile *Oleispira antarctica* in *E. coli* allowed the transgenic strain to grow well at temperatures below 4 °C (Ferrer et al. 2003). Similarly, metabolic pathway analyses of mesophiles vs. psychrophiles uncovered that as T decreases, psychrophiles increase ATP production while mesophiles decrease ATP production (Parry and Shain 2011). The researchers noted that in the purine biosynthetic pathways of the two classes of organisms, psychrophiles were enriched in de novo AMP-synthesizing enzymes, whereas mesophiles were enriched in AMP-degrading enzymes. By genetic manipulation, they engineered an *E. coli* strain lacking its native AMP nucleosidase while expressing the AMP-generating PurA enzyme from the psychrophile *Psychrobacter cryohalolentis* and observed that the engineered *E. coli* strain grew 70% faster at low T (Parry and Shain 2011). Thus it is clear from the above experiments that relatively few and simple genetic changes can profoundly alter the minimum growth T of an *E. coli*, common mesophilic bacterium.

4.2 Low Pressure (P)

We have undertaken experiments with the aim of exploring the evolution to growth at low P using the Gram-positive spore-forming bacterium *Bacillus subtilis*. We had previously shown that *B. subtilis* could grow (albeit slowly) at 5 kPa but that its growth essentially ceased at P's at or below ~2.5 kPa (Schuerger and Nicholson 2006). To analyze how *B. subtilis* responded to low P, we compared its global transcriptional pattern (i.e., transcriptome) when cultivated at either ~101.3 kPa or at 5 kPa (Waters et al. 2014). We found that incubation of *B. subtilis* at low P resulted in significant alteration in the expression of 10 regulons and most notably resulted in upregulation of 86 transcripts involved in the general stress response (GSR) regulon (Waters et al. 2014). Transcription of GSR genes is controlled by RNA polymerase containing the sigma-B factor ($Esig^B$), and we showed that expression of the GSR gene *ctc* was induced at low P in an $Esig^B$-dependent manner (Waters et al. 2014).

We were interested to see if *B. subtilis* could evolve to improve growth at low P, using the near-inhibitory low P of 5 kPa as the selective factor (Nicholson et al. 2010). We propagated a wild-type laboratory strain of *B. subtilis*, called WN624, in rich liquid (LB) medium for 1000 generations at 27 °C and 5 kPa, making frozen glycerol stocks from samples of the population at 50-generation intervals for later study. Over the course of 1000 generations at low P, the population exhibited a stepwise evolution to better growth at 5 kPa. We isolated a strain from the 1000-generation culture, called strain WN1106, which could grow to higher cell density at 5 kPa than the ancestral strain WN624. In pairwise competition experiments, low P-evolved strain WN1106 readily outcompeted the ancestral strain WN624 at 5 kPa, but not at Earth-normal pressure (~101.3 kPa). The growth advantage of WN1106 over ancestral strain WN624 at 5 kPa was not due to its better growth under oxygen limitation, because both strains grew equally under oxygen-limited conditions (Nicholson et al. 2010).

We suspected that the enhanced low-P growth capability of strain WN1106 was likely due to mutation(s) occurring in its genome during 1000 generations of propagation at 5 kPa. To identify such mutations, we subjected strains WN624 and WN1106 to whole-genome sequencing and compared their genome sequences (Waters et al. 2015). We found only eight mutations in the genome of WN1106; seven single-nucleotide polymorphisms (SNPs) in the *fliI*, *parC*, *resD*, *ytoI*, *yvlD*, *bacD*, and *walK* genes; and one in-frame, nine-nucleotide deletion in the *rnjB* gene (Waters et al. 2015). We were particularly interested in the *rnjB* deletion mutation, which we call *rnjBΔ9*, because *rnjB* encodes the RNase J2 subunit of the *B. subtilis* RNA degradosome, a multi-subunit enzyme complex that governs global RNA turnover in *B. subtilis* [reviewed recently in Cho (2017)]. Furthermore, *B. subtilis* strains carrying a complete deletion of the *rnjB* gene were viable and did not display an observable phenotype under normal laboratory conditions. To test the role of *rnjB* in low-P growth, we constructed genetically identical strains of *B. subtilis* carrying either the wild-type $rnjB^+$ or the mutant *rnjBΔ9* gene and competed them at P's of ~101.3 or 5 kPa and at T's of 20 °C, 25 °C, or 30 °C. We found that the competition outcomes depended on the particular combination of T and P used. At 20 °C the mutant strain was less fit than wild-type at both pressures, and at 30 °C, the mutant was more fit than wild-type at both pressures. Only at 25 °C (close to the T at which the original evolution experiment was conducted) was the mutant more fit at low P and less fit at standard P (H. Nguyen and W.L. Nicholson, unpublished data)—highlighting the inextricable linkage between T and P in performing environmental experiments. We currently are working to understand how the *rnjBΔ9* mutation differentially affects gene expression at 25 °C and either ~101.3 or 5 kPa, by comparing the wild-type and mutant transcriptomes under these environmental conditions using RNA-seq technology (experiment in progress).

4.3 UV and Ionizing Radiation

Numerous studies have been performed to test the resistance of various microorganisms to conditions simulating the UV and ionizing radiation environment of Mars, but only one study has been published that actually tests the question of whether cells could evolve higher resistance to UV, again using *B. subtilis* as the test organism (Wassmann et al. 2010). In this study, vegetative wild-type *B. subtilis* strain 168 cells were subjected to periodic episodes of selection by daily exposure of stationary phase cultures to polychromatic UV (200–400 nm), followed by dilution into fresh medium and regrowth. Cells were exposed to 69 cycles of UV treatment over the course of 700 generations, and it was shown that the evolving populations exhibited a statistically significant 3- to 4-fold increase in resistance of vegetative cells to both polychromatic (200–400 nm) and monochromatic (254-nm) UV (Wassmann et al. 2010). Vegetative cells of the UV-adapted strain also exhibited significantly increased resistance to ionizing radiation (X-rays; ~8-fold), high osmolarity (1 M NaCl; ~7-fold), desiccation (33% RH; ~5-fold), and hydrogen peroxide (10 mM; ~4-fold), but not to wet heat (55 °C) (Wassmann et al. 2010). Taken together, the data suggest that, in response to the single stressor of UV, *B. subtilis* may have evolved a generalized upregulated resistance to a variety of environmental stresses relevant to enhanced growth in the martian environment. Unfortunately, follow-up studies have not appeared in the literature, and at present, the molecular mechanism for this effect is unknown.

4.4 High Salinity and Osmolarity

The issue of high osmolarity is relevant to growth in the martian environment, because under the low-T, low-P martian atmospheric regime, only water loaded with ionic solutes is stable in liquid form. Various sites on Mars are thought to be the remains of rivers, seas, and lake beds from which water has evaporated, leaving behind once-dissolved solutes as precipitates. Analyzing the mineral composition at various sites on Mars indicates that evaporitic salts of K^+, Na^+, Mg^{+2}, or Ca^{+2} paired with Cl^-, SO_4^{-2}, or ClO_4^- dominate, suggesting that liquid water on Mars was briny. Microbes need special adaptations for growth at high salinities and osmolarities, as described below.

All cells possess semipermeable membranes through which water, but not most other molecules, passes freely. To maintain proper turgor pressure in environments of ever-changing osmotic strength, cells must actively adjust their intracellular osmotic potential to prevent dehydration or rupture [reviewed in Hoffmann and Bremer (2017); Detkova and Boltyanskaya (2007)]. They do so by accumulating and/or expelling ions (usually K^+ and Na^+) or by producing, accumulating, and/or expelling a number of compounds collectively known as compatible solutes (proline, glycine betaine, choline, among several others) (Hoffmann and Bremer 2017). Using these mechanisms, some extreme halophiles can grow at NaCl concentrations approaching saturation (above 5 M).

As noted above, UV selection of *B. subtilis* resulted in a concomitant increase of its resistance to high osmolarity (1 M NaCl), due to an unknown mechanism (Wassmann et al. 2010). A search of the literature revealed that studies in which microorganisms were subjected to experimental evolution specifically directed at increasing their growth at increased osmotic potentials are lacking. However, some evolutionary insights can be gained by studying the properties of macromolecules from halophiles vs. mesophiles. The proteins of archaeal halophiles typically have an increased content of acidic amino acids and exhibit increased negative surface charges, which compensate for their extreme ionic environments (Reed et al. 2013). Regarding protein-DNA interactions, the binding of archaeal promoters by the TATA-box-binding protein (TBP) is an essential step in initiation of transcription. In mesophilic archaea, the affinity of TBP for DNA decreases with increasing salt concentration, but in the halophilic archaeon *Pyrococcus woesei*, TBP binds its cognate DNA better at higher salt concentrations (Bergqvist et al. 2003). This effect could be attributed to only three amino acid changes in TBP, suggesting that the important phenotype of halophilicity in TBP, thus global transcription at high osmolarity, could be rapidly acquired in evolutionary time (Bergqvist et al. 2003). Valuable information regarding the possible growth of Earth microbes in martian environments could be gained by future experiments directed toward evolution of terrestrial microbes to conditions of increased osmolarity, particularly in regard to perchlorate salts.

4.5 Low or Absent Organic Nutrients

Evolution of life toward the utilization of increasingly scarce nutrients (termed *oligotrophy*) started on Earth billions of years ago [reviewed in Raven et al. (2005)]. Scarcity of energy, carbon, and nitrogen likely drove the early evolution of processes such as photosynthesis and nitrogen fixation. The resulting oxygenation of the Earth via photosynthesis in turn led to scarcity of soluble phosphorus and iron, thus driving the evolution of various strategies for phosphorus and iron dissolution, uptake, and scavenging among evolving prokaryotes (Raven et al. 2005). On present-day Earth, oligotrophy is a common condition facing microbes in nutrient-poor environments, and even in nutrient-rich environments, nutrients can become limiting through competition; in particular, phosphorus and nitrogen are recognized to be limiting nutrients in most freshwater, oceanic, and terrestrial environments (Guildford and Hecky 2000). The technical difficulties encountered to date both in identifying and quantifying the actual, particularly organic, nutrients present in martian soils have hampered development of experiments to test the possible evolution of Earth microbes under simulated Mars conditions, and to date no directed studies have appeared in the literature.

5 Conclusions and Perspectives

From the discussions presented above, we have seen that experimental evolution can be a valuable tool in understanding how Earth life might adapt to some of the environmental parameters currently present on Mars. In addition, we have seen that *some* Earth microorganisms already possess *some* of the adaptations necessary for successful growth in martian environments. At this point it is perhaps instructional to summarize the myriad environmental extremes present on Mars, what attributes would be needed to enable survival and growth, and which Earth microbes possess such attributes, presented in Table 10.3.

In examining Table 10.3, it should be kept in mind that in order to grow in the environment of Mars, a microorganism must possess *all* the attributes listed *simultaneously*—and to date no such single organism has been found. At present, Earth's continuous permafrosts seem to be the most promising environments for isolation of microbes with the potential to grow on Mars. Permafrosts exhibit several Mars-like properties: permanent cold, limited access to liquid water, low nutrient availability, high salinity in brine inclusions and cryopegs, and low oxygen availability. As discussed in Sect. 3, we were able to isolate numerous bacteria from a Siberian permafrost borehole capable of growing under the T, P, and AC regime of the Mars atmosphere—but on hydrated rich medium (Table 10.2). Recently, an exciting report appeared in the literature showing that methanogenic archaea isolated from permafrost could (1) convert CO_2 to CH_4 in the presence of high concentrations of perchlorate and (2) utilize perchlorate as an electron acceptor in anaerobic methane oxidation (Shcherbakova et al. 2015). However, this study was not performed under simulated martian T, P, and AC conditions, and the media used contained vitamins and other components not known to be present in the martian environment.

Nevertheless, these results are encouraging and contribute significantly toward our understanding of the limits of life and the ability of Earth microbes to grow in extraterrestrial environments. I am optimistic that many of the gaps identified and discussed in this review will be filled in near-future studies on this fascinating topic.

Table 10.3 Attributes of Earth microbes potentially enabling them to inhabit Mars environments

Challenge	Attribute	Representative genera	Reviewed in
Low temperature	Psychrophily/ psychrotolerance	*Arthrobacter*, *Halobacterium*, *Hyphomonas*, *Pseudomonas*, *Psychrobacter*, *Sphingomonas*	D'Amico et al. (2006), Casanueva et al. (2010)
Low pressure	Hypobarophilicity	See Table 10.2	This chapter.
UV radiation	Resistance mechanisms	Spores: *Bacillus*, *Clostridium*, *Paenibacillus*, etc. Vegetative cells: *Deinococcus*, *Janibacter*	Nicholson et al. (2000), Friedberg et al. (2006)
Ionizing radiation	Resistance mechanisms	*Bacillus* (spores), *Deinococcus*, *Chelatococcus*, *Corbulabacter*, *Spirosoma*, *Geodermatophilus*, *Hymenobacter*, *Planococcus*	Munteanu et al. (2015), Friedberg et al. (2006), Rainey et al. (2005)
High osmolarity	Osmophily/ halophily	Archaea: numerous genera Bacteria: *Chromohalobacter*, *Halomonas*	Detkova and Boltyanskaya (2007)
Perchlorate salts	Redox chemistry	*Azospirillum*, *Ideonella*, *Proteus*, *Pseudomonas*, *Wolinella*	Van der Zee and Cervantes (2009), Coates et al. (1999)
Low nutrient levels	Oligotrophy	Wide diversity, in the mainly uncharacterized "rare biosphere"	Egli (2010)
Lack of organic carbon	Autotrophy	*Methanobacterium*, *Methanosarcina*, *Nitrosomonas*, *Thiobacillus*	Wood et al. (2004)

And who knows? Perhaps in the not-so-distant future, many of the questions posed here will be answered through laboratory study of bona fide martian microbes.

Acknowledgments The author thanks Andrew Schuerger and Patricia Fajardo-Cavazos for helpful discussions and Hoang Nguyen for communication of data before publication. This work has been supported over the years by grants from the NASA Exobiology (NNA04CI35A, NNX08AO15G) and Planetary Protection (NNA05CS68G, NNA06CB58G) programs.

References

Bergqvist S, Williams MA, O'Brien R, Ladbury JE (2003) Halophilic adaptation of protein-DNA interactions. Biochem Soc Trans 31:677–680. https://doi.org/10.1042/bst0310677

Berry BJ, Jenkins DG, Schuerger AC (2010) Effects of simulated Mars conditions on the survival and growth of *Escherichia coli* and *Serratia liquefaciens*. Appl Environ Microbiol 76 (8):2377–2386. https://doi.org/10.1128/AEM.02147-09

Casanueva A, Tuffin M, Cary C, Cowan DA (2010) Molecular adaptations to psychrophily: the impact of 'omic' technologies. Trends Microbiol 18(8):374–381. https://doi.org/10.1016/j.tim.2010.05.002

Cho KH (2017) The structure and function of the Gram-positive bacterial RNA degradosome. Front Microbiol 8:154. https://doi.org/10.3389/fmicb.2017.00154

Coates JD, Michaelidou U, Bruce RA, O'Connor SM, Crespi JN, Achenbach LA (1999) Ubiquity and diversity of dissimilatory (per)chlorate-reducing bacteria. Appl Environ Microbiol 65 (12):5234–5241

Cockell CS, Catling DC, Davis WL, Snook K, Kepner RL, Lee P, McKay CP (2000) The ultraviolet environment of Mars: biological implications past, present, and future. Icarus 146(2):343–359

Cockell CS, Bush T, Bryce C, Direito S, Fox-Powell M, Harrison JP, Lammer H, Landenmark H, Martin-Torres J, Nicholson N, Noack L, O'Malley-James J, Payler SJ, Rushby A, Samuels T, Schwendner P, Wadsworth J, Zorzano MP (2016) Habitability: a review. Astrobiology 16 (1):89–117. https://doi.org/10.1089/ast.2015.1295

Collins MA, Buick RK (1989) Effect of temperature on the spoilage of stored peas by *Rhodotorula glutinis*. Food Microbiol 6:135–142

Crawford RL (2005) Microbial diversity and its relationship to planetary protection. Appl Environ Microbiol 71(8):4163–4168. https://doi.org/10.1128/AEM.71.8.4163-4168.2005

D'Amico S, Collins T, Marx JC, Feller G, Gerday C (2006) Psychrophilic microorganisms: challenges for life. EMBO Rep 7(4):385–389. https://doi.org/10.1038/sj.embor.7400662

Detkova EN, Boltyanskaya YV (2007) Osmoadaptation of haloalkaliphilic bacteria: role of osmoregulators and their possible practical application. Microbiology 76(5):511–522. https://doi.org/10.1134/s0026261707050013

Eigenbrode JL, Summons RE, Steele A, Freissinet C, Millan M, Navarro-González R, Sutter B, McAdam AC, Franz HB, Glavin DP, Archer PD, Mahaffy PR, Conrad PG, Hurowitz JA, Grotzinger JP, Gupta S, Ming DW, Sumner DY, Szopa C, Malespin C, Buch A, Coll P (2018) Organic matter preserved in 3-billion-year-old mudstones at Gale crater, Mars. Science 360 (6393):1096–1101

Egli T (2010) How to live at very low substrate concentration. Water Res 44(17):4826–4837. https://doi.org/10.1016/j.watres.2010.07.023

Ehlmann BL, Mustard JF, Murchie SL (2010) Geologic setting of serpentine deposits on Mars. Geophys Res Lett 37:5. https://doi.org/10.1029/2010gl042596

Fajardo-Cavazos P, Schuerger A, Nicholson W (2007) Testing interplanetary transfer of bacteria between Earth and Mars as a result of natural impact phenomena and human spaceflight activities. Acta Astronaut 60(4-7):534–540

Ferrer M, Chernikova TN, Yakimov MM, Golyshin PN, Timmis KN (2003) Chaperonins govern growth of *Escherichia coli* at low temperatures Chaperonins govern growth of *Escherichia coli* at low temperatures. Nat Biotechnol 21(11):1266–1267. https://doi.org/10.1038/nbt1103-1266

Freissinet C, Glavin DP, Mahaffy PR, Miller KE, Eigenbrode JL, Summons RE, Brunner AE, Buch A, Szopa C, Archer PD, Franz HB, Atreya SK, Brinckerhoff WB, Cabane M, Coll P, Conrad PG, Des Marais DJ, Dworkin JP, Fairen AG, Francois P, Grotzinger JP, Kashyap S, ten Kate IL, Leshin LA, Malespin CA, Martin MG, Martin-Torres FJ, McAdam AC, Ming DW, Navarro-Gonzalez R, Pavlov AA, Prats BD, Squyres SW, Steele A, Stern JC, Sumner DY, Sutter B, Zorzano MP, Team MSLS (2015) Organic molecules in the Sheepbed Mudstone, Gale Crater, Mars. J Geophys Res Planets 120(3):495–514. https://doi.org/10.1002/2014je004737

Friedberg EC, Walker GC, Siede W, Wood RD, Schultz RA, Ellenberger T (2006) DNA repair and mutagenesis, 2nd edn. ASM Press, Washington, DC

Gaillard F, Michalski J, Berger G, McLennan SM, Scaillet B (2013) Geochemical reservoirs and timing of sulfur cycling on Mars. Space Sci Rev 174(1–4):251–300. https://doi.org/10.1007/s11214-012-9947-4

Glavin DP, Freissinet C, Miller KE, Eigenbrode JL, Brunner AE, Buch A, Sutter B, Archer PD, Atreya SK, Brinckerhoff WB, Cabane M, Coll P, Conrad PG, Coscia D, Dworkin JP, Franz HB, Grotzinger JP, Leshin LA, Martin MG, McKay C, Ming DW, Navarro-Gonzalez R, Pavlov A, Steele A, Summons RE, Szopa C, Teinturier S, Mahaffy PR (2013) Evidence for perchlorates and the origin of chlorinated hydrocarbons detected by SAM at the Rocknest aeolian deposit in Gale Crater. J Geophys Res Planets 118(10):1955–1973. https://doi.org/10.1002/jgre.20144

Guildford SJ, Hecky RE (2000) Total nitrogen, total phosphorus, and nutrient limitation in lakes and oceans: is there a common relationship? Limnol Oceanogr 45(6):1213–1223

Hassler DM, Zeitlin C, Wimmer-Schweingruber RF, Ehresmann B, Rafkin S, Eigenbrode JL, Brinza DE, Weigle G, Böttcher S, Böhm E, Burmeister S, Guo J, Köhler J, Martin C, Reitz G, Cucinotta FA, Kim MH, Grinspoon D, Bullock MA, Posner A, Gómez-Elvira J, Vasavada A, Grotzinger JP, Team MS (2014) Mars' surface radiation environment measured with the Mars Science Laboratory's Curiosity rover. Science 343(6169):1244797. https://doi.org/10.1126/science.1244797

Hecht MH, Kounaves SP, Quinn RC, West SJ, Young SM, Ming DW, Catling DC, Clark BC, Boynton WV, Hoffman J, Deflores LP, Gospodinova K, Kapit J, Smith PH (2009) Detection of perchlorate and the soluble chemistry of martian soil at the Phoenix lander site. Science 325 (5936):64–67. https://doi.org/10.1126/science.1172466

Heller R, Armstrong J (2014) Superhabitable worlds. Astrobiology 14(1):50–66. https://doi.org/10.1089/ast.2013.1088

Hoffmann T, Bremer E (2017) Guardians in a stressful world: the Opu family of compatible solute transporters from *Bacillus subtilis*. Biol Chem 398(2):193–214. https://doi.org/10.1515/hsz-2016-0265

Horikoshi K, Antranikian G, Bull AT, Robb FT, Stetter KO (eds) (2011) Extremophiles handbook. Springer, Berlin

Horneck G (1993) Responses of *Bacillus subtilis* spores to the space environment: results from experiments in space. Orig Life Evol Biosph 23(1):37–52

Kempf M, Chen F, Kern R, Venkateswaran K (2005) Recurrent isolation of hydrogen peroxide-resistant spores of *Bacillus pumilus* from a spacecraft assembly facility. Astrobiology 5 (3):391–405

Kral TA, Altheide TS, Lueders AE, Schuerger AC (2011) Low pressure and desiccation effects on methanogens: implications for life on Mars. Planet Space Sci 59:264–270

La Duc M, Nicholson W, Kern R, Venkateswaran K (2003) Microbial characterization of the Mars Odyssey spacecraft and its encapsulation facility. Environ Microbiol 5(10):977–985

La Duc M, Kern R, Venkateswaran K (2004) Microbial monitoring of spacecraft and associated environments. Microb Ecol 47(2):150–158. https://doi.org/10.1007/s00248-003-1012-0

La Duc M, Dekas A, Osman S, Moissl C, Newcombe D, Venkateswaran K (2007) Isolation and characterization of bacteria capable of tolerating the extreme conditions of clean room environments. Appl Environ Microbiol 73(8):2600–2611. https://doi.org/10.1128/AEM.03007-06

Link L, Sawyer J, Venkateswaran K, Nicholson W (2004) Extreme spore UV resistance of *Bacillus pumilus* isolates obtained from an ultraclean spacecraft assembly facility. Microb Ecol 47 (2):159–163

Melosh H (1984) Impact ejection, spallation, and the origin of meteorites. Icarus 59(2):234–260

Melosh HJ (1989) Impact cratering: a geologic process. Oxford University Press, New York

Mileikowsky C, Cucinotta F, Wilson J, Gladman B, Horneck G, Lindegren L, Melosh J, Rickman H, Valtonen M, Zheng J (2000) Natural transfer of viable microbes in space – 1. From Mars to Earth and Earth to Mars. Icarus 145(2):391–427

Mileikowsky C, Cucinotta FA, Wilson JW, Gladman B, Horneck G, Lindegren L, Melosh HJ, Rickman H, Valtonen M, Zheng JQ (2002) Natural transfer of viable microbes in space. Part 1: From Mars to Earth and Earth to Mars. Icarus 145:391–427

Millan M, Szopa C, Buch A, Coll P, Glavin DP, Freissinet C, Navarro-Gonzalez R, Francois P, Coscia D, Bonnet JY, Teinturier S, Cabane M, Mahaffy PR (2016) In situ analysis of martian regolith with the SAM experiment during the first mars year of the MSL mission: identification of organic molecules by gas chromatography from laboratory measurements. Planet Space Sci 129:88–102. https://doi.org/10.1016/j.pss.2016.06.007

Mumma MJ, Villanueva GL, Novak RE, Hewagama T, Bonev BP, Disanti MA, Mandell AM, Smith MD (2009) Strong release of methane on Mars in northern summer 2003. Science 323 (5917):1041–1045. https://doi.org/10.1126/science.1165243

Munteanu A, Uivarosi V, Andries A (2015) Recent progress in understanding the molecular mechanisms of radioresistance in Deinococcus bacteria. Extremophiles 19(4):707–719. https://doi.org/10.1007/s00792-015-0759-9

Newcombe DA, Schuerger AC, Benardini JN, Dickinson D, Tanner R, Venkateswaran K (2005) Survival of spacecraft-associated microorganisms under simulated martian UV irradiation. Appl Environ Microbiol 71(12):8147–8156. https://doi.org/10.1128/aem.71.12.8147-8156.2005

Nicholson WL (2009) Ancient micronauts: interplanetary transport of microbes by cosmic impacts. Trends Microbiol 17(6):243–250. https://doi.org/10.1016/j.tim.2009.03.004

Nicholson WL, Munakata N, Horneck G, Melosh HJ, Setlow P (2000) Resistance of *Bacillus* endospores to extreme terrestrial and extraterrestrial environments. Microbiol Mol Biol Rev 64 (3):548–572. https://doi.org/10.1128/mmbr.64.3.548-572.2000

Nicholson WL, Schuerger AC, Setlow P (2005) The solar UV environment and bacterial spore UV resistance: considerations for Earth-to-Mars transport by natural processes and human spaceflight. Mutat Res 571(1–2):249–264. https://doi.org/10.1016/j.mrfmmm.2004.10.012

Nicholson W, Schuerger A, Race M (2009) Migrating microbes and planetary protection. Trends Microbiol 17(9):389–392. https://doi.org/10.1016/j.tim.2009.07.001

Nicholson WL, Fajardo-Cavazos P, Fedenko J, Ortiz-Lugo JL, Rivas-Castillo A, Waters SM, Schuerger AC (2010) Exploring the low-pressure growth limit: evolution of *Bacillus subtilis* in the laboratory to enhanced growth at 5 kilopascals. Appl Environ Microbiol 76 (22):7559–7565. https://doi.org/10.1128/aem.01126-10

Nicholson WL, McCoy L, Kerney K, Ming DW, Golden DC, Schuerger AC (2012) An aqueous extract of Mars analogue soil from the Phoenix landing site does not inhibit spore germination or growth of model spacecraft contaminants *Bacillus subtilis* 168 and *B. pumilus* SAFR-032. Icarus 220:904–910

Nicholson WL, Krivushin K, Gilichinsky D, Schuerger AC (2013) Growth of *Carnobacterium* spp. from permafrost under low pressure, temperature, and anoxic atmosphere has implications for Earth microbes on Mars. Proc Natl Acad Sci USA 110(2):666–671. https://doi.org/10.1073/pnas.1209793110

O'Leary M (2008) Anaxagoras and the origin of panspermia theory. iUniverse Press, Bloomington

Parry BR, Shain DH (2011) Manipulations of AMP metabolic genes increase growth rate and cold tolerance in *Escherichia coli*: implications for psychrophilic evolution. Mol Biol Evol 28 (7):2139–2145. https://doi.org/10.1093/molbev/msr038

Rainey FA, Ray K, Ferreira M, Gatz BZ, Nobre F, Bagaley D, Rash BA, Park MJ, Earl AA, Shank NC, Small AM, Henk MC, Battista JR, Kampfer P, da Costa MS (2005) Extensive diversity of ionizing-radiation-resistant bacteria recovered from Sonoran desert soil and description of nine new species of the genus Deinococcus obtained from a single soil sample. Appl Environ Microbiol 71(11):7630. https://doi.org/10.1128/aem.71.11.7630.2005

Raven JA, Andrews M, Quigg A (2005) The evolution of oligotrophy: implications for the breeding of crop plants for low input agricultural systems. Ann Appl Biol 146(3):261–280. https://doi.org/10.1111/j.1744-7348.2005.040138.x

Reed CJ, Lewis H, Trejo E, Winston V, Evilia C (2013) Protein adaptations in archaeal extremophiles. Archaea Int Microbiol J 2013:14. https://doi.org/10.1155/2013/373275

Rettberg P, Anesio A, Baker V, Baross J, Cady SL, Foreman CM, Hauber E, Gabriele-Ori G, Pearce D, Rennó N, Ruvkun G, Sattler B, Saunders MP, Wagner D, Westall F (2015) Review of the MEPAG report on Mars special regions. National Academies Press, Washington, DC

Rummel J (2001) Planetary exploration in the time of astrobiology: protecting against biological contamination. Proc Natl Acad Sci USA 98(8):2128–2131

Rummel JD, Beaty DW, Jones MA, Bakermans C, Barlow NG, Boston PJ, Chevrier VF, Clark BC, de Vera JP, Gough RV, Hallsworth JE, Head JW, Hipkin VJ, Kieft TL, McEwen AS, Mellon MT, Mikucki JA, Nicholson WL, Omelon CR, Peterson R, Roden EE, Sherwood Lollar B, Tanaka KL, Viola D, Wray JJ (2014) A new analysis of Mars "Special Regions": findings of the second MEPAG Special Regions Science Analysis Group (SR-SAG2). Astrobiology 14 (11):887–968. https://doi.org/10.1089/ast.2014.1227

Schuerger AC (2004) Microbial ecology of the surface exploration of Mars with human-operated vehicles. In: Cockell CS (ed) Martian expedition planning. Univelt Publishers, Santa Barbra, CA, pp 363–386

Schuerger AC, Nicholson WL (2006) Interactive effects of hypobaria, low temperature, and CO_2 atmospheres inhibit the growth of mesophilic *Bacillus* spp. under simulated martian conditions. Icarus 185(1):143–152. https://doi.org/10.1016/j.icarus.2006.06.014

Schuerger AC, Nicholson WL (2016) Twenty species of hypobarophilic bacteria recovered from diverse soils exhibit growth under simulated martian conditions at 0.7 kPa. Astrobiology 16 (12):964–976. https://doi.org/10.1089/ast.2016.1587

Schuerger A, Richards J, Hintze P, Kern R (2005) Surface characteristics of spacecraft components affect the aggregation of microorganisms and may lead to different survival rates of bacteria on Mars landers. Astrobiology 5(4):545–559

Schuerger A, Richards J, Newcombe D, Venkateswaran K (2006) Rapid inactivation of seven *Bacillus* spp. under simulated Mars UV irradiation. Icarus 181(1):52–62. https://doi.org/10.1016/j.icarus.2005.10.008

Schuerger AC, Golden DC, Ming DW (2012) Biotoxicity of Mars soils: 1. Dry deposition of analog soils on microbial colonies and survival under martian conditions. Planet Space Sci 72 (1):91–101

Schuerger AC, Ulrich R, Berry BJ, Nicholson WL (2013) Growth of *Serratia liquefaciens* under 7 mbar, 0 °C, and CO_2-enriched anoxic atmospheres. Astrobiology 13(2):115–131. https://doi.org/10.1089/ast.2011.0811

Sephton MA, Lewis JMT, Watson JS, Montgomery W, Garnier C (2014) Perchlorate-induced combustion of organic matter with variable molecular weights: implications for Mars missions. Geophys Res Lett 41(21):7453–7460. https://doi.org/10.1002/2014gl062109

Shcherbakova V, Oshurkova V, Yoshimura Y (2015) The effects of perchlorates on the permafrost methanogens: implication for autotrophic life on Mars. Microorganisms 3(3):518–534. https://doi.org/10.3390/microorganisms3030518

Smith DJ, Schuerger AC, Davidson MM, Pacala SW, Bakermans C, Onstott TC (2009) Survivability of *Psychrobacter cryohalolentis* K5 under simulated martian surface conditions. Astrobiology 9(2):221–228. https://doi.org/10.1089/ast.2007.0231

Stern JC, Sutter B, Freissinet C, Navarro-Gonzalez R, McKay CP, Archer PD, Buch A, Brunner AE, Coll P, Eigenbrode JL, Fairen AG, Franz HB, Glavin DP, Kashyap S, McAdam AC, Ming DW, Steele A, Szopa C, Wray JJ, Martin-Torres FJ, Zorzano MP, Conrad PG, Mahaffy PR, Team MSLS (2015) Evidence for indigenous nitrogen in sedimentary and aeolian deposits from the Curiosity rover investigations at Gale crater, Mars. Proc Natl Acad Sci USA 112 (14):4245–4250. https://doi.org/10.1073/pnas.1420932112

Stillman DE, Grimm RE (2011) Dielectric signatures of adsorbed and salty liquid water at the Phoenix landing site, Mars. J Geophys Res Planets 116:11. https://doi.org/10.1029/2011je003838

Stolper DA, Revsbech NP, Canfield DE (2010) Aerobic growth at nanomolar oxygen concentrations. Proc Natl Acad Sci USA 107(44):18755–18760. https://doi.org/10.1073/pnas.1013435107

Toner JD, Catling DC (2016) Water activities of $NaClO_4$, $Ca(ClO_4)(2)$, and $Mg(ClO_4)(2)$ brines from experimental heat capacities: water activity $>$ 0.6 below 200 K. Geochim Cosmochim Acta 181:164–174. https://doi.org/10.1016/j.gca.2016.03.005

Toner JD, Catling DC, Light B (2014) The formation of supercooled brines, viscous liquids, and low-temperature perchlorate glasses in aqueous solutions relevant to Mars. Icarus 233:36–47. https://doi.org/10.1016/j.icarus.2014.01.018

Van der Zee FR, Cervantes FJ (2009) Impact and application of electron shuttles on the redox (bio) transformation of contaminants: a review. Biotechnol Adv 27(3):256–277. https://doi.org/10.1016/j.biotechadv.2009.01.004

Venkateswaran K, Satomi M, Chung S, Kern R, Koukol R, Basic C, White D (2001) Molecular microbial diversity of a spacecraft assembly facility. Syst Appl Microbiol 24(2):311–320

Waite JH, Combi MR, Ip WH, Cravens TE, McNutt RL, Kasprzak W, Yelle R, Luhmann J, Niemann H, Gell D, Magee B, Fletcher G, Lunine J, Tseng WL (2006) Cassini ion and neutral mass spectrometer: enceladus plume composition and structure. Science 311(5766):1419–1422. https://doi.org/10.1126/science.1121290

Wassmann M, Moeller R, Reitz G, Rettberg P (2010) Adaptation of *Bacillus subtilis* cells to Archean-like UV climate: relevant hints of microbial evolution to remarkably increased radiation resistance. Astrobiology 10(6):605–615. https://doi.org/10.1089/ast.2009.0455

Waters SM, Robles-Martínez JA, Nicholson WL (2014) Exposure of *Bacillus subtilis* to low pressure (5 kPa) induces several global regulons including the *sigB*-mediated General Stress Response. Appl Environ Microbiol 80(16):4788–4794. https://doi.org/10.1128/AEM.00885-14

Waters SM, Zeigler DR, Nicholson WL (2015) Experimental evolution of enhanced growth by *Bacillus subtilis* at low atmospheric pressure: genomic changes revealed by whole-genome sequencing. Appl Environ Microbiol 81(21):7525–7532. https://doi.org/10.1128/AEM.01690-15

Webster CR, Mahaffy PR, Atreya SK, Flesch GJ, Mischna MA, Meslin PY, Farley KA, Conrad PG, Christensen LE, Pavlov AA, Martín-Torres J, Zorzano MP, McConnochie TH, Owen T, Eigenbrode JL, Glavin DP, Steele A, Malespin CA, Archer PD, Sutter B, Coll P, Freissinet C, McKay CP, Moores JE, Schwenzer SP, Bridges JC, Navarro-Gonzalez R, Gellert R, Lemmon MT, Team MS (2015) Mars atmosphere. Mars methane detection and variability at Gale crater. Science 347(6220):415–417. https://doi.org/10.1126/science.1261713

Webster CR, Mahaffy PR, Atreya SK, Moores JE, Flesch GJ, Malespin C, McKay CP, Martinez G, Smith CL, Martin-Torres J, Gomez-Elvira J, Zorzano M-P, Wong MH, Trainer MG, Steele A, Archer D Jr, Sutter B, Coll PJ, Freissinet C, Meslin P-Y, Gough RV, House CH, Pavlov A, Eigenbrode JL, Glavin DP, Pearson JC, Keymeulen D, Christensen LE, Schwenzer SP, Navarro-Gonzalez R, Pla-García J, Rafkin SCR, Vicente-Retortillo Á, Kahanpää H, Viudez-Moreiras D, Smith MD, Harri A-M, Genzer M, Hassler DM, Lemmon M, Crisp J, Sander SP, Zurek RW, Vasavada AR (2018) Background levels of methane in Mars' atmosphere show strong seasonal variations. Science 360:1093–1096. https://doi.org/10.1126/science.aaq0131

Wood AP, Aurikko JP, Kelly DP (2004) A challenge for 21st century molecular biology and biochemistry: what are the causes of obligate autotrophy and methanotrophy? FEMS Microbiol Rev 28(3):335–352. https://doi.org/10.1016/j.femsre.2003.12.001

Chapter 11
The Role of Phage in the Adaptation of Bacteria to New Environmental Niches

Veronica Casas and Stanley Maloy

1 Introduction

As the most abundant biological entity on earth, bacterial viruses (bacteriophages or phages) play significant roles in the adaptation of bacteria to environmental and animal niches. Numerous studies investigating the interplay between phages and host in natural, animal, and laboratory models have demonstrated that phages profoundly influence the evolution of their bacterial hosts. In this chapter we discuss the three major ways phages mediate this process: (1) by the selective pressure they exert on their host as a result of lysis ("predation"), (2) by the generalized transduction of genetic material between bacteria, and (3) as lysogens that provide genetic traits carried on the phage to the bacterial genome (lysogenic conversion).

The basic dynamics of phage-bacteria interactions are well-studied, but we will briefly review them here as a matter of introduction. When phages and their bacterial hosts interact in the environment, there are multiple fates possible for the phages, bacterial hosts, and bacterial community as a whole (Fig. 11.1). The simplest outcome is predation, infection, and lysis of the susceptible bacterial population—releasing many copies of progeny phages capable of continued rounds of lytic infection (Fig. 11.1a). The key element in these predator-prey interactions is that pressure is created for selection of bacteria that are resistant to phage infection and spared from lysis. Relatively free of competition from the original susceptible bacterial population, these phage-resistant bacteria outgrow the previous population of phage-sensitive bacteria and present a new target for phages. Survival of the original phage population within this environment will now depend on selection of

V. Casas (✉) · S. Maloy (✉)
Department of Biology, San Diego State University, San Diego, USA

Center for Microbial Sciences, San Diego, USA
e-mail: vcasas@sdsu.edu; smaloy@sdsu.edu

P. H. Rampelotto (ed.), *Molecular Mechanisms of Microbial Evolution*, Grand Challenges in Biology and Biotechnology,
https://doi.org/10.1007/978-3-319-69078-0_11

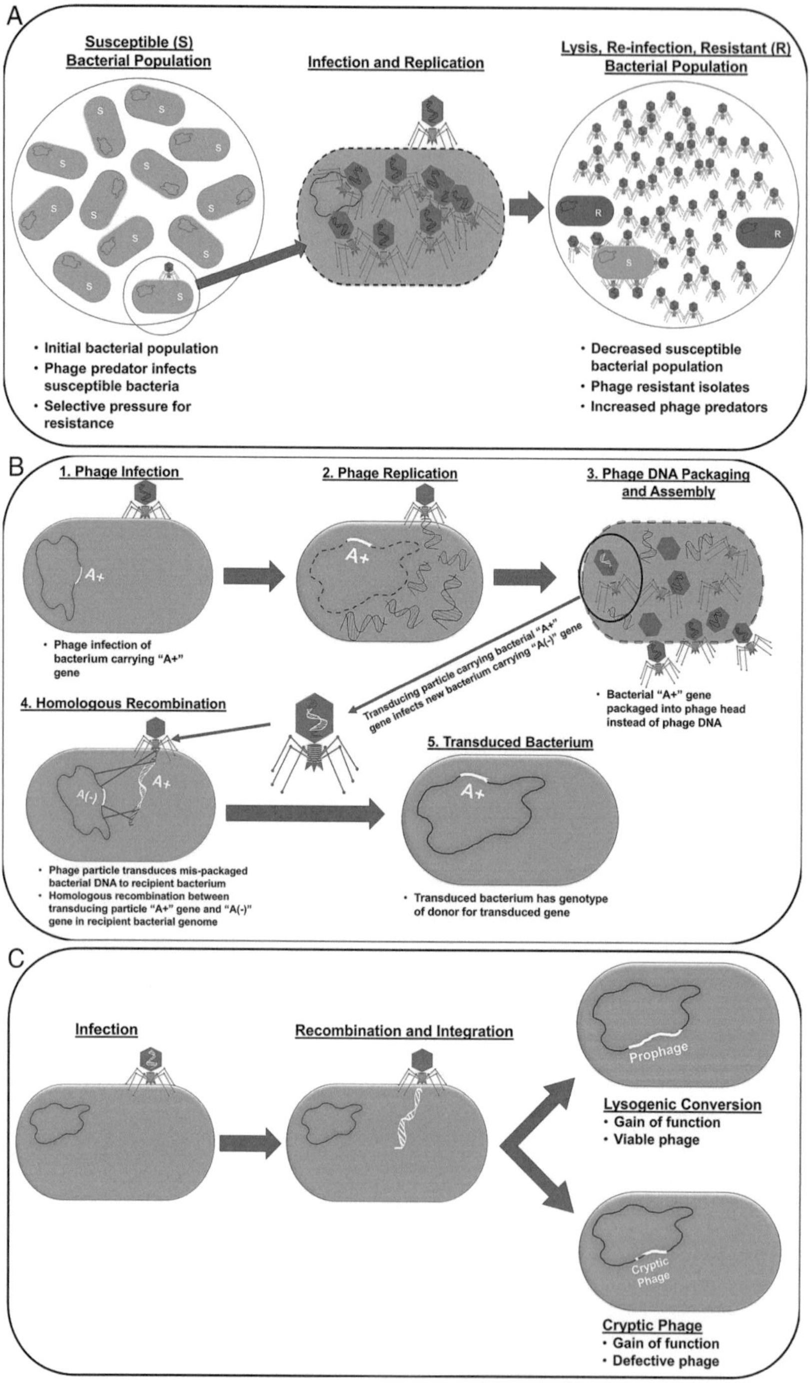

Fig. 11.1 Outcomes of phage and bacterial host interactions. (**a**) Outcomes of phage predation on a susceptible bacterial population. (**b**) Outcomes of transduction of genes between bacterial hosts via mis-packaging of bacterial DNA into phage head. (**c**) Outcomes of lysogeny via integration

phages that are capable of infecting the emerging resistant bacterial clones (Fig. 11.1a). The result is a microbial population that is coevolving due to the strong selective pressures provided by the predator-prey associations (Horne 1970; Lenski and Levin 1985; Lenski 1988).

The relationship between phages and bacteria in any given niche is a complex natural selection system that results in an "arms race" that drives coevolution of phages and the bacterial hosts (Horne 1970; Lenski and Levin 1985; Buckling et al. 2006; Calendar 2006; Pal et al. 2007; Buckling and Hodgson 2007; Morgan et al. 2010). While it may be surprising to consider that the immediate propagation of phage particles in a lytic infection could lead to bacterial adaptation in specialized niches, the sacrifice of one can benefit the many. While the individual infected cell does not directly benefit from phage predation, the population's overall fitness is enhanced by the repeated exposure to those selective pressures (Lenski 1988; Buckling and Rainey 2003; Buckling et al. 2006; Buckling and Hodgson 2007; Pal et al. 2007). Studies of the predator-prey relationship between bacteria and phages have shown that there is an increase in fitness and survival of bacterial populations exposed to high selective pressure from phage predation (Scanlan et al. 2011; Morgan et al. 2010; Shapiro et al. 2016; Gorter et al. 2016; McGee et al. 2014; Goldhill and Turner 2014; O'Brien et al. 2013; Koskella and Meaden 2013; Paterson et al. 2010).

Fitness test results from a continuous culture model system studying the effects of M13 predation on *Escherichia coli* indicate that infected bacteria reproduce faster and to higher density than uninfected bacteria (Shapiro et al. 2016). When the predator-prey interactions were allowed to continue, through both vertical and horizontal transmission, both lytic and lysogenic interactions resulted in increased bacterial growth and cell density (Shapiro et al. 2016). Examples of acquired phenotypes that allow evasion of phage predation include those that reduce adsorption, affect phage DNA processing, and inhibit production of phage particles, packaging of phage DNA, or lysis (Brown et al. 2013; Buckling and Brockhurst 2012; Hall et al. 2011; Kashiwagi et al. 2011). Additionally, increased fitness is manifest in a subpopulation of bacteria that acquire resistance to other phages through mechanisms such as adaptation of the bacterial CRISPR-Cas immune system (Al-Attar et al. 2011; Levin and Guttman 2010).

Another way phages contribute to the adaptation of bacteria to new environmental niches is by promoting exchange of chromosomal or plasmid DNA between bacteria via transduction (Fig. 11.1b). The result of mis-packaging bacterial DNA into a phage capsid, generalized transduction can transfer essentially any region of a bacterial genome to another bacterium (Perez et al. 2009). When the transducing

Fig. 11.1 (continued) of whole phage genome into bacterial host genome. Lysogenic conversion produces a gain of function to bacterial host and continues to exist as a viable phage in prophage form. Over generations and replication of bacterial genome, unused phage-specific genes are lost from prophage genome and render it a cryptic phage incapable of producing viable phage particles

phage particle infects the next host, the bacterial DNA within its capsid can integrate into the chromosome through homologous recombination. Transduction of genes between bacteria can provide a bacterial recipient with an arsenal of genetic and phenotypic traits that may have selective advantage in particular environmental niches (Fig. 11.1b). The prolific genetic exchange between bacteria and phages, through transduction in particular, has contributed to the highly mosaic nature of both their genomes (Yoshida et al. 2015; Kelly et al. 2013; Timms et al. 2010; Morris et al. 2008; Smith et al. 2012; Summer et al. 2005; Aziz et al. 2005; Casjens and Thuman-Commike 2011; Ernst et al. 2003). The resulting genetic alterations influence the bacterial host's ability to adapt and survive in its environment. These adaptations include acquisition of various traits such as antibiotic resistance, new virulence properties, resistance to predation from other phages or protozoa, or by providing enhanced or novel metabolic properties (Colomer-Lluch et al. 2011b; Hawkey and Jones 2009; Karmali 2017; Plunkett et al. 1999; Waldor and Mekalanos 1996; Groman 1953; Arnold and Koudelka 2014; Lainhart et al. 2009; Waksman 1961). These adaptations will be discussed in further detail in subsequent sections of this chapter.

Finally, phages are capable of altering the bacterial genome via lysogeny, providing genes carried on the phage genome that can alter the phenotype of the infected bacteria (Fig. 11.1c). In lysogeny, the phage genome (or prophage) is integrated into the bacterial genome, or persists as a plasmid, repressing the lytic functions and allowing phage DNA replication to be coordinated with the genome of the bacterial host. These quiescent prophages can shift to lytic growth capable of producing viable phage particles upon induction. A common phenotype conferred by lysogens is resistance to related phages due to immunity (expression of a repressor that shuts down incoming phages) or superinfection exclusion (modification of the host to prevent entry or inactivate incoming phages; Maloy et al. 1996). When the integrated prophage encodes genetic traits that change the physiology of the bacterial host, lysogenic conversion occurs. Pathogens such as *E. coli* and *Vibrio cholerae* are examples of bacterial hosts that have undergone lysogenic conversion due to infection by phages carrying exotoxin genes (Waldor and Mekalanos 1996; O'Brien et al. 1984; Strockbine et al. 1986). These bacteria have benefited from this conversion as they have become better adapted to causing disease in an animal host and maintaining their proliferation in the environment. Lysogeny is not without its own drawbacks, however, and maintaining inactive prophage genes has a fitness cost. Unless there is a clear advantage in maintaining the entire phage genome, prophage genes can become an energetic burden for the bacterial host and can lead to selection for reduction in phage genes without a selective advantage. With the majority of phage genes deleted, only a cryptic prophage remains and is incapable of producing viable phage particles upon induction. Oftentimes bacteria can accumulate these cryptic prophages or phage remnants as pseudogenes, and while their direct benefit is not currently understood, it is hypothesized that these extra genes might provide a selective advantage or fitness benefit (Nedialkova et al. 2016; Bondy-Denomy and Davidson 2014).

The interactions between phages and their bacterial hosts happen frequently and are a major factor driving bacterial evolution. When genetic alterations to the bacterial host's genome result in acquisition of enhanced virulence traits, understanding the intricacies of the phage-bacterial host relationship has implications for human health by understanding how this relationship promotes the evolution of pathogens.

2 Antibiotic Resistance

Emergence of highly antibiotic-resistant pathogens has become a global public health concern. Ever since antibiotics were introduced as chemotherapy against bacterial infections, it became clear that bacteria quickly evolved resistance. Shortly after the use of penicillin as a chemotherapeutic in 1943, penicillin-resistant strains of *Staphylococcus* had already been identified (Ventola 2015; Benvenis and Davies 1973; Cohen et al. 1972; Davies and Devies 2010). As new antibiotics were introduced to address development of resistance, the window between introduction and emergence of resistant organisms was initially on the order of 10 years, but that window has become increasingly smaller, now spanning only 1–2 years on average (Ventola 2015). The extensive misuse of antibiotics in human medicine, coupled with the excessive non-clinical use of antibiotics in livestock and food production, releases a large input of antibiotics into the environment via wastewater treatment plants and other sources (Kraushaar et al. 2017; Wei et al. 2011; Kemper 2008; Kay et al. 2004; Miao et al. 2004; Jjemba 2002a). This provides strong selective pressure for the evolution of antibiotic-resistant bacteria within natural environments (Colomer-Lluch et al. 2011a). Phage-mediated transduction plays a key role in the transfer of antibiotic resistance between bacteria. Much of the spread of this antibiotic resistance between bacteria can be attributed to mobile genetic elements such as plasmids and phages. Recently, phage-encoded antibiotic resistance genes have been identified in several human- or animal-associated ecosystems, suggesting phages are key players in the transfer of antibiotic resistance genes in the environment (Lee and Park 2016; Haaber et al. 2016; Colombo et al. 2016; Calero-Caceres and Muniesa 2016; Subirats et al. 2016; Ross and Topp 2015; Colomer-Lluch et al. 2014).

As humans and animals continue to encroach on natural ecosystems, the influx of human or animal waste in these locations results in a mixing of bacteria and phages from the human/animal microbiome with bacteria and phages from the environment (Colomer-Lluch et al. 2011a; Chee-Sanford et al. 2009). Coupled with widespread use of antibiotics, this creates a hotbed for multiple selective pressures creating environmental bacteria that are resistant to a variety of different agents (Zhao et al. 2010; Kemper 2008; Focazio et al. 2008; Sarmah et al. 2006; Jjemba 2002b). Within the phagocytic cells in animals and protozoa in nature, bacteria face similar environmental challenges, including pH changes, antimicrobial peptides, oxidative agents, and phagocytic cells. Many of the conditions that select for survival in contaminated environments also select for increased virulence.

An example of this confluence of events occurs in the Ganges river-estuarine environment of South Bengal. Cholera infections have been endemic in this region for decades and are tied to a seasonal fluctuation of *Vibrio* spp. and their phages (Mookerjee et al. 2015). The seasonal shifts of these bacteria in the water are closely tied to blooms of phytoplankton associated with changes in temperature and increase in nutrients from runoff. *V. parahaemolyticus* and *V. vulnificus* strains isolated from this highly dynamic environment were resistant to β-lactam antibiotics and also carried virulence genes associated with cholera such as *toxR*, thermostable direct hemolysin (*tdh*), and hemolysin (*vvh*) (Mookerjee et al. 2015). The transfer of virulence genes carried on phages, high rates of transduction, and mutability in these aquatic environments, combined with the ability of *V. parahaemolyticus* and *V. cholerae* to persist in rivers and estuarine environments, provide a "perfect storm" for evolution of new *Vibrio* pathogens in this cholera-endemic region of the world.

2.1 Examples of Phage-Encoded Antibiotic Resistance Genes

It has been known since the 1960s that bacteria can acquire antibiotic resistance via conjugation of plasmids that contain resistance cassettes (Benvenis and Davies 1973; Cohen et al. 1972; Cohen and Miller 1970; Jacob and Hobbs 1974). Likewise, acquisition of antibiotic resistance via transformation with naked DNA is a common mechanism for acquisition of antibiotic resistance in "naturally transformable" bacteria. Metagenomic studies have provided evidence that phages also play a major role in transfer of antibiotic-resistant genes (Chambers et al. 2015; Colomer-Lluch et al. 2011a; Modi et al. 2013; Colomer-Lluch et al. 2014). These findings have implications on how we approach research and treatment of antibiotic-resistant organisms. With the reservoir of antibiotic resistance genes expanding, it is important to understand the role phages play in spreading antibiotic resistance and the dynamics of the phage-bacteria interactions that lead to emergence of novel antibiotic-resistant organisms of clinical significance (Haaber et al. 2016).

Since they were first discovered (Colomer-Lluch et al. 2011a, 2014), more phage-encoded antibiotic resistance genes have been identified in a wide variety of environments. Among the first discovered were the phage-encoded β-lactamase (*bla*) genes (discussed in more detail in Sect. 2.2 below) (Colomer-Lluch et al. 2011a; Uyaguari et al. 2011). Since then, phages have also been found to carry genes that code for antibiotic efflux pumps, antimicrobial peptides, and resistance to aminoglycosides, macrolides, polymyxin-B, streptomycin, sulfonamides, and quinolones (Colomer-Lluch et al. 2011a; Lupo et al. 2012; Fancello et al. 2011; Looft et al. 2014; Goh et al. 2013).

Investigations into the human microbiome and clinically associated samples indicate that phage-encoded antibiotic resistance genes are abundant in these environments (Subirats et al. 2016; Solheim et al. 2011). Multi-locus sequence analysis of *Enterococcus faecalis* clinical strains from blood, feces, urine, and wounds revealed various phage-associated virulence factors, including vancomycin resistance (*vanB*) (Solheim et al. 2011). The combination of these genes makes *E. faecalis* a formidable pathogen in the hospital environment. The composition of the hospital wastewater viriome parallels the results of individual clinical pathogenic genomes indicating a high proportion of class A and D β-lactamases in the phage DNA fraction versus bacterial DNA fraction (Subirats et al. 2016). A survey of the dsDNA viral fraction associated with human skin showed a co-occurrence of lysogenic phages and antibiotic resistance genes in these skin metagenomes (Hannigan et al. 2015). These results suggest that these phages may confer resistance to skin bacteria and provide them with an advantage over their non-lysogenic counterparts (Sommer et al. 2009). Likewise, genomic analysis of *Helicobacter pylori* associated with gastrointestinal ulcers identified a lysogenic phage with the gene homologous to the multidrug-resistant protein D (*emrD*) (Fan et al. 2016). As more human-associated samples and clinical isolates are analyzed, it becomes clear that the variety of micro-niches within the human body provides a fertile ground for the presence and dissemination of antibiotic-resistant genes via the phage gene reservoir.

The animal microbiome (livestock, animal products, animal waste, etc.) provides a similar reservoir of phage-encoded antibiotic resistance genes. Particularly significant is the influence that treating animals with antibiotics has on the presence of resistance genes in the animal gut. In the fecal metagenome of dairy cows treated with ceftiofur—a third-generation cephalosporin antibiotic used throughout the agricultural industry—there was a greater abundance of antibiotic resistance genes identified when compared to the fecal metagenome of dairy cows *without* antibiotic treatment (Chambers et al. 2015). Moreover, these metagenomes contained a high proportion of phage- and prophage-like sequences, suggesting a high potential for horizontal transfer of the antibiotic resistance genes within the metagenome (Chambers et al. 2015). In the phage metagenomes from aquaculture wastewater, there was an abundance of antibiotic resistance genes encoding efflux pump regulating proteins (16–19%) as well as macrolide resistance (Colombo et al. 2016). In areas where gastrointestinal diseases are common, such as India, *Salmonella* phages isolated from raw sewage and river water carried both polymyxin-B and penicillin resistance-associated genes (Karpe et al. 2016). As antimicrobials continue to be used in animal husbandry and veterinary medicine, the bacteria and phages present in the animal gut will continue to be exposed to the selective pressures imposed by these antimicrobials, providing a fertile environment for exchange of resistance genes between bacteria and phages.

This genetic exchange is not limited to the animal gut but also occurs in the local environment that is exposed to animal waste and by-products abundant in antimicrobials (Kraushaar et al. 2017; Wei et al. 2011; Kemper 2008; Kay et al. 2004; Miao et al. 2004; Jjemba 2002a). Genes for resistance to streptomycin, sulfamethazine, aminoglycosides, and β-lactam antibiotics have been found in phage DNA fractions isolated from agricultural soils that had been mixed with animal manure or solid biological waste by-products ("biosolids") produced during wastewater treatment (Ross and Topp 2015). In this study, when soil-derived bacteria were co-incubated with phages derived from biosolids, coliforms resistant to cefoxitin (second-generation cephalosporin) and sulfamethazine (sulfonamide) developed in high numbers when compared to controls not co-incubated with the phages (Ross and Topp 2015). The practices of adding biosolids directly to agricultural soils to fertilize crops and the use of antibiotics to supplement farm animal feed provide strong selection for horizontal exchange of antibiotic resistance genes, potentially leading to novel antibiotic-resistant organisms and pathogens that are adapted to persist in environments like these. Additionally, since phages are capable of high rates of transduction (Baugher et al. 2014; Cornick et al. 2006; Petridis et al. 2006; Jiang and Paul 1998; Chiura 1997; Rohwer and Thurber 2009) and are typically more tolerant of common environmental stresses and wastewater treatment processes than are bacteria (Dumke et al. 2006; Haramoto et al. 2006; McLaughlin and Rose 2006; Moce-Llivina et al. 2003; Tanji et al. 2003; Hawkey and Jones 2009; Colomer-Lluch et al. 2011a, 2014; Faruque et al. 2000; Weinbauer and Suttle 1999; Jiang and Paul 1996), the reservoir of phage-encoded antibiotic resistance genes in natural environments associated with human and animal waste may play a particularly important role in the evolution of antibiotic-resistant organisms.

2.2 *Beta-Lactamases*

One of the best-studied examples of phage-encoded antibiotic resistance genes involves the family of genes associated with resistance to β-lactam antibiotics (Colomer-Lluch et al. 2011a). Over two-thirds of the antibiotics used in human medicine are β-lactams, providing a strong selection for bacteria that are resistant to this class of antibiotics, often by acquisition of genes that encode for β-lactamase (*bla*) (Hawkey and Jones 2009; Uyaguari et al. 2011). These antibiotic resistance genes were initially demonstrated to be acquired via plasmid-mediated lateral gene transfer between bacteria (Hawkey and Jones 2009; Uyaguari et al. 2011). However, recent studies have shown that certain *bla* genes, such as *blaCTXM-1*, are also carried by phages (Colomer-Lluch et al. 2011a; Hawkey and Jones 2009; Lachmayr et al. 2009).

Phages can withstand harsh environmental conditions, as compared to their bacterial hosts, and on some occasions their induction may actually be stimulated by these same environmental conditions (Moce-Llivina et al. 2003; Sinton et al. 2002). Antimicrobial agents used in household, commercial, and industrial settings provide some of these adverse conditions that induce phage replication. In fact, transduction of antibiotic resistance genes has been shown to be preceded by induction of a lysogenic phage from its bacterial host. A recent study found that compounds such as EDTA and sodium citrate (common additives in antimicrobial products) can induce lysogens and lead to the release of phages carrying antibiotic-resistant genes into the environment (Colomer-Lluch et al. 2014). The harsh chemical and physical treatments used to remove bacteria in wastewater treatment facilities are not always as effective at removing the sewage-associated phages. Given their ability to persist in harsh environments, it is no surprise that phage-encoded antibiotic-resistant genes have been consistently detected in sewage-impacted environments (Colomer-Lluch et al. 2011a; Lachmayr et al. 2009; Uyaguari et al. 2011). Examination of phage fractions isolated from raw wastewater and exposed to a range of temperatures showed the persistence of beta-lactamase genes (Calero-Caceres and Muniesa 2016). The reservoir of resistant genes carried by phages in these types of environments promotes the evolution and survival of resistant bacteria. Comparison of pre- and post-treatment samples from several wastewater treatment plants demonstrated that the concentration of antibiotic-resistant bacteria was greater in post-treatment samples (Lachmayr et al. 2009; Uyaguari et al. 2011). Antibiotic resistance genes are found not only in the wastewater collected directly from a treatment plant, but they have also been found in high concentrations in the phages isolated from surrounding environments (Karpe et al. 2016; Colomer-Lluch et al. 2011a; Lachmayr et al. 2009; Uyaguari et al. 2011). In samples gathered from areas where water flows after sewage treatment, β-lactamase genes were found in greater concentrations and in greater proportion than other antibiotic resistance genes when using 16S rDNA genes as a proxy for bacterial counts (Lachmayr et al. 2009). The ratio for post-treatment samples was one order of magnitude greater than the pre-treatment samples, suggesting a greater portion of bacteria that survive wastewater treatment carry beta-lactamase genes. Since this reservoir is teeming with β-lactamase genes, there is extremely high potential for rampant transfer of these genes. In turn, the environment downstream of sewage treatment plants becomes an area with an overwhelming influx of potent virulence factors carried by the bacteria and phages released from the treated wastewater. In addition to altering the composition of the bacterial communities in this environment, the mixing of transducing phages with human and environmental bacteria may lead to the creation of a reservoir of antibiotic-resistant genes in the surrounding environment.

This reservoir of phage-encoded antibiotic resistance genes and their transfer between microbes is not limited to the environment in and around wastewater treatment plants. Antibiotics produced by bacteria in nature are used as defense mechanisms against other bacteria and predators but also to secure their place and resources in a particular niche. It should come as no surprise that genes for resistance against one of the most prevalent antibiotics are so widespread. Genomic and metagenomic evidence suggests phage-encoded *bla* genes are present in the natural and agricultural environment, clinical and laboratory settings, and in the food supply (Lee and Park 2016; Marti et al. 2016; Krahn et al. 2016; Subirats et al. 2016). In wild-type *Staphylococcus aureus* isolates, the gene for metallo-β-lactamase was transduced from phage TEM123 (Lee and Park 2016). These transductants were also more highly resistant to β-lactam antibiotics than their *S. aureus* S133 counterparts (Lee and Park 2016). Metagenomic analysis of the phage DNA fraction from hospital wastewater samples showed a higher concentration of antibiotic-resistant genes in phage DNA versus bacterial DNA. These analyses also revealed that class D β-lactamase genes *bla*(OXA-10), *bla*(OXA-58), and *bla*(OXA-24) were particularly prevalent in phage DNA (Subirats et al. 2016). *S. aureus* phages have been shown to transduce antibiotic resistance genes at frequencies of ~10E-06 cfu/pfu between clinical strains, suggesting the dynamics of adaptation in clinical settings may occur at faster rates than in nature (Varga et al. 2012, 2016).

Considering antibiotics are routinely administered in hospital and clinical settings, it is not surprising that this environment is a hotbed for selection of antibiotic-resistant bacteria. Metagenomic analyses of phages from various hospital environments indicated a high proportion of phages with β-lactamase genes (Haaber et al. 2016; Subirats et al. 2016). Understanding the genetic interactions between phages and bacteria from hospital environments becomes increasingly important as new clinical strains of antibiotic resistant bacteria emerge. In a genomic study of *Acinetobacter baumannii* clinical isolates, the β-lactam-resistant isolate R2090 was able to transfer the NDM-1 *bla* gene it was carrying to an antibiotic-sensitive *A. baumannii* clinical isolate, CIP 70.10 (Krahn et al. 2016). *Acinetobacter* spp. are naturally transformable at a high frequency and can acquire resistance genes via plasmids or DNA fragments. However, in this instance, the donor R2090 *A. baumannii* isolate did not carry a plasmid, and generalized transduction via phages between the two isolates was likely the mode of transfer of the NDM-1 *bla* gene. Genomic comparison of the R2090 and CIP 70.10 isolates located three putative prophages present in the R2090 genome that could have been induced and errantly packaged the chromosomal NDM-1 *bla* DNA for transfer to the CIP 70.10 isolate (Mookerjee et al. 2015). While this is a possible mechanism, evidence of phage transfer needs to be empirically demonstrated to distinguish generalized transduction from transformation with chromosomal DNA. Nevertheless, this example provides a scenario where the hospital environment becomes ripe for emergence of new antibiotic-resistant pathogens. Association of antibiotic-resistant genes with phage metagenomes could be due to either packaging of these genes in transducing particles or association with the phage genome per se. Distinguishing between these two possibilities requires analysis of sequences adjacent to the antibiotic-resistant genes, to determine if they are contained within a phage genome.

2.3 Impact of Phages on Antibiotic Resistance in Cystic Fibrosis Microbiome

Chronic infections of cystic fibrosis (CF) patients exemplify adaptation of pathogens to specialized niches via acquisition of virulence properties, such as antibiotic resistance. The lung mucosa of CF patients provides an ideal environment for establishment of chronic infections, mostly dominated by *Pseudomonas aeruginosa* (Costerton et al. 1983; Ciofu et al. 2013; Govan and Deretic 1996; Lam et al. 1980; Potts et al. 1995; Sharma et al. 2014; Goerke and Wolz 2010; Willner and Furlan 2010; Hall-Stoodley et al. 2004). These infections are hard to treat, because the bacteria are frequently in biofilms and are often resistant to multiple antibiotics (Rolain et al. 2011; Fancello et al. 2011). Analyses of the lung microbiome of CF patients have shown that this complex environment harbors an abundance of both antibiotic-resistant bacteria and phages, and antibiotic resistance genes are prevalent in the phages (Rolain et al. 2011). Additionally, simply harboring multiple cryptic prophages can offer a distinct selective advantage for *P. aeruginosa* in a multi-species environment where competition for resources and space is strong—such as the complex microbial community in CF lungs (Burns et al. 2015).

Whether the antibiotic-resistant genes are carried on the phages or phages are simply the vehicle for transfer of resistance genes, the role they play in establishment of chronic *P. aeruginosa* infection of CF patients is beginning to become clear. Mutations in the prophages present in the CF-adapted Liverpool strain of *P. aeruginosa* (LESB58) have been shown to reduce the virulence of this opportunistic pathogen outside of the CF lung. Competitive index assays in the rat chronic infection model demonstrated a 3–70-fold decrease in virulence (Lemieux et al. 2016). The effects of these mutations on production levels of phages in this strain were also measured. The difference in production levels of phages induced spontaneously or deliberately was statistically significant in the mutants when compared to the wild type, suggesting prophages may play a role in the virulence of *P. aeruginosa* in CF lungs (Lemieux et al. 2016). Overwhelming evidence shows *P. aeruginosa* strains adapt relatively quickly to the selective pressures of multiple antibiotic treatments delivered to CF patients, and the genome of the *P. aeruginosa* within the CF lung is in constant flux (Latino et al. 2014; Sharma et al. 2014).

These selective and adaptive responses are not limited to *P. aeruginosa*. Presence of multiple prophages in the genomes of pathogens is a common trait (Table 11.1). The phages of *S. aureus* are also capable of modulating their host population so that it survives better within the CF lung environment. In a changing environment exposed to selective pressures from the immune system, therapeutics, and competition with the normal inhabitants of the lung microbiome, *Staphylococcus* phages are induced at higher frequencies, yielding greater opportunities for genetic exchange (Goerke and Wolz 2010). In this circumstance, the bacterial host persists longer because of the adaptive benefits provided by the phages. Altogether, the evolutionary dynamics between bacterial pathogen and phages contribute to the rapid development of increasingly potent opportunistic pathogens in the lungs of CF patients (Sharma et al. 2014).

3 Virulence

In addition to the selective advantage conferred by transfer of antibiotic resistance genes to bacteria, phages also transfer virulence genes responsible for the disease symptoms associated with particular pathogens (Boyd 2012; Wagner and Waldor 2002). Since the discovery of the phage-encoded diphtheria toxin gene (Freeman 1951; Groman 1953) in 1951, many more exotoxin genes have been found associated with phages (Table 11.1)—including the genes for Shiga-like toxin (*stx*) of

Table 11.1 Pathogenic bacteria and temperate phage

Bacterial host(s)	Phage(s)	Pathogenic properties	Disease(s)	References
Bordetella pertussis	*Bordetella* phage	Pertussis toxin (mobilized via generalized transduction)	Pertussis (whooping cough)	Karataev et al. (1988), Lapaeva et al. (1980)
Brucella inopinata (O1), other *Brucella* spp., *Ochrobactrum* spp.	BiPBO1, *Brucella* phages	Broad host range of phage enhances interspecies transfer of genetic information	Brucellosis—fever, multiple organ disease, opportunistic nosocomial infections	Hammerl et al. (2014), Hammerl et al. (2016)
Clostridium botulinum	Neurotoxin C and D phage, CEb	*Clostridium botulinum* type C neurotoxin (*botxn/C1*)	Botulism	Eklund and Poysky (1974), Inoue and Iida (1970), Zhou et al. (1993)
Clostridium difficile	CD phages ϕCD119	Repressor and regulator of PaLoc genes (*repR*)	Nosocomial diarrhea and colitis	Govind et al. (2011), Popoff and Bouvet (2013)
Corynebacterium diphtheria	Phage β	Diphtheria exotoxin (*dtx*)	Diphtheria	Freeman (1951), Groman (1953)
Escherichia coli	stx phages (933 J, H19B, 933 W) lambda, P1,	*Shigella*-like exotoxin (*stx*)	Enterohemorrhagic diarrhea, hemorrhagic colitis, hemolytic uremic syndrome	Calderwood et al. (1987), Strockbine et al. (1986), Barondess and Beckwith (1990), Plunkett et al. (1999)
Pseudomonas aeruginosa	*Pseudomonas* phages	Cytotoxins, O-antigen variation, fitness adaptation to human lung environment	Opportunistic infections (nosocomial, cystic fibrosis lung patients)	Lemieux et al. (2016), Rolain et al. (2011)
Salmonella enterica sv. typhimurium	P22, Gifsy, Fels	Increased fitness in enteric environments of animal hosts	Diarrhea, systemic disease	Brüssow et al. (2004)
Staphylococcus aureus	Phage 13, TSST phage, ϕETA, ϕ80	Staphylococcal enterotoxin A, superantigen (*sea*), toxic shock syndrome toxin (*tsst1*), exfoliative toxin A	Food poisoning, toxic shock syndrome, scalded skin syndrome	Betley and Mekalanos (1985), Coleman et al. (1989), Lindsay et al. (1998), Yamaguchi et al. (2000)

(continued)

Table 11.1 (continued)

Bacterial host(s)	Phage(s)	Pathogenic properties	Disease(s)	References
Streptococcus pyogenes (Group A *Streptococcus*, GAS)	GAS phage, phage T12	Exfoliative endotoxin A (*eta*), scarlet fever, toxic shock syndrome toxin (tsst1), hyaluronidase (*hylP*)	Necrotizing fasciitis, puerperal fever, STSS (streptococcal toxic shock syndrome), SSTI (soft skin tissue infection), scarlet fever	Johnson and Schlievert (1984), Weeks and Ferretti (1984), Yu and Ferretti (1991)
Vibrio cholerae	Ctxϕ, TCPϕ	Cholera exotoxin (ctx)	Cholera	Boyd and Waldor (1999), Boyd et al. (2000a, b), Waldor and Mekalanos (1996)
Vibrio parahaemolyticus (O3:K6)	ϕ237, VP882	ORF8 (putative adhesive property)	Gastroenteritis, wound infections, septicemia	Lan et al. (2009), Mookerjee et al. (2015), Okuda et al. (1997), Nasu et al. (2000)

Carrying multiple temperate phages, as viable or cryptic phages, in their genome is a common characteristic of many pathogenic bacteria. Some of the temperate phages contribute directly to the virulence of their bacterial host via virulence genes carried by the phage and integrated into the bacterial genome, while others may contribute indirectly to virulence through increase in bacterial host fitness and survivability in their animal host environment

E. coli, cholera toxin (*ctx*) of *V. cholerae*, and multiple toxins of *S. aureus* (Barondess and Beckwith 1990; Betley and Mekalanos 1985; Boyd and Waldor 1999; Lindsay et al. 1998; Paton and Paton 1996; Plunkett et al. 1999; Strockbine et al. 1986; Yamaguchi et al. 2000; Pearson et al. 1993; Davis and Waldor 2003). Transduction of virulence genes between clinical and natural hosts has been demonstrated in laboratory and natural model systems (Nyambe et al. 2016a, b; Tozzoli et al. 2014; Solheim et al. 2013; Petridis et al. 2006; Campos et al. 2003; O'Shea and Boyd 2002). In metagenomic studies investigating the presence of phage-encoded exotoxin genes in the environment, it was determined that these virulence genes are widespread in aquatic and terrestrial environments (Casas et al. 2006, 2010). These studies also revealed that these genes can be found in alternative, non-cognate hosts—that is, in bacteria that are not usually associated with the toxin-associated disease (Casas et al. 2006, 2010, 2011). Recent epidemic outbreaks of increasingly pathogenic strains of *E. coli* are examples of the outcome of this rampant genetic exchange occurring between phages and their bacterial hosts in the environment (Karmali 2017; Tozzoli et al. 2014; Beutin et al. 2013; Fruth et al. 2015).

3.1 Exotoxin Genes

3.1.1 Shiga-Like Toxin of *Escherichia coli*

Exotoxins are potent virulence factors that are often encoded by phages. Since it was first discovered that the Shiga-like toxins of *E. coli* were transferred by phages, *stx*-encoding phages have been found in numerous environments (Table 11.1; O'Brien et al. 1984; Barondess and Beckwith 1990; Calderwood et al. 1987; Plunkett et al. 1999; Strockbine et al. 1986; Muniesa and Jofre 2004; Yan et al. 2011; Muniesa et al. 2004; Park et al. 2007, 2013). *E. coli* carrying the *stx* gene have become increasingly virulent through acquisition of other horizontally transferred traits. In the 2011 outbreak of enterohemorrhagic *E. coli* infections that originated in Germany, the *E. coli* O104:H4 strain responsible for the outbreak had acquired not only the Shiga toxin from an *stx*-phage but had also acquired antibiotic resistance and enhanced adhesion properties via plasmids (Fig. 11.2; Altmann et al. 2011; Casas and Maloy 2011; Buchholz et al. 2011; Mora et al. 2011; Beutin et al. 2013; Frank et al. 2011; L'Abee-Lund et al. 2012).

As a normal inhabitant of the animal intestine, *E. coli* is highly adapted to this environment. However, in order to persist and be capable of infecting new hosts, *E. coli* must be able to survive in other environments once it is shed from its host. Evidence that pathogenic *E. coli* can readily survive outside the animal host are provided by metagenomic studies investigating microbiomes of environments like agricultural soils, livestock waste, dairy farms, food sources, sewage treatment

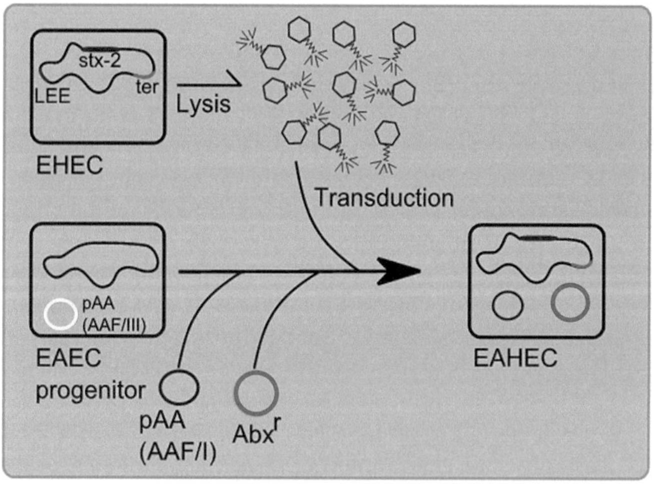

Fig. 11.2 Evolution of *Escherichia coli* O104:H4 Germany outbreak strain. Acquisition of multiple virulence traits like enteroaggregative factors, antibiotic resistance, and Shiga toxin production created a very potent novel human pathogen that caused 852 cases of hemolytic uremic syndrome, including 32 deaths in Europe and the Americas (Center for Disease Control and Prevention 2011; Copyright Future Medicine, Casas and Maloy 2011)

plants, oceans, rivers, and lakes (Allue-Guardia et al. 2011; Allue-Guardia et al. 2014; Beutin et al. 2013; Bonanno et al. 2016; Casas et al. 2011; Garcia-Aljaro et al. 2006, 2009; Tozzoli et al. 2014). Many of the *E. coli* from these environments have been shown to carry the phage-encoded *stx* gene. While the benefits of carrying phage-encoded toxin genes are not well-understood, one hypothesis is that carrying the toxin-coding prophage provides a selective advantage in bacterial survival and fitness in the environments where *E. coli* is found (Los et al. 2013; Veses-Garcia et al. 2015; Gamage et al. 2004).

Surprisingly, much like antibiotics are a weapon against other bacteria competing for space and resources in an environment, exotoxins can also be used as chemical weapons against eukaryotic predators. Production of Shiga toxin in *E. coli* EDL933 strain gives it protection against the ciliate predator *Tetrahymena thermophila* and kills it in co-culture. Enzyme and mutation experiments demonstrated that this killing protection was a direct result of the production of the phage-encoded Shiga toxin and EDL933 strains carrying the *stx* gene displayed a growth advantage over those absent the *stx* gene (Lainhart et al. 2009; Steinberg and Levin 2007).

3.1.2 Cholera Toxin

Vibrio cholerae is another organism that acquires virulence genes from phages. The cholera toxin gene (*ctx*) is transferred to *V. cholerae* via the ctxϕ via binding to the toxin-coregulated pilus (Table 11.1; Boyd and Waldor 1999; Waldor and Mekalanos 1996; Boyd et al. 2000a, b; Boyd and Waldor 2002; Davis and Waldor 2003). The acquisition of virulence factors via phages has been instrumental in the evolution of *V. cholerae* as a pathogen (Yeroshenko and Smirnova 2004; Das et al. 2016; Kim et al. 2015). Cholera is endemic in many regions of Southeast Asia. Many of the pathogenic *V. cholerae* strains isolated from these areas have multiple phage-associated genes within their genomes (Boyd et al. 2000a, b). In a study of 274 *V. cholerae* genomes that represented different niches of space, time, and environment, components from phages and prophages were present at high frequencies (Dutilh et al. 2014). The results from this comparative study provide strong evidence supporting the role of phages and other mobile genetic elements in the evolution and adaptation of bacteria to specialized niches.

The ctxϕ is not exclusively present in *V. cholerae* strains, however. In a study of four *V. mimicus* isolates, multiple copies of ctxϕ and the toxin-coregulated pilus-encoding phage VPIϕ were detected (Boyd et al. 2000a, b). This suggests that *V. mimicus* may be an intermediary host in the retention and dissemination of the *ctx* gene. The aquatic-associated reservoirs of *Vibrio* spp. impose selective pressures on survival such as salinity, ultraviolet radiation, pH, dissolved oxygen levels, nutrient, and human waste input. Despite these selective pressures, pathogenic strains of *V. cholerae* continue to persist in these dynamic environments (Das et al. 2016; Davis et al. 2000; Payne et al. 2004). In fact, the biotic and abiotic factors might be positively influencing the transfer of *ctx* in the ocean. Environmental vibriophages isolated from Newport Bay, California, were able to infect

V. cholerae O1 strains and transfer *ctx* genes to non-O1 *V. cholerae* strains (Choi et al. 2010). The reservoir of phage-encoded exotoxin genes has been found widespread in the environment (Casas et al. 2006). Given the abundance of the phages carrying exotoxin genes in the environment, it is not surprising that they are able to transfer these genes to multiple bacterial hosts—sometimes even environmental bacteria previously not known to carry these genes. If ctxϕ is able to persist in aquatic environments, this suggests that it is capable of finding suitable *Vibrio* and sometimes non-*Vibrio* bacterial hosts within this environment. This suggests that transfer of the phage-encoded *ctx* gene could offer some fitness benefit to the bacterial host in order for the bacterium to maintain the gene within its genome.

3.1.3 Toxins of *Staphylococcus aureus*

S. aureus has a wide arsenal of virulence factors. Many of these virulence factors are encoded and transferred via plasmids, but several are acquired from phages (Table 11.1; Xia and Wolz 2014; Freer and Arbuthnott 1982). The *sea* gene was the first phage-encoded virulence gene discovered in *S. aureus* (Betley and Mekalanos 1985), and since then more *S. aureus* virulence factors encoded by phages have been discovered, including exfoliative enterotoxin A (*eta*), Panton-Valentine leukocidin (*pvl*), β-hemolysin (*hlb*) staphylokinase (*sak*), and toxic shock syndrome toxin (*tsst1*) (Betley and Mekalanos 1985; Lindsay et al. 1998; Yamaguchi et al. 2000; Xia and Wolz 2014; Helbin et al. 2012; Coleman et al. 1989; McGavin et al. 2012). Phages can also contribute to *S. aureus* pathogenicity by aiding in the expression of genes located on pathogenicity islands (SaPIs) (Novick et al. 2010). The nu Sa beta genomic island of *S. aureus* can be induced and transferred to other *S. aureus* strains via prophages present in its genome. This phage-aided induction of SaPIs can also allow for interspecies transfer of virulence factors and niche adaptation (Helbin et al. 2012). In hospital settings, nosocomial infections with opportunistic and antibiotic-resistant bacteria create an environment that selects for evolution of new pathogens. In silico analyses of hospital-acquired methicillin-resistant *S. aureus* (HA-MRSA) CC30 strains indicate these bacteria evolve to persist in this environment. Together with single nucleotide polymorphisms (SNP) that affect tryptophan metabolism, iron acquisition, and toxin-antitoxin addiction module, the phage-associated genes for SaPI2, SaPI4, and *tsst1* are also involved in adaptation of this strain to the hospital environment (McGavin et al. 2012).

3.1.4 Toxin Genes in Other Hosts

Molecular studies of the genomes of potent and clinically relevant bacteria continue to expand our knowledge of other phage-encoded toxin genes and the pathogens that carry them. Some of these pathogens are listed in Table 11.1 and include *Streptococcus pyogenes*, *Clostridium difficile*, *C. botulinum*, *Enterobacter cloacae*,

A. baumannii, *Bartonella* spp., and *Yersinia pestis*. Comparative genomic analysis indicates the most striking differences in genomic composition of Group A *S. pyogenes* (GAS) strains can be attributed to the differences in the lysogens they carry (Cleary et al. 2016). Many of the GAS toxins are phage-encoded (Zabriskie 1964; Cleary et al. 1998). The GAS phage T12 carries the gene for the erythrogenic toxin A (*speA*)—one of the major virulence factors of GAS (Johnson and Schlievert 1984; Weeks and Ferretti 1984; Yu and Ferretti 1991). One GAS strain, AP53.2, acquired multiple exotoxins from phages: the genes for erythrogenic toxin K (*speK*) and streptococcal phospholipase A (*slaA*). These phage-encoded toxins enhance the virulence of AP53.2 (Bao et al. 2016). The highly pathogenic plague microbe, *Yersinia pestis*, is infected with an unstable filamentous phage (YpfΦ) that alters its pathogenicity in animal models, and despite its instability, maintenance in the *Y. pestis* genome suggests it confers some advantage in natural ecosystems (Derbise 2014; Derbise 2007). Finally, a homolog of the zonula occludens toxin gene (*zot*) of the plant pathogen *Ralstonia solanacearum* is carried by phage PE226 (Murugaiyan et al. 2011).

Clostridium species also carry phage-encoded toxin genes. Both *C. difficile* and *C. botulinum* contain toxin-encoding lysogens in their genomes that influence their virulence (Popoff and Bouvet 2013; Govind et al. 2011, Inoue and Iida 1970; Zhou et al. 1993; Eklund and Poysky 1974). *C. difficile* is responsible for a chronic gastrointestinal infection in which toxin production and virulence are encoded by the phiCD119 lysogen (Govind et al. 2011). Moreover, phiCD119 has been shown to be transferred in the intestine during the infection process (Govind et al. 2011). The botulinum toxin gene (*btx*) of *C. botulinum* is a potent virulence factor. Genetic diversity of botulism toxins is due to the multiple phages that carry this gene (Popoff and Bouvet 2013; Sugiyama 1980). Lysogenic conversion of *C. botulinum* and *C. butyricum* strains has been observed in laboratory settings (Eklund and Poysky 1974; Inoue and Iida 1970; Zhou et al. 1993). Genomic comparisons of pathogenic *Clostridium* strains suggest horizontal transfer of the *btx* gene to non-toxigenic *Clostridium* strains (Popoff and Bouvet 2013).

Traditionally, phage-encoded exotoxin genes have been found associated with only one known pathogen, its cognate host—e.g., ctxϕ with *V. cholerae*. However, phage-encoded genes have also been found in non-cognate, nonpathogenic bacteria. For instance, the phage-encoded *stx* gene that is usually associated with *E. coli* strains has been found in alternative bacteria from different environments, including *Enterobacter cloacae*, *Enterococcus* spp., and *Pseudomonas* spp. (Paton and Paton 1996; Casas et al. 2010; Casas et al. 2011). Likewise, the phage-encoded *sea* gene commonly associated with *S. aureus* has been found in *Pseudomonas* spp. isolated from an ambient environment (Casas et al. 2010). Additionally, a cholera-like toxin gene has been found in a putative prophage of *Bartonella bacilliformis* and is conserved between isolates that originated from cow, moose, dog, and kangaroo samples (Guy et al. 2013). While experiments directly testing fitness impact of phage-encoded exotoxin genes in natural environments are still required, conservation of this phage-associated toxin gene across multiple animal environments suggests it may influence host fitness and survival in these specialized niches by an as yet unknown mechanism.

Another set of toxic proteins of phage origin are the bacteriocins produced primarily by Gram-negative bacteria. The bacteriocins of *Pseudomonas aeruginosa* are known as pyocins and include proteins such as F- and R-type pyocins (Michel-Briand and Baysse 2002). Pyocins, and similar proteins, are thought to have evolved from once infective phages due to their physiological similarities including induction by UV and mitomycin C, receptor binding, and bactericidal action through protein translocation and lysis. The most compelling evidence supporting the phage origin of pyocins is their physical structure. F-pyocins have a structure similar to flexible noncontractile phage tails, while R-pyocin structure resembles the inflexible and contractile phage tails (Michel-Briand and Baysse 2002). The importance of these phage-associated bacteriocins in bacterial fitness and niche adaptation is demonstrated in their conservation in multiple species across the bacterial and archaeal domains. In a comparative analysis of whole-microbial genome sequences, phage tail-like genetic elements were identified in Gram-negative and Gram-positive bacteria and archaea (Sarris et al. 2014). These phage-associated bacteriocin genes included F- and R-pyocins, Type VI secretion systems (T6SS), the anti-feeding phage toxin (Afp) of the grass grub pathogen, *Serratia entomophila*, and the *Photorhabdus luminescens* virulence cassette (PVC) (Sarris et al. 2014).

While it is clear that phage-encoded exotoxin genes do play a role in pathogenicity, it is still not well-understood why bacteria have evolved to continue to replicate and bear the fitness burden of carrying these genes or if they also have a function separate from increased virulence. Why non-cognate and nonpathogenic environmental bacteria carry phage-encoded exotoxin genes is even more perplexing. Presumably, in these bacteria, the virulence genes play a different role not associated with human or animal pathogenicity. One possible function of these genes could be as antimicrobial weapons when competing for space and resources in a particular niche (see Sect. 4 for discussion). Another possibility is that maintaining the prophage in its genome protects the bacterium from infection by lytic phages (see Sect. 4.2 for discussion).

3.1.5 Secretion Systems

Secretion systems also play a role in bacterial host infections. Many of these secretion systems are found on pathogenicity islands, but some are also found on or mobilized by phages. A resident prophage mobilizes the nu Sa beta genomic island of *S. aureus* (Moon et al. 2015). The human pathogen *Bartonella bacilliformis* contains Type III, IV, and V secretion systems encoded by phages (Guy et al. 2013). In *Salmonella enterica* sv. enteritidis, the Type III secretion system 2 (TTSS2) is encoded by a phage "moron" gene SEN1140 (Vishwakarma et al. 2012). Acquisition of this phage-encoded TTSS2 is necessary for early colonization and infection in the human gut (Vishwakarma et al. 2012). A Type III secretion system effector encoded by SopEϕ plays a role in virulence of *Salmonella* strains that carry this prophage. Inflammation during infection promotes the transfer of SopEϕ to non-lysogenic *Salmonella* strains (Diard et al. 2017).

3.2 Other Properties to Enhance Overall Virulence

In addition to toxins, bacteriocins, and secretion systems, additional phage-encoded virulence properties play a role in niche adaptation of pathogens. In the emerging pathogenic strains of *S. pneumoniae* serotype 22F (ST433, ST698), enhanced virulence and clinical significance may be due to presence of intact prophages and numerous phage insertions (Cleary et al. 2016). The phage-derived *sda1* gene of the invasive M1 T1 clone of Group A *Streptococcus* encodes a DNase that degrades neutrophil DNA extracellular tags and allows it to escape phagocytic ingestion (Wang et al. 2013a, b). Phage-related elements are responsible for enhanced virulence and niche adaptation in *Enterococcus faecalis* clonal complex 2 (CC2) strains (Solheim et al. 2011). The previously innocuous commensal organism is now responsible for many hospital-acquired infections. Genome comparison analyses of various clinical *E. faecalis* CC2 strains show that much of the adaptation of *E. faecalis* to its hospital environment is due to phage-derived genetic elements in its genome. These various phage-related genes are hypothesized to confer enhanced fitness and pervasiveness in clinical and hospital settings. For example, phage-encoded helicase and MutT proteins may provide the bacterial lysogen with protection from oxidative DNA damage within the infected host (Solheim et al. 2011).

4 Defense Adaptations

Bacteria encounter many competitors and predators in their environment, including other bacteria, protists, and phages. In response, bacteria have evolved various defense mechanisms specifically adapted to evade these predators. Production of small extracellular molecules like antibiotics and exotoxins is an adaptation to defend against other bacteria competing for space and resources as well as to protect against protist predation (Arnold and Koudelka 2014; Lainhart et al. 2009; Waksman 1961). Since phages are the major predators of bacteria, it is not surprising that they have evolved multiple processes for resisting infection (Fig. 11.3). Phage resistance can occur through the altering of cell surface receptors to which phages bind, expression of restriction/modification systems that destroy the genome of incoming phages, immunity or superinfection exclusion by resident prophages, and adaptive

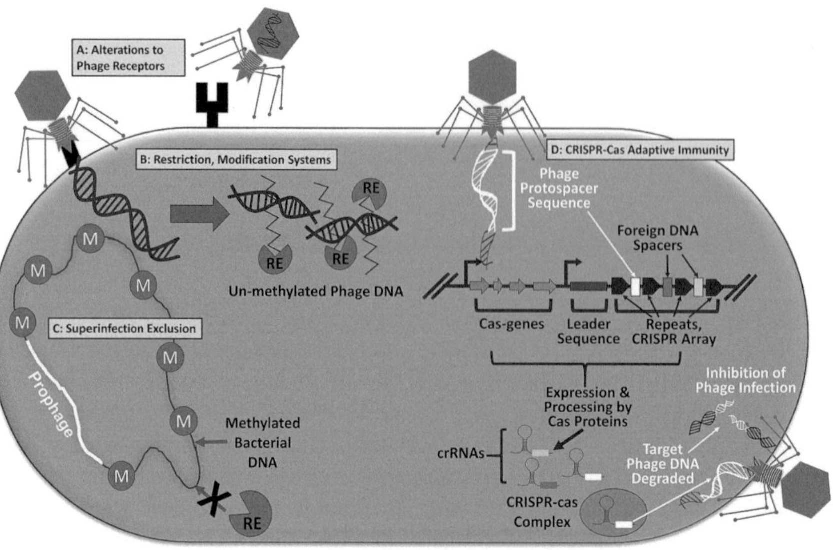

Fig. 11.3 Bacterial mechanisms of phage resistance. (**a**) Alteration of phage receptors renders phage incapable of binding to bacterial surface and prevents infection. (**b**) Bacterial resistance and modification systems protect bacterial DNA from degradation while targeting foreign phage DNA for destruction. Phage is unable to replicate within the bacterial host. (**c**) Superinfection exclusion prevents infection by multiple phages when a prophage is integrated into bacterial host chromosome. (**d**) The CRISPR-Cas adaptive immune response recognizes signature phage protospacer sequences that are integrated into the CRISPR array locus on the bacterial genome, expressed and processed by Cas proteins into crRNAs and together with other Cas proteins form the CRISPR-Cas complex that specifically targets the foreign phage DNA, degrades the phage DNA, and prevents production of new phage particles. *RE* restriction enzyme, *M* methylation site, *CRISPR* clustered regularly interspaced short palindromic repeats, *Cas* CRISPR associated

immunity provided by the CRISPR-Cas system (Koonin et al. 2017; Mai-Prochnow 2015; Derbise 2014; Chouikha et al. 2010).

4.1 Phage Resistance

The challenges bacterial hosts face in response to phage infection, while seemingly disastrous to the individual bacterium, can be beneficial to the bacterial population as a whole. The selective pressures placed on the host through phage infection provide opportunities for evolution of novel physiological characteristics that give the host an advantage over other bacteria within the local ecosystem. Increased fitness and survival within an environment can occur through acquisition of resistance factors or by exploiting novel metabolic properties that allow them to metabolize an underutilized resource within that environment (Burns et al. 2015).

In the natural environment, repeated exposure to and infection by phages leads to development of resistance. Consequently, adaptation to this phage predation permits these bacteria to survive and compete in their local environmental niche. For example, in the leaves of the horse chestnut tree, bacteria will adapt across time and space to their phage predators, such that they become resistant to non-evolved phages regardless of which tree or leaf the phages originated (Koskella 2013). Rapid coevolution of bacteria induced by the selective pressure of phage predation in a natural soil microbial community provides further insight into the fluctuating dynamics of this relationship. The extent to which resistance develops can be influenced by nutrient availability. In the natural soil environment, resistance to phages is directly proportional to nutrient levels—when nutrients are high, so is phage resistance (Gomez et al. 2015). In this study, phage/host coevolution dynamics and resistance development did reach a maximum sustainable resistance level. Bacterial hosts indeed became resistant to evolved phage predators, and even though there was a maximum level of resistance developed, the antagonistic relationship in the soil environment showed that phages influenced bacterial community structures over time (Gomez and Buckling 2011).

To better understand how phages and bacteria interact in nature and adapt to their local environment, laboratory systems are used to model these interactions and determine the factors that most impact resistance during coevolution. Mutations leading to changes in cell surface structure or proteins preventing adsorption of phages to their bacterial hosts are very common in this microbial "arms race." Phage resistance dynamics between *E. coli* B and phage T3 were tested in vitro in a chemostat system. In this system, the first round of phage resistance developed by selection of cells with a mutation in the *waaG* gene responsible for LPS assembly and resulted in decreased adsorption of phage T3 (Perry et al. 2015). Phages subsequently evolved to infect both the ancestral and evolved bacterial hosts. After the third round of coevolution, *E. coli* B was resistant to both the original phage T3 population as well as the evolved second-generation phage T3 population (Perry et al. 2015). A similar outcome is observed between *E. coli* and phage λ. Initially, *E. coli* became resistant to λ via a mutation of the LamB receptor, but subsequently the phages evolved to infect these bacteria via increased adsorption affinity to the LamB receptor and expression of a new OmpF receptor (Meyer et al. 2012). Phage resistance does not come without consequences, however (Koskella et al. 2012; Buckling et al. 2006). Mutations in bacterial host genomes leading to resistant phenotypes had deleterious consequences for host fitness when coevolving with single or multiple phage types; however, absence of predation dense bacterial growth is observed (Koskella et al. 2012). The longer bacteria coevolve with phage predators, the greater the negative impact on fitness (as measured by resistance to ancestral and evolving phage). This effect is amplified when mutations in the bacterial host ("mutation load") are high (Buckling et al. 2006). Despite the transient fitness consequences, predation still drives selection for bacteria that have evolved resistance to phages in their local environment and aids in adaptation through synergistic epistasis of the bacterial genome (Vasse et al. 2015).

4.2 CRISPR-Cas Adaptive Immunity

The CRISPR-Cas adaptive immunity of bacteria is response mechanism to phage-bacterial dynamics. Like any adaptive immune system, continued exposure of an organism to microbial threats increases the repertoire of antimicrobial molecules generated by the immune system and available to the host. Evasion of phage predators expands via the CRISPR-Cas system and is driven by the constant exposure of the bacterium to phage infection, which increases the number and diversity of spacer regions targeting foreign phage DNA (Heler et al. 2014). *S. pyogenes* generates spacers via Cas9 cleavage of protospacer adjacent motifs (PAM) in the viral genome (Heler et al. 2014). As the bacterium accumulates more of these spacer regions, the organism's ability to adapt to potential threats increases proportionally. In a system exposing cells to replication-defective phages, these cells quickly develop their CRISPR-Cas anti-phage repertoire, in direct proportion to the concentration of phages to which they are exposed (Hynes et al. 2014).

Even in the face of a highly dynamic system, where phages and bacteria are rapidly evolving defense systems against each other, the bacterial CRISPR-Cas system has the ability to use preexisting phage spacers to genetically respond to similar invaders. Immunity is provided against these evolved phages by incorporating mutations and deletions into preexisting spacers "priming" the immune system to evade the newly evolved phages (Fineran et al. 2014). The high level of adaptability of this "priming" allows the host to avoid predation and generates an increasingly diverse and rapid immune response.

There are many examples of how phages drive the development and fine tuning of the bacterial CRISPR-Cas system. Evidence that evolution of CRISPR-Cas systems can provide protection to subsequent infection of genetically divergent phages can be found in the Type II-C CRISPR-Cas system in *Bordetella* species, *B. pseudohinzii*. In this bacterium, the Type II-C CRISPR system transcribes CRISPR RNAs (crRNA) that have homology to *B. hinzii* prophages, but it does not itself contain any prophages (Ivanov et al. 2015).

Evolution of effective and diverse CRISPR-Cas systems are especially important in industrial environments where large-scale processes are sometimes disrupted by phage infection and rapid lysis of the production bacteria. In the dairy industry, for instance, it is important that fermentation strains are resistant to phage predation, but also maintain genome qualities key to their use as industrial producers. *S. thermophilus,* used in the production of yogurt, is able to rapidly generate novel spacers in their CRISPR1 and CRISPR3 loci in response to phage infection (Wei et al. 2015; Horvath and Barrangou 2011; Picozzi et al. 2012). Having a greater understanding of the dynamics of adaptive immunity in industrial settings allows for a more targeted approach to reduce production losses due to phage infection (Al-Attar et al. 2011). As more CRISPR-Cas systems are discovered and analyzed, it is evident that this adaptive immunity is driven by phage predation and infection (Heler et al. 2014). The CRISPR-Cas system is a compelling example of the role of phages in driving bacterial evolution and the numerous practical applications that arise from understanding phage-bacterial interactions.

5 Fitness Adaptations: Animal and Natural Environments

5.1 Adaptation to the Animal Environment

Bacterial fitness is associated with an organism's ability to adapt to its environmental niche in order to proliferate. Because of the high rates of transfer of bacterial genes via phages, much of this adaptation is associated with phages. Phage-inducible chromosomal islands (PICI) can use temperate phages as helpers to increase their spread and survival of their host (Penadés and Christie 2015). Adaptation to diverse environments and increased pathogenicity in pathogens like *Vibrio* spp. and *H. pylori* are highly influenced by genetic exchange between phages and bacteria (Fernandez-Gonzalez and Backert 2014; Hazen et al. 2010).

Adaptation to the animal intestinal tract is one example of phages driving bacterial evolution. In a mouse study investigating the effects of a "Western diet" on the spatial distribution and diversity of the associated phage and bacterial metagenomes, the microbiome was dominated by *Caudovirales* temperate phages and their associated *Bacilli*, *Negativicutes*, and *Bacteroidia* bacterial hosts (Kim and Bae 2016). The temperate phages from the *Bacteroidia* class carried genes that provided adaptations to environmental stresses and provided specific functions for specialized niche adaptation (Kim and Bae 2016). Phage-acquired genotypes are also found in bacteria adapted to other animal environments. *Lactobacillus johnsonii* strain DPC6026 has restriction modification and phage-adapted CRISPR loci that enhance its adaptation to the porcine intestinal tract (Guinane et al. 2011). Inflammation from persistent *C. difficile* infections in the gut also influences phage and bacterial dynamics. Stress from the inflammation response and bacterial lysis lead to activation of *Clostridium* phages during infection. This increased abundance appears to contribute to severity of disease, as the viriome of a healthy human fecal transplant donor was characterized by low phage abundance. This low phage abundance was maintained in healthy transplant recipients for up to 5 years. Additionally, the composition of the gut microbiome appears to be directly influenced by the phages, as both phages and bacterial host were present in the healthy microbiome of the fecal transplant recipient (Broecker et al. 2016).

Other phage-encoded genes contributing to host adaptation in animal environments are found in Group A *Streptococcus pyogenes* (GAS) and *Klebsiella pneumoniae*. *S. pyogenes* carries a prophage-like chromosomal island (SpyCI) controlling DNA mismatch repair and is implicated in promoting survival and adaption to novel environments or environmental changes (Scott et al. 2012).

5.2 Adaptation to the Natural Environment

Phage-derived genes can also impact bacterial hosts in the natural environment. Terrestrial and aquatic environments each present unique challenges to the microbial communities that inhabit them. The physical and chemical composition of each environment as well as the availability of resources and mobility within these environments is distinct. The dynamics of the competitive and selective pressures, therefore, are also distinct.

In the soil and arable ecosystem, the rich microbial diversity is paralleled with rich nutrient resources, making competition for these resources the driving force behind selection in this environment. Any selective advantage that can increase survivability and fitness within this environment is beneficial to the bacterium. Many of these advantages can come by way of interactions with phages or the long-term consequences of these interactions. In a model phage/host system using a soil-adapted psychrophilic *Pseudomonas fluorescens* strain and ϕ2, the temperature at which both coevolved had a statistically significant influence on increased resistance and infectivity (Gorter et al. 2016). Density of bacterial growth was greater at any temperature when they were coevolving with phage than when they were not (Gorter et al. 2016). This suggests that in soil-adapted phages and bacteria, the selective pressure of phage predation positively influences bacterial fitness and growth.

Increased fitness and infectivity of a bacterium can also be affected by predation in the plant environment. In the "phytosphere"—the plant-associated micro-niche which includes the plant itself, adjacent soil, and rhizosphere—transduction of genes that allow for niche adaptation and prolonged survival may indirectly lead to increased virulence of enteric human pathogens. An example of this in a plant environment is the increased transmissibility of pathogens like *E. coli* and *Salmonella* from plants and plant products to which they have adapted via horizontal transfer of genes supplying enhanced abilities to survive in the more varied conditions of the phytosphere (van Overbeek et al. 2014). This indirectly enhances virulence because increased colonization, stress resistance, and nutrient acquisition and utilization within the phytosphere allow them to persist in this environment, in turn providing greater opportunities for encounters with their animal host (van Overbeek et al. 2014).

In contrast to the rich and diverse terrestrial environment, the oligotrophic aquatic environment presents greater challenges for access to nutrients and resources. However, the aquatic environment is a more dynamic and quickly changing system of genetic exchange between phages and their bacterial hosts—allowing for more rapid evolution and adaptation in this niche (Motlagh et al. 2017). The photosynthetic genes carried by cyanophages and transferred to their *Synechococcus* and *Prochlorococcus* hosts are a striking example of this phenomenon (Paul et al. 2002; Puxty et al. 2015; Dammeyer et al. 2008). These genes are involved in photosystem electron transfer, pigment biosynthesis, and carbon metabolism in the light-independent reactions of photosynthesis (Dammeyer et al. 2008; Puxty et al. 2015). Cyanophages and their hosts are particularly well-adapted to their estuarine and ocean environments. Metagenomic studies of cyanopodophages from these environments indicate that regardless of origin, the phages share a core genome interspersed with variant genes that may confer their bacterial hosts with phenotypes that enhance growth or survival in specialized niches. These accessory genes are related to the environmental origin of the host and host adaptation to that environment, as well as genes for photosynthesis (Huang et al. 2015).

In addition to transferring genes for photo-adaptation and persistence in an oligotrophic ocean, phages can also provide genes necessary for survival in marine niches. *Vibrio* spp. can carry multiple mobile genetic elements that enhance their virulence and survival. *Vibrio* phage KVP40 carries genes encoding enzymes involved in pyridine nucleotide metabolism and NAD biosynthesis (Fan et al. 2016; Yun Lee et al. 2017). With resources scarce and often specialized in the different strata and zones of the ocean, phage-acquired genes that contribute to energy production and metabolism could prove advantageous. In a metagenomic study investigating the photic and aphotic viriomes of the Pacific Ocean, results indicated that phages carry a variety of auxiliary genes that may benefit their marine bacterial hosts and permit niche adaptation (Hurwitz et al. 2015). Genes associated with photosynthesis were prevalent, but also present, and spatially distributed according to ocean depth and photic zone, were those associated with amino acid transport and metabolism, carbohydrate transport and metabolism, DNA metabolism, and iron-sulfur protein biogenesis with likely critical functions in electron transfer and enzyme catalysis (Hurwitz et al. 2015).

Myoviridae, *Siphoviridae*, and *Podoviridae* from deep-sea hydrothermal vents also carry genes advantageous to their bacterial hosts (Wang and Zhang 2010). Phages infecting the deep-sea marine bacterium SUP05, found in hydrothermal vents, carry the α-subunit (*rdsrA*) and the γ-subunit (*rdsrC*) genes—essential components of the reverse dissimilatory sulfite reductase (*RDSR*) complex (Anantharaman et al. 2014). As sulfur is quite abundant in deep-sea hydrothermal vents, the RDSR complex supports sulfur oxidation and metabolism for the ubiquitous SUP05 marine bacterium—offering a fitness advantage in this severely nutrient-poor environment (Anantharaman et al. 2014). A comparative metagenomic study of the microbial communities of deep-sea hydrothermal systems provides compelling evidence on the benefits of lysogeny to bacterial hosts. Viral, bacterial, and archaeal metagenomes from the Hulk hydrothermal vent in the Main Endeavour Field on the Juan de Fuca Ridge were isolated and analyzed for gene content and composition, presence of prophages and prophage-associated genes, and overall genome plasticity to determine the modes of interaction in this stratified, extreme environment (Anderson et al. 2014). A rich variety of auxiliary genes for metabolism and energy production were detected in the viral and cellular (bacteria and archaea) fractions. Fragment recruitment analysis and gene annotation revealed phage-associated genes for hydrogenases, cofactors, vitamins, pigments, and amino acid biosynthesis. Most importantly with regard to niche adaptation, there was an abundance of genes for sulfur and methane oxidation, providing alternative sources for energy metabolism (Anderson et al. 2014). Altogether, these examples of niche adaptation to extreme environments demonstrate that the relationship between phages and their hosts is

Table 11.2 Phage-encoded genes that modify host metabolism

Bacterial host	Environmental niche	Phage(s)	Phage gene function	References
Antibiotic resistance				
Staphylococcus aureus	Urban runoff, untreated sewage, hospital wastewater	Staphylococcal phages, *S. aureus* phage phiJB, *S. aureus* phage TEM123	β-lactamases (*blaCTXM*, *mecA*), metallo-beta-lactamase (*blaTEM*)	Subirats et al. (2016), Lee and Park (2016)
Acinetobacter baumannii	Nosocomial infection	*A. baumannii* phages	β-lactamase (*NDM-1 bla*)	Krahn et al. (2016)
Agricultural waste-associated bacteria	Aquaculture wastewater	Environmental phages	Efflux pump regulating proteins, macrolide resistance (erythromycin, telithromycin, clarithromycin)	Colombo et al. (2016)
Salmonella enterica sv. typhimurium	Raw sewage and river water	*Salmonella* phages	Polymyxin-B resistance, penicillin resistance	Karpe et al. (2016)
Environmental bacteria	Raw sewage and river water	Environmental phage	Quinolone resistance (*qnrA, qnrS*)	Colomer-Lluch et al. (2011a), Colomer-Lluch et al. (2014)
Helicobacter pylori	Human-associated wastewater	*H. pylori* phage	Multidrug-resistant protein D (*emrD)*	Fan et al. (2016)
Human skin-associated bacteria	Skin phage fraction DNA	Skin-associated phage	Antibiotic resistance	Hannigan et al. (2015)
Dairy cow fecal-associated bacteria	Dairy cattle feces	Various fecal phage	Ceftiofur resistance (cephalosporin)	Chambers et al. (2015)
Vibrio cholerae, *V. parahaemolyticus*	River and estuary	Vibrio phages	Enhanced and multidrug resistance	Mookerjee et al. (2015), Rolain et al. (2011)
Pseudomonas aeruginosa, *S. aureus*	Cystic fibrosis viriome	*P. aeruginosa* LESB58 prophages, Staphylococcal phages	Efflux pumps, fluoroquinolone resistance, β-lactamases	Rolain et al. (2011), Fancello et al. (2011), Lemieux et al. (2016), Willner et al. (2009), Willner and Furlan (2010)
Resource adaptation				
Prochlorococcus spp.	Marine oceanic environment	Cyanophage	Photosynthesis, alternative carbon metabolism, phosphate stress response (*psbA, hliP, talc, phoH, pstS)*	Dammeyer et al. (2008)

(continued)

Table 11.2 (continued)

Bacterial host	Environmental niche	Phage(s)	Phage gene function	References
Synechococcus spp.	Marine oceanic environment	Cyanophage	Photosynthesis, electron transfer, carbon fixation and metabolism, niche adaptation	Puxty et al. (2015), Huang et al. (2015)
Vibrio spp.	River and estuary	Vibriophage KVP40	NAD^+ scavenging, pyridine nucleotide metabolism (*nadV, natV*)	Yun Lee et al. (2017)

Select examples of phage-encoded genes which benefit host metabolism and allow for increased viability in the animal host and the natural environment

more mutualistic than parasitic with both the phages and bacteria benefitting from these genetic exchanges.

6 Conclusions

As the primary predators of bacteria, it is not surprising that phages provide the selective pressure for evolution of their hosts in different environments. A great deal is known about the dynamics of *E. coli* and some coliphages, providing model systems to understand how interactions with phages drive evolution of the bacterial host. As the bacterial host evolves, the phages also evolve to adapt to changes in their host. The outcomes from phage-bacteria interactions described in this chapter—virulence, metabolic traits, antibiotic resistance, phage and antibiotic resistance—suggest that the phage-bacteria relationship might actually be more mutualistic than parasitic (Obeng et al. 2016). These interactions yield not only more fit bacteria but also increased fitness of the phage predator. Dissecting the complexities of the relationship between phages and their bacterial hosts, along with what drives this dynamic relationship, is critical for understanding how phages contribute to the overall diversity and fitness of the ecosystem. Because these interactions can modulate virulence and antibiotic resistance, bacteria-phage interactions also have profound impacts on the evolution of human pathogens.

Phage predation leading to bacterial lysis contributes to the population dynamics and composition of bacterial communities by providing selective pressure for bacterial evolution (Fig. 11.1a). Phage predation, lysis, and evolution events of the lytic cycle provide a mechanism for evolution of resistant bacterial populations with enhanced capacity of resisting phage infections. These adaptations are not without fitness costs and do not offer resistance to all future phage infections. Pleiotropy of resistance mutations results in trade-offs between phage resistance or bacterial growth and fitness. Therefore, in the "arms race" between phages and bacteria, coevolution is not an infinite cycle. Factoring in an environment constantly in flux and the bacterial community as a whole, phage-resistant mechanisms provide an overall benefit in the local environment.

While the lytic cycle of infection is fluid and constantly changing, the lysogenic cycle is much more enduring. Either through generalized or specialized transduction of bacterial and phage genes, the bacterial genome is changed indefinitely during lysogeny. The result is genetically altered bacteria with enhanced phenotypes and fitness (Fig. 11.1b and c). Lysogenic conversion is responsible for contributing to the augmented virulence of many human pathogens (Table 11.1).

Improved fitness within animal environments is often associated with increased pathogenicity. Acquisition of phage-encoded virulence factors such as toxin and antibiotic resistance genes not only enhances the bacterium's ability to cause disease but invariably increases the ability of the bacterium to persist in these animal environments (Tables 11.1 and 11.2). *Pseudomonas* spp. and *E. coli* spp. are examples of bacteria whose increased fitness in specialized niches can be partly attributed to presence of phage-derived traits in their genomes. In the natural environment, phage-acquired genes can boost the metabolic capabilities of their bacterial hosts (Table 11.2). The relationship between cyanophages and their *Synechococcus* and *Prochlorococcus* marine hosts is a classic example of the beneficial outcomes of integration of phage-derived genes into a bacterial host. In an oligotrophic environment, these genes provide the bacterial host a mechanism for carbon metabolism that makes them particularly well-suited for surviving in this aquatic niche.

There is clear evidence for the considerable contribution of phages to their bacterial host's ability to survive in specialized niches. We have seen that phages drive bacterial community dynamics through the selective pressures of predation. We have also seen how phage-mediated transfer of bacterial genes through generalized transduction is a major contributor to the mosaicism of bacterial genomes and the ability to adapt to new environmental niches. Finally, through lysogenic conversion, phage-encoded genes can generate increasingly virulent human pathogens as well as provide a mechanism for the evolution of novel human pathogens. Altogether, the impact of phages on bacterial populations extends beyond a mere predator-prey relationship and has a more global impact on their survival in specialized niches. Given the complex role of phages in modifying and modulating ecosystems from humans, animals, and the environment, there remain many questions about how this occurs and what selective conditions drive phage-host interactions. Addressing these challenges will require interdisciplinary teams using novel tools for quantifying phage, host, and environmental determinants in real time.

References

Al-Attar S, Westra ER, van der Oost J, Brouns SJJ (2011) Clustered regularly interspaced short palindromic repeats (CRISPRs): the hallmark of an ingenious antiviral defense mechanism in prokaryotes. Biol Chem 392:277–289

Allue-Guardia A, Garcia-Aljaro C, Muniesa M (2011) Bacteriophage-encoding cytolethal distending toxin type v gene induced from nonclinical *Escherichia coli* isolates. Infect Immun 79:3262–3272

Allue-Guardia A, Martinez-Castillo A, Muniesa M (2014) Persistence of infectious shiga toxin-encoding bacteriophages after disinfection treatments. Appl Environ Microbiol 80:2142–2149

Altmann M, Wadl M, Altmann D, Benzler J, Eckmanns T, Krause G, Spode A, an der Heiden M (2011) Timeliness of surveillance during outbreak of shiga toxin-producing *Escherichia coli* infection, Germany, 2011. Emerg Infect Dis 17:1906–1909

Anantharaman A, Duhaime MB, Breier JA, Wendt KA, Toner BM, Dick GJ (2014) Sulfur oxidation genes in diverse deep-sea viruses. Science 344(6185):757–760. https://doi.org/10.1126/science.1252229

Anderson RE, Sogin ML, Baross JA, Uversky VN (2014) Evolutionary strategies of viruses, bacteria and archaea in hydrothermal vent ecosystems revealed through metagenomics. PLoS One 9(10):e109696

Arnold JW, Koudelka GB (2014) The Trojan Horse of the microbiological arms race: phage-encoded toxins as a defence against eukaryotic predators. Environ Microbiol 16(2):454–466

Aziz RK, Edwards RA, Taylor WW, Low DE, McGeer A, Kotb M (2005) Mosaic prophages with horizontally acquired genes account for the emergence and diversification of the globally disseminated M1T1 clone of Streptococcus pyogenes. J Bacteriol 187(10):3311–3318

Bao YJ, Liang Z, Mayfield JA, Donahue DL, Carothers KE, Lee SW, Ploplis VA, Castellino FJ (2016) Genomic characterization of a pattern D *Streptococcus pyogenes* emm53 isolate reveals a genetic rationale for invasive skin tropicity. J Bacteriol 198:1712–1724

Barondess J, Beckwith J (1990) A bacterial virulence determinant encoded by lysogenic coliphage lambda. Nature 346:871–874

Baugher JL, Durmaz E, Klaenhammer TR (2014) Spontaneously induced prophages in *Lactobacillus gasseri* contribute to horizontal gene transfer. Appl Environ Microbiol 80(11):3508–3517. https://doi.org/10.1128/aem.04092-13

Benveniste R, Davies J (1973) Mechanisms of antibiotic resistance in bacteria. Annu Rev Biochem 42(1):471–506

Betley M, Mekalanos J (1985) Staphylococcal enterotoxin A is encoded by phage. Science 229 (4709):185–187

Beutin L, Hammerl JA, Reetz J, Strauch E (2013) Shiga toxin-producing *Escherichia coli* strains from cattle as a source of the Stx2a bacteriophages present in enteroaggregative *Escherichia coli* O104:H4 strains. Int J Med Microbiol 303(8):595–602. https://doi.org/10.1016/j.ijmm.2013.08.001

Bonanno L, Petit MA, Loukiadis E, Michel V, Auvraya F (2016) Heterogeneity in induction level, infection ability, and morphology of shiga toxin-encoding phages (stx phages) from dairy and human shiga toxin-producing *Escherichia coli* O26:H11 Isolates. Appl Environ Microbiol 82:2177–2186

Bondy-Denomy J, Davidson AR (2014) When a virus is not a parasite: the beneficial effects of prophages on bacterial fitness. J Microbiol 52(3):235–242

Boyd EF (2012) Bacteriophage-encoded bacterial virulence factors and phage-pathogenicity island interactions. In: Lobocka M, Szybalski WT (eds) Advances in virus research: Bacteriophages, Pt A, vol 82. Elsevier Academic Press, San Diego, pp 91–118. https://doi.org/10.1016/b978-0-12-394621-8.00014-5

Boyd E, Waldor M (1999) Alternative mechanism of cholera toxin acquisition by *Vibrio cholerae*: generalized transduction of CTX phi by bacteriophage CP-T1. Infect Immun 67(11):5898–5905

Boyd EF, Waldor MK (2002) Evolutionary and functional analyses of variants of the toxin-coregulated pilus protein TcpA from toxigenic *Vibrio cholerae* nonO1/non-O139 serogroup isolates. Microbiology 148:1655–1666

Boyd E, Heilpern A, Waldor M (2000a) Molecular analyses of a putative CTX phi precursor and evidence for independent acquisition of distinct CTX phi s by toxigenic *Vibrio cholerae*. J Bacteriol 182(19):5530–5538

Boyd EF, Moyer KE, Shi L, Waldor MK (2000b) Infectious CTX Phi, and the *Vibrio* pathogenicity island prophage in *Vibrio mimicus*: evidence for recent horizontal transfer between *V. mimicus* and *V. cholerae*. Infect Immun 68:1507–1513

Brown CJ, Millstein J, Williams CJ, Wichman HA (2013) Selection affects genes involved in replication during long-term evolution in experimental populations of the bacteriophage φ. PLoS One 8(3):e60401. https://doi.org/10.1371/journal.pone.0060401

Broecker F, Klumpp J, Moelling K (2016) Long-term microbiota and virome in a Zurich patient after fecal transplantation against Clostridium difficile infection. In: Moelling K (ed) Nutrition and the microbiome, vol 1372, pp 29–41

Brüssow H, Canchaya C, Hardt W-D (2004) Phages and the evolution of bacterial pathogens: from genomic rearrangements to lysogenic conversion. Microbiol Mol Biol Rev 68(3):560–602

Buchholz U, Bernard H, Werber D, Bohmer MM, Remschmidt C, Wilking H, Delere Y, an der Heiden M, Adlhoch C, Dreesman J, Ehlers J, Ethelberg S, Faber M, Frank C, Fricke G, Greiner M, Hohle M, Ivarsson S, Jark U, Kirchner M, Koch J, Krause G, Luber P, Rosner B, Stark K, Kuhne M (2011) German outbreak of *Escherichia coli* O104:H4 associated with sprouts. N Engl J Med 365:1763–1770

Buckling A, Brockhurst M (2012) Bacteria–virus coevolution. In: Soyer O (ed) Evolutionary systems biology. Advances in experimental medicine and biology, vol 751. Springer, New York

Buckling A, Hodgson DJ (2007) Short-term rates of parasite evolution predict the evolution of host diversity. J Evol Biol 20(5):1682–1688. https://doi.org/10.1111/j.1420-9101.2007.01402.x

Buckling A, Rainey PB (2003) The role of parasites in sympatric and allopatric host diversification. Nature 421(6920):294–294. https://doi.org/10.1038/nature01349

Buckling A, Wei Y, Massey RC, Brockhurst MA, Hochberg ME (2006) Antagonistic coevolution with parasites increases the cost of host deleterious mutations. Proc R Soc B Biol Sci 273 (1582):45–49

Burns N, James CE, Harrison E (2015) Polylysogeny magnifies competitiveness of a bacterial pathogen. Evol Appl 8(4):346–351

Calderwood S, Auclair F, Donohue-RolfE A, Keusch G, Mekalanos J (1987) Nucleotide sequence of the shiga-like toxin genes of *Escherichia coli*. Proc Natl Acad Sci USA 84(13):4364–4368. https://doi.org/10.1073/pnas.84.13.4364

Calendar R (2006) The Bacteriophages, 2nd ed. Edited by Richard Calendar and Stephen T. Abedon. Oxford University Press, Oxford

Calero-Caceres W, Muniesa M (2016) Persistence of naturally occurring antibiotic resistance genes in the bacteria and bacteriophage fractions of wastewater. Water Res 95:11–18. https://doi.org/10.1016/j.watres.2016.03.006

Campos J, Martinez E, Marrero K, Silva Y, Rodriguez BL, Suzarte E, Ledon T, Fando R (2003) Novel type of specialized transduction for CTXÂ or its satellite phage RS1 mediated by filamentous phage VGJÂ in Vibrio cholerae. J Bacteriol 185(24):7231–7240

Casas V, Maloy SR (2011) Role of bacteriophage-encoded exotoxins in the evolution of bacterial pathogens. Future Microbiol 6(12):1461–1473

Casas V, Miyake J, Balsley H, Roark J, Telles S, Leeds S, Zurita I, Breitbart M, Bartlett D, Azam F, Rohwer F (2006) Widespread occurrence of phage-encoded exotoxin genes in terrestrial and aquatic environments in Southern California. FEMS Microbiol Lett 261(1):141–149. https://doi.org/10.1111/j.1574-6968.2006.00345.x

Casas V, Magbanua J, Sobrepeña G, Kelley ST, Maloy SR (2010) Reservoir of bacterial exotoxin genes in the environment. Int J Microbiol 2010:754368. https://doi.org/10.1155/2010/754368

Casas V, Sobrepena G, Rodriguez-Mueller B, AhTye J, Maloy SR (2011) Bacteriophage-encoded shiga toxin gene in atypical bacterial host. Gut Pathogens 3:7

Casjens SR, Thuman-Commike PA (2011) Evolution of mosaically related tailed bacteriophage genomes seen through the lens of phage P22 virion assembly. Virology 411(2):393–415

Chambers L, Yang Y, Littier H, Ray P, Zhang T, Pruden A, Strickland M, Katharine K, Mark Ibekwe A (2015) Metagenomic analysis of antibiotic resistance genes in dairy cow feces following therapeutic administration of third generation cephalosporin. PLoS One 10(8): e0133764

Chee-Sanford JC, Mackie RI, Koike S, Krapac IG, Lin YF, Yannarell AC, Maxwell S, Aminov RI (2009) Fate and transport of antibiotic residues and antibiotic resistance genes following land application of manure waste. J Environ Qual 38(3):1086–1108. https://doi.org/10.2134/jeq2008.0128

Chiura HX (1997) Generalized gene transfer by virus-like particles from marine bacteria. Aquat Microb Ecol 13(1):75–83. https://doi.org/10.3354/ame013075

Chouikha I, Charrier L, Filali S, Derbise A, Carniel E (2010) Insights into the infective properties of YpfΊ¦, the Yersinia pestis filamentous phage. Virology 407(1):43–52

Choi S, Dunams D, Jiang SC (2010) Transfer of cholera toxin genes from O1 to non-O1/O139 strains by vibriophages from California coastal waters. J Appl Microbiol 108:1015–1022

Ciofu O, Hansen CR, Høiby N (2013) Respiratory bacterial infections in cystic fibrosis. Curr Opin Pulm Med 19(3):251–258

Cleary PP, LaPenta D, Vessela R, Lam H, Cue D (1998) A globally disseminated M1 subclone of Group A *Streptococci* differs from other subclones by 70 kilobases of prophage DNA and capacity for high-frequency intracellular invasion. Infect Immun 66:5592–5597

Cohen SN, Miller CA (1970) Non-chromosomal antibiotic resistance in bacteria II: molecular nature of R-factors isolated from *Proteus mirabilis* and *Escherichia coli*. J Mol Biol 50:671–687

Cohen SN, Chang ACY, Hsu L (1972) Nonchromosomal antibiotic resistance in bacteria: genetic transformation of Escherichia coli by R-factor DNA. Proc Natl Acad Sci U S A 69(8):2110–2114

Coleman D, Sullivan D, Russell R, Arbuthnott J, Carey B, Pomeroy H (1989) *Staphylococcus aureus* bacteriophages mediating the simultaneous lysogenic conversion of .beta.-lysin, staphylokinase and enterotoxin A: molecular mechanism of triple conversion. J Gen Microbiol 135:1679–1698

Colombo S, Arioli S, Guglielmetti S, Lunelli F, Mora D (2016) Virome-associated antibiotic-resistance genes in an experimental aquaculture facility. FEMS Microbiol Ecol 92(3). https://doi.org/10.1093/femsec/fiw003

Colomer-Lluch M, Imamovic L, Jofre J, Muniesa M (2011a) Bacteriophages carrying antibiotic resistance genes in fecal waste from cattle, pigs, and poultry. Antimicrob Agents Chemother 55 (10):4908–4911

Colomer-Lluch M, Jofre J, Muniesa M, Aziz R (2011b) Antibiotic resistance genes in the bacteriophage DNA fraction of environmental samples. PLoS One 6(3):e17549

Colomer-Lluch M, Jofre J, Muniesa M (2014) Quinolone resistance genes (*qnrA* and *qnrS*) in bacteriophage particles from wastewater samples and the effect of inducing agents on packaged antibiotic resistance genes. J Antimicrob Chemother 69(5):1265–1274. https://doi.org/10.1093/jac/dkt528

Cornick NA, Helgerson AF, Mai V, Ritchie JM, Acheson DWK (2006) In vivo transduction of an Stx-encoding phage in ruminants. Appl Environ Microbiol 72(7):5086–5088. https://doi.org/10.1128/aem.00157-06

Costerton JW, Lam J, Lam K, Chan R (1983) The role of the microcolony mode of growth in the pathogenesis of *Pseudomonas aeruginosa* infections. Rev Infect Dis 5:s867–S873

Dammeyer T, Bagby S, Sullivan M, Chisholm S, Frankenberg-Dinkel N (2008) Efficient phage mediated pigment biosynthesis in oceanic cyanobacteria. Curr Biol 18(6):442–448

Das B, Pazhani GP, Sarkar A, Mukhopadhyay AK, Nair GB, Ramamurthy T (2016) Molecular evolution and functional divergence of *Vibrio cholerae*. Curr Opin Infect Dis 29:520–527

Davies J, Davies D (2010) Origins and evolution of antibiotic resistance. Microbiol Mol Biol Rev 74(3):417–433

Davis BM, Waldor MK (2003) Filamentous phages linked to virulence of *Vibrio cholerae*. Curr Opin Microbiol 6:35–42

Davis BM, Moyer KE, Boyd EF, Waldor MK (2000) CTX prophages in classical biotype *Vibrio cholerae*: Functional phage genes but dysfunctional phage genomes. J Bacteriol 182:6992–6998

Derbise A (2014) Ypf Phi: a filamentous phage acquired by Yersinia pestis. Front Microbiol 5:701

Diard M, Bakkeren E, Cornuault JK, Moor K, Hausmann A, Sellin ME, Loverdo C, Aertsen A, Ackermann M, De Paepe M, Slack E, Hardt WD (2017) Inflammation boosts bacteriophage transfer between *Salmonella* spp. Science 355:1211–1215

Dumke R, Schroter-Bobsin U, Jacobs E, Roske I (2006) Detection of phages carrying the Shiga toxin 1 and 2 genes in waste water and river water samples. Lett Appl Microbiol 42:48–53

Dutilh BE, Thompson CC, Vicente ACP, Marin MA, Lee C, Silva GGZ, Schmieder R, Andrade BGN, Chimetto L, Cuevas D, Garza DR, Okeke IN, Aboderin AO, Spangler J, Ross T, Dinsdale EA, Thompson FL, Harkins TT, Edwards RA (2014) Comparative genomics of 274 *Vibrio cholerae* genomes reveals mobile functions structuring three niche dimensions. BMC Genomics 15:11

Eklund M, Poysky F (1974) Interconversion of type C and D strains of *Clostridium botulinum* by specific bacteriophages. Appl Environ Microbiol 27:251–258

Ernst RK, D'Argenio DA, Ichikawa JK, Bangera MG, Selgrade S, Burns JL, Hiatt P, McCoy K, Brittnacher M, Kas A, Spencer DH, Olson MV, Ramsey BW, Lory S, Miller SI (2003) Genome mosaicism is conserved but not unique in Pseudomonas aeruginosa isolates from the airways of young children with cystic fibrosis. Environ Microbiol 5(12):1341–1349

Fan X, Li Y, He R, Li Q, He W (2016) Comparative analysis of prophage-like elements in *Helicobacter* sp genomes. PeerJ 4:e2012

Fancello L, Desnues C, Raoult D, Rolain JM (2011) Bacteriophages and diffusion of genes encoding antimicrobial resistance in cystic fibrosis sputum microbiota. J Antimicrob Chemother 66(11):2448–2454

Faruque SM, Asadulghani, Rahman MM, Waldor MK, Sack DA (2000) Sunlight-induced propagation of the lysogenic phage encoding cholera toxin. Infect Immun 68(8):4795–4801. https://doi.org/10.1128/iai.68.8.4795-4801.2000

Fernandez-Gonzalez E, Backert S (2014) DNA transfer in the gastric pathogen *Helicobacter pylori*. J Gastroenterol 49:594–604

Fineran PC, Gerritzen MJH, Suarez-Diez M, Kunne T, Boekhorst J, van Hijum S, Staals RHJ, Brouns SJJ (2014) Degenerate target sites mediate rapid primed CRISPR adaptation. Proc Natl Acad Sci USA 111:e1629–E1638

Focazio MJ, Kolpin DW, Barnes KK, Furlong ET, Meyer MT, Zaugg SD, Barber LB, Thurman ME (2008) A national reconnaissance for pharmaceuticals and other organic wastewater contaminants in the United States II: Untreated drinking water sources. Sci Total Environ 402 (2–3):201–216. https://doi.org/10.1016/j.scitotenv.2008.02.021

Frank C, Werber D, Cramer JP, Askar M, Faber M, an der Heiden M, Bernard H, Fruth A, Prager R, Spode A, Wadl M, Zoufaly A, Jordan S, Kemper MJ, Follin P, Muller L, King LA, Rosner B, Buchholz U, Stark K, Krause G, Team HUSI (2011) Epidemic profile of shiga-toxin-producing *Escherichia coli* O104:H4 outbreak in Germany. N Engl J Med 365:1771–1780

Freeman VJ (1951) Studies on the virulence of bacteriophage-infected strains of *Corynebacterium diphtheriae*. J Bacteriol 61(6):675–688

Freer JH, Arbuthnott JP (1982) Toxins of *Staphylococcus aureus*. Pharmacol Ther 19:55–106

Fruth A, Prager R, Tietze E, Rabsch W, Flieger A (2015) Molecular epidemiological view on Shiga toxin-producing *Escherichia coli* causing human disease in Germany: diversity, prevalence, and outbreaks. Int J Med Microbiol 305(7):697–704. https://doi.org/10.1016/j.ijmm.2015.08.020

Gamage SD, Patton AK, Hanson JF, Weiss AA (2004) Diversity and host range of Shiga toxin-encoding phage. Infect Immun 72(12):7131–7139

Garcia-Aljaro C, Muniesa M, Jofre J, Blanch AR (2006) Newly identified bacteriophages carrying the stx Shiga toxin gene isolated from *Escherichia coli* strains in polluted waters. FEMS Microbiol Lett 258:127–135

Garcia-Aljaro C, Muniesa M, Jofre J, Blanch AR (2009) Genotypic and phenotypic diversity among induced, stx(2)-carrying bacteriophages from environmental *Escherichia coli* strains. Appl Environ Microbiol 75:329–336

Goerke C, Wolz C (2010) Adaptation of *Staphylococcus aureus* to the cystic fibrosis lung. Int J Med Microbiol 300(8):520–525. https://doi.org/10.1016/j.ijmm.2010.08.003

Goh S, Hussain H, Chang BJ, Emmett W, Riley TV, Mullany P (2013) Phage phi C2 mediates transduction of Tn6215, encoding erythromycin resistance, between *Clostridium difficile* strains. MBio 4(6):e00840-13. https://doi.org/10.1128/mBio.00840-13

Goldhill DH, Turner PE (2014) The evolution of life history trade-offs in viruses. Curr Opin Virol 8:79–84

Gomez P, Bennie J, Gaston KJ, Buckling A (2015) The impact of resource availability on bacterial resistance to phages in soil. PLoS One 10(4):e0123752. https://doi.org/10.1371/journal.pone.0123752

Gomez P, Buckling A (2011) Bacteria-phage antagonistic coevolution in soil. Science 332 (6025):106–109

Gorter FA, Scanlan PD, Buckling A (2016) Adaptation to abiotic conditions drives local adaptation in bacteria and viruses coevolving in heterogeneous environments. Biol Lett 12(2):20150879

Govan JRW, Deretic V (1996) Microbial pathogenesis in cystic fibrosis: mucoid *Pseudomonas aeruginosa* and *Burkholderia cepacia*. Microbiol Rev 60:539–574

Govind R, Fralick JA, Rolfe RD (2011) In vivo lysogenization of a *Clostridium difficile* bacteriophage Phi CD119. Anaerobe 17(3):125–129

Groman N (1953) The relation of bacteriophage to the change of *Corynebacterium* diphtheriae from avirulence to virulence. Science 117:297–299

Guinane CM, Kent RM, Norberg S, Hill C, Fitzgerald GF, Stanton C, Ross RP (2011) Host specific diversity in *Lactobacillus johnsonii* as evidenced by a major chromosomal inversion and phage resistance mechanisms. PLoS One 6(4):e18740

Guy L, Nystedt B, Toft C, Zaremba-Niedzwiedzka K, Berglund EC, Granberg F, Naslund K, Eriksson AS, Andersson SGE (2013) A gene transfer agent and a dynamic repertoire of secretion systems hold the keys to the explosive radiation of the emerging pathogen *Bartonella*. PLoS Genet 9:22

Haaber J, Leisner JJ, Cohn MT, Catalan-Moreno A, Nielsen JB, Westh H, Penadés JR, Ingmer H (2016) Bacterial viruses enable their host to acquire antibiotic resistance genes from neighbouring cells. Nat Commun 7:13333

Hall AR, Scanlan PD, Morgan AD, Buckling A (2011) Host-parasite coevolutionary arms races give way to fluctuating selection. Ecol Lett 14(7):635–642

Hall-Stoodley L, Costerton JW, Stoodley P (2004) Bacterial biofilms: from the natural environment to infectious diseases. Nat Rev Microbiol 2(2):95–108. https://doi.org/10.1038/nrmicro821

Hammerl JA, Al Dahouk S, Nöckler K, Göllner C, Appel B, Hertwig S (2014) F1 and Tbilisi are closely related Brucellaphages exhibiting some distinct nucleotide variations which determine the host specificity. Genome Announc 2(1):e01250-01213

Hammerl JA, Göllner C, AlDahouk S, Nöckler K, Reetz J, Hertwig S (2016) Analysis of the first temperate broad host range Brucellaphage (BiPBO1) isolated from *B. inopinata*. Front Microbiol 7:24. https://doi.org/10.3389/fmicb.2016.00024

Hannigan GD, Meisel JS, Tyldsley AS, Zheng Q, Hodkinson BP, SanMiguel AJ, Minot S, Bushman FD, Grice EA (2015) The Human skin double-stranded DNA virome: topographical and temporal diversity, genetic enrichment, and dynamic associations with the host microbiome. mBio 6(5):e01578-15. https://doi.org/10.1128/mBio.01578-15

Haramoto E, Katayama H, Oguma K, Yamashita H, Tajima A, Nakajima H, Ohgaki S (2006) Seasonal profiles of human Noroviruses and indicator bacteria in a wastewater treatment plant in Tokyo, Japan. Water Sci Technol 54:301–308

Hawkey PM, Jones AM (2009) The changing epidemiology of resistance. J Antimicrob Chemother 64(Supplement 1):i3–i10

Hazen TH, Pan L, Gu JD, Sobecky PA (2010) The contribution of mobile genetic elements to the evolution and ecology of *Vibrios*. FEMS Microbiol Ecol 74:485–499

Helbin WM, Polakowska K, Miedzobrodzki J (2012) Phage-related virulence factors of *Staphylococcus aureus*. Postepy Mikrobiologii 51:291–298

Heler R, Marraffini LA, Bikard D (2014) Adapting to new threats: the generation of memory by CRISPR-Cas immune systems. Mol Microbiol 93:1–9

Horne MT (1970) Coevolution of *Escherichia coli* and bacteriophages in chemostat culture. Science 168:992–993

Horvath P, Barrangou R (2011) Protection against foreign DNA. In: Storz G, Hengge R (eds) Bacterial stress responses, 2nd edn. ASM Press, Washington, DC, pp 333–348

Huang S, Zhang S, Jiao N, Chen F (2015) Comparative genomic and phylogenomic analyses reveal a conserved core genome shared by estuarine and oceanic cyanopodoviruses. PLoS One 10(11): e0142962. https://doi.org/10.1371/journal.pone.0142962

Hurwitz BL, Brum JR, Sullivan MB (2015) Depth-stratified functional and taxonomic niche specialization in the 'core' and 'flexible' Pacific Ocean virome. ISME J 9:472–484

Hynes AP, Villion M, Moineau S (2014) Adaptation in bacterial CRISPR-Cas immunity can be driven by defective phages. Nat Commun 5:4399. https://doi.org/10.1038/ncomms5399

Inoue K, Iida H (1970) Conversion of toxigenicity in *Clostridium botulinum* type C. Jpn J Microbiol 14(1):87–89

Ivanov YV, Shariat N, Register KB, Linz B, Rivera I, Hu K, Dudley EG, Harvill ET (2015) A newly discovered *Bordetella* species carries a transcriptionally active CRISPR-Cas with a small Cas9 endonuclease. BMC Genomics 16:863. https://doi.org/10.1186/s12864-015-2028-9

Jacob AE, Hobbs SJ (1974) Conjugal transfer of plasmid-borne multiple antibiotic-resistance in *Streptococcus-faecalis* var zymogenes. J Bacteriol 117:360–372

Jiang SC, Paul JH (1996) Occurrence of lysogenic bacteria in marine microbial communities as determined by prophage induction. Mar Ecol Prog Ser 142(1–3):27–38

Jiang SC, Paul JH (1998) Gene transfer by transduction in the marine environment. Appl Environ Microbiol 64(8):2780–2787

Jjemba PK (2002a) The potential impact of veterinary and human therapeutic agents in manure and biosolids on plants grown on arable land: a review. Agric Ecosyst Environ 93(1–3):267–278. https://doi.org/10.1016/s0167-8809(01)00350-4

Jjemba PK (2002b) The effect of chloroquine, quinacrine, and metronidazole on both soybean plants and soil microbiota. Chemosphere 46(7):1019–1025

Johnson L, Schlievert P (1984) Group A *streptococcal* phage T12 carries the structural gene for pyrogenic exotoxin type A. Mol Gen Genet 194(1–2):52–56

Kashiwagi A, Yomo T, Casadesús J (2011) Ongoing phenotypic and genomic changes in experimental coevolution of RNA bacteriophage QÎ² and Escherichia coli. PLoS Genet 7(8): e1002188

Karmali MA (2017) Emerging public health challenges of shiga toxin-producing *Escherichia coli* related to changes in the pathogen, the population, and the environment. Clin Infect Dis 64 (3):371–376. https://doi.org/10.1093/cid/ciw708

Karpe YA, Kanade GD, Pingale KD, Arankalle VA, Banerjee K (2016) Genomic characterization of Salmonella bacteriophages isolated from India. Virus Genes 52(1):117–126

Karataev GI, Moskivina IL, Ryabinina OP, Miller GG, Mebel SM, Lapaeva IA (1988) Isolation and characterization of bacteriophage from the vaccine strain Tohama Phase I. Mol Genet Microbiol Virol 4:22–25

Kay P, Blackwell PA, Boxall ABA (2004) Fate of veterinary antibiotics in a macroporous tile drained clay soil. Environ Toxicol Chem 23(5):1136–1144. https://doi.org/10.1897/03-374

Kelly WJ, Altermann E, Lambie SC, Leahy SC (2013) Interaction between the genomes of Lactococcus lactis and phages of the P335 species. Front Microbiol 4:257

Kemper N (2008) Veterinary antibiotics in the aquatic and terrestrial environment. Ecol Indic 8 (1):1–13. https://doi.org/10.1016/j.ecolind.2007.06.002

Kim MS, Bae JW (2016) Spatial disturbances in altered mucosal and luminal gut viromes of diet-induced obese mice. Environ Microbiol 18:1498–1510

Kim EJ, Lee CH, Nair GB, Kim DW (2015) Whole-genome sequence comparisons reveal the evolution of *Vibrio cholerae* O1. Trends Microbiol 23:479–489

Koonin EV, Makarova KS, Wolf YI (2017) Evolutionary genomics of defense systems in archaea and bacteria. Annu Rev Microbiol 71(1):233–261

Koskella B (2013) Phage-mediated selection on microbiota of a long-lived host. Curr Biol 23:1256–1260

Koskella B, Meaden S (2013) Understanding bacteriophage specificity in natural microbial communities. Viruses 5(3):806–823. PMC. Web. 20 July 2018

Koskella B, Lin DM, Buckling A, Thompson JN (2012) The costs of evolving resistance in heterogeneous parasite environments. Proc R Soc B Biol Sci 279(1735):1896–1903

Krahn T, Wibberg D, Maus I, Winkler A, Bontron S, Sczyrba A, Nordmann P, Puehler A, Poirel L, Schlueter A (2016) Intraspecies transfer of the chromosomal *Acinetobacter baumannii bla* (NDM-1) carbapenemase gene. Antimicrob Agents Chemother 60(5):3032–3040

Kraushaar B, Hammerl JA, Kienol M, Heinig ML, Sperling N, Thanh MD, Reetz J, Jakel C, Fetsch A, Hertwig S (2017) Acquisition of virulence factors in livestock-associated MRSA: lysogenic conversion of CC398 strains by virulence gene-containing phages. Sci Rep 7. https://doi.org/10.1038/s41598-017-02175-4

L'Abee-Lund TM, Jorgensen HJ, O'Sullivan K, Bohlin J, Ligard G, Granum PE, Lindback T (2012) The highly virulent 2006 Norwegian EHEC O103:H25 outbreak strain is related to the 2011 German O104:H4 outbreak strain. PLoS One 7(3):e31413. https://doi.org/10.1371/journal.pone.0031413

Lachmayr KL, Kerkhof LJ, DiRienzo AG, Cavanaugh CM, Ford TE (2009) Quantifying nonspecific TEM beta-lactamase (*bla*(TEM)) genes in a wastewater stream. Appl Environ Microbiol 75:203–211

Lainhart W, Stolfa G, Koudelka G (2009) Shiga toxin as a bacterial defense against a eukaryotic predator, *Tetrahymena thermophila*. J Bacteriol 191(16):5116–5122

Lam J, Chan R, Lam K, Costerton JW (1980) Production of mucoid microcolonies by *Pseudomonas aeruginosa* within infected lungs in cystic fibrosis. Infect Immun 28(2):546–556

Lan S-F, Huang C-H, Chang C-H, Liao W-C, Lin I-H (2009) Characterization of a new plasmid-like prophage in a pandemic. Appl Environ Microbiol 75(9):2659–2667

Lapaeva IA, Mebel SM, Pereverzev NA, Sinyashina LN (1980) *Bordetella pertussis* bacteriophage. Zh Mikrobiol Epidemiol Immunobiol 5:85–90

Latino L, Essoh C, Blouin Y, Thien HV, Pourcel C (2014) A novel *Pseudomonas aeruginosa* bacteriophage, Ab31, a chimera formed from temperate phage PAJU2 and *P. putida* lytic phage AF: characteristics and mechanism of bacterial resistance. PLoS One 9(4):e93777. https://doi.org/10.1371/journal.pone.0093777

Levin BR, Guttman DS (2010) Nasty viruses, costly plasmids, population dynamics, and the conditions for establishing and maintaining CRISPR-mediated adaptive immunity in bacteria. PLoS Genet 6(10):e1001171

Lee YD, Park JH (2016) Phage conversion for beta-lactam antibiotic resistance of *Staphylococcus aureus* from foods. J Microbiol Biotechnol 26(2):263–269

Lee JY, Li ZQ, Miller ES (2017) Vibrio phage KVP40 encodes a functional NAD+ salvage pathway. J Bacteriol 199(9):e00855-16. https://doi.org/10.1128/JB.00855-16

Lemieux A-A, Jeukens J, Kukavica-Ibrulj I, Fothergill JL, Boyle B, Laroche J, Tucker NP, Winstanley C, Levesque RC (2016) Genes required for free phage production are essential for *Pseudomonas aeruginosa* chronic lung infections. J Infect Dis 213(3):395–402

Lenski RE (1988) Dynamics of interactions between bacteria and virulent bacteriophage. Adv Microb Ecol 10:1–44

Lenski RE, Levin BR (1985) Constraints on the coevolution of bacteria and virulent phage: a model, some experiments, and predictions for natural communities. Am Nat 125(4):585–602

Lindsay J, Ruzin A, Ross H, Kurepina N, Novick R (1998) The gene for toxic shock toxin is carried by a family of mobile pathogenicity islands in *Staphylococcus aureus*. Mol Microbiol 29 (2):527–543

Looft T, Allen HK, Cantarel BL, Levine UY, Bayles DO, Alt DP, Henrissat B, Stanton TB (2014) Bacteria, phages and pigs: the effects of in-feed antibiotics on the microbiome at different gut locations. ISME J 8(8):1566–1576. https://doi.org/10.1038/ismej.2014.12

Los JM, Los M, Wegrzyn A, Wegrzyn G (2013) Altruism of shiga toxin-producing *Escherichia coli*: recent hypothesis versus experimental results. Front Cell Infect Microbiol 3:166. https://doi.org/10.3389/fcimb.2012.00166

Lupo A, Coyne S, Berendonk TU (2012) Origin and evolution of antibiotic resistance: the common mechanisms of emergence and spread in water bodies. Front Microbiol 3:13

Mai-Prochnow A, Hui JGK, Kjelleberg S, Rakonjac J, McDougald D, Rice SA (2015) Big things in small packages: the genetics of filamentous phage and effects on fitness of their host. FEMS Microbiol Rev 39(4):465–487. https://doi.org/10.1093/femsre/fuu007

Maloy SR, Stewart VJ, Taylor RK (1996) Genetic analysis of pathogenic bacteria: a laboratory manual. Cold Spring Harbor Laboratory Press, Cold Spring Harbor, NY

Marti R, Muniesa M, Schmid M, Ahrens CH, Naskova J, Hummerjohann J (2016) Short communication: heat-resistant *Escherichia coli* as potential persistent reservoir of extended-spectrum beta-lactamases and Shiga toxin-encoding phages in dairy. J Dairy Sci 99(11):8622–8632. https://doi.org/10.3168/jds.2016-11076

McGavin MJ, Arsic B, Nickerson NN (2012) Evolutionary blueprint for host- and niche-adaptation in *Staphylococcus aureus* clonal complex CC30. Front Cell Infect Microbiol 2:48. https://doi.org/10.3389/fcimb.2012.00048

McGee LW, Aitchison EW, Brian Caudle S, Morrison AJ, Zheng L, Yang W, Rokyta DR, Worobeg M (2014) Payoffs, not tradeoffs, in the adaptation of a virus to ostensibly conflicting selective pressures. PLoS Genet 10(10):e1004611

McLaughlin MR, Rose JB (2006) Application of *Bacteroides fragilis* phage as an alternative indicator of sewage pollution in Tampa Bay, Florida. Estuar Coasts 29(2):246–256

Meyer JR, Dobias DT, Weitz JS, Barrick JE, Quick RT, Lenski RE (2012) Repeatability and contingency in the evolution of a key innovation in phage lambda. Science 335:428–432

Miao XS, Bishay F, Chen M, Metcalfe CD (2004) Occurrence of antimicrobials in the final effluents of wastewater treatment plants in Canada. Environ Sci Technol 38(13):3533–3541. https://doi.org/10.1021/es030653q

Michel-Briand Y, Baysse C (2002) The pyocins of Pseudomonas aeruginosa. Biochimie 84(5–6):499–510

Morris P, Marinelli LJ, Jacobs-Sera D, Hendrix RW, Hatfull GF (2008) Genomic characterization of mycobacteriophage giles: evidence for phage acquisition of host DNA by illegitimate recombination. J Bacteriol 190(6):2172–2182

Moce-Llivina L, Muniesa M, Pimenta-Vale H, Lucena F, Jofre J (2003) Survival of bacterial indicator species and bacteriophages after thermal treatment of sludge and sewage. Appl Environ Microbiol 69:1452–1456

Modi SR, Lee HH, Spina CS, Collins JJ (2013) Antibiotic treatment expands the resistance reservoir and ecological network of the phage metagenome. Nature 499:219–222

Mookerjee S, Batabyal P, Sarkar M, Palit A (2015) Seasonal prevalence of enteropathogenic *Vibrio* and their phages in the riverine estuarine ecosystem of South Bengal. PLoS One 10(9):13

Moon BY, Park JY, Hwang SY, Robinson DA, Thomas JC, Fitzgerald JR, Park YH, Seo KS (2015) Phage-mediated horizontal transfer of a *Staphylococcus aureus* virulence-associated genomic island. Sci Rep 5(1):9784. https://doi.org/10.1038/srep09784

Mora A, Herrera A, Lopez C, Dahbi G, Mamani R, Pita JM, Alonso MP, Llovo J, Bernardez MI, Blanco JE, Blanco M, Blanco J (2011) Characteristics of the Shiga-toxin-producing enteroaggregative *Escherichia coli* O104:H4 German outbreak strain and of STEC strains isolated in Spain. Int Microbiol 14:121–141

Morgan AD, Bonsall MB, Buckling A (2010) Impact of bacterial mutation rate on coevolutionary dynamics between bacteria and phages. Evolution 64(10):2980–2987. https://doi.org/10.1111/j.1558-5646.2010.01037.x

Motlagh AM, Bhattacharjee AS, Coutinho FH, Dutilh BE, Casjens SR, Goel RK (2017) Insights of phage-host interaction in hypersaline ecosystem through metagenomics analyses. Front Microbiol 8:15

Muniesa M, Jofre J (2004) Abundance in sewage of bacteriophages infecting *Escherichia coli* O157:H7. Public Health Microbiol Methods Protocols 268:79–88

Muniesa M, Serra-Moreno R, Jofre J (2004) Free Shiga toxin bacteriophages isolated from sewage showed diversity although the *stx* genes appeared conserved. Environ Microbiol 6:716–725

Murugaiyan S, Bae JY, Wu J, Lee SD, Um HY, Choi HK, Chung E, Lee JH, Lee SW (2011) Characterization of filamentous bacteriophage PE226 infecting *Ralstonia solanacearum* strains. J Appl Microbiol 110:296–303

Nasu H, Iida T, Sugahara T, Yamaichi Y, Park K, Yokoyama K, Makino K, Shinagawa H, Honda T (2000) A filamentous phage associated with recent pandemic *Vibrio parahaemolyticus* O3:K6 strains. J Clin Microbiol 38(6):2156–2161

Nedialkova LP, Sidstedt M, Koeppel MB, Spriewald S, Ring D, Gerlach RG, Bossi L, Stecher B (2016) Temperate phages promote colicin-dependent fitness of serovar Typhimurium. Environ Microbiol 18(5):1591–1603

Novick RP, Christie GE, Penades JR (2010) The phage-related chromosomal islands of Gram-positive bacteria. Nat Rev Microbiol 8:541–551

Nyambe S, Burgess C, Whyte P, Bolton D (2016a) Survival studies of a temperate and lytic bacteriophage in bovine faeces and slurry. J Appl Microbiol 121(4):1144–1151

Nyambe S, Burgess C, Whyte P, Bolton D (2016b) The survival of a temperate vtx bacteriophage and an anti-verocytotoxigenic O157 lytic phage in water and soil samples. Zoonoses Public Health 63(8):632–640

Obeng N, Pratama AA, van Elsas JD (2016) The significance of mutualistic phages for bacterial ecology and evolution. Trends Microbiol 24(6):440–449

O'Brien AD, Newland JW, Miller SF, Holmes RK, Smith HW, Formal SB (1984) Shiga-like toxin-converting phages from Escherichia coli strains that cause hemorrhagic colitis or infantile diarrhea. Science 226(4675):694–696

O'Brien S, Rodrigues AMM, Buckling A (2013) The evolution of bacterial mutation rates under simultaneous selection by interspecific and social parasitism. Proc R Soc B Biol Sci 280 (1773):20131913

Okuda J, Ishibashi M, Hayakawa E, Nishino T (1997) Emergence of a Unique O3:K6 Clone of *Vibrio parahaemolyticus*. J Clin Microbiol 35(12):3150–3155

O'Shea YA, Fidelma Boyd E (2002) Mobilization of the pathogenicity island between isolates mediated by CP-T1 generalized transduction. FEMS Microbiol Lett 214(2):153–157

Pal C, Macia MD, Oliver A, Schachar I, Buckling A (2007) Coevolution with viruses drives the evolution of bacterial mutation rates. Nature 450(7172):1079–1081. https://doi.org/10.1038/nature06350

Park MO, Ikenaga H, Watanabe K (2007) Phage diversity in a methanogenic digester. Microb Ecol 53:98–103

Park D, Stanton E, Ciezki K, Parrell D, Bozile M, Pike D, Forst SA, Jeong KC, Ivanek R, Dopfer D, Kaspar CW (2013) Evolution of the stx2-encoding prophage in persistent bovine *Escherichia coli* O157:H7 strains. Appl Environ Microbiol 79:1563–1572

Paterson S, Vogwill T, Buckling A, Benmayor R, Spiers AJ, Thomson NR, Quail M, Smith F, Walker D, Libberton B, Fenton A, Hall N, Brockhurst MA (2010) Antagonistic coevolution accelerates molecular evolution. Nature 464(7286):275–278

Paton A, Paton J (1996) *Enterobacter cloacae* producing a shiga-like toxin II-related cytotoxin associated with a case of hemolytic-uremic syndrome. J Clin Microbiol 34(2):463–465

Paul JH, Sullivan MB, Segall AM, Rohwer F (2002) Marine phage genomics. Comp Biochem Physiol B Biochem Mol Biol 133:463–476

Payne M, Oakey J, Owens L (2004) The ability of two different *Vibrio* spp. bacteriophages to infect *Vibrio harveyi, Vibrio cholerae* and *Vibrio mimicus*. J Appl Microbiol 97:663–672

Pearson GDN, Woods A, Chiang SL, Mekalanos JJ (1993) Ctx genetic element encodes a site-specific recombination system and an intestinal colonization factor. Proc Natl Acad Sci USA 90:3750–3754

Penadés JR, Christie GE (2015) The phage-inducible chromosomal islands: a family of highly evolved molecular parasites. In: Enquist LW (ed) Annual review of virology, vol 2, pp 181–201. https://doi.org/10.1146/annurev-virology-031413-085446

Perez G, Thierauf A, Maloy S (2009) Generalized transduction. In: Bacteriophages: methods in molecular biology. Humana Press, New York, pp 267–286

Perry EB, Barrick JE, Bohannan BJM (2015) The molecular and genetic basis of repeatable coevolution between *Escherichia coli* and bacteriophage T3 in a laboratory microcosm. PLoS One 10(6):e0130639

Petridis M, Bagdasarian M, Waldor MK, Walker E (2006) Horizontal transfer of Shiga toxin and antibiotic resistance genes among Escherichia coli strains in house fly (Diptera: Muscidae) gut. J Med Entomol 43(2):288–295

Picozzi C, Volponi G, Vigentini I, Grassi S, Foschino R (2012) Assessment of transduction of *Escherichia coli* stx2-encoding phage in dairy process conditions. Int J Food Microbiol 153:388–394

Plunkett G, Rose D, Durfee T, Blattner F (1999) Sequence of shiga toxin 2 phage 933w from Escherichia coli O157:H7: Shiga toxin as a phage late-gene product. J Bacteriol 181 (6):1767–1778

Popoff MR, Bouvet P (2013) Genetic characteristics of toxigenic *Clostridia* and toxin gene evolution. Toxicon 75:63–89

Potts SB, Roggli VL, Spock A (1995) Immunohistologic quantification of *Pseudomonas aeruginosa* in the tracheobronchial tree from patients with cystic-fibrosis. Pediatr Pathol Lab Med 15:707–721

Puxty R, Millard A, Evans D, Scanlan D (2015) Shedding new light on viral photosynthesis. Photosynth Res 12(1):71–97

Rohwer F, Thurber RV (2009) Viruses manipulate the marine environment. Nature 459 (7244):207–212. https://doi.org/10.1038/nature08060

Rolain J, Fancello L, Desnues C, Raoult D (2011) Bacteriophages as vehicles of the resistome in cystic fibrosis. J Antimicrob Chemother 66(11):2444–2447

Ross J, Topp E (2015) Abundance of antibiotic resistance genes in bacteriophage following soil fertilization with dairy manure or municipal biosolids, and evidence for potential transduction. Appl Environ Microbiol 81(22):7905–7913. https://doi.org/10.1128/aem.02363-15

Sarmah AK, Meyer MT, Boxall ABA (2006) A global perspective on the use, sales, exposure pathways, occurrence, fate and effects of veterinary antibiotics (VAs) in the environment. Chemosphere 65(5):725–759. https://doi.org/10.1016/j.chemosphere.2006.03.026

Sarris PF, Ladoukakis ED, Panopoulos NJ, Scoulica EV (2014) A phage tail-derived element with wide distribution among both prokaryotic domains: a comparative genomic and phylogenetic study. Genome Biol Evol 6(7):1739–1747

Scanlan PD, Hall AR, Lopez-Pascua LDC, Buckling A (2011) Genetic basis of infectivity evolution in a bacteriophage. Mol Ecol 20:981–989

Scott J, Nguyen SV, King CJ, Hendrickson C, McShan WM (2012) Phage-like *Streptococcus pyogenes* chromosomal islands (SpyCI) and mutator phenotypes: control by growth state and rescue by a SpyCI-encoded promoter. Front Microbiol 3:317. https://doi.org/10.3389/fmicb.2012.00317

Shapiro JW, Williams E, Turner PE (2016) Evolution of parasitism and mutualism between filamentous phage M13 and *Escherichia coli*. PeerJ 4:e2060. https://doi.org/10.7717/peerj.2060

Sharma P, Gupta SK, Rolain JM (2014) Whole genome sequencing of bacteria in cystic fibrosis as a model for bacterial genome adaptation and evolution. Expert Rev Anti-Infect Ther 12 (3):343–355. https://doi.org/10.1586/14787210.2014.887441

Sinton LW, Hall CH, Lynch PA, Davies-Colley RJ (2002) Sunlight inactivation of fecal indicator bacteria and bacteriophages from waste stabilization pond effluent in fresh and saline waters. Appl Environ Microbiol 68:1122–1131

Smith DL, Rooks DJ, Fogg PCM, Darby AC, Thomson NR, McCarthy AJ, Allison HE (2012) Comparative genomics of Shiga toxin encoding bacteriophages. BMC Genomics 13(1):311

Solheim M, Brekke MC, Snipen LG, Willems RJL, Nes IF, Brede DA (2011) Comparative genomic analysis reveals significant enrichment of mobile genetic elements and genes encoding surface structure-proteins in hospital-associated clonal complex 2 *Enterococcus faecalis*. BMC Microbiol 11:3. https://doi.org/10.1186/1471-2180-11-3

Solheim HT, Sekse C, Urdahl AM, Wasteson Y, Nesse LL (2013) Biofilm as an environment for dissemination of *stx* genes by transduction. Appl Environ Microbiol 79:896–900

Sommer MOA, Dantas G, Church GM (2009) Functional characterization of the antibiotic resistance reservoir in the human microflora. Science 325:1128–1131

Strockbine N, Marques L, Newland J et al (1986) Two toxin-converting phages from *Escherichia coli* O157:H7 strain 933 encode antigenically distinct toxins with similar biological activities. Infect Immun 53(1):135–140

Steinberg KM, Levin BR (2007) Grazing protozoa and the evolution of the Escherichia coli O157: H7 Shiga toxin-encoding prophage. Proc R Soc B Biol Sci 274(1621):1921–1929

Subirats J, Sanchez-Melsio A, Borrego CM, Balcazar JL, Simonet P (2016) Metagenomic analysis reveals that bacteriophages are reservoirs of antibiotic resistance genes. Int J Antimicrob Agents 48(2):163–167. https://doi.org/10.1016/j.ijantimicag.2016.04.028

Sugiyama H (1980) *Clostridium botulinum* neurotoxin. Microbiol Rev 44(3):419–448

Summer EJ, Gonzalez CF, Bomer M, Carlile T, Embry A, Kucherka AM, Lee J, Mebane L, Morrison WC, Mark L, King MD, LiPuma JJ, Vidaver AK, Young R (2005) Divergence and mosaicism among virulent soil phages of the Burkholderia cepacia complex. J Bacteriol 188 (1):255–268

Tanji Y, Mizoguchi K, Yoichi M, Morita M, Kijima N, Kator H, Unno H (2003) Seasonal change and fate of coliphages infected to *Escherichia coli* O157:H7 in a wastewater treatment plant. Water Res 37:1136–1142

Timms AR, Cambray-Young J, Scott AE, Petty NK, Connerton PL, Clarke L, Seeger K, Quail M, Cummings N, Maskell DJ, Thomson NR, Connerton IF (2010) Evidence for a lineage of virulent bacteriophages that target Campylobacter. BMC Genomics 11(1):214

Tozzoli R, Grande L, Michelacci V, Ranieri P, Maugliani A, Caprioli A, Morabito S (2014) Shiga toxinconverting phages and the emergence of new pathogenic Escherichia coli: a world in motion. Front Cell Infect Microbiol 4:80

Uyaguari MI, Fichot EB, Scott GI, Norman RS (2011) Characterization and quantitation of a novel β-lactamase gene found in a wastewater treatment facility and the surrounding coastal ecosystem. Appl Environ Microbiol 77(23):8226–8233

van Overbeek LS, van Doorn J, Wichers JH, van Amerongen A, van Roermund HJW, Willemsen PTJ (2014) The arable ecosystem as battleground for emergence of new human pathogens. Front Microbiol 5:104

Varga M, Kuntova L, Pantucek R, Maslanova I, Ruzickova V, Doskar J (2012) Efficient transfer of antibiotic resistance plasmids by transduction within methicillin-resistant *Staphylococcus aureus* USA300 clone. FEMS Microbiol Lett 332:146–152

Varga M, Pantucek R, Ruzickova V, Doskar J (2016) Molecular characterization of a new efficiently transducing bacteriophage identified in methicillin-resistant *Staphylococcus aureus*. J Gen Virol 97:258–267

Vasse M, Torres-Barceló C, Hochberg ME (2015) Phage selection for bacterial cheats leads to population decline. Proc R Soc B Biol Sci 282(1818):20152207

Ventola CL (2015) The antibiotic resistance crisis: part 1: causes and threats. Pharm Ther 40(4):277–283

Veses-Garcia M, Liu X, Rigden DJ, Kenny JG, McCarthy AJ, Allison HE (2015) Transcriptomic analysis of shiga-toxigenic bacteriophage carriage reveals a profound regulatory effect on acid resistance in *Escherichia coli*. Appl Environ Microbiol 81(23):8118–8125. https://doi.org/10.1128/aem.02034-15

Vishwakarma V, Periaswamy B, Pati NB, Slack E, Hardt WD, Suar M (2012) A novel phage element of *Salmonella enterica* serovar enteritidis p125109 contributes to accelerated type III secretion system 2-dependent early inflammation kinetics in a mouse colitis model. Infect Immun 80:3236–3246

Wagner PL, Waldor MK (2002) Bacteriophage control of bacterial virulence. Infect Immun 70:3985–3993

Waksman S (1961) Role of Antibiotics in Nature. Perspect Biol Med 4(3):271

Waldor MK, Mekalanos JJ (1996) Lysogenic conversion by a filamentous phage encoding cholera toxin. Science 272(5270):1910–1914

Wang YQ, Zhang XB (2010) Genome analysis of deep-sea thermophilic phage D6E. Appl Environ Microbiol 76:7861–7866

Wang CH, Chuan CN, Kuo HT, Zheng PX, Tsou CC, Wang SY, Tsai PJ, Chuang WJ, Lin YS, Liu CC, Wu JJ (2013a) Peroxide responsive regulator perR of Group A *Streptococcus* is required for the expression of phage-associated DNAse Sda1 under oxidative stress. PLoS One 8(12): e81882. https://doi.org/10.1371/journal.pone.0081882

Wang QY, Kan BA, Wang RB (2013b) Isolation and characterization of the new mosaic filamentous phage VFJ phi of *Vibrio cholerae*. PLoS One 8:9

Weeks C, Ferretti J (1984) The gene for type A *Streptococcal exotoxin* (erythrogenic toxin) is located in bacteriophage T12. Infect Immun 46(2):531–536

Wei RC, Ge F, Huang SY, Chen M, Wang R (2011) Occurrence of veterinary antibiotics in animal wastewater and surface water around farms in Jiangsu Province, China. Chemosphere 82 (10):1408–1414. https://doi.org/10.1016/j.chemosphere.2010.11.067

Wei YZ, Chesne MT, Terns RM, Terns MP (2015) Sequences spanning the leader-repeat junction mediate CRISPR adaptation to phage in *Streptococcus thermophilus*. Nucleic Acids Res 43:1749–1758

Weinbauer MG, Suttle CA (1999) Lysogeny and prophage induction in coastal and offshore bacterial communities. Aquat Microb Ecol 18:217–225

Willner D, Furlan M (2010) Deciphering the role of phage in the cystic fibrosis airway. Virulence 1 (4):309–313. https://doi.org/10.4161/viru.1.4.12071

Willner D, Furlan M, Haynes M, Schmieder R, Angly F, Silva J, Tammadoni S, Nosrat B, Conrad D, Rohwer F (2009) Metagenomic analysis of respiratory tract DNA viral communities in cystic fibrosis and non-cystic fibrosis individuals. PLoS One 4(10):e7370

Xia GQ, Wolz C (2014) Phages of *Staphylococcus aureus* and their impact on host evolution. Infect Genet Evol 21:593–601

Yamaguchi T, Hayashi T, Takami H, Nakasone K, Ohnishi M, Nakayama K, Yamada S, Sugai M (2000) Phage conversion of exfoliative toxin A production in *Staphylococcus aureus*. Mol Microbiol 38(4):694–705

Yan YX, Shi YB, Cao DM, Meng XP, Xia LM, Sun JH (2011) Prevalence of *stx* phages in environments of a pig farm and lysogenic infection of the field *E. coli* O157 isolates with a recombinant converting phage. Curr Microbiol 62:458–464

Yeroshenko GA, Smirnova NI (2004) Role of temperate bacteriophage 139 in changing cholera toxin production in *Vibrio cholerae* classical biovar. Russ J Genet 40:348–355

Yoshida M, Yoshida-Takashima Y, Nunoura T, Takai K (2015) Identification and genomic analysis of temperate *Pseudomonas* bacteriophage PstS-1 from the Japan trench at a depth of 7000 m. Res Microbiol 166:668–676

Yu C, Ferretti J (1991) Molecular characterization of new group A streptococcal bacteriophages containing the gene for streptococcal erythrogenic toxin (speA). Mol Gen Genet 231:161–168

Zabriskie J (1964) The role of temperate bacteriophage in the production of erythrogenic toxin by Group A *Streptococci*. J Exp Med 119:761–780

Zhao L, Dong YH, Wang H (2010) Residues of veterinary antibiotics in manures from feedlot livestock in eight provinces of China. Sci Total Environ 408(5):1069–1075. https://doi.org/10.1016/j.scitotenv.2009.11.014

Zhou Y, Sugiyama H, Johnson E (1993) Transfer of neurotoxigenicity from *Clostridium butyricum* to a nontoxigenic *Clostridium botulinum* type E-like strain. Appl Environ Microbiol 59 (11):3825–3831

Chapter 12
Clonally Evolving Pathogenic Bacteria

Sofia Hauck and Martin C. J. Maiden

1 Clonality in Bacteria

A population is a group of organisms that occupy the same niche and that share an evolutionary path, adapting to their environment through the process of natural selection. Members of a population are closely related and united by a recent common ancestor or ancestral population. Despite these similarities, bacterial populations are rarely composed of single clones that are genetically identical. Indeed, genetic variation within a population is a necessary condition for evolution by means of natural selection to occur. A population of identical clones cannot adapt, as all of its members have genomes that are equally suited to any set of conditions (Nielsen 2005).

Natural selection results in changes in the frequency of genotypes according to their fitness, as higher rates of reproduction and survival increase the abundance of the fittest genotypes within the population. Likewise, lower rates of success reduce the frequency of the least fit genotypes, sometimes to the point of elimination. This process therefore reduces the genetic diversity of the population over time, as it can only maintain or decrease diversity. Diversity is not completely reduced by natural selection because it is balanced by the introduction of new variants through mutation and recombination. All genetic variation ultimately arises from mutations, but sexual processes can introduce mutations from other populations via horizontal gene

S. Hauck · M. C. J. Maiden (✉)
Department of Zoology, University of Oxford, Oxford, UK
e-mail: martin.maiden@zoo.ox.ac.uk

P. H. Rampelotto (ed.), *Molecular Mechanisms of Microbial Evolution*, Grand Challenges in Biology and Biotechnology,
https://doi.org/10.1007/978-3-319-69078-0_12

transfer (HGT) as well as causing the reassortment of genes within a population (Feil and Spratt 2001).

Sexual recombination in bacteria differs from that of plants and animals in both frequency and range of source genetic material. Bacterial HGT occurs widely and promiscuously, with genetic material sourced throughout the domain, rather than exclusively from other members of the same species (Cohan 2002). Unlike most plants and animals, which recombine as a matter of course during each reproductive cycle, HGT in bacteria occurs at rates and extents that vary widely by species and by ecological situation (Hanage et al. 2006a). Across the bacterial domain, a wide spectrum of recombination rates has been observed, from very high rates that erase most clonal signal to complete clonality (Spratt and Maiden 1999). In a fundamentally non-clonal population, genes are the evolutionary units, as recombination is so frequent that it overwhelms the non-random associations among genetic variants in a genome, collectively named "linkage disequilibrium." A recombination rate that is ten times higher than the mutation rate may be enough for the linkage caused by shared descent to break down entirely (Cohan 1994). When this is the case, each variant of a particular gene can exist in a range of genomic backgrounds and therefore evolve independently. This is not an hypothetical extreme as it occurs in *Helicobacter pylori*, a human-specific pathogen that causes chronic gastric infections (Suerbaum et al. 1998), which has been found to exchange up to a tenth of its genome in one multi-strain infection over a period of 4 years (Cao et al. 2015).

At the opposite end of the spectrum is full clonality, which is a consequence of limited or no HGT (Tibayrenc and Ayala 2002). HGT can occur occasionally without affecting a clonal population structure, as long as it is infrequent enough to remain beneath the "clonality threshold." This is the point at which recombination is too rare to act as a homogenising force within the population and prevent the divergence of distinct branches (Tibayrenc and Ayala 2015). In the extreme and unusual case of a fully clonal population, where no recombination can be observed, genes cannot be exchanged among individuals and therefore each genome evolves independently. The phylogeny of a population in these conditions fits the "Russian doll" model, where, with increasing resolution, each branch of the species phylogeny is subdivided into smaller distinct and diverging branches (Tibayrenc and Ayala 2016).

2 The Clonal Paradigm

This article will focus on fully clonal populations, where HGT is entirely absent or sufficiently rare that it has essentially no impact on the evolution, population structure, and ecology of the organism in question during the timeframe under consideration (Shapiro 2016). The first major consequence of limited or no HGT is the inability to repair mutations that arise from time to time. Bacteria can use a variety of mechanisms for repairing DNA damage, including nucleotide or base excision repair and mismatch repair. These mechanisms can lower the rate of mutation by reversing damage as it happens, but without HGT, bacteria cannot repair mutations that have been copied to both stands of DNA (van der Veen and Tang 2015). Two other major consequences follow from a lack of HGT: the absence of a mechanism whereby different mutations can be combined to make novel variants and the inability to acquire genetic material from other organisms. All three of these affect the diversity of a population. In the absence of HGT, a population is both limited in its ability to generate new genetic diversity and to maintain existing diversity, as it is frequently subject to purges (Namouchi et al. 2012) by means of bottlenecks when population sizes are small or genome-wide sweeps when selection is strong (Smith et al. 2006). This lack of diversity has consequences for the adaptability and long-term viability of a population.

Populations that do not participate in HGT typically exhibit very low genetic diversity, although a lack of diversity is also a feature of relatively young populations, as over time mutations will accumulate even in the absence of HGT. The term "clonal" can describe both of these traits, the genealogical concept of reproduction through clonal descent and the genetic concept of identical clones forming all members of a population. This state of low diversity within the population is referred to as genetic monomorphism. The confusion between these definitions is particularly pronounced because these two traits often, but not necessarily, co-exist in bacteria (Tibayrenc and Ayala 2012).

From the perspective of evolutionary and molecular epidemiological studies, the low diversity of monomorphic bacteria can present problems for the identification of epidemiologically relevant variants for practical purposes. When typing resolution is not sufficient to distinguish similar isolates, the crucial task of tracking pathogen transmission becomes more challenging (Kohl et al. 2014). Improvements in the speed, cost, and accuracy of whole-genome sequencing (WGS) made in the 10 years prior to writing this article ensured that these bacteria could be examined at an unprecedented level of detail, covering virtually all of the existing genetic variation. Consequently, fully clonal, monomorphic pathogen species have been some of the first to be analysed in-depth by this technology, precisely because their low diversity simplifies large-scale genomics studies (Achtman 2008). This research has made it possible to discover the common traits of fully clonal bacteria.

Most known examples of highly clonal pathogens are lineages nested within a more diverse phylogeny of nonpathogenic organisms (Achtman 2008). An example of this is the agent of tuberculosis, *Mycobacterium tuberculosis*, which is perhaps the best studied example of a fully clonal bacterium. It is one of nine lineages within the *Mycobacterium tuberculosis* complex, each of which is host-specific, though some spillover in hosts does exist (Ghodbane et al. 2014). The complex as a whole is considered to be one species, due to its high degree of relatedness (99% or more average nucleotide identity), but individual species names are still used to refer to each subspecies, or ecotype, within it (Djelouadji et al. 2011). There are a number of parallel examples elsewhere in the domain including the following: *Yersinia pestis*, the agent of plague, which is best considered a specific clone of *Yersinia pseudotuberculosis* as it is nested within this more diverse species (Rasmussen et al. 2015);

Bordetella pertussis, which causes whooping cough, is a monophyletic branch of the *B. bronchiseptica* species (Diavatopoulos et al. 2005); and *Bacillus anthracis*, responsible for anthrax, is an equivalent branch of the *B. cereus* phylogeny (Okinaka et al. 2006). Not all clonal populations are given species names, as can be observed with *Salmonella enterica* subspecies *enterica* serovar Typhi, the agent of typhoid fever (Didelot et al. 2011).

In the absence of wholly sexual reproduction, which is used to define animal species, and the widespread occurrence of HGT in bacteria from a variety of sources, precise definitions of bacterial species remain difficult, contingent, and to a degree controversial (Achtman and Wagner 2008). For pathogenic bacteria, the importance of accurately circumscribing disease-causing microbes has led to pathogenicity being used as the defining trait of a species. However, pathogenicity, like many other phenotypic traits, correlates poorly with genotype, as is becoming increasing clear as genotyping methods improve. A species name that may not be "accurate," that is, not defining a distinct monophyletic cluster of genotypic and phenotypic traits, may nonetheless still be used in cases where distinguishing dangerous strains from harmless genetic relatives can be a life-or-death concern. This concept of a *nomen periculosum* is one example of the many complications in the field of bacterial taxonomy (Stackebrandt et al. 2002).

The complex nature of phenotypic and genotypic clusters in bacteria, both in defining clear clusters by either set of criteria or in correlating genotype to phenotype, has meant that a universal bacterial species concept is yet to be achieved. While the search for a unifying definition continues, "species" remains a term without one particular meaning, largely used as a label of convenience (Bapteste and Boucher 2009). Lineages are given names "dependent on the level of divergence thought appropriate" and only "where this serves a useful purpose" (Lan and Reeves, 2001). Clonal populations present particular problems in defining species clusters, as the absence of recombination means that each lineage is "irreversibly set" on divergence from other members of its species (Hanage et al. 2006b). For this reason, "species," "clone," "serovar," or any other terms for a specific phylogenetic group of bacteria are used here as per the convention for that species.

Despite HGT not being a mechanism in the maintenance of clonal lineages, it is often key to the evolutionary events that have led to them becoming distinct from their parental populations. HGT events can be catalysts for the expansion of a population to a new niche and form critical junctions in the evolutionary path of a species (Wiedenbeck and Cohan 2011). A clear example of this has been observed in the history of *Bacillus anthracis*, the agent of anthrax. It diverged from *Bacillus cereus* after acquiring the pXO1 and pXO2 plasmids, which includes the anthrax toxin genes implicated in the high fatality rate of the disease (Kolstø et al. 2009). After this acquisition, which occurred perhaps 20,000 years ago, *B. anthracis*'s unique pathobiology has caused it to specialise into a novel ecological niche (Keim et al. 2009). The evolutionary history of *S. enterica* serovar Typhi displays a similar pattern. *S. enterica* serovar Typhi lost the promiscuous HGT that is characteristic of the *S. enterica* species, changed niche as it transformed from a gastroenteric infection to an invasive one and underwent functional gene loss and genome degradation in the process (Holt et al. 2008).

Irreversible specialisation is the only possibility for fully clonal populations, due to the loss of adaptability that comes with restricted HGT (Moran and Wernegreen 2000). Parasitism itself is a form of specialisation, and the extreme host specificity and obligate pathogenicity evident in many fully clonal species is an even more extreme version of this specialization (Shapiro 2016). While the possibility of a discovery bias caused by the overrepresentation of pathogens in bacterial research cannot be discounted, it appears that all persistent fully clonal bacteria are pathogenic (Achtman 2012).

Many of the most deadly bacterial pathogens are species with low genetic diversity that fit the fully clonal paradigm (Namouchi et al. 2012). For example, the etiological agent of plague, *Y. pestis*, was responsible for the Black Death pandemic that killed nearly a third of Europeans in the fourteenth century, and this organism remains the cause of small but often fatal outbreaks in the modern era (Gage and Kosoy 2005). Meanwhile, *M. tuberculosis* is the most deadly bacterial pathogen in the world, infecting 100 million new patients and causing 1.5 million deaths a year, and appears to have coevolved with humans (Niemann et al. 2016). *S. enterica* serovar Typhi causes 200,000 deaths a year and may have been the cause of the plague of Athens in 430 B.C. (Galan 2016).

These species are genetically monomorphic despite their wide geographic range and long history as human pathogens, stretching over tens of thousands of years. While this is not substantial stretch of time in evolutionary terms, it is enough for analyses of mutation rate to be made across species and for phylogeographic signals to emerge within species (Comas et al. 2013). A comparison between modern *Y. pestis* strains and DNA recovered from a Black Death victim showed that few genetic changes occurred in the interceding 660 years, with only 10 cases of altered gene order and only 97 chromosomal sites and 6 plasmid single-nucleotide sites differing from modern strains (Bos et al. 2011). For *M. tuberculosis*, whole-genome alignments find that even the most distant strains that can be isolated from humans differ at most in 1800 single nucleotide polymorphisms or SNPs (Coscolla and Gagneux 2014). In contrast, over 15,000 SNP differences have been found between isolates of *Staphylococcus aureus* (Planet et al. 2017), and *Campylobacter jejuni* isolates can differ by 13,000 SNPs even within one outbreak (Moffatt et al. 2016).

Lifestyle stages with slow or no growth, common to many clonal pathogenic bacteria, may further exacerbate their low genetic diversity, by prolonging generation times and slowing evolutionary rates (Ohta 2011). *M. tuberculosis* can enter a granuloma phase in infected individuals, resulting in essentially asymptomatic carriage (Ford et al. 2011). *Bacillus anthracis* can survive as a spore in dry soil for decades, potentially centuries, in which time it does not replicate and therefore does not accumulate mutations (Rasko et al. 2011). This does not seem to be a necessary condition, however, as *Bordetella pertussis* is an obligate human pathogen with no chronic carrier state (Park et al. 2012). The source-sink evolutionary dynamics caused by carrier stages in their lifestyles may also reduce variability by stabilising the genome, as adaptations that are advantageous in the infectious phase confer no fitness benefits for invasive disease. This may be the case in *Yersinia pestis*, which constantly circulates in rodent populations between its infrequent outbreaks in humans (Ayyadurai et al. 2008) and in *S. enterica* serovar Typhi, whose reservoir is human asymptomatic carriers that may shed the bacteria for decades over their lifetimes (Holt et al. 2008).

3 Limiting the Rate of Recombination

To date, pathogens with restricted host ranges are among the best studied clonally evolving populations. A host represents a highly controlled and relatively uniform environment, and the restriction to a given host range deepens the extreme level of specialization these populations have undergone. As specialization increases, newly introduced genes are more likely to disturb existing favorable interactions in the genome, an effect named recombinational load (Michod et al. 2008). Because of this, it is advantageous for clonally evolving bacteria to actively suppress their rate of HGT so as to reduce this load. The realised rate is at a maximum equivalent to the potential rate, which is determined by the available mechanisms for HGT, but which in practice is restricted by ecological conditions of the recipient bacteria (Yahara et al. 2016).

DNA can be transferred between bacteria by three mechanisms, all of which transfer a small amount of DNA from a donor to a recipient: transduction, conjugation, and transformation. Briefly, transduction involves infectious virions (most often bacteriophages) as vectors, while in conjugation the vectors are self-replicating plasmids or other mobile genetic elements. Bacteria can protect themselves against these parasitic elements by means of endonucleases, enzymes that cleave foreign DNA. With these enzymes, invading DNA sequences can be destroyed before or after they recombine with the bacterial chromosome and consequently prevent cases of successful HGT from being passed on to subsequent generations (Cohan 2002). Restriction endonucleases recognise and cleave DNA at specific "target" sequences, while the more recently discovered CRISPR-Cas system detects the clustered regularly interspaced short palindromic repeats (CRISPR) after which it is named. The high specificity of the targets in both restriction and CRISPR systems is key to their ability to protect bacteria from "foreign" DNA without damaging their own genomes.

In contrast, transformation depends on a recipient with functional and active competency machinery to complete the uptake of DNA from the environment. This requires a set of highly specialised loci that are often spread throughout the genome and which are conserved across a species (Croucher et al. 2016). Because expressing this machinery carries a high cost, competence is typically regulated so that it is transiently expressed only under specific circumstances. When the loss of one of the genes occurs, relaxed selection on the others ensures that they are quickly lost. These losses often signal a point of no return for the lineages in which they occur, resulting in their extinction, albeit in the long-term perspective of hundreds of millions of years (Redfield et al. 2006).

Even in the presence of the possibility of HGT, for any of these mechanisms to succeed at facilitating transfer between two bacteria, the donor and recipient must be in close proximity to one another, so that the DNA may be passed from one to the other. Spatial isolation may therefore be a barrier to gene flow (Polz et al. 2013) and is the basis for the "starving sex hypothesis" that posits clonality as a passive process, occurring due to the absence of opportunity for mating (Tibayrenc and Ayala 2016). This may in fact be the case in many pathogenic bacteria, for which transmission events often mean only a small founding population infects each new host (Bergstrom et al. 1999).

The limitation of small founding populations can prevent the impact of recombination in yet another way: if the genetic distance between host and recipient is too small, the exchange of DNA is inconsequential and indeed unobservable. After DNA is exchanged, homologous recombination is a necessary step for genetic material to be incorporated into the genome and subsequently heritable. This requires that the donor DNA be paired with a chromosomal sequence with which it has some sequence identity, at least on its flanking sequences. If the donor sequence and host sequence are identical, homologous recombination may indeed be occurring, but the like-for-like exchange leaves no trace in a case of "invisible sex" (Tibayrenc and Ayala 2015). If both donor and recipient bacteria are descendants of a recent transmission event, they may be too closely related to recombine in ways that can be observed. This may in fact be the most common form of HGT within bacteria, a mechanism that is advantageous due to its utility in DNA repair (Feil and Spratt 2001).

With the high resolution possible by WGS analyses, data from multiple samples of a single patient have shown that multi-clonal infections are more common than previously thought (Votintseva et al. 2014). For *M. tuberculosis*, DNA recovered from human remains indicates that multiple strain infections were commonplace in eighteenth century Europe (Kay et al. 2015). It is also likely that any colonising bacteria are not alone in their niche, as every area of the human body has an associated microbiome (Huttenhower et al. 2012); however, physical proximity to distantly related bacteria may not lead to HGT as the frequency of transformation is reduced exponentially as sequence divergence increases. Therefore, the rate of successful HGT is reduced as the genetic distance between a host and a recipient increases and the homology between their genome sequences weakens (Cohan 1994). Genetic distance between bacteria can also lower the efficacy of plasmids and bacteriophages as vectors of HGT as their host ranges are limited to similar bacteria (Cohan 2002).

In this way, recombination may not occur, even when potential mechanisms are present and the donor and recipient bacteria are physically close, if the bacteria are either too distantly related (no HGT will occur) or too closely related (HGT is non-observable). HGT is a major force in the evolution of the bacteria domain as a whole, with estimates that up to 20% of genes across the domain have been recently mobilised by this method. Bacteria rapidly adapt to changes in their environment using this method for gene flow to swap the genes in their accessory genome. In a fully clonal population, HGT is not available as a mechanism for increasing genetic diversity within a species, and so mutation remains as the only option (Feil and Spratt 2001).

4 Mutation and the Evolution of Clonal Pathogens

Mutations arise from DNA damage or copying errors and are therefore random with respect to gene function. Some gene features, such as homopolymeric tracts, may locally raise the rate of mutation and therefore be found primarily in genes where this effect is beneficial, but even mutations such as this occur randomly (Orsi et al. 2010). According to Ohta's nearly neutral theory model, the majority of mutations have little or no effect on phenotype, with a small selective coefficient if any, and out of these, most are slightly deleterious. A minority of mutations will be fatal, and an even smaller number will confer a strong positive effect on fitness (Ohta 1992). Obligate pathogens, by the nature of the specialization inherent in their ecology, can be particularly likely to incur fitness costs from most mutations (Achtman and Wagner 2008). The consequence of this is that point mutations, by and large, generate mutations that negatively impact the fitness of the bacterium in which they arise.

Hyper-mutating bacteria, those that are defective in DNA proof-reading mechanisms, can develop new traits, such as antibiotic resistance, at faster rates, but this comes at the cost of their overall fitness of the organism as a whole (Woodford and Ellington 2007). This is because deleterious mutations cannot be easily removed from the genome in the absence of HGT. In strictly clonal organisms, with no HGT, linkage disequilibrium is so strong that the entire genome is the evolutionary unit and natural selection can only remove a deleterious mutation through a strong purifying selective pressure that causes a genome-wide sweep. By this mechanism, however, other deleterious mutations can become fixed in a population by hitchhiking with beneficial mutations during selective sweeps, as the sudden decrease in population size causes their mild negative effect to be counteracted by the overwhelmingly positive effect acting upon the allele selected for by the sweep (Fay and Wu 2000). Population bottlenecks, like sweeps, can remove or fix mutations; however, in this case, due to the small effective population size, it is random genetic drift, rather than natural selection, which is the predominant force. Consequently, such bottlenecks often fix deleterious mutations that actually reduce fitness.

Distinguishing between sweeps and bottlenecks, especially in retrospect, can be difficult, due to their similar effects on population structure, in that they both cause a sudden and severe reduction in the effective population size (Smith et al. 2006). The survival of a particular variant through a bottleneck may not be entirely decided by random genetic drift when more virulent genotypes lead to higher bacterial load within the host and a consequently higher likelihood of transmission to a new host.

This may explain the bias of clonal lineages to high disease-causing capacity and greater severity or virulence (Coscolla and Gagneux 2014).

Genome-wide sweeps bear the cost of high genetic load (Nielsen 2005), which can be mitigated if the effective population size is large, or by increasing the rate of HGT, which weakens linkage disequilibrium (Parkhill et al. 2003). As neither of these compensatory options is possible in fully clonal pathogens, purifying selection pressure is relaxed and ineffective in removing deleterious mutations. While strongly deleterious mutations, such as those that cause cell death, are removed immediately from a population, the smaller the negative impact on fitness that a mutation has, the slower will be the speed at which it is removed from the population. This time lag is especially marked when the pathogen population is expanding to a new niche, as is the case in a recently emerged or emerging pathogen species. The rate of non-synonymous mutations remains constant relative to the time of divergence while the synonymous mutation rate decreases (Rocha et al. 2006).

The small effective population sizes of specialised pathogens leave them particularly vulnerable to random genetic drift. Transmission events are in effect population bottlenecks, sudden decreases in population size that lead to only a minority of the population surviving to reproduce further, purging diversity and barring variants from future generations. Mutations that increase in prevalence due to genetic drift are the basis for one of the most widely used methods of pathogen classification, multilocus sequence typing (MLST). This approach indexes housekeeping genes that are key to the survival of bacteria and form part of the core genome that is typically present in all members of a species. Their sequences can be investigated for neutral (or nearly neutral) mutations that reveal phylogenetic relationships (Maiden et al. 1998, 2013).

Deleterious mutations may be fixed by such mechanisms in a clonally evolving population, as a consequence of population bottlenecks or through hitchhiking when genome-wide selective sweeps occur. Combined with relaxed purifying selective pressure, deleterious mutations can accumulate in a process known as Muller's ratchet (Gordo and Charlesworth 2000). During population bottlenecks, small population sizes weaken purifying selection so that random drift becomes the predominant evolutionary force and deleterious mutations can become fixed by chance. As these mutations are effectively random and can cause genes to lose their function, purifying selective pressure on them is further relaxed, causing gene loss and genome degradation. This has been observed in the emergence of *B. pertussis*, an obligate human pathogen, from the generalist *B. bronchiseptica*, which has a much broader-host specificity and can infect a variety of mammals and birds (Gross et al. 2010). The novel host specificity of *B. pertussis* has been accompanied by and perhaps driven by gene loss. In this case, gene loss followed host restriction, most likely due to the removal of genes that are no longer needed for growth and survival in the environment between host infections (Cummings et al. 2004). This can give the bacteria a fitness advantage by streamlining its genome, removing parts which may have associated costs but no longer incur fitness benefits, but reduction in genome size may instead be a consequence of genome degradation due to Muller's rachet (Mooi 2010). In the case of *B. pertussis*, the reduction in genome size has been accompanied by a massive increase in insertion sequence copy number and a consequent loss of genome structure, which suggest that this has not simply been a streamlining process and that some element of degradative evolution has also occurred (Bart et al. 2014).

A similar situation is observed in *M. tuberculosis*, where gene loss is associated with increased virulence and transmission (Djelouadji et al. 2011). *Mycobacterium leprae*, the agent of leprosy, is an even more extreme example from within the same genus. It is a fully clonal organism, marked by gene decay so extreme that half the genes in its chromosome have become inactive since the bottleneck that formed the emergence of the species (Cole et al. 2001). Homologous recombination, through its origin as a tool for DNA repair, is also an effective way to excise parasitic DNA from the genome. Species that do not participate in HGT are therefore left vulnerable to insertion elements and other forms of parasitic DNA, which may accelerate the degradation of the genome (Croucher et al. 2016). The pattern forms a general syndrome of genome evolution, where host restriction is followed by pseudogene formation, gene loss, and increasing numbers of insertion elements (Moran and Plague 2004). This syndrome can even be observed in the parallel evolution of independent clonal lineages. *S. enterica* serovars Paratyphi A and Typhi shared genes by HGT in the recent past, but their subsequent evolution shows evidence of convergence. Both serovars have evolved to a greater degree of host restriction and to cause systemic disease through the loss of function in many of the same genes (Holt et al. 2009).

5 Examples of Adaptive Evolution in Clonal Pathogens

As we have seen, strictly clonal pathogens have limited means by which they can adapt to changes in their environment, being constrained in their ability to generate new variants and to reassort these variants by HGT, even within the same species. Nevertheless, when sufficiently strong, selective pressures can impact the evolution of these pathogens and lead to adaptation based on single mutations that arise in the population. Most often this is observed following positive periodic selection, where a strong selective pressure, following, for example, a change to the environment, causes a genome-wide sweep that removes any isolates that do not have the allele or alleles most advantageous to the new environment (Shapiro 2016). In pathogenic bacteria, such events are typically associated with changes caused by medical interventions including the introduction of specific treatments.

B. pertussis has been a significant cause of whooping cough morbidity and mortality in children worldwide, but while the implementation of effective vaccines successfully reduced the disease burden, the bacterium may still be circulating in the population as a milder infection in adults (Guiso 2014). Whole-cell vaccines for *B. pertussis* are highly effective but often cause mild side effects and occasionally severe ones. For this reason they were replaced by acellular vaccines in the late 1980s and early 1990s, after which a worldwide selective sweep was observed in the circulating *B. pertussis* strains. The previously dominant *ptxP1* organisms were replaced with strains carrying the *ptxP3* allele, which is associated with an increase in the expression of pertussis toxin. The resulting change in population composition caused phenotypic consequences beyond the pertussis toxin genes, as other mutations hitchhiked with the *ptxP3* variant (de Gouw et al. 2014).

The limited amount of genetic diversity present in clonally evolving populations may belie their adaptive potential. The few existing points of variability in a genetically monomorphic population can have a disproportionally high impact on their biology (Reiling et al. 2013). In *M. tuberculosis*, whole-genome sequencing has revealed that the species can adapt when faced with sufficiently strong selection pressure, as is the case with the emergence of antibiotic resistance in the face of treatment. The difficulty in achieving therapeutic doses of antibiotics in the intracellular environment of *M. tuberculosis* compounds the problem of resistance in this species, as therapy with a single antibiotic is rarely effective and the use of combination therapies accelerates the emergence of multidrug-resistant strains (Moreno-Gamez et al. 2015). Despite the need for each resistance-conferring mutation to occur sequentially within each lineage, resistance is commonplace, and multidrug resistance is a growing concern for tuberculosis treatment worldwide (Hershberg et al. 2008).

Hypermutability can enhance the ability of clonal linages to adapt rapidly and may therefore be actively selected in clonal pathogen populations. This appears to have been the case for the W-Beijing lineage of *M. tuberculosis*, a genotype which is associated with high virulence, in terms of the rate of activation of tuberculosis disease in patients (Merker et al. 2015). This phenotype enhances the spread of tuberculosis in large dense urbanised human populations but would be less advantageous in less dense human populations where the chronic latent stage, rather than the acute symptomatic stage, would be favoured. The mutations promoting virulence in the W-Beijing lineage have probably played a role in the pandemic spread of this lineage during the twentieth century (Hanekom et al. 2011). The genetic basis for this increased pathogenicity is variable throughout the lineage, with the common factor appearing to be the relaxation of purifying selection on the genes that control replication and DNA repair, leading to a higher rate of mutations. This increase in genomic variability provides an evolutionary advantage under stressful conditions and can compensate for the loss of adaptability caused by the lack of horizontal gene transfer (Vultos Dos et al. 2008).

S. enterica serovar Typhi has also adapted to the pressure imposed by antibiotic use, with 15 mutations that confer fluoroquinolone resistance being observed at the *gyrA* gene in just a decade. This accumulation of adaptations as a consequence of strong positive selection is in stark contrast to the evidence of neutral evolution in the remainder of *S. typhi* genome: genetic drift is the predominant force throughout the genome, with no observed difference in the selective forces in genes associated with housekeeping functions or pathogenicity (Roumagnac et al. 2006).

6 Consequences of Clonality

As we have discussed, clonal population structures can lead to specialization and vice versa, leading to the evolution of a highly specialised bacterial clonal pathogen, which has lost much of its ancestors' metabolic capacity. The ecological and evolutionary consequences of the interaction between clonality and specialisation are a continuous cycle of small effective population sizes that relax purifying pressure and increase the dominance of genetic drift, causing deleterious mutations which may arise to accumulate in the genome. These lead to genome degradation and genetic isolation, which in turn reduces the genetic diversity of a population through sweeps and bottlenecks. Specialising to a specific niche, under these circumstances, can become the only possible evolutionary path that is available to the organism in question, however short-lived that path may ultimately be.

Sexual reproduction is costly for the organisms that engage in it. They must bear the cost of searching and competing for a mate, of producing male offspring, and of sharing only half their genes with any offspring. Why sexual reproduction is widely maintained despite these costs is an "evolutionary puzzle" (Lehtonen et al. 2012). In vertebrates, clonality is always a short-lived strategy that inevitably leads to extinction, though it can be rescued by even occasional recombination (Avise 2015). In bacteria, parasexual processes, mediated by gene transfer and recombination mechanisms, may have evolved primarily as a DNA repair mechanism, belying their far-reaching evolutionary consequences as a method of genetic exchange. The ability of HGT to generate diversity and allow genes, rather than entire genomes, to become the evolutionary units is critical for the long-term success of bacterial species. As with vertebrates, even occasional recombination seems to prevent the complementary processes of clonality and specialisation, by markedly improving the viability of populations over that of a strictly clonal lifestyle (Tibayrenc and Ayala 2012).

When considering the time scales over which clonality can emerge as a stable trait of a given highly specialised pathogen, it is necessary to "distinguish between short term emergence of clonal complexes and the more long-term evolutionary history of a bacterial population" (Feil and Spratt 2001). The concept of clonal lineages emerging from a context with a high frequency of recombination, due to small population sizes that lead to genetic isolation, predates genome sequencing (Levin 1981). Expanding upon this theory led to the epidemic clone model, which posits that even in populations with rampant recombination, successful clones will occasionally cause epidemic waves (Maynard Smith et al. 1993). In this model, a freely recombining population, comprising many variants that participate in HGT, occasionally gives rise to a clonal lineage that successfully emerges as a transmissible pathogen (Turner and Feil 2007).

Are the clonal lineages that survive over millennia fundamentally different from those that survive only until the next genotype in clonal replacement takes over? It is possible that the various pandemics of *Vibrio cholerae*, the cause of the diarrheal disease cholera, are an intermediate between short-lived epidemic clones and more long-lived clonal species such as *M. tuberculosis*. Sublineages that define each epidemic are observed over the space of a few decades but are replaced by successive waves of new sublineages. These lineages can also develop into hypermutators, again with the pattern of fast adaptive evolution to a specific pressure matched by a decrease in overall fitness (Didelot et al. 2015).

In *Salmonella enterica*, two time scales of evolution have been proposed, based on studies of the clonal *S. typhi* serovar. Evolution to antibiotic resistance occurs in clinically observable time, affecting transmission and infections, with a much slower process of genetic drift fixing neutral mutations gradually, resulting in "two distinct epidemiological dynamics" being present in one population structure (Roumagnac et al. 2006). This two-level evolution may explain, at least in part, the predominant nature of clonal evolution being one of stability over thousands of years and marked only by the slow accumulation of mostly neutral mutations but overlaid with the rapid adaptation in response to the strong selection pressures that are the consequence of antibiotic use or vaccine implementation.

Intriguingly, in another serovar of this species, *S. enterica* serovar Typhimurium, a novel clone appears to currently be emerging as a highly specialised pathogen in observable time. The ST313 sublineage of *S. typhimurium* causes invasive disease rather than the gastroenteritis typical of this serovar. Further, the disease that it causes is associated with much higher mortality, with up to half of adult cases resulting in death. The ST313 clone has, through a HGT event, gained a virulence-associated plasmid that bears many antibiotic resistance genes, and this seems to have enabled a change in its ecology that has resulted in a shift in its pathology. The change to an invasive disease niche has subsequently led to genome degradation, including loss of many of the same genes that are now absent from *S. typhi* (Kingsley et al. 2009), in an apparent example of convergent evolution.

Strictly clonal, obligate pathogens of one or a few host species can be seen as a quirk of evolution. They have emerged as specialised clones of large, more adaptable, and diverse populations, but their invasion of this novel niche, with its consequent process of specialization, comes with the associated costs of loss of adaptability and perhaps leading to a process of degradative evolution that inevitably leads to extinction, even if the host species survives. This process of extinction may take very many host generations and, for a number of human pathogens, this seems to have been ongoing for many millennia, sufficient for global distribution of these seemingly evolutionarily doomed pathogens. Given the high virulence that these pathogens can attain, a fuller understanding of the evolutionary path taken by these bacteria will aid us in containing the long-standing threats that they pose and that may be posed by novel pathogens that take the same evolutionary path as a consequence of changes in human ecology. In this respect, it is worth noting as a final comment, that our own species is a relative newcomer in terms of worldwide distribution and ecological success (indeed we are unique among the family Hominidae in this respect), and it is therefore unsurprising that the obligate clonal pathogens of humans are also of recent date, with some of them, a notable example being *M. tuberculosis* evolving alongside *Homo sapiens*, with the fate of both species inexorably linked (Soares et al. 2012) (Fig. 12.1).

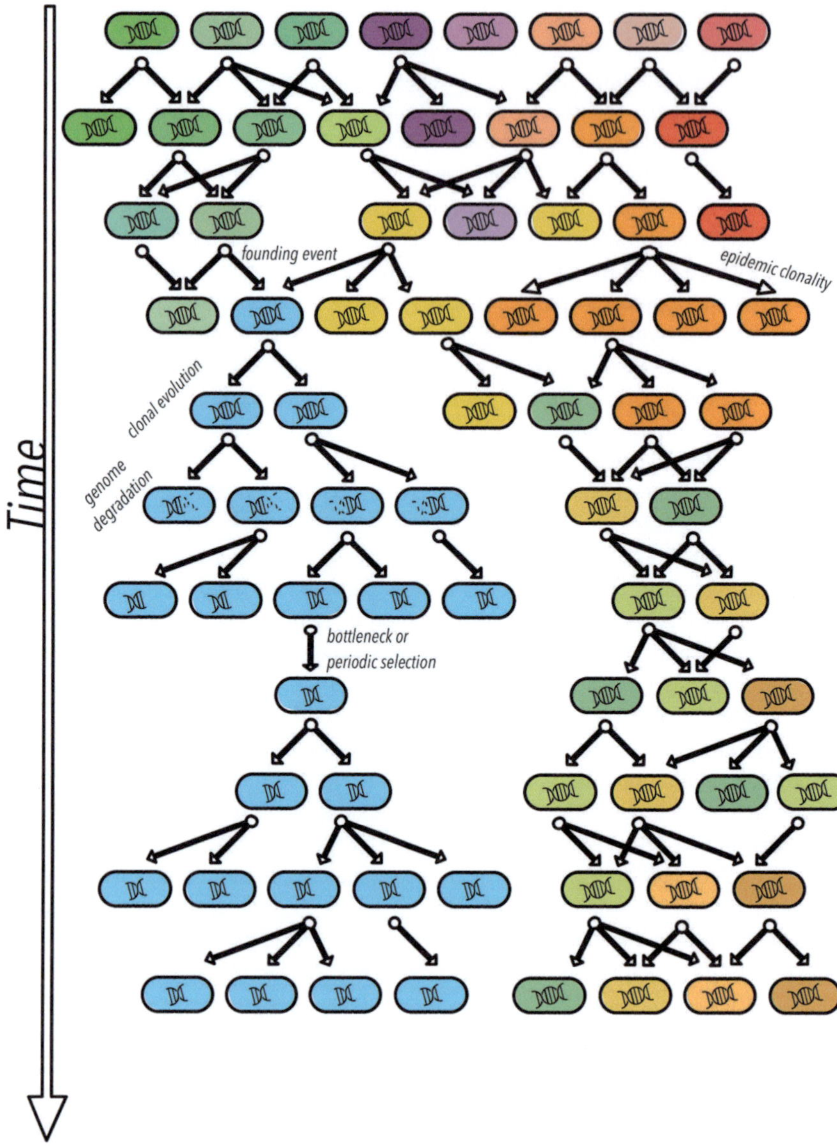

Fig. 12.1 Illustration of bacterial reproduction and evolution, including epidemic clonality and fully clonal evolution. Each oval represents one bacterium and its genome, and each row represents one sampling time period, so that the bacteria in a row are descendants of those in the row above, separated by many generations. At first the population is evolving as is typical of most bacteria, with a mix of clonal descent and horizontal gene transfer. Therefore most bacteria contain genetic material from more than one ancestor, though not all bacteria contribute genetically to further generations. Between the third and fourth time points, one bacterium is particularly successful and rapidly increases in abundance, in an example of *epidemic clonality*. This is a temporary trait, however, and by the fifth time point onward, the evolutionary processes occurring in its descendants have returned to the patterns typical of the ancestral population. A different bacterium within the

References

Achtman M (2008) Evolution, population structure, and phylogeography of genetically monomorphic bacterial pathogens. Annu Rev Microbiol 62:53–70. https://doi.org/10.1146/annurev.micro.62.081307.162832
Achtman M (2012) Insights from genomic comparisons of genetically monomorphic bacterial pathogens. 367:860–867. https://doi.org/10.1098/rstb.2011.0303
Achtman M, Wagner M (2008) Microbial diversity and the genetic nature of microbial species. Nat Rev Microbiol 6:431–440. https://doi.org/10.1038/nrmicro1872
Avise JC (2015) Evolutionary perspectives on clonal reproduction in vertebrate animals. Proc Natl Acad Sci U S A 112:8867–8873. https://doi.org/10.1073/pnas.1501820112
Ayyadurai S, Houhamdi L, Lepidi H et al (2008) Long-term persistence of virulent *Yersinia pestis* in soil. Microbiol Immunol 154:2865–2871. https://doi.org/10.1099/mic.0.2007/016154-0
Bapteste E, Boucher Y (2009) Epistemological impacts of horizontal gene transfer on classification in microbiology. Methods Mol Biol 532:55–72. https://doi.org/10.1007/978-1-60327-853-9_4
Bart MJ, Harris SR, Advani A et al (2014) Global population structure and evolution of *Bordetella pertussis* and their relationship with vaccination. MBio 5:e01074–e01074–14. https://doi.org/10.1128/mBio.01074-14
Bergstrom CT, McElhany P, Real LA (1999) Transmission bottlenecks as determinants of virulence in rapidly evolving pathogens. Proc Natl Acad Sci U S A 96:5095–5100
Bos KI, Schuenemann VJ, Golding GB et al (2011) A draft genome of *Yersinia pestis* from victims of the Black Death. Nature 478:506–510. https://doi.org/10.1038/nature10549
Cao Q, Didelot X, Wu Z et al (2015) Progressive genomic convergence of two *Helicobacter pylori* strains during mixed infection of a patient with chronic gastritis. Gut 64:554–561. https://doi.org/10.1136/gutjnl-2014-307345
Cohan FM (1994) Genetic exchange and evolutionary divergence in prokaryotes. Trends Ecol Evol 9:175–180. https://doi.org/10.1016/0169-5347(94)90081-7
Cohan FM (2002) Sexual isolation and speciation in bacteria. Genetica 116:359–370. https://doi.org/10.1023/A:1021232409545
Cole ST, Eiglmeier K, Parkhill J et al (2001) Massive gene decay in the leprosy bacillus. Nature 409:1007–1011. https://doi.org/10.1038/35059006
Comas I, Coscolla M, Luo T et al (2013) Out-of-Africa migration and Neolithic coexpansion of *Mycobacterium tuberculosis* with modern humans. Nat Genet 45:1176–1182. https://doi.org/10.1038/ng.2744
Coscolla M, Gagneux S (2014) Consequences of genomic diversity in *Mycobacterium tuberculosis*. Semin Immunol 26:431–444. https://doi.org/10.1016/j.smim.2014.09.012
Croucher NJ, Mostowy R, Wymant C et al (2016) Horizontal DNA transfer mechanisms of bacteria as weapons of intragenomic conflict. PLoS Biol 14:e1002394. https://doi.org/10.1371/journal.pbio.1002394
Cummings CA, Brinig MM, Lepp PW (2004) Bordetella species are distinguished by patterns of substantial gene loss and host adaptation. J Bacteriol. https://doi.org/10.1128/JB.186.5.1484

Fig. 12.1 (continued) population becomes reproductively isolated in the fourth time point, after an HGT event that led to its emergence (the *founding event*). The descendants of this bacterium undergo *clonal evolution*, with no HGT observable anywhere in the lineage, so that every bacterium is descended from solely one bacterium in a previous time point. This lineage undergoes *genome degradation* leading to gene loss and eventually to a smaller genome. A *bottleneck or periodic selection* event dramatically reduces the population size of the clonally evolving lineage between the seventh and eighth time points, reducing the diversity of the population and fixing the gene loss in the genome

de Gouw D, Diavatopoulos D, Heuvelman K et al (2014) Differentially expressed genes in *Bordetella pertussis* strains belonging to a lineage which recently spread globally. PLoS One 9:e84523. https://doi.org/10.1371/journal.pone.0084523

Diavatopoulos DA, Cummings CA, Schouls LM, Brinig MM, Relman DA, Mooi FR (2005) *Bordetella pertussis*, the causative agent of whooping cough, evolved from a distinct, human-associated lineage of *B. bronchiseptica*. PLoS Pathog 1(4):e45

Didelot X, Bowden R, Street T et al (2011) Recombination and population structure in *Salmonella enterica*. PLoS Genet 7:e1002191–e1002191. https://doi.org/10.1371/journal.pgen.1002191

Didelot X, Pang B, Zhou Z et al (2015) The role of China in the global spread of the current cholera pandemic. PLoS Genet 11:e1005072. https://doi.org/10.1371/journal.pgen.1005072

Djelouadji Z, Raoult D, Drancourt M (2011) Palaeogenomics of *Mycobacterium tuberculosis*: epidemic bursts with a degrading genome. Lancet Infect Dis 11:641–650. https://doi.org/10.1016/S1473-3099(11)70093-7

Fay JC, Wu CI (2000) Hitchhiking under positive Darwinian selection. Genetics 155:1405–1413

Feil EJ, Spratt BG (2001) Recombination and the population structures of bacterial pathogens. Annu Rev Microbiol 55:561–590. https://doi.org/10.1146/annurev.micro.55.1.561

Ford CB, Lin PL, Chase MR et al (2011) Use of whole genome sequencing to estimate the mutation rate of *Mycobacterium tuberculosis* during latent infection. Nat Genet 43:482–486. https://doi.org/10.1038/ng.811

Gage KL, Kosoy MY (2005) Natural history of plague: perspectives from more than a century of research. Annu Rev Entomol 50:505–528. https://doi.org/10.1146/annurev.ento.50.071803.130337

Galan JE (2016) Typhoid toxin provides a window into typhoid fever and the biology of *Salmonella typhi*. Proc Natl Acad Sci U S A 113:6338–6344. https://doi.org/10.1073/pnas.1606335113

Ghodbane R, Mba Medie F, Lepidi H et al (2014) Long-term survival of tuberculosis complex mycobacteria in soil. Microbiol Rev 160:496–501. https://doi.org/10.1099/mic.0.073379-0

Gordo I, Charlesworth B (2000) The degeneration of asexual haploid populations and the speed of Muller's ratchet. Genetics 154:1379–1387. https://doi.org/10.1016/S0960-9822(02)00448-7

Gross R, Keidel K, Schmitt K (2010) Resemblance and divergence: the "new" members of the genus Bordetella. Med Microbiol Immunol 199:155–163. https://doi.org/10.1007/s00430-010-0148-z

Guiso N (2014) *Bordetella pertussis*: why is it still circulating? J Infect 68(Suppl 1):S119–S124. https://doi.org/10.1016/j.jinf.2013.09.022

Hanage WP, Fraser C, Spratt BG (2006a) The impact of homologous recombination on the generation of diversity in bacteria. J Theor Biol 239:210–219. https://doi.org/10.1016/j.jtbi.2005.08.035

Hanage WP, Fraser C, Spratt BG (2006b) Sequences, sequence clusters and bacterial species. Philos Trans R Soc B 361:1917–1927. https://doi.org/10.1098/rstb.2006.1917

Hanekom M, Gey Van Pittius NC, McEvoy C et al (2011) *Mycobacterium tuberculosis* Beijing genotype: a template for success. Tuberculosis 91:510–523. https://doi.org/10.1016/j.tube.2011.07.005

Hershberg R, Lipatov M, Small PM et al (2008) High functional diversity in *Mycobacterium tuberculosis* driven by genetic drift and human demography. PLoS Biol 6:e311. https://doi.org/10.1371/journal.pbio.0060311

Holt KE, Parkhill J, Mazzoni CJ et al (2008) High-throughput sequencing provides insights into genome variation and evolution in *Salmonella typhi*. Nat Genet 40:987–993. https://doi.org/10.1038/ng.195

Holt KE, Teo YY, Li H et al (2009) Detecting SNPs and estimating allele frequencies in clonal bacterial populations by sequencing pooled DNA. Vaccine 25:2074–2075. https://doi.org/10.1093/bioinformatics/btp344

Huttenhower C, Gevers D, Knight R et al (2012) Structure, function and diversity of the healthy human microbiome. Nature 486:207–214. https://doi.org/10.1038/nature11234

Kay GL, Sergeant MJ, Zhou Z et al (2015) Eighteenth-century genomes show that mixed infections were common at time of peak tuberculosis in Europe. Nat Commun 6:6717. https://doi.org/10.1038/ncomms7717

Keim P, Gruendike JM, Klevytska AM et al (2009) The genome and variation of *Bacillus anthracis*. Mol Aspects Med 30:397–405. https://doi.org/10.1016/j.mam.2009.08.005

Kingsley RA, Msefula CL, Thomson NR et al (2009) Epidemic multiple drug resistant *Salmonella typhimurium* causing invasive disease in sub-Saharan Africa have a distinct genotype. Genome Res 19:2279–2287. https://doi.org/10.1101/gr.091017.109

Kohl TA, Diel R, Harmsen D et al (2014) Whole genome based *Mycobacterium tuberculosis* surveillance: a standardized, portable and expandable approach. J Clin Microbiol 52:2479–2486. https://doi.org/10.1128/JCM.00567-14

Kolstø A-B, Tourasse NJ, Økstad OA (2009) What sets *Bacillus anthracis* apart from other Bacillus species? Annu Rev Microbiol 63:451–476. https://doi.org/10.1146/annurev.micro.091208.073255

Lan R, Reeves PR (2001) When does a clone deserve a name? A perspective on bacterial species based on population genetics. Trends Microbiol 9:419–424

Lehtonen J, Jennions MD, Kokko H (2012) The many costs of sex. Trends Ecol Evol 27:172–178. https://doi.org/10.1016/j.tree.2011.09.016

Levin BR (1981) Periodic selection, infectious gene exchange and the genetic structure of *E. coli* populations. Genetics 99:1–23

Maiden MCJ, Bygraves JA, Feil E et al (1998) Multilocus sequence typing: a portable approach to the identification of clones within populations of pathogenic microorganisms. Proc Natl Acad Sci U S A 95:3140–3145. https://doi.org/10.1073/pnas.95.6.3140

Maiden MCJ, van Rensburg MJJ, Bray JE et al (2013) MLST revisited: the gene-by-gene approach to bacterial genomics. Nat Rev Microbiol 11:728–736. https://doi.org/10.1038/nrmicro3093

Maynard Smith J, Smith NH, O'Rourke M, Spratt BG (1993) How clonal are bacteria? Proc Natl Acad Sci U S A 90:4384–4388. https://doi.org/10.1902/jop.2011.110063

Merker M, Blin C, Mona S et al (2015) Evolutionary history and global spread of the *Mycobacterium tuberculosis* Beijing lineage. Nat Genet 47:242–249. https://doi.org/10.1038/ng.3195

Michod RE, Bernstein H, Nedelcu AM (2008) Adaptive value of sex in microbial pathogens. Infect Genet Evol 8:267–285. https://doi.org/10.1016/j.meegid.2008.01.002

Moffatt CRM, Greig A, Valcanis M et al (2016) A large outbreak of *Campylobacter jejuni* infection in a university college caused by chicken liver pâté, Australia, 2013. Epidemiol Infect 144:2971–2978. https://doi.org/10.1017/S0950268816001187

Mooi F (2010) *Bordetella pertussis* and vaccination: the persistence of a genetically monomorphic pathogen. Infect Genet Evol 10:36–49. https://doi.org/10.1016/j.meegid.2009.10.007

Moran NA, Plague GR (2004) Genomic changes following host restriction in bacteria. Curr Opin Genet Dev 14:627–633. https://doi.org/10.1016/j.gde.2004.09.003

Moran NA, Wernegreen JJ (2000) Lifestyle evolution in symbiotic bacteria: insights from genomics. Trends Ecol Evol 15:321–326

Moreno-Gamez S, Hilla AL, Rosenbloom DIS et al (2015) Imperfect drug penetration leads to spatial monotherapy and rapid evolution of multidrug resistance. Proc Natl Acad Sci U S A 112: E2874–E2883. https://doi.org/10.1073/pnas.1424184112

Namouchi A, Didelot X, Schöck U et al (2012) After the bottleneck: Genome-wide diversification of the *Mycobacterium tuberculosis* complex by mutation, recombination, and natural selection. Genome Res 22:721–734. https://doi.org/10.1101/gr.129544.111

Nielsen R (2005) Molecular signatures of natural selection. Annu Rev Genet 39:197–218. https://doi.org/10.1146/annurev.genet.39.073003.112420

Niemann S, Merker M, Kohl T, Supply P (2016) Impact of genetic diversity on the biology of *Mycobacterium tuberculosis* complex strains. Microbiol Spectr. https://doi.org/10.1128/microbiolspec.TBTB2-0022-2016

Ohta T (1992) The nearly neutral theory of molecular evolution. Annu Rev Ecol Syst 23:263–286. https://doi.org/10.2307/2097289

Ohta T (2011) Near-neutrality, robustness, and epigenetics. Genome Biol Evol 3:1034–1038. https://doi.org/10.1093/gbe/evr012

Okinaka R, Pearson T, Keim P (2006) Anthrax, but not *Bacillus anthracis*? PLoS Pathog 2:e122

Orsi RH, Bowen BM, Wiedmann M (2010) Homopolymeric tracts represent a general regulatory mechanism in prokaryotes. BMC Genomics 11(1):102

Park J, Zhang Y, Buboltz AM et al (2012) Comparative genomics of the classical Bordetella subspecies: the evolution and exchange of virulence-associated diversity amongst closely related pathogens. BMC Genomics 13:545–545. https://doi.org/10.1186/1471-2164-13-545

Parkhill J, Sebaihia M, Preston A et al (2003) Comparative analysis of the genome sequences of *Bordetella pertussis*, *Bordetella parapertussis* and *Bordetella bronchiseptica*. Nat Genet 35:32–40. https://doi.org/10.1038/ng1227

Planet PJ, Narechania A, Chen L et al (2017) Architecture of a Species: phylogenomics of *Staphylococcus aureus*. Trends Microbiol 25:153–166. https://doi.org/10.1016/j.tim.2016.09.009

Polz MF, Alm EJ, Hanage WP (2013) Horizontal gene transfer and the evolution of bacterial and archaeal population structure. Trends Genet 29:170–175. https://doi.org/10.1016/j.tig.2012.12.006

Rasko DA, Worsham PL, Abshire TG et al (2011) *Bacillus anthracis* comparative genome analysis in support of the Amerithrax investigation. Proc Natl Acad Sci U S A 108:5027–5032. https://doi.org/10.1073/pnas.1016657108

Rasmussen S, Allentoft ME, Nielsen K et al (2015) Early divergent strains of *Yersinia pestis* in Eurasia 5,000 years ago. Cell 163:571–582. https://doi.org/10.1016/j.cell.2015.10.009

Redfield RJ, Findlay WA, Bosse J et al (2006) Evolution of competence and DNA uptake specificity in the Pasteurellaceae. BMC Evol Biol 6:82. https://doi.org/10.1186/1471-2148-6-82

Reiling N, Homolka S, Walter K et al (2013) Clade-specific virulence patterns of *Mycobacterium tuberculosis* complex strains in human primary macrophages and aerogenically infected mice. MBio 4:e00250–e00213. https://doi.org/10.1128/mBio.00250-13

Rocha E, Maynard Smith J, Hurst LD et al (2006) Comparisons of dN/dS are time dependent for closely related bacterial genomes. J Theor Biol 239:226–235. https://doi.org/10.1016/j.jtbi.2005.08.037

Roumagnac P, Weill F-X, Dolecek C et al (2006) Evolutionary history of *Salmonella typhi*. Science 314:1301–1304. https://doi.org/10.1126/science.1134933

Shapiro BJ (2016) How clonal are bacteria over time? Curr Opin Microbiol 31:116–123. https://doi.org/10.1016/j.mib.2016.03.013

Smith NH, Gordon SV, de la Rua-Domenech R et al (2006) Bottlenecks and broomsticks: the molecular evolution of *Mycobacterium bovis*. Nat Rev Microbiol 4:670–681. https://doi.org/10.1038/nrmicro1472

Soares P, Alshamali F, Pereira JB et al (2012) The expansion of mtDNA Haplogroup L3 within and out of Africa. Mol Biol Evol 29:915–927. https://doi.org/10.1093/molbev/msr245

Spratt BG, Maiden MCJ (1999) Bacterial population genetics, evolution and epidemiology. Philos Trans R Soc B 354:701–710. https://doi.org/10.1098/rstb.1999.0423

Stackebrandt E, Frederiksen W, Garrity GM et al (2002) Report of the ad hoc committee for the re-evaluation of the species definition in bacteriology. Int J Syst Evol Microbiol 52:1043–1047. https://doi.org/10.1099/00207713-52-3-1043

Suerbaum S, Smith JM, Bapumia K et al (1998) Free recombination within *Helicobacter pylori*. Proc Natl Acad Sci U S A 95:12619–12624

Tibayrenc M, Ayala FJ (2002) The clonal theory of parasitic protozoa: 12 years on. Trends Parasitol 18:405–410

Tibayrenc M, Ayala FJ (2012) Reproductive clonality of pathogens: a perspective on pathogenic viruses, bacteria, fungi, and parasitic protozoa. Proc Natl Acad Sci U S A 109:E3305–E3313. https://doi.org/10.1073/pnas.1212452109

Tibayrenc M, Ayala FJ (2015) How clonal are Neisseria species? The epidemic clonality model revisited. Proc Natl Acad Sci U S A 112:8909–8913. https://doi.org/10.1073/pnas.1502900112

Tibayrenc M, Ayala FJ (2016) Is predominant clonal evolution a common evolutionary adaptation to parasitism in pathogenic parasitic protozoa, fungi, bacteria, and viruses? Adv Parasitol 97:243–325

Turner KME, Feil EJ (2007) The secret life of the multilocus sequence type. Int J Antimicrob Agents 29:129–135

van der Veen S, Tang CM (2015) The BER necessities: the repair of DNA damage in human-adapted bacterial pathogens. Nat Rev Microbiol 13:83–94. https://doi.org/10.1038/nrmicro3391

Votintseva AA, Miller RR, Fung R et al (2014) Multiple-strain colonization in nasal carriers of *Staphylococcus aureus*. J Clin Microbiol 52:1192–1200. https://doi.org/10.1128/JCM.03254-13

Vultos Dos T, Mestre O, Rauzier J et al (2008) Evolution and diversity of clonal bacteria: the paradigm of *Mycobacterium tuberculosis*. PLoS One 3:e1538–e1538. https://doi.org/10.1371/journal.pone.0001538

Wiedenbeck J, Cohan FM (2011) Origins of bacterial diversity through horizontal genetic transfer and adaptation to new ecological niches. FEMS Microbiol Rev 35:957–976. https://doi.org/10.1111/j.1574-6976.2011.00292.x

Woodford N, Ellington MJ (2007) The emergence of antibiotic resistance by mutation. Clin Microbiol Infect 13:5–18. https://doi.org/10.1111/j.1469-0691.2006.01492.x

Yahara K, Didelot X, Jolley KA et al (2016) The landscape of realized homologous recombination in pathogenic bacteria. Mol Biol Evol 33:456–471. https://doi.org/10.1093/molbev/msv237

Chapter 13
A Case for the Evolution from Commensalism to Pathogenicity and Possibly Back Again: Lessons Learned from the Human-Adapted *Neisseria* Species

Lauren L. Priniski and H. Steven Seifert

1 The *Neisseria* spp.

In 1879, Albert Neisser observed the presence of Gram-negative diplococci within urethral exudates from people with gonorrhea (Liu et al. 2015). The organism was *Neisseria gonorrhoeae* (*Ng*), the sole causative agent of the sexually transmitted infection, gonorrhea. There are an estimated 106 million new cases of gonorrhea annually, worldwide (WHO 2012). In females, *Ng* typically colonizes the cervix and, if untreated, can ascend into the fallopian tubes and result in pelvic inflammatory disease with the sequelae of pain and infertility (Edwards and Butler 2011). In males, *Ng* normally colonizes the urethra and often results in painful urination. While the dogma in the field is that males are usually symptomatic and females are usually asymptomatic (WHO 2012; Judson 1990), it is not clear that this difference in symptoms reflects the true colonization status in infected populations (Walker and Sweet 2011). Symptomatic infection of both sexes results in a purulent discharge comprising polymorphonuclear leukocytes (PMNs) and diplococci, and this discharge is more obvious from the male urethra (Rice et al. 2017). It is recommended that all patients that test positive for *Ng* are treated with antibiotics (Workowski et al. 2015). On rare occasions, *Ng* can disseminate from the genital tract to the heart, skin, or joints, resulting in dermatitis, endocarditis, or arthritis (Edwards and Apicella 2004; Wiesner and Thompson 1980). *Ng* can also colonize the nasopharynx and rectum of both men and women, observed more commonly in the population of men who have sex with men (MSM) (Fairley et al. 2017). Colonization of the noncanonical sites is usually asymptomatic (Kent et al. 2005; Kinghorn 2010).

L. L. Priniski · H. S. Seifert (✉)
Northwestern University Feinberg School of Medicine, Chicago, IL, USA
e-mail: h-seifert@northwestern.edu

P. H. Rampelotto (ed.), *Molecular Mechanisms of Microbial Evolution*, Grand Challenges in Biology and Biotechnology,
https://doi.org/10.1007/978-3-319-69078-0_13

Neisseria meningitidis (*Nm*) is typically an asymptomatic colonizer of the nasopharynx (similar to the commensal *Neisseria* spp.) and is resident in approximately 10% of the human population (Cartwright et al. 1987). In contrast to *Ng*, there is usually not active screening for *Nm* colonization, and treatment is only necessary in the case of invasive disease (Klughammer et al. 2017). Invasive disease occurs when the organism transits from the nasopharynx into the brain or bloodstream and can present as bacterial meningitis or bacteremia (meningococcemia), respectively. About 1.2 million cases of meningococcal meningitis occur annually worldwide (Rouphael and Stephens 2012). The primary clinical manifestations of meningococcal meningitis are headache, fever, and nausea. The disease can be rapidly diagnosed by the presence of PMNs in the cerebrospinal fluid (MacNeil and Cohn 2011). About 20% of patients with meningococcal meningitis also have meningococcemia (MacNeil and Cohn 2011). Transmission occurs by person-to-person spread of the organism within throat secretions through actions such as coughing or kissing (Stephens 2009). Systemic infection appears to be a dead end for *Nm* because transmission is from the throat. The two populations at highest risk for meningococcal disease are young children aged 0–3 and young adults aged 14–24 (Cartwright et al. 1987). Children under the age of 3 are more susceptible to a variaty of organisms causing bacterial menigitis due to the loss of maternal immunity before their immunity develops (Goldschneider et al. 1969). Some meningococcal strains are more likely to asymptomatically colonize humans, whereas others appear to be more likely to cause disease (Yazdankhah et al. 2004), but the reasons for this remain controversial. The differences between carriage and invasive disease are not well understood but are influenced by polysaccharide capsule structure, presence of genetic elements, and host susceptibility (Tzeng et al. 2016).

The genus *Neisseria* is in the family *Neisseriaceae*, in the phylum of *Proteobacteria* and contains a core genome of 896 genes, which are present in all species (Marri et al. 2010). There are ten other commensal *Neisseria* species that regularly colonize humans. These commensal species are *Neisseria lactamica* (*Nl*), *Neisseria polysaccharea* (*Np*), *Neisseria cinerea*, *Neisseria subflava*, *Neisseria flavescens*, *Neisseria perflava*, *Neisseria mucosa*, *Neisseria sicca*, *Neisseria elongata*, and *Neisseria weaveri*. It should be noted that *N. flavescens*, *N. mucosa*, *N. weaveri*, *N. sicca*, and *N. cinerea* are not human-specific as members of these species also colonize other animals (Liu et al. 2015; Bennet et al. 2014). It is not clear if the human and animal colonizers of these species are distinct lineages or whether these *Neisseria* spp. have a broader host range. Colonization by the nonpathogenic *Neisseria* spp. species rarely results in pathology (Workowski et al. 2015) in either humans or animal hosts (Liu et al. 2015). The best-studied human commensal, *Nl*, can leave the nasopharynx to cause bacterial meningitis, but the incidence of these cases is extremely low (Liu et al. 2015). Interestingly, *Nm* and *Nl* exhibit an inverse pattern of colonization. *Nl* carriage starts shortly after birth, peaks after 1–2 years, and then declines with age (Bennett et al. 2005). Current efforts aim to test the hypothesis that *Nl* colonization protects from *Nm* colonization (Deasy et al. 2015). Notably, *Nl* causes B-cell-dependent mitogenic proliferative response that produces polyclonal IgM. This does not occur in *Nm* colonization and is hypothesized to shield *Nl* from the host adaptive response allowing for colonization (Vaughan et al. 2009).

Historically, *Neisseria* spp. have been distinguished based on phenotypes such as carbohydrate utilization or enzyme substrate tests; however, these tests do not always allow for the differentiation of species within the genus (Dossett et al. 1985). The advent of molecular analyses, including genome sequencing, has made possible a species-level phylogenetic tree allowing for analysis of the relationship between *Neisseria* spp. (Fig. 13.1) (Bennett et al. 2012). Bennett et al. estimated that *Ng* and *Nm* split about 100,000 years ago (Bennett et al. 2010), although without an accurate molecular "clock," this number should be considered a best guess rather than a firm estimate. This high-throughput sequencing has allowed for the discovery of new species and analysis of genes within each species (Weyand et al. 2016; Marri et al. 2010). These advances in sequencing technology and bioinformatics allow scientists to broadly compare genomic content within species and compare whether colonization phenotypes reflect genotypes.

Bacteria that establish persistence with hosts possess a variety of factors and mechanisms that allow for colonization, growth, immune evasion, and transmission. Throughout the study of these pathogens, the term *virulence factor* has been used to describe any attribute of the bacteria that aids in any part of the disease process; however, many of these factors are also present in commensal species (Pirofski and Casadevall 2012; Casadevall and Pirofski 2014). Throughout this chapter, we will discuss which factors are specific to the pathogens (virulence factors) and those that are *core colonization determinants* for both pathogens and commensals and speculate on how these pathogens evolved to cause their respective diseases through the acquisition of specific factors. We will present the case that the presence of many shared factors in both pathogens argues that a pathogenic ancestor arose from a commensal progenitor in the genital tract in response to a shift to this distinct anatomical location.

In general, pathogen evolution is thought of as the acquisition of virulence factors through plasmids, phage, and/or other mobile genetic elements (Finlay and Falkow 1997). These mobile genetic elements can be passed between species and allow bacteria to gain new abilities, some of which will be beneficial in the bacterial lifestyle, so these elements will be maintained in the population. Frequently, many factors are found together in what has been termed pathogenicity islands (Lee and Walker 1996). Pathogenicity islands are important for virulence in many different bacterial pathogens such as *Yersinia* spp., *Shigella* spp., *Pseudomonas aeruginosa*, and others (Schmidt and Hensel 2004). *Ng* contains one genetic island that is thought to enhance dissemination and will be mentioned below; however much is not known about the evolution of these pathogens. The pathogenic *Neisseria* are unique because both have evolved in the human host for thousands of years. Many of the other factors discussed in this chapter are located throughout the genome and their origin unknown.

The natural competence for transformation shared by all *Neisseria* spp. makes genetic exchange between the species possible, even though there are genetic barriers to transfer such as restriction modification systems (Stein et al. 1995),

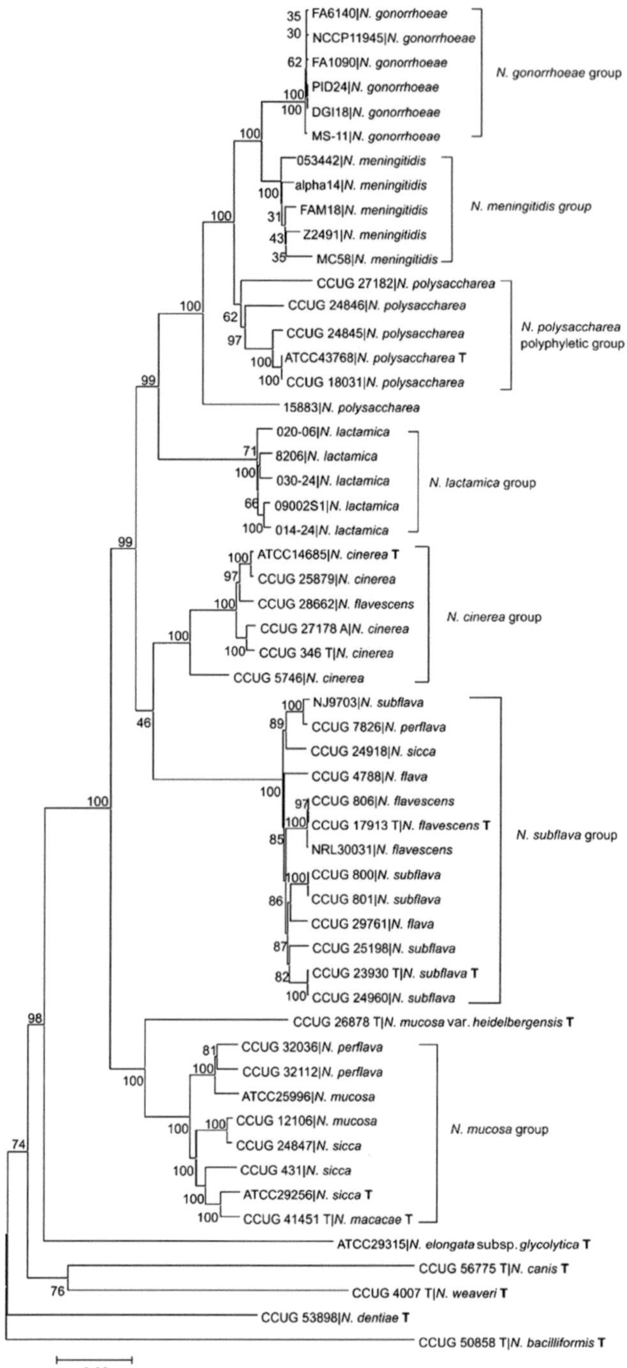

Fig. 13.1 Phylogenetic tree of *Neisseria* species. This tree is based on the comparison of 53 ribosomal genes within each *Neisseria* isolate shown and was constructed using the neighbor-joining method (Bennett et al. 2012). This method identified seven species groups, which mostly follow

CRISPR (Zhang et al. 2013), and physical limitations for transfer, between *Ng* and the other *Neisseria* spp. due to different sites of colonization. Still, there is clear evidence of HGT between the species (Marri et al. 2010; Davis et al. 2001), and this sometimes leads to ambiguity in assigning certain traits to a single *Neisseria* spp. Genomic sequencing of many *Neisseria* spp. has found that virulence genes have gone through intra- and interspecies recombination and the genetic repertoire of specific species can be dynamic (Marri et al. 2010). Because of this high rate of HGT, it has been observed that select isolates of a species possess orthologs of a gene found predominately in another species. Yet, the gene is not broadly represented within that entire species. In these cases, the penetrance of the gene in the population defines whether we consider it a trait of the organism for the purpose of this review. In the case of genes that are present in major subsets of the species isolates, but that are not found in all isolates, we consider the gene to be part of that species genetic makeup and presume that there are compensating genetic changes that allow the loss of that trait in a few lineages. The few cases where these compensating changes are known will be discussed. It should be noted that there is potential for sampling bias in the *Neisseria* spp. literature because there is much less known about many of the commensal *Neisseria* spp. than the pathogenic species, so some of the differences may change with more genome sequencing and phenotypic studies.

2 Colonization Determinants Shared by the Human-Adapted *Neisseria* spp.

The human-adapted *Neisseria* spp. share a variety of factors that allow for colonization of the human host. Some determinants are only shared between the commensals and *Nm*, presumably because they colonize the nasopharyngeal niche. This list of factors is not comprehensive and only includes factors with a likely role in the evolution of *Neisseria* spp. pathogenesis. The core colonization determinants and virulence factors are summarized in Table 13.1. We assume that most of these factors were present when the ancestral organism first colonized humans or evolved before the pathogenic ancestor split from the commensals.

Fig. 13.1 (continued) other species classifications of previous studies (Tonjum 2005). Of note, the pathogenic *Neisseria* are clustered together, and *Np* and *Nl* are the next most closely related species. This figure is reproduced under a Creative Commons Attribution License and is used with the permission of the authors

Table 13.1 Summary of *Neisseria* colonization and virulence factors

Distribution	Factor	Function	Reference
Core colonization determinants			
Most *Neisseria*	Iron systems	High-affinity membrane iron receptors	Harrison et al. (2013a)
Most *Neisseria*	LOS	Endotoxin, important for phagocytosis by neutrophils	Stein et al. (2011)
Most *Neisseria*	Pilus	Mediates Attachment to host cells, aggregation and motility, PMN defense	Wormann et al. (2014)
Pathogenesis-related factors			
Nm, *Ng*	Av Systems	Mechanism of altering the sequence and/or expression of antigens	Rotman and Seifert (2014)
Nm, *Ng*, *Nl*, *Np*[a], *Nf*[b]	PV systems	Mechanism of ON/OFF expression of genes	Jordan et al. (2005)
Nm, *Ng*, *Nl*, *Np*, *Nf*	Opa	Mediates attachment and invasion, bacterial aggregation	Marri et al. (2010)
Nm, *Ng*, *Nl*, *Np*, *Nc*[c]	Maf	Polymorphic toxin/immunity protein system used in bacterial competition	Jamet et al. (2015)
Nm, *Ng*, *Nl*	IgA protease	Secreted protease, cleaves IgA and LAMP1	Marri et al. (2010)
Nm, *Ng*	TdfF	Iron uptake important for intracellular survival	Marri et al. (2010)
Nm-specific factors			
Nm	Capsule	Required for infection, antibody, and complement resistance	Harrison et al. (2013b)
Nm	PorA	Attachment and serum resistance	Massari et al. (2003)
Nm	NadA	Important for adhesion and invasion	Capecchi et al. (2005)
Nm	Opc	Adhesin, mediates attachment to extracellular matrix proteins, integrins	Virji (2009)
Nm, *Nc*	NalP	Secreted IgA protease	van Ulsen et al. (2003)
Nm	TpsA/B	Toxin/antitoxin system used in bacterial competition	Arenas et al. (2013)
Nm, *Nl*	CRISPR	Restricts DNA uptake, allows for protection from foreign DNA	Zhang et al. (2013)
Nm	MDA	Filamentous prophage encoded island	Bille (2005)
Ng-specific factors			
Ng	PG release	Stimulates immune system	Cloud-Hansen et al. (2008)
Ng, *Nm*[a]	GGI	Type IV secretion system, DNA secretion, stimulate immune system	Dillard and Seifert (2001)

Other *Neisseria* spp. contain some of the *Ng*- and *Nm*-specific factors. These are usually only found in a few isolates and determined based on sequence alignment. Each case is further described in the respective section

[a]Function not tested or unknown

[b]*Nf* = *N. flavescens*

[c]*Nc* = *N. cinerea*

2.1 *Iron Acquisition*

The ability to scavenge iron from the host is a core colonization determinant of all bacteria that colonize animals, and in the *Neisseria* spp., the systems used diverge in different species. None of the *Neisseria* spp. secrete siderophores; instead, they utilize high-affinity receptors/transporters that capture iron-binding proteins from the host (Hagen and Cornelissen 2006). There are up to five iron acquisition systems used by *Neisseria* spp.: transferrin-binding proteins TbpA and TbpB, lactoferrin-binding proteins LbpA and LbpB, transporter proteins FrpB/FetA, hemoglobin receptor Hmbr, and haptoglobin-hemoglobin receptor HpuAB (Perkins-Balding et al. 2004; Saleem et al. 2013). HmbR is an outer membrane protein that binds and transports hemoglobin (Miyoshi et al. 2003). This receptor is encoded on a genomic island that is found more commonly in *Nm* disease isolates relative to *Nm* carriage isolates (Harrison et al. 2009; Kahler et al. 2001). HmbR is also found in certain commensal species (*Nl*, *Np*, *N. subflava*, *N. cinerea*, and *N. mucosa*). While *Ng* contains an ortholog of the *hmbR* gene, it has a premature stop codon, which results in a nonfunctional protein (Harrison et al. 2013a; Evans et al. 2010). The distribution of a functional HmbR receptor gene suggests that it is important for nasopharyngeal colonization but is detrimental for colonization of the urogenital tract. As it is unclear whether any of the *Neisseria* spp. encounter hemoglobin at their mucosal colonization sites (Schryvers and Stojiljkovic 1999), there may be another unrealized function for this protein during nasopharyngeal colonization.

Another iron acquisition system, *hpuAB*, encodes a TonB-dependent hemoglobin and/or haptoglobin receptor that can import both iron and heme into the bacteria and is found in both the pathogens and commensals (Harrison et al. 2013a; Perkins-Balding et al. 2004). Variable expression of *hpuAB* by phase variation occurs in the pathogenic *Neisseria* and the commensals, *Nl* and *Np* (Harrison et al. 2013a). The other commensals do not phase vary *hpuAB*, but the reasons for this difference are unclear. The other three iron systems do not differ in any systematic way in the different *Neisseria* spp. All *Neisseria* spp. require iron uptake systems to obtain iron from the limiting environment of the human host, and although there are some differences in expression, iron acquisition appears to be a shared colonization determinant.

2.2 *LOS*

Lipooligosaccharide (LOS) is a surface-exposed molecule on the outer membrane of all *Neisseria* spp. *Neisseria* spp. LOS is similar to the lipopolysaccharide of other Gram-negative bacteria, except it is missing the polymeric O-antigen. Although all *Neisseria* spp. have a common lipid A and an inner core of 3-deoxy-D-manno-octulosonic acid (KDO) and heptose, the number of KDOs and heptose differs as well as additions to the core structure (Bartley and Kahler 2014). Interestingly, there are substitutions to the backbone of lipid A that are seen in *Nm*, *Ng*, *Nl*, and some *Np* isolates that create a heterogeneity in the LOS structure that is not seen in other commensals (John et al. 2012). Additionally, unlike most commensal *Neisseria* spp., *Nm*, *Ng*, *Nl*, and *N. subflava* have the ability to add sialic acid to the LOS using host-derived NeuNAc (Serino and Virji 2000).

2.3 *Pilus*

A type IV pilus (Tfp) is expressed on all the *Neisseria* spp. and has defined roles in colonization, including twitching motility, transformation, and adherence to host cells (Cahoon and Seifert 2011; Rudel et al. 1992; Sparling 1966; Wolfgang et al. 1998; Park et al. 2001). Expression of a functional pilus contributes to *Ng* resistance to oxidative and non-oxidative killing by PMNs (Stohl et al. 2013). There are two types of Tfp: class I pili can undergo antigenic variation (see below) and are expressed on *Ng* and a majority of *Nm* isolates, and class II pili cannot undergo antigenic variation and are expressed on the commensal species and a subset of *Nm* isolates (Aho et al. 2000; Davies et al. 2014).

2.4 *Summary of* Neisseria *spp. Colonization Factors*

We propose that all of the factors discussed above that are shared widely within the *Neisseria* spp. be considered colonization factors and not discussed as virulence factors. However, there are differences in expression of some of the core colonization determinants between the pathogens and commensals that may influence pathogenicity, and these will be discussed in the next sections.

3 Determinants Shared by the Pathogenic *Neisseria* spp.

This section will explore the factors and mechanisms specific to the pathogens. We propose that many of these factors and mechanisms were acquired by the common ancestor of the pathogens before *Ng* and *Nm* diverged into different species. We cannot currently determine the order in which these adaptations occurred, and while it is possible that some of these acquisitions occurred independently in each pathogenic species, we consider this possibility unlikely.

3.1 Antigenic and Phase Variation

The pathogenic *Neisseria* spp. utilize sophisticated genetic systems to allow expression of multiple antigenic determinants of important surface-exposed antigens that create stochastic diversity within bacterial populations. These alterations provide subpopulations that can evade immune surveillance mechanisms or alter functional properties of the bacteria. The two main types of diversity generation systems are phase variation and antigenic variation.

Phase variation is the simplest of these diversity-generation systems and is defined as a reversible process, which allows high-frequency changes in gene product expression between limited numbers of defined expression states. Phase variation usually acts between two distinct states, either ON/OFF, high/low, or ON variant-1/ON variant-2. Most phase variation in *Neisseria* spp. is of the ON/OFF type, mediated by changes in polynucleotide repeats (PNRs) present in either the promoter of a gene or within the coding sequence of the gene (Snyder et al. 2001). Changes in nucleotide repeat number are mediated by slipped-strand mispairing that occurs in the PNRs during DNA replication (van der Woude 2011). The pathogenic *Neisseria* spp. and several other host-restricted bacteria have multiple genes whose expression is modulated by phase variation. These collectively have been coined "contingency loci" (Moxon et al. 2006) since their changes in expression are not dependent on environmental or developmental cues (Murphy et al. 1989). Experimental theory has postulated that phase variation can be selected for in organisms that cannot sense environmental cues or do not encounter many different environments (Wolf et al. 2005). In experiments using *Saccharomyces cerevisiae*, phase variation rates correspond to environmental change rates, i.e., when an environment changes, frequently phase variation rates are high (Acar et al. 2005). Certain commensal *Neisseria* spp. also contain a few phase variable genes, and their occurrence will be noted in the individual descriptions, but overall their occurrence in different isolates is much lower than in the pathogenic *Neisseria* spp. About 100 genes are phase variable within the pathogenic *Neisseria* spp., with individual strains having around 80 phase variable genes (Jordan et al. 2005; Saunders 2000; Snyder et al. 2001). Recent analysis of an *Nm* strain predicted 277 phase variable genes, although only 26 have been confirmed (Siena et al. 2016). It is unclear what specific selective pressures promote the evolution of such a large set of phase variable genes or how differential expression of most of these genes alters the biology of the pathogens. Moreover, there have not been any studies that suggest that subsets of the phase variable genes are expressed in coordination. The large number of phase variable genes in the two pathogenic species and their occasional occurrence in the commensal species clearly indicates that the pathogens rely on many different subpopulations with different sets of expressed gene products being available for selection and suggest that these stochastic changes are as important as environmental cues for adaptation.

Antigenic variation (Av) refers to more complex diversity-generation systems that use a high-frequency, reversible process that results in the expression of many different forms of a gene product. Av can occur through multigene phase variation or through recombination-based systems that vary the coding sequence of a gene product (Cahoon and Seifert 2011). While the name antigenic variation implies these systems have evolved for immune evasion, many Av systems alter functional properties of the organism, and similar diversity-generation systems occur in free-living organisms that never encounter an immune system (Simon and Schmidt 2007). Both pathogenic *Neisseria* spp. use Av to express different repertoires of three surface-exposed antigens (Tfp, LOS, and Opa). Though Tfp and LOS are core colonization determinants and found in all *Neisseria* spp., Av processes do not occur in any of the commensal species. A few other gene systems also have the ability to create multiple forms of the gene product in a stochastic manner and will be discussed individually below. The presence of many stochastically regulated genes in the pathogenic *Neisseria* supports the idea that these organisms face strong purifying selection at the population level rather than at the individual organism level. The idea is that in a heterogeneous population, although some bacteria will not survive, other subsets can survive and grow. It is also possible that the extensive evolution of phase and Av systems in the pathogens is due to interactions with a wider range of environments, but considering the host restriction, we consider this to be less likely.

3.1.1 Pilus Av

Although the Tfp is a core colonization determinant found in many *Neisseria* spp., the process of Tfp Av is specific to the pathogens. In strains expressing class I pili, the Tfp structure varies through extensive diversity in the coding sequence of the major pilin subunit, PilE (Hagblom et al. 1985). During pilin Av, a portion of a silent copy sequence replaces part of the *pilE* gene in a nonreciprocal, homologous recombination process (Fig. 13.2a) (Haas and Meyer 1986; Meyer et al. 1984). Any portion of the recombining silent copy can be transferred into the *pilE* locus, resulting in an altered *pilE* and then PilE protein sequence, while the silent copy gene sequence remains unchanged in these reactions (Fig. 13.2b). Some pilin Av events can result PilE molecules that are inefficiently assembled into the pilus. Other events can cause the recombination of a premature stop codon in *pilE* which results in nonpiliated bacteria (Criss et al. 2005). Pilin Av requires many conserved recombination and repair factors including RecA, RecOR, and RecG (Sechman et al. 2005; Mehr and Seifert 1997, 1998). Moreover, loss of other recombination and repair factors, RecQ, RecJ, and Rep, lowers the Av frequency, but does not block the process (Mehr and Seifert 1998; Skaar et al. 2002; Kline and Seifert 2005; Rotman et al. 2016). Since these factors are present in most bacterial species that do not undergo a similar recombination-based, diversity-generation process, it was uncertain what unique processes in the pathogenic *Neisseria* spp. allow these high-frequency gene conversion reactions to occur. However, two findings have revealed processes unique to these organisms.

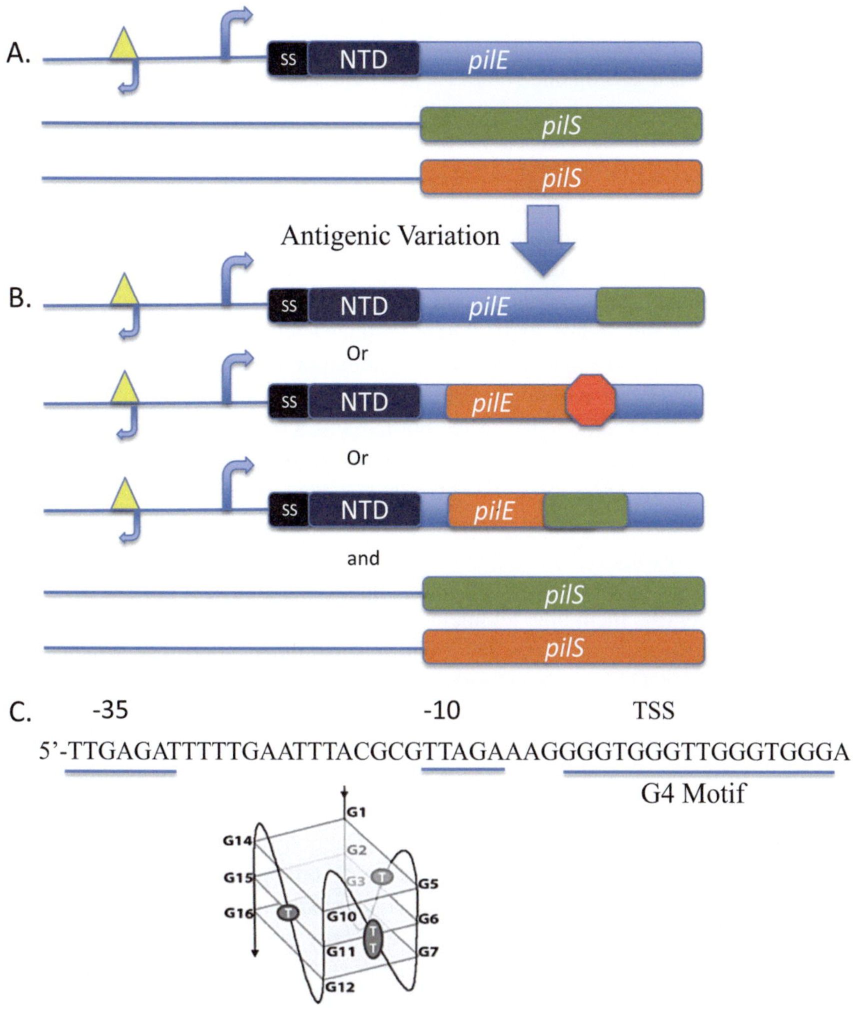

Fig. 13.2 Pilin Av. (**a**) The *pilE* gene contains a G4 motif and associated sRNA upstream of the promoter, which are both required for Av. The *pilE* gene contains a signal sequence (SS), a conserved N-terminal domain (NTD), and a variable C-terminal domain. The *pilS* copies have microhomology to the *pilE* gene but lack the G4 motif, promoter, and N-terminal domain (Obergfell and Seifert 2015). (**b**) Nonreciprocal recombination occurs between *pilE* and one or more *pilS* copy. Recombination occurs at sites of microhomology in the C-terminal region. Some silent copies include a frameshift. If this region recombines into the expression locus, it can cause an alternate reading frame and can cause a premature stop codon resulting in a nonpiliated bacteria. In all Av events, the *pilS* sequences remain unchanged. (**c**) G4 motif and associated sRNA. All four guanine tracts are required for Av and fold into a parallel quadruplex (Kuryavyi et al. 2012). Transcription of the sRNA begins within the second G tract (Cahoon and Seifert 2013)

The first unique property of the pathogenic *Neisseria* spp. linked to pilin Av is that the pathogens have two identical copies of their genome per cell (Tobiason and Seifert 2006a). The pathogenic *Neisseria* spp. are similar to many other bacterial species that have been shown to have multiple copies of their genome (Pecoraro et al. 2011; Robson et al. 1984; Hansen 1978; Angert and Clements 2004; Bresler et al. 1998; Kitten and Barbour 1992; Minton 1994; Nagpal et al. 1989; Komaki and Ishikawa 1999). While diploid, it is not possible to incorporate two different markers into the *Ng* genome, showing that they are genetically haploid. As such, they are designated as homogeneous diploid organisms (Tobiason and Seifert 2010). *Nl* does not undergo pilin antigenic variation and only contains a single-chromosome copy per cell (Tobiason and Seifert 2010). These findings lead to the hypothesis that pilin Av requires two chromosomes and that recombination occurs between the diploid chromosomes (Tobiason and Seifert 2006b; Howell-Adams and Seifert 2000). This hypothesis has never been directly tested but remains a reasonable model to explain the high-frequency recombination occurring between multiple *pilS* copies and *pilE*.

The second unique property of the pathogenic *Neisseria* spp. that allows for pilin Av is the presence of a conserved sequence located upstream of *pilE* that can form an alternate DNA structure called a guanine quadruplex (G4) (Fig. 13.2c) (Cahoon and Seifert 2009, 2011). G4 structures have been suggested to play specific roles in many molecular processes, including telomere maintenance, oncogene regulation, and recombination (Rhodes and Lipps 2015; Maizels and Gray 2013). The sequence of the *pilE*-associated G4 is unique in the *Ng* chromosome, and none of the other G4-forming sequences present in the chromosome are located near a *pilS* locus (Cahoon and Seifert 2009). Mutation of any of the individual GC base pairs necessary to form the *pilE* G4 structure prevented pilin antigenic variation, while mutation of any of the TA base pairs within the sequence, which are not required to form the structure, did not prevent pilin Av (Cahoon and Seifert 2009). It was subsequently discovered that pilin Av also depends on a promoter located adjacent to the G4-forming sequence that initiates transcription within the G4 sequence resulting in a small, noncoding RNA that can only function *in cis* (Fig. 13.2c) (Cahoon and Seifert 2013). All of these genetic studies led to the conclusion that formation of the G4 structure was required to initiate pilin Av.

There are about 18 *Ng* silent *pilS* copies that reside in various loci in distinct locations in the genome (Hamrick et al. 2001). There are 4–11 silent copies in *Nm* isolates, generally within one locus localized near the *pilE* gene. *Nm* also shows a lower frequency of pilin Av compared to some *Ng* strains (Tettelin et al. 2000; Helm and Seifert 2010). Some pilin Av events in *Ng* cause fewer or no pili, resulting in a phase variation event (Koomey et al. 1987; Criss et al. 2005). Whether pilus phase variation (i.e., little to no pili) through Av recombination also occurs in *Nm* has not been established in the literature, so this aspect may be *Ng* specific. As a necessary condition of any phase variation system, piliation can be restored when a functional *pilS* copy sequence recombines into *pilE*. Pilus ON/OFF phase variation can also be mediated by changes in a PNR sequence in the pilus assembly gene, *pilC* (Jonsson et al. 1991). The antigenically variable class I pilin gene is expressed by *Ng* and most

Nm isolates, while the commensal *Neisseria* spp. and a subset of *Nm* isolates express class II pilin that does not undergo Av (Helm and Seifert 2010; Davies et al. 2014).

Whole genome sequencing of *Nm* isolates has shown that *Nm* containing class II pili still contain two or three silent copies at the locus where the class I pilin gene is found in class I pilus strains, but the adjacent class I *pilE* has been deleted (Wormann et al. 2014). They also described two independent molecular events leading to class II pili. These findings suggest the hypothesis that the class II pilin expressing *Nm* used to have the class I pilin, but it was lost after the class II *pilE* was acquired from commensal *Neisseria* spp. Since the class II pilins of *Nm* are most similar to *Nl* and *Np*, these are the most probable donor species as their *pilE* genes are closest to *Nm* class II pilins (Wormann et al. 2014). The conversion of class I to class II pilus strains is a more recent evolutionary event. What is not known is whether there was selection for class II pilus transfer to *Nm* or whether the gene transfer was a neutral event promoting other changes to maintain colonization. *Nm* expressing non-variable class II pili can still colonize and cause invasive disease, but changes in the pili can affect cell signaling in epithelial cells, and non-variable pili cannot alter signaling (Miller et al. 2014). It appears that altered pili are advantageous to *Ng* and class II containing *Nm* cannot avoid immune detection or alter cell signaling by pilin Av, so other mechanisms may exist for these functions.

Phase-Variable Pilin Glycosylation

Both neisserial pathogens and *Nl* possess an O-linked glycosylation system that can modify several proteins including PilE. Pilin glycosylation has been shown to be important for *Ng* adhesion and invasion of human cervical epithelium (Takahashi et al. 2012; Jennings et al. 2011; Bartley et al. 2013). Four of the genes in the glycosylation pathway are phase variable in some strains (Power et al. 2003; Aas et al. 2007; Borud et al. 2011). The genes in this glycosylation system are similar among *Neisseria* spp. but the total repertoire of modified proteins is different between the species (Borud et al. 2010). Not much is known about commensal *Neisseria* spp. glycosylation. *N. elongata* does possess a similar glycosylation system but lacks the phase-variable genes, produces different glycoforms from the *Neisseria* pathogens, and does not have a glycosylated pilus (Anonsen et al. 2015). Pilin glycosylation may be specific to neisserial pathogens and their close relatives, although this has not been thoroughly investigated. Interestingly, *Nm* that express the less prevalent class II pilin show increased levels of glycosylation of pilin that masks the surface of the pilus fiber from antibody recognition, which might functionally substitute for pilin Av (Gault et al. 2015).

3.1.2 Opa

Opacity-associated (Opa) proteins are a family of surface-exposed proteins expressed on the *Neisseria* spp. Some but not all Opa proteins confer an opaque colony appearance to *Neisseria* spp. grown on agar medium (Swanson 1978; King and Swanson 1978). Opa proteins are expressed on both the pathogenic *Neisseria* spp. and certain commensals (*Nl*, *Np*, and *N. flavescens*) (Marri et al. 2010). Different Opa proteins bind to different host receptors to mediate adherence and also can induce signal transduction pathways upon binding (Chen and Gotschlich 1996; Chen et al. 1997; Gray-Owen et al. 1997; Virji et al. 1996; Bos et al. 1997). The *Ng* genome contains ~11 different Opa-encoding genes, whereas *Nm* has only 4–5 Opa proteins (Bhat et al. 1991; Virji et al. 1992; Marri et al. 2010). Each Opa protein can phase vary a pentameric repeat (CTCCT) located at the start of each gene (Stern et al. 1986). Opa phase variation occurs in the two pathogens and *Nl*, which has two phase-variable Opa genes (Stern et al. 1986). While not yet tested, *Np* and *N. flavescens* also contain the pentameric repeat, so phase variation probably also occurs in these commensals (Marri et al. 2010).

Opa proteins bind to human carcinoembryonic, antigen-related cell adhesion molecules (CEACAMs) (Virji et al. 1999). Due to their basic nature, Opa proteins have also been shown to bind heparan sulfate proteoglycans and sialic acids (de Vries et al. 1998; Moore et al. 2005). Opa phase variation has been shown both in vitro and during human infection. CEACAM1 and CEACAM6 are located on most tissues, and expression is upregulated during inflammation. CEACAM5 and CEACAM3 are specific to epithelial cells and neutrophils, respectively (Hammarstrom 1999). Different Opa variants bind to diverse receptors located on different host cells. Interaction with different receptors can change the host cell response such as bacterial uptake, inflammation, or stimulation of an oxidative burst (reviewed in Sadarangani et al. 2011). Therefore, changing the expression of receptors by phase variation alters neisserial interactions with the host.

There are three commensal *Neisseria* spp. that express Opa orthologs that interact with human CEACAM1, similar to some Opa proteins from *Ng* and *Nm* (Marri et al. 2010). However, these commensal orthologs have been shown to have lower affinity to human CEACAM1, allowing *Nm* Opa to outcompete for binding to the human receptors (Toleman et al. 2001). The existence of this limited Opa repertoire in the commensal species suggests that the expression of a single Opa adhesin is a colonization determinant, but that the ability to express multiple Opa adhesins is correlated with pathogenesis (Sarantis and Gray-Owen 2012). The use of human CEACAM expressing mouse strains has allowed researchers to use study *Ng* and *Nm* infection in a mouse model, further demonstrating the critical role of Opa-CEACAM interactions (Johswich et al. 2013; Sarantis and Gray-Owen 2012).

Based on their broad representation in the human-restricted *Neisseria* spp. and the clear functional differences between Opa paralogs, we assume that the selective pressure that promoted the evolution of the Opa gene families in the pathogenic species was mainly based on their varying functional properties rather than immune evasion, although both types of selection could have been operating. The ability of a heterogeneous population to interact with the host in different ways may aid in survival and persistence. The tissue-specific localization of different CEACAM proteins on the epithelium or PMNs may have provided the need for *Ng* and *Nm* to maintain a heterogeneous population expressing different Opa proteins to allow some bacteria to spread to other sites, survive intracellularly, or maintain colonization at the mucosal surface.

3.1.3 LOS

LOS is involved in both immune evasion and functional changes, including attachment to host cell and tissues, and as an immune modulator to stimulate inflammation. *Nm* and *Ng* LOS are both a strong activators of TLR4 and subsequent inflammatory response in host cells (Zughaier et al. 2004). Conversely, commensal *Neisseria* spp. largely lack the specific modifications that result in immune stimulation (Zughaier et al. 2004). Therefore, the LOS is one of the main immunomodulatory molecules that differentiates the pathogens from the commensal *Neisseria* spp. Since the change from immunomodulatory to commensal is fairly simple, this could have been one of the major changes that produced the alteration from commensal to pathogen (see Sect. 6).

This branched oligosaccharide of LOS is attached to the cell surface through lipid A. Phase variation occurs in the genes coding for multiple glycosyltransferases resulting in differential glycosylation patterns and a heterogeneous population of bacteria. There are six glycosyltransferases involved in LOS synthesis, four of which can independently phase vary (Jennings et al. 1999; Snyder et al. 2001; Gotschlich 1994). For example, the *lgtA* gene, which encodes a β1,3-*N*-acetylgalactosamine transferase, contains a 12-nucleotide guanine tract within the coding sequence. This long stretch of guanines will frameshift when a base is added or deleted by slipped-strand mispairing during replication (Danaher et al. 1995; Yang and Gotschlich 1996). Therefore, all populations will have LtgA expressing variants and LtgA-non-expressing variants with each subset having a distinct LOS structure. There are 12 known LOS immunotypes that result from phase variation of the different glycosyltransferases (Scholten et al. 1994).

There are no known homopolymeric repeats in the glycosyltransferases of commensal species, so they are predicted to be invariant (Stein et al. 2011). It is likely that the variation in LOS types is, therefore, one of the major changes that allowed the pathogenic species to evolve from commensal to immune stimulating and host damaging. Since the evolution of LOS variation is relatively simple compared to the other Av systems, it is likely that this was one of the earliest adaptations that started the transition from the commensal ancestor to the pathogen's precursor.

Similarly to changes in Opa expression, changes in LOS sialylation have been shown to alter *Neisseria* spp. interactions with antibodies and immune cells (Bayliss et al. 2008; van Vliet et al. 2009). While the frequency of phase variation is low enough that glycosylation patterns are usually stable, one study did determine that two LOS variants can be expressed on the same cell using immuno-TEM. This result demonstrated that the variation in LOS structure can vary at both the population and single-cell level (Burch et al. 1997).

There is another alteration to LOS that is also likely to have contributed to the development of pathogenesis. Phosphoethanolamine, a TLR4 agonist, is added to lipid A by LptA (Cox et al. 2003). Most commensals do not contain *lptA*, with the exception of two *Nl* strains. Purified lipid A from *Nl* has high inflammatory stimulation compared to other commensals, although the whole bacteria do not when cultured with ThP1 monocytes. This result indicates that *Nl* has stimulatory lipid A, similar to *Nm*, but possesses additional mechanisms to reduce inflammatory signaling (John et al. 2012).

Overall, it seems important for the *Neisseria* pathogens but not the commensals to change their LOS structure and stimulate an immune response. This phase variation occurs in multiple genes in the locus allowing for the creation of a variety of structures and Av. LOS, Opa, and Tfp are all surface-exposed antigenic determinants, and the variation of all three allows for the formation of a heterogeneous population of bacteria, which can allow for immune evasion during host response and altered interactions with host cells.

3.1.4 Maf

Nm and *Ng* contain multiple gene families encoding multiple adhesins family (Maf) toxins. The Maf genomic island contains both a toxin and immunity protein, MafB and MafI, respectively (Zhang et al. 2012). These Maf genes have similarity to *Cdi* genes, which are found in many different bacterial species and mediate interbacterial species competition (Arenas et al. 2015; Aoki et al. 2010). Fifteen strains of pathogenic *Neisseria* spp. were analyzed, and all were found to have multiple Maf loci. In fact, Maf genes make up 2% of the genome of the pathogens (Jamet et al. 2015). In contrast, *Nl*, *N. cinerea*, and *Np* each contain only one Maf island, while in other commensals, the Maf island is incomplete or absent (Jamet et al. 2015). The reasons why there are more Maf islands in the pathogenic species are not known. Bacteria that outcompete surrounding bacteria can better establish a niche and increase their population density within the host. However, most commensal bacterial species limit their growth to help limit their detection by the innate immune system. Perhaps the Maf gene families allow for increased growth and an inflammatory response. Alternatively, the Maf toxins could show unknown activities against host cells that are important for pathogenesis.

A commonality in the Maf genetic island is the variance in *mafB* genes. This Av is thought to occur through the recombination of the 5′ end of the *mafB* gene, sometimes resulting in a nonfunctional gene (Jamet and Nassif 2015). The variation in the *mafB* gene sequence in these genomic islands is present in both *Nm* and *Ng*. The diverse 5′ ends of the genes result in different carboxy-terminal domains (Jamet et al. 2015). Changing the carboxy-terminal domain of the toxin protein may allow for a variety of functions or targets including other bacteria and even eukaryotic cells (Jamet et al. 2015). Though the research of *Neisseria* spp. Maf proteins is not yet extensive, Maf presents a new antigenically variable set of proteins that appear to be important for pathogenesis due to their high prevalence in the pathogenic genomes.

3.2 IgA Protease

Many mucosal pathogens secrete a protease that targets the most common antibody found in human mucosa, immunoglobulin A1 (Mulks and Shoberg 1994). IgA acts by binding the bacteria to mucin, which results in mechanical clearing of the bacteria using mucus and cilia on the epithelia (Childers et al. 1989). All pathogenic *Neisseria* spp. secrete one of the two IgA1 proteases. The two proteases cleave at different bonds in the human IgA (Pohlner et al. 1987; Plaut et al. 1975). IgA protease is not usually found in commensals with the exception of an intact Iga2 gene in some *Nl* genomes (Marri et al. 2010).

Experimental human challenge infections with IgA protease-deficient *Ng* did not result in any difference in rate of infection or symptoms, indicating that it is not required for experimental urethritis in males (Johannsen et al. 1999). IgA protease may be important for extended infection or in female infection, which were not tested in the human challenge model. IgA1 protease can also act intracellularly to cleave lysosome-associated membrane protein 1 (LAMP1) and reduce the levels of other lysosomal proteins, which aids in intracellular survival of the pathogens (Lin et al. 1997; Ayala et al. 1998). Loss of these lysosomal proteins presumably causes destabilization of the lysosome and allows for *Ng* to survive intracellularly and possibly transit through the epithelium creating a new niche. Why the IgA1 protease was selected for within the pathogenic species and not most commensal species is not explained by its known targets, and therefore, it is likely that the pathogenesis-associated function(s) of this protease is unknown.

3.3 Pathogen-Specific Metabolism

While most of the iron acquisition functions are expressed in both the pathogens and the commensal species, one outer membrane-specific transporter, TdfF, is only found in the pathogens (Marri et al. 2010). As TdfF is essential for intracellular survival and gene expression of *tdfF* is iron regulated, it is thought to be important for intracellular iron import (Hagen and Cornelissen 2006). The diversity of iron acquisition systems in *Neisseria* spp. has been used to group similar species together since the genetically similar species tend to possess the same systems (Marri et al. 2010). Overall, many of the iron acquisition systems are shared among species, but *Ng* and *Nm* have a few differences that allow for them to occupy the intracellular niche.

In addition to iron metabolism, host lactate utilization is a key metabolic factor for pathogenic *Neisseria* spp. Both *Neisseria* pathogens preferentially incorporate lactate carbon into fatty acids and the LOS. When glucose is not present, lactate is used for both gluconeogenesis and energy production via the TCA cycle. However, lactate is not used in gluconeogenesis when glucose is present (Leighton et al. 2001). Lactate enhances LPS sialylation, which can increase serum resistance

(Smith et al. 2001). Additionally, increased oxygen consumption caused by lactate utilization can help the bacteria survive oxygen-dependent killing by the host immune system (Britigan et al. 1988; Simons et al. 2005). *Ng* mutants lacking lactose dehydrogenases have decreased survival in anaerobic environments and inside neutrophils (Atack et al. 2014). Recently, an *Nm* isolate has emerged as a urogenital pathogen. This clone acquired a functional nitrification system from *Ng*, which allows the bacteria to grow anaerobically (Tzeng et al. 2017). Host lactate can also induce microcolony dispersal in both pathogens (Sigurlásdóttir et al. 2017). *Nl* as is its name is the only *Neisseria* spp. that utilizes lactose to make acid, which is used in diagnostic test to identify this species (Knapp 1988). Overall, the ability to metabolize lactose and use as a signal allows the *Neisseria* pathogens to better survive and spread in the host.

3.4 Summary of Pathogen-Specific Factors

It is clear there are many pathogen-specific factors and mechanisms that allow the pathogens to modulate the immune response and possibly adapt to a broader sets of environments. The number of features that are shared between the pathogens (albeit with clear differences) strongly supports the phylogenetically supported concept of a common pathogenic ancestor.

4 Determinants Specific to *N. meningitidis*

Nm has many factors that are not found in the other species. We presume that many of these factors contribute to *Nm*'s ability to leave the nasal pharynx and go systemic, but since this is a dead-end process, their evolution is more likely the result of contributions to other processes including transmission and survival. It is important to realize that *Nm* is usually an asymptomatic colonizer, similar to other *Neisseria* commensals, and that only some sequence types have an increased probability to go systemic. The difference between strains that can cause invasive disease rather than just carriage is not fully understood. *Nm* utilizes many different adhesins to bind to the nasopharyngeal surface highlighting the importance of this step to *Nm* colonization and pathogenesis. *Nm* also contains a two-partner secretion system that aids in attachment and can act in bacterial fratricide.

4.1 Polysaccharide Capsule

The polysaccharide capsule of *Nm* is a major feature specific to this pathogen. There are 12 defined capsular polysaccharides expressed by *Nm*, which are used to define

Nm isolates into serogroups. Six serogroups have been found to cause the majority of invasive disease (Harrison et al. 2013b). Each serogroup has a distinct capsule operon. Since capsule operons have a lower GC content than the surrounding genome, it is likely that these capsule variants were obtained through independent HGT events (Harrison et al. 2013b). The most efficient meningococcal vaccines are based on polysaccharide capsule protein conjugates (Frasch et al. 2012). However, the serogroup B capsule is a poor immunogen, because the capsule structure closely resembles polysialic acid moieties found on human tissue (Sadarangani and Pollard 2010; Zimmer and Stephens 2006). This lack of immunogenicity of the MenB capsule has led to the creation of two commercially available MenB protein vaccines that both contain the factor H binding protein, Bexsero® and Trumemba® (Crum-Cianflone and Sullivan 2016). Wide-scale vaccination is likely to provide selective pressure for the establishment of new serotypes in the population. It will be interesting to determine whether vaccination also causes a shift in the clones that produce invasive disease (Borrow et al. 2017).

The capsule has anti-adherent properties due to the masking of multiple adhesins (Stephens et al. 1993; Virji et al. 1993a, b). The presence of a capsule can also result in avoidance of opsonization and complement dependent killing, which enhances survival of *Nm* in the bloodstream (Spinosa et al. 2007). Capsule interactions with complement are complex. Depending on the structure, each serotype can reduce or increase complement activation of both the classic and/or alternative pathway (Lewis and Ram 2014).

Capsule expression can change through four distinct mechanisms. First, expression of capsule biosynthesis genes is regulated at the transcriptional level by a two-component system MisR/S and the translational level by altered mRNA folding in different temperatures (Tzeng et al. 2004; Loh et al. 2013). Second, ON/OFF phase variation of the biosynthesis genes by slipped-strand mispairing occurs in the B capsule type (Hammerschmidt et al. 1996). Third, genetic exchange of capsule biosynthetic operons can occur. This was demonstrated by meningococcal isolates with similar genetic markers but different capsule types. An example of this was shown through the observation of outbreak isolates, capsule type switched with the exchange of a polysialyltransferase allele from one strain to another, probably through horizontal DNA exchange (Swartley et al. 1997). Fourth, there is an insertion element, IS1301, in the first gene of the capsular biosynthesis operon, *ccsA*. When it is excised, capsule production is ON, and when it is present, capsule production is OFF. Different isolates possess either the ON or OFF version of this gene, and OFF can change to ON if the IS1301 element excises. This insertion was found in 10–50% of bacteria recovered from infection assays to epithelial cells, which may show that reduced capsule is beneficial during some stages of infection (Hammerschmidt et al. 1996). Whether this phenotype change is found during human infection is not known and may only be an in vitro artifact. Insertion element excision is not considered a canonical phase variation, because it is low frequency and involves a transposable element. *Ng* and commensal *Neisseria* spp. do not have copies of the IS1301 element (Tzeng et al. 2016).

4.2 Porins

Porin forms a voltage-gated, outer membrane pore that allows ion exchange from the exterior to the periplasm (Haines et al. 1988). *Ng* and *Nm* have one of the two alleles of PorB (*porB1A* and *porB1B* in *Ng* and *porB2* and *porB3* in *Nm*) (Derrick et al. 1999). In *Ng*, one allele, *porB1A*, is associated with disseminated disease (Brunham et al. 1985; Cannon et al. 1983). *Nm* PorB is a potent TLR2 agonist, stimulating proinflammatory cytokines and an immune response (Massari et al. 2002; Toussi et al. 2012). Porin binds complement factors to repress complement-mediated signaling. Notably, this serum resistance is only observed with human serum as animal serum kills *Ng* (Ngampasutadol et al. 2008). This specificity to resist killing by human serum most likely occurred at the level of mutation of the existing porin gene rather than gene transfer.

Though PorB is a core colonization determinant and shared among all *Neisseria* spp., it does not function identically in different *Neisseria* spp. Surprisingly, *Nm* PorB has a much higher affinity for TLR2 compared to *Nl* PorB, and *Nm* PorB induces higher production of proinflammatory cytokines (Toussi et al. 2012). These attributes all suggest that the immune-stimulatory aspect of *Nm* PorB is one of the traits that differentiates *Nm* from the commensal species.

PorA, an *Nm*-specific factor, is a class 1 porin of the Gram-negative superfamily of porins (Frasch et al. 1985; Tommassen et al. 1990). *Nm* PorA also contributes to resistance to the complement system through binding to the negative regulator C4BP, which inhibits the classic complement pathway (Jarva et al. 2005). The PorA porin is only expressed in *Nm* and is one molecule used to classify lineages of *Nm*. There is a conserved *porA* pseudogene in *Ng* (Feavers and Maiden 1998), suggesting that expression of this porin was lost after the split between *Ng* and *Nm* due to PorA being disadvantageous in the genital tract. However, why this particular porin might be selected against in a specific location has not been elucidated.

Nm PorA also undergoes phase variation with PNRs occurring in both the *porA* promoter and within the coding region of PorA (van der Ende et al. 2000). Variable levels of PorA expression have been detected in *Nm* isolates from carriers and patients with invasive disease (Alcala et al. 2004; van der Ende et al. 2000). This suggests that in *Nm*, PorA expression is selected against in some conditions, but in other times, PorA provides an advantage.

4.3 NadA

Neisseria adhesion protein A (NadA) is expressed on the outer membrane of *Nm*. This protein promotes adhesion and invasion of tissue culture cells when expressed on *Escherichia coli* cell surfaces. Deletion of this protein also leads to reduced adherence of *Nm* to epithelial cells in vitro (Capecchi et al. 2005). The NadA gene is mostly conserved between *Nm* isolates, with three major alleles, but the gene is only found in about 50% of *Nm* isolates. In general, the presence of *nadA* tends to correlate with hyperinvasive strains. While NadA is one of the antigens in the Bexsero® MenB vaccine, the variability in expression between strains suggests it is not the most important component. There is no homolog of *nadA* in *Ng*, *Nl*, and *Np*, although the surrounding genes in the locus are highly conserved in these species. One strain of *N. cinerea* contains an intact *nadA* gene, and two other isolates have fragments of *nadA*, but expression was not tested (Comanducci et al. 2004; Muzzi et al. 2013). NadA expression is regulated by the transcriptional repressor NadR, although transcriptional repression can be alleviated by the human catabolite 4-HPA (Schielke et al. 2009; Liguori et al. 2016). NadA is also phase variable. An ATAA repeat in the promoter region can vary in length and, therefore, change binding of the NadR repressor and subsequent expression of NadA. The global regulator, IHF, can also bind this AT-rich region between the operators and alter the repression of NadA (Metruccio et al. 2009). The multilevel regulation of NadA expression is proposed to allow for the population to have different expression levels of this adhesion protein, so a subset of the population could invade host tissue, while another subset remains extracellular (Metruccio et al. 2009).

4.4 Opc

Another *Nm*-specific adhesin is Opc, a surface-exposed protein that can bind to extracellular matrix proteins, integrins, and heparin sulfate proteoglycan (Virji 2009). Opc mediates attachment and invasion of epithelial cells (Virji et al. 1992, 1993a). Opc is mostly conserved in different *Nm* lineages, but a few clonal complexes lack the *opc* gene, and these lineages are more often isolated from patients with meningococcemia than from patients with meningococcal meningitis (Unkmeir et al. 2002). Whether this difference is causal or represents another aspect of these clonal complexes is not known. Opc undergoes up/down phase variation with a PNR sequence within the promoter region. Slipped-strand mispairing results in a varying number of cytosine residues between the −10 and −35 sequences and alters the promoter strength and levels of protein expressed (Sarkari et al. 1994). It is not clear why the stochastic modulation of Opc levels is advantageous, but again it seems valuable to have subpopulations of *Nm* with different surface protein expression that can be selected for by functional or immunological pressure.

4.5 NalP

NalP is a serine protease specific to *Nm* that can mediate its own transport using an autotransporter mechanism (van Ulsen et al. 2003). This protease was initially identified due to its homology with a serine protease in *Serratia marcescens*, but *NalP* also contains a lipoprotein motif (van Ulsen et al. 2003). IgA1 and human C3 (a subunit of the complement system) are among the substrates cleaved by NalP, but it does not degrade rabbit or mouse C3. A *nalP* mutant survives less well in human serum, due to complement-mediated killing (Del Tordello et al. 2014). *nalP* can undergo phase variation via a cytosine PNR found within the coding sequence of the gene suggesting that there are conditions when NalP expression is detrimental to *Nm* (Turner et al. 2002).

4.6 Nm-*Specific Two-Partner Secretion System*

Nm possesses three distinct type Va secretion systems (TPS) similar to Maf that are located on separate genomic islands (Arenas et al. 2013; Leo et al. 2012). In general, TpsA is translocated across the inner membrane by the Sec system and then across the outer membrane by its partner protein TpsB. The C-terminal domain of TspA is transported into neighboring cells by the BamA transporter and can then inhibit growth by acting as a DNase or RNase (Aoki et al. 2008). The third protein in this system is TpsI, which is an immunity protein that acts in the donor cell to bind TpsA and inhibit its activities (Aoki et al. 2010). TPS system 1 is ubiquitous in *Nm* and promotes biofilm formation, adherence to epithelial cells in vitro, intracellular survival and escape from epithelial cells, and bacterial fratricide (Neil and Apicella 2009; Schmitt et al. 2007; Tala et al. 2008). TPS systems 2 and 3 are associated with hyperinvasive strains. The TpsA of systems 2 and 3 may act in killing neighboring bacteria that lack the immunity protein, but this has not been tested (Poole et al. 2011). The genomic island that contains system 1 *tspA*, *tspB*, and *tspI* also contains one to five alternative *tpsC* cassettes, which have sequence similarity to *tpsA*, but lacks the signal sequence of *tpsA*. Each of these alternative *tpsC* cassettes can recombine into the carboxy terminus of *tpsA* (Arenas et al. 2013). The function for this variation through recombination is not currently known but could function to increase the repertoire of alternate toxin proteins.

4.7 *CRISPR*

Nm possesses clustered regularly interspaced short palindromic repeat (CRISPR) loci. These loci are found in bacteria and archaea and can restrict foreign DNA to which the organism has been previously exposed (Marraffini 2015; Burstein et al. 2017). The *Nm* CRISPR system interferes with natural transformation when the transforming DNA matches one of the spacers (Jansen et al. 2002; Zhang et al. 2013). Unlike other bacterial CRISPR systems, the type II-C system expressed by *Nm* has individual promoters to individually express each CRISPR RNA without RNA processing.

The functional CRISPR system is found in six *Nm* strains and *Nl*, but not in *Ng* (Zhang et al. 2013; Grissa et al. 2008). *Ng* does contain one spacer and a few *cas* type

I genes, but each gene contains frameshift mutations resulting in a loss of function (Zhang et al. 2013). There is also evidence of putative CRISPR systems in *N. sicca*, *N. weaveri*, *N. cinerea*, and *N. mucosa* (Louwen et al. 2014; Zhang 2017). The spacer sequences in *Nm* match the DNA sequences found in some commensal *Neisseria* spp., indicating that *Nm* is often exposed to commensal-derived DNA while inhabiting the same niche. It is proposed that the *Nm* CRISPR system might help maintain the genomic isolation of *Nm* from the neighboring commensal *Neisseria* spp. Since *Ng*'s niche is not shared with any closely related species, it would not need a way to limit genetic exchange. While a recent report demonstrated that the *Nm* CRISPR system alters the interactions of *Nm* with epithelial cells in culture, the mechanistic basis for this observation has not been elucidated (Price et al. 2015).

4.8 MDA Phage

The 8 kb meningococcal disease-associated (MDA) island is associated with invasive meningococcal disease (Bille et al. 2005, 2008). The island encodes a filamentous prophage, which is expressed and secreted via PilQ, the pore for the Tfp (Meyer et al. 2016). Recently, it was discovered that the MDA phage increases colonization on epithelial cells, by increasing pilus bundle formation and aggregation (Bille et al. 2017). The authors proposed that increased aggregation can increase overall bacterial populations during colonization of the nasopharynx and lead to dissemination. Even though the island is associated with invasive disease, the phage was seen to have no effect on the septicemia stage of infection in mouse model (Bille et al. 2017).

5 Determinants Specific to *N. gonorrhoeae*

Ng has only a few species-specific determinants, and these factors either evolved after the split from the common pathogen ancestor or were lost in *Nm* after *Ng* and *Nm* diverged. A key characteristic of gonorrhea is a robust influx of PMNs to the genital tract (Wiesner and Thompson 1980). One hypothesis is that the ability to recruit PMNs to the site of infection was the initiating event that differentiated the ancestral pathogenic *Neisseria* spp. from the precursor commensal organism and gaining the ability to release peptidoglycan (PG) fragments extracellularly could be one of the first adaptations developed to life as a sexually transmitted infection (see Sect. 6).

5.1 Peptidoglycan Release

While PG is a key component in the bacterial cell, it is one of the core pathogen molecular patterns that can trigger the host immune response when released. Addition of *Ng* PG monomers to cell culture media causes ciliated cells to die, which phenocopies what is observed in *Ng* infection of the human fallopian tubes (McGee et al. 1981; Melly et al. 1984). Killing of ciliated cells could possibly allow *Ng* access to subepithelial tissue and cause more severe disease. The release of similar PG fragments that contribute to pathogenesis is also found in *Bordetella pertussis* and *Helicobacter pylori* (Cookson et al. 1989; Viala et al. 2004). In all these cases, PG fragments stimulate the release of cytokine and chemokines, through stimulation of NOD1 signaling (Girardin et al. 2003).

Lytic transglycosylases, LtgA and LtgD, are responsible for PG release in *Ng* through cleavage of the *N*-acetylmuramic acid B-1, 4-*N*-acetylglucosamine linkage in PG. A double *lgtA/lgtD* mutant was found to not release PG monomers, yet has no growth defect (Cloud and Dillard 2002; Cloud-Hansen et al. 2008). The lack of a growth defect suggests that the core function of LtgA and LtgD is to produce these PG monomers to stimulate an immune response. Additionally, two important proteins in PG fragment release are the anhydromuramyl-specific permease and amidases, AmpG and AmpD, respectively. These proteins function to recycle PG in *Ng*. PG recycling occurs at a rate similar to that observed in *E. coli* (85%) but produces unique fragments (Garcia and Dillard 2008; Uehara and Park 2007; Girardin et al. 2003).

Nm recycles 96% of PG monomers and has been shown to release less proinflammatory monomers than *Ng*. Though the *ampG* orthologs are 97% similar, the level of PG recycling is threefold higher in *Nm* compared to *Ng*. Experiments demonstrated that *Nm* AmpG is twice as efficient as *Ng* AmpG at recycling PG. Due to the reduced PG recycling, *Ng* releases much more tripeptide than other fragments, which is an agonist of NOD1. *Nm* supernatants are also less stimulatory to hNOD1-dependent NF-KB activation in tissue culture (Woodhams et al. 2013). These differences may help reveal why *Nm* is usually an asymptomatic colonizer, while *Ng* more often results in increased inflammation. As *Ng* can survive and spread during inflammation, this creates a more distinct niche for the *Ng* colonization. *Nm* can cause inflammatory disease in the genital tract: however we cannot compare the two species because the rate of *Nm* disease and colonization is not known at this location.

5.2 Gonococcal Genetic Island and the Ng Type IV Secretion System

The gonococcal genetic island (GGI) is a chromosomal locus that contains 62 open reading frames, 23 of which encode genes similar to the F-plasmid conjugation system and a type IV secretion system (Hamilton et al. 2005). It is likely that the GGI originated from a conjugal plasmid that was transferred into *Ng* and then integrated into the chromosome using a site-specific recombination (Hacker and Kaper 2000; Hamilton et al. 2005). The GGI functions to secrete chromosomal DNA out of the bacterial cell, where it can be recognized by other *Ng* and used for DNA transformation (Dillard and Seifert 2001). All *Neisseria* spp. preferentially uptake species-specific, self-DNA that contains a specific DNA uptake sequence (DUS). This sequence varies slightly among species, and recognition of the specific DUS promotes transformation of self over foreign DNA (Elkins et al. 1991; Goodman and Scocca 1988; Donati et al. 2016).

Analysis of 115 clinical isolates showed that 80% of the isolates contained GGI genes, suggesting that the presence of the GGI is not necessary for disease. There are multiple variants of the GGI, defined by distinct *traG* alleles and the presence of the *altA* gene. Although no causation has been established, one GGI variant was found more often in isolates from disseminated infection (Dillard and Seifert 2001).

While some *Nm* strains were found to have large deletions in part of the island, a few maintained an intact island (Woodhams et al. 2012; Hamilton et al. 2005). The intact *Nm* GGI does not secrete DNA or affect the association with human cells, so it is unknown what, if any, role these GGI play in *Nm* pathogenesis (Woodhams et al. 2012). We therefore define the GGI as an *Ng*-specific factor.

Whether the main advantage the GGI provides *Ng* is in HGT or to provide DNA to the host and cause inflammation remains to be determined. Since many type IV secretion systems also transport proteins outside the bacterial cell (Alvarez-Martinez and Christie 2009), it is possible that protein effectors important for *Ng* pathogenesis are also secreted by the GGI. Further studies are needed to understand how these processes influence bacterial pathogenesis and survival.

6 How Might the Pathogenic *Neisseria* spp. Evolved from a Commensal Ancestor?

As discussed in the previous sections, the pathogenic and commensal *Neisseria* spp. share many characteristics and factors that are important for colonization. *Nm* and commensals also share a niche in the nasopharynx, though there may be differences in the spatial distribution of species within the nasopharynx (Donati et al. 2016). Whether slight location differences within the nasopharynx dictate unique traits is unknown. The two pathogenic *Neisseria* spp. are more similar to each other than to any commensal *Neisseria* spp.—both on the level of DNA sequence similarity and in shared phenotypic traits. Based on these shared characteristics, we propose that there was a common pathogenic ancestor that arose from a commensal predecessor, as opposed to independent evolutionary events for each pathogen. Our overriding assumption that underlies the following discussion is that the ability to be seen as

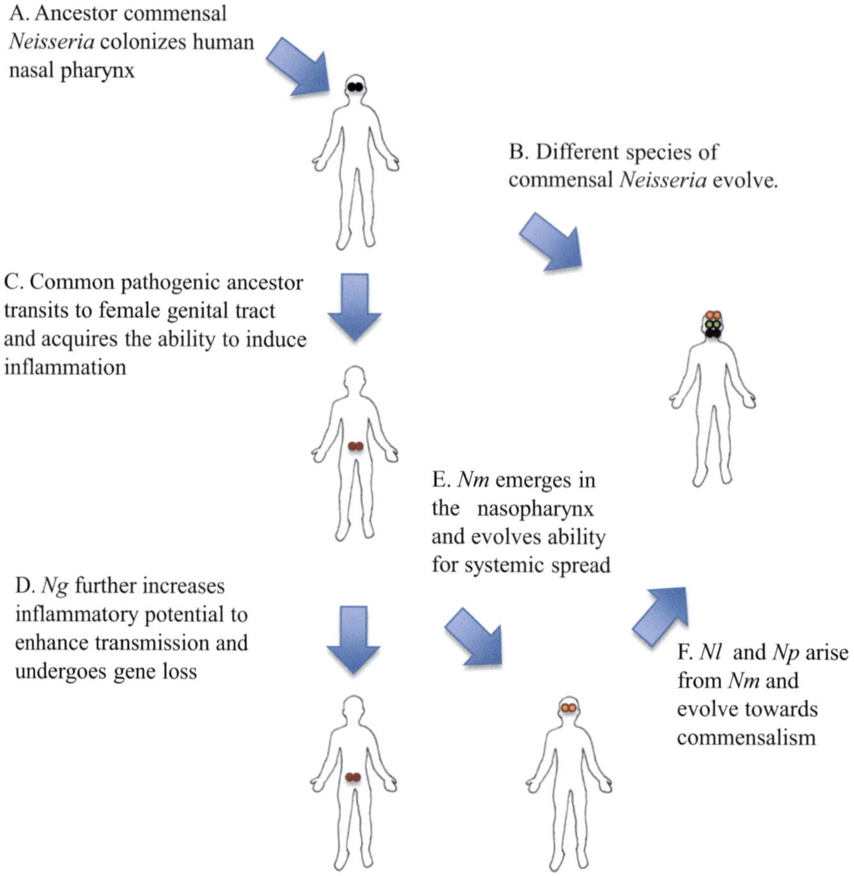

Fig. 13.3 Preferred pathways for the evolution of the human-restricted *Neisseria.* (**a**) We propose that the human-restricted *Neisseria* first colonized the nasopharynx of a human or humanoid. (**b**) This initial colonizer then produced many of the other commensal *Neisseria* species. (**c**) Colonization of the female genital tract from the nasal pharynx by the common ancestral organism promotes the switch from a commensal state that did not stimulate the innate immune system to an organism with pathogenic potential. (**d**) *Ng* continues to evolve separately due, mainly, through gene loss and the enhance ability to induce PMN inflammation. (**e**) *Nm* evolves further in the nasopharynx and gains more unique factors that allow for systemic spread and disease. (**f**) The closely related commensals, *Nl* and *Np*, gain or lose factors to evolve toward commensalism

foreign within the host and elicit an innate immune response was the event that differentiated the common pathogenic ancestor from the commensal ancestor.

There are two main scenarios that could explain the evolution of the pathogenic ancestor from a commensal. Our favored hypothesis is that the pathogenic potential evolved when the common ancestor relocated from the nasopharynx to the genital tract of a female (Fig. 13.3). While the different environment of the genital tract likely selected for additional colonization properties, this hypothesis assumes that it was the ability to promote transmission through sexual contact that supplied strong selective pressure, because efficient transmission ensures spread of the species into the population and continuation of the infection cycle. We make this assumption since transfer of the organism from the female genital tract to the male genital tract has natural physical barriers that can be overcome by utilizing the motile PMN cell population for transit. This model is supported by the fact that the pathogenic *Neisseria* spp. express many functions that allow them to modulate PMN antimicrobial functions and survival (Criss and Seifert 2012). To produce the *Nm* lineage, this progenitor could have subsequently relocated to the nasopharynx or, alternatively, could have transferred the acquired trait(s) through DNA transformation from the common predecessor to a nasopharynx-localized commensal, but we consider this option less likely (see below). The transit of the common pathogen ancestor back to the nasopharynx is supported by the phylogenetic analyses of the *Neisseria* genus (Fig. 13.1). Regardless of the details of these transitions that are lost in evolutionary history, this hypothesis is based on the assumption that colonizing a new niche was the alteration that provided strong selective pressure for the evolution or acquisition of new traits.

The main alternative model for the evolution of the pathogenic predecessor posits that the ability to induce PMN-based inflammation was acquired by the pathogenic ancestor in the nasopharynx and that a clone of the progenitor later transited to the genital tract to begin the *Ng* lineage. We consider this proposal less probable as it seems unlikely that induction of inflammation in the nasopharynx would be beneficial for the ancestral commensal, but not for any of the other commensal species present in the same locale. However, this argument brings up a clear contradiction—if there were no advantage to *Nm* to induce inflammation, it likely would have been lost after nasopharyngeal colonization was restored. Alternatively, if the basis for the selection was strong, these trait(s) should have been transferred into other commensals. Therefore, if this hypothesis is correct, the ability to induce inflammation must have provided a selective advantage to the *Nm* ancestor within the nasopharynx but only in combination with other traits that evolved in the genital tract and are not found in the commensals. It is also possible that some of the inflammation-inducing traits were transferred to one or more of the commensal *Neisseria* spp. but that the ability to go systemic, survive in the human bloodstream, and cross the blood-brain barrier was not. The pathogenic *Neisseria* spp. contain multiple immunostimulatory factors such LOS but also multiple mechanisms of avoiding immune killing, such as IgA protease, and multiple variation mechanisms. The chances of a commensal gaining all the traits necessary to be pathogenic are low. This partial acquisition of traits would be consistent with the idea that the ability to cause systemic disease provides no obvious advantage to *Nm* and is an evolutionary dead end.

The third neisserial evolution possibility is the commensal species have evolved from a common pathogenic ancestor. The ability to alter many of the surface-exposed structures is one of the most notable pathogen-associated traits, and therefore the acquisition of these systems most probably occurred in the ancestor of the pathogens. It is striking that *N. elongata*, *N. sicca*, *N. mucosa*, *N. subflava*, *N. flavescens*, *N. cinerea*, *Np*, and *Nl* all have a class II *pilE* gene that cannot undergo pilin Av and have only one chromosome but carry two to five *pilS* copies (Marri et al.

2010). The presence of *pilS* copies is the most widespread of the pathogen-related, genetic signatures; however since there is no pilin Av in these species, this argues that these *pilS* copies are residual and all of these commensals evolved from the common ancestor that could undergo pilin Av but lost this ability in the past. Since acquiring multiple *pilS* copies is a necessary priming event to allow for pilin Av to occur, duplication events, resulting in multiple *pilS* copies, must have occurred during the evolution of the pathogens. Then, pressures from the immune system may have selected for a more developed Av system in the pathogens, particularly in *Ng*. This hypothesis would require the commensal *Neisseria* to lose the Av systems and many of the inflammation-inducing characteristics. It is possible that there are multiple commensal-specific factors that were gained after the split that suppress inflammation and cellular invasion. Based on one study, there were only four genes found uniquely in all commensals, and all were classified as hypothetical proteins (Marri et al. 2010). The supporting evidence for this theory is the presence of residual *pilS* copies in many commensals. These commensals may have simply horizontally acquired these pathogen-specific *pilS* copies from *Nm* in a selection neutral event, but this scenario is less likely. The existence of the residual pilS copies in the commensal organisms is the best evidence for the hypothesis that the commensals are actually less virulent derivatives of the pathogens, in contrast to the hypothesis that the pathogens evolved from commensals. This idea is supported by the hypothesis that long-term host adaptation results in reduced virulence (Didelot et al. 2016).

7 The Evolutionary Events That Differentiate the Pathogens and Commensals Are Complex

Regardless of how the pathogens initially arose, there were clearly new traits that both pathogens acquired after the split that, presumably, arose in response to selective pressures specific to each anatomical niche. Diversity generation, likely aided by the maintenance of two chromosomes, results in a heterogeneous population, which allows for evasion of the adaptive immune system and different interactions with host cells. The pathogens also have more toxin-antitoxin systems that can not only aid in bacterial fratricide but also aid in attachment and biofilm formation. Acquisition of intracellular iron allows them to survive and replicate inside host cells, potentially avoid immune surveillance, and occupy a different niche than that of the commensal *Neisseria* spp. The ability to cleave IgA in the host is another specific trait of pathogenic *Neisseria* spp., which could modulate the immune response to benefit pathogen survival or by cleaving alternate targets to alter host cell function. It is likely that these factors were acquired earlier by the common pathogen ancestor prior to speciation.

Nm's location in the nasopharynx now places it in competition with closely related bacteria within same niche. *Nm* possesses a CRISPR system that not only protects against phage but also restricts HGT through limiting transformation. *Nm* also expresses multiple factors to aid in attachment and bacterial competition that are specific to the pathogen and may help *Nm* maintain colonization around many similar species. Arguably the most important virulence factor for *Nm* is the capsule, which masks antigenic factors and aids in attachment. It is probable that acquisition of the capsule was a critical step in the separate evolution of *Nm* after it split from *Ng*.

Ng inhabits a niche that does not contain any related species, and DUS-mediated restriction of foreign DNA is sufficient to protect *Ng* from deleterious DNA. *Ng* has fewer species-specific factors, but *Ng* has the greatest capacity for Av, harboring many more *pilS* gene copies and more Opa genes than *Nm*. *Ng* also stimulates the most robust PMN inflammatory response, leading to pelvic inflammatory disease and can result in infertility. It has been postulated that infertility may lead to an increase in sexual partners, which could aid in the spread of disease (Mackey and Immerman 2003).

Today, there is increasing research on both alternate modes of transmission of pathogens and alternate sites of colonization. Clinical studies of MSM have observed colonization and transmission of *Ng* from the genital tract, anus, and nasopharyngeal tract. Additionally, *Nm* can also colonize the urogenital tract and cause urethritis in men and pelvic inflammatory disease in women (Hagman et al. 1991; McKenna et al. 1993). Urogenital *Nm* appears to be spread sexually (Urra et al. 2005). *Nm* would appear the same as *Ng* using standard Gram stain tests, so the prevalence and historical occurrence of this type of colonization is presumably underappreciated (Bazan et al. 2016). There is a recent clone of *Nm* that has evolved to be an efficient urogenital pathogen. This *Nm* strain has lost encapsulation and acquired a *norB-aniA* gene cassette that promotes anaerobic growth (Tzeng et al. 2017; Taha et al. 2016). It is therefore likely that these alternate transmissions are not new events but rather are just newly appreciated. Currently, urethral *Nm* can be treated similarly to *Ng* (Workowski 2015), but new research is needed to determine whether these strains isolated from alternate colonization sites have any other specific adaptations to growth and transmission from this alternative niche.

Sequencing and phenotypic analysis has linked *Nl* to the common pathogenic ancestor since it shares more of virulence-associated traits compared with other commensals—such as Opa, Maf, CRISPR, IgA protease, and limited phase variation (Schoen et al. 2008). However, *Nl* has since lost the disease-causing traits or gained anti-inflammatory factors that promote commensalism. However, other analyses show that *Np* is the commensal most closely related to *Nm*, based on both 16s sequencing and core gene sequence comparisons (Fig. 13.1) (Bennett et al. 2012). *Np* may also have shared the common pathogenic ancestor, because it contains Maf and has a phase-variable iron uptake system. *Np* does not appear to share as many factors with *Nm*, but this might be due simply to it having not been studied as thoroughly as *Nl*. Nevertheless, both of these commensals probably separated from the common ancestor most recently before speciation.

8 Conclusions

What is obvious from this discussion is that we are unable or unwilling to definitively determine how the human-restricted *Neisseria* spp. evolved. Both *Ng* and *Nm* are exquisitely adapted to maintain colonization in their respective niches in the human host. It is clear that they have evolved from a common commensal ancestor, but as discussed in this chapter, what the order of events is and how much HGT contributed to the evolution of these organisms is difficult to discern. Recently, *Neisseria* spp. has evolved to survive different antibiotic treatments, which is a major public health concern. Appreciating how pathogens gain factors and how each aspect is important for colonization and disease can increase understanding of *Neisseria* biology and the factors that might influence the evolution of host damage (i.e., pathogenesis) and commensalism might occur. This discussion has not included ideas of how the human host has evolved over this time period and how this would influence the evolution of the host-adapted organisms.

References

Aas FE, Vik A, Vedde J, Koomey M, Egge-Jacobsen W (2007) *Neisseria gonorrhoeae* O-linked pilin glycosylation: functional analyses define both the biosynthetic pathway and glycan structure. Mol Microbiol 65(3):607–624

Acar M, Becskei A, van Oudenaarden A (2005) Enhancement of cellular memory by reducing stochastic transitions. Nature 435(7039):228–232. https://doi.org/10.1038/nature03524

Aho EL, Keating AM, McGillivray SM (2000) A comparative analysis of pilin genes from pathogenic and nonpathogenic *Neisseria* species. Microb Pathog 28(2):81–88. https://doi.org/10.1006/mpat.1999.0325

Alcala B, Salcedo C, Arreaza L, Abad R, Enriquez R, De La Fuente L, Uria MJ, Vazquez JA (2004) Antigenic and/or phase variation of PorA protein in non-subtypable *Neisseria meningitidis* strains isolated in Spain. J Med Microbiol 53(Pt 6):515–518

Alvarez-Martinez CE, Christie PJ (2009) Biological diversity of prokaryotic type IV secretion systems. Microbiol Mol Biol Rev 73(4):775–808. https://doi.org/10.1128/MMBR.00023-09

Angert ER, Clements KD (2004) Initiation of intracellular offspring in Epulopiscium. Mol Microbiol 51(3):827–835

Anonsen JH, Vik A, Borud B, Viburiene R, Aas FE, Kidd SW, Aspholm M, Koomey M (2015) Characterization of a unique tetrasaccharide and distinct glycoproteome in the O-linked protein glycosylation system of *Neisseria elongata* subsp. glycolytica. J Bacteriol 198(2):256–267. https://doi.org/10.1128/JB.00620-15

Aoki SK, Malinverni JC, Jacoby K, Thomas B, Pamma R, Trinh BN, Remers S, Webb J, Braaten BA, Silhavy TJ, Low DA (2008) Contact-dependent growth inhibition requires the essential outer membrane protein BamA (YaeT) as the receptor and the inner membrane transport protein AcrB. Mol Microbiol 70(2):323–340. https://doi.org/10.1111/j.1365-2958.2008.06404.x

Aoki SK, Diner EJ, de Roodenbeke CT, Burgess BR, Poole SJ, Braaten BA, Jones AM, Webb JS, Hayes CS, Cotter PA, Low DA (2010) A widespread family of polymorphic contact-dependent toxin delivery systems in bacteria. Nature 468(7322):439–442. https://doi.org/10.1038/nature09490

Arenas J, Schipper K, van Ulsen P, van der Ende A, Tommassen J (2013) Domain exchange at the 3′ end of the gene encoding the fratricide meningococcal two-partner secretion protein A. BMC Genomics 14:622. https://doi.org/10.1186/1471-2164-14-622

Atack JM, Ibranovic I, Ong CL, Djoko KY, Chen NH, Vanden Hoven R, Jennings MP, Edwards JL, McEwan AG (2014) A role for lactate dehydrogenases in the survival of *Neisseria gonorrhoeae* in human polymorphonuclear leukocytes and cervical epithelial cells. J Infect Dis 210(8):1311–1318. https://doi.org/10.1093/infdis/jiu230

Arenas J, de Maat V, Caton L, Krekorian M, Herrero JC, Ferrara F, Tommassen J (2015) Fratricide activity of MafB protein of *N. meningitidis* strain B16B6. BMC Microbiol 15:156. https://doi.org/10.1186/s12866-015-0493-6

Ayala P, Lin L, Hopper S, Fukuda M, So M (1998) Infection of epithelial cells by pathogenic Neisseriae reduces the levels of multiple lysosomal constituents. Infect Immun 66 (10):5001–5007

Bartley S, Kahler CM (2014) The glycome of *Neisseria* spp.: how does this relate to pathogenesis? In: Davies JK, Kahler CM (eds) Pathogenic *Neisseria*: genomics, molecular biology and disease intervention. Caister Academic Press, Norfolk, pp 115–145

Bartley SN, Tzeng YL, Heel K, Lee CW, Mowlaboccus S, Seemann T, Lu W, Lin YH, Ryan CS, Peacock C, Stephens DS, Davies JK, Kahler CM (2013) Attachment and invasion of *Neisseria meningitidis* to host cells is related to surface hydrophobicity, bacterial cell size and capsule. PLoS One 8(2):e55798. https://doi.org/10.1371/journal.pone.0055798

Bayliss CD, Hoe JC, Makepeace K, Martin P, Hood DW, Moxon ER (2008) *Neisseria meningitidis* escape from the bactericidal activity of a monoclonal antibody is mediated by phase variation of *lgtG* and enhanced by a mutator phenotype. Infect Immun 76(11):5038–5048. https://doi.org/10.1128/IAI.00395-08

Bazan JA, Peterson AS, Kirkcaldy RD, Briere EC, Maierhofer C, Turner AN, Licon DB, Parker N, Dennison A, Ervin M, Johnson L, Weberman B, Hackert P, Wang X, Kretz CB, Abrams AJ, Trees DL, Del Rio C, Stephens DS, Tzeng YL, DiOrio M, Roberts MW (2016) Notes from the field: increase in *Neisseria meningitidis*-associated urethritis among men at two sentinel clinics – Columbus, Ohio, and Oakland County, Michigan, 2015. MMWR Morb Mortal Wkly Rep 65 (21):550–552. https://doi.org/10.15585/mmwr.mm6521a5

Bennet J, Bratcher HB, Brehony C, Harrison OB, Maiden CJ (2014) The genus *Neisseria*. In: Rosenberg E (ed) The prokaryotes–alphaproteobacteria and betaproteobacteria. Springer, Berlin. https://doi.org/10.1007/978-3-642-30197-1_241

Bennett JS, Griffiths DT, McCarthy ND, Sleeman KL, Jolley KA, Crook DW, Maiden MC (2005) Genetic diversity and carriage dynamics of *Neisseria lactamica* in infants. Infect Immun 73 (4):2424–2432. https://doi.org/10.1128/IAI.73.4.2424-2432.2005

Bennett JS, Bentley SD, Vernikos GS, Quail MA, Cherevach I, White B, Parkhill J, Maiden MC (2010) Independent evolution of the core and accessory gene sets in the genus *Neisseria:* insights gained from the genome of *Neisseria lactamica* isolate 020-06. BMC Genomics 11:652. https://doi.org/10.1186/1471-2164-11-652

Bennett JS, Jolley KA, Earle SG, Corton C, Bentley SD, Parkhill J, Maiden MC (2012) A genomic approach to bacterial taxonomy: an examination and proposed reclassification of species within the genus *Neisseria*. Microbiology 158(Pt 6):1570–1580. https://doi.org/10.1099/mic.0.056077-0

Bhat KS, Gibbs CP, Barrera O, Morrison SG, Jahnig F, Stern A, Kupsch EM, Meyer TF, Swanson J (1991) The opacity proteins of *Neisseria gonorrhoeae* strain MS11 are encoded by a family of 11 complete genes. Mol Microbiol 5(8):1889–1901

Bille E, Zahar JR, Perrin A, Morelle S, Kriz P, Jolley KA, Maiden MC, Dervin C, Nassif X, Tinsley CR (2005) A chromosomally integrated bacteriophage in invasive meningococci. J Exp Med 201(12):1905–1913

Bille E, Ure R, Gray SJ, Kaczmarski EB, McCarthy ND, Nassif X, Maiden MC, Tinsley CR (2008) Association of a bacteriophage with meningococcal disease in young adults. PLoS One 3(12): e3885. https://doi.org/10.1371/journal.pone.0003885

Bille E, Meyer J, Jamet A, Euphrasie D, Barnier JP, Brissac T, Larsen A, Pelissier P, Nassif X (2017) A virulenceassociated filamentous bacteriophage of *Neisseria meningitidis* increases host-cell colonisation. PLoS Pathog 13(7):e1006495. https://doi.org/10.1371/journal.ppat.1006495

Borrow R, Alarcon P, Carlos J, Caugant DA, Christensen H, Debbag R, De Wals P, Echaniz-Aviles G, Findlow J, Head C, Holt D, Kamiya H, Saha SK, Sidorenko S, Taha MK, Trotter C, Vazquez Moreno JA, von Gottberg A, Safadi MA, Global Meningococcal I (2017) The Global Meningococcal Initiative: global epidemiology, the impact of vaccines on meningococcal disease and the importance of herd protection. Expert Rev Vaccines 16(4):313–328. https://doi.org/10.1080/14760584.2017.1258308

Borud B, Aas FE, Vik A, Winther-Larsen HC, Egge-Jacobsen W, Koomey M (2010) Genetic, structural, and antigenic analyses of glycan diversity in the O-linked protein glycosylation systems of human *Neisseria* species. J Bacteriol 192(11):2816–2829. https://doi.org/10.1128/JB.00101-10

Borud B, Viburiene R, Hartley MD, Paulsen BS, Egge-Jacobsen W, Imperiali B, Koomey M (2011) Genetic and molecular analyses reveal an evolutionary trajectory for glycan synthesis in a bacterial protein glycosylation system. Proc Natl Acad Sci U S A 108(23):9643–9648. https://doi.org/10.1073/pnas.1103321108

Bos MP, Grunert F, Belland RJ (1997) Differential recognition of members of the carcinoembryonic antigen family by Opa variants of *Neisseria gonorrhoeae*. Infect Immun 65 (6):2353–2361

Bresler V, Montgomery WL, Fishelson L, Pollak PE (1998) Gigantism in a bacterium, *Epulopiscium fishelsoni*, correlates with complex patterns in arrangement, quantity, and segregation of DNA. J Bacteriol 180(21):5601–5611

Britigan BE, Klapper D, Svendsen T, Cohen MS (1988) Phagocyte-derived lactate stimulates oxygen consumption by *Neisseria gonorrhoeae.* An unrecognized aspect of the oxygen metabolism of phagocytosis. J Clin Investig 81:318–324

Brunham RC, Plummer F, Slaney L, Rand F, DeWitt W (1985) Correlation of auxotype and protein I type with expression of disease due to *Neisseria gonorrhoeae*. J Infect Dis 152:339–343

Burch CL, Danaher RJ, Stein DC (1997) Antigenic variation in *Neisseria gonorrhoeae*: production of multiple lipooligosaccharides. J Bacteriol 179(3):982–986

Burstein D, Harrington LB, Strutt SC, Probst AJ, Anantharaman K, Thomas BC, Doudna JA, Banfield JF (2017) New CRISPR-Cas systems from uncultivated microbes. Nature 542 (7640):237–241. https://doi.org/10.1038/nature21059

Cahoon LA, Seifert HS (2009) An alternative DNA structure is necessary for pilin antigenic variation in *Neisseria gonorrhoeae*. Science 325(5941):764–767. https://doi.org/10.1126/science.1175653

Cahoon LA, Seifert HS (2011) Focusing homologous recombination: pilin antigenic variation in the pathogenic *Neisseria*. Mol Microbiol 81(5):1136–1143. https://doi.org/10.1111/j.1365-2958.2011.07773.x

Cahoon LA, Seifert HS (2013) Transcription of a cis-acting, noncoding, small RNA is required for pilin antigenic variation in *Neisseria gonorrhoeae*. PLoS Pathog 9(1):e1003074. https://doi.org/10.1371/journal.ppat.1003074

Cannon JG, Buchanan TM, Sparling PF (1983) Confirmation of association of protein I serotype of *Neisseria gonorrhoeae* with ability to cause disseminated infection. Infect Immun 40:816–819

Capecchi B, Adu-Bobie J, Di Marcello F, Ciucchi L, Masignani V, Taddei A, Rappuoli R, Pizza M, Arico B (2005) *Neisseria meningitidis* NadA is a new invasin which promotes bacterial adhesion to and penetration into human epithelial cells. Mol Microbiol 55(3):687–698. https://doi.org/10.1111/j.1365-2958.2004.04423.x

Cartwright KA, Stuart JM, Jones DM, Noah ND (1987) The Stonehouse survey: nasopharyngeal carriage of meningococci and *Neisseria lactamica*. Epidemiol Infect 99(3):591–601

Casadevall A, Pirofski LA (2014) Microbiology: ditch the term pathogen. Nature 516 (7530):165–166. https://doi.org/10.1038/516165a

Chen T, Gotschlich EC (1996) CGM1a antigen of neutrophils, a receptor of gonococcal opacity proteins. Proc Natl Acad Sci U S A 93(25):14851–14856

Chen T, Grunert F, Medina-Marino A, Gotschlich EC (1997) Several carcinoembryonic antigens (CD66) serve as receptors for gonococcal opacity proteins. J Exp Med 185(9):1557–1564

Childers NK, Bruce MG, McGhee JR (1989) Molecular mechanisms of immunoglobulin A defense. Annu Rev Microbiol 43:503–536. https://doi.org/10.1146/annurev.mi.43.100189.002443

Cloud KA, Dillard JP (2002) A lytic transglycosylase of *Neisseria gonorrhoeae* is involved in peptidoglycan-derived cytotoxin production. Infect Immun 70(6):2752–2757

Cloud-Hansen KA, Hackett KT, Garcia DL, Dillard JP (2008) *Neisseria gonorrhoeae* uses two lytic transglycosylases to produce cytotoxic peptidoglycan monomers. J Bacteriol 190 (17):5989–5994. https://doi.org/10.1128/JB.00506-08

Comanducci M, Bambini S, Caugant DA, Mora M, Brunelli B, Capecchi B, Ciucchi L, Rappuoli R, Pizza M (2004) NadA diversity and carriage in *Neisseria meningitidis*. Infect Immun 72 (7):4217–4223. https://doi.org/10.1128/IAI.72.7.4217-4223.2004

Cookson BT, Tyler AN, Goldman WE (1989) Primary structure of the peptidoglycan-derived tracheal cytotoxin of *Bordetella pertussis*. Biochemistry 28(4):1744–1749

Cox AD, Wright JC, Li J, Hood DW, Moxon ER, Richards JC (2003) Phosphorylation of the lipid A region of meningococcal lipopolysaccharide: identification of a family of transferases that add phosphoethanolamine to lipopolysaccharide. J Bacteriol 185(11):3270–3277

Criss AK, Seifert HS (2012) A bacterial siren song: intimate interactions between *Neisseria* and neutrophils. Nat Rev Microbiol 10(3):178–190. https://doi.org/10.1038/nrmicro2713

Criss AK, Kline KA, Seifert HS (2005) The frequency and rate of pilin antigenic variation in *Neisseria gonorrhoeae*. Mol Microbiol 58(2):510–519. https://doi.org/10.1111/j.1365-2958.2005.04838.x

Crum-Cianflone N, Sullivan E (2016) Meningococcal vaccinations. Infect Dis Ther 5(2):89–112. https://doi.org/10.1007/s40121-016-0107-0

Danaher RJ, Levin JC, Arking D, Burch CL, Sandlin R, Stein DC (1995) Genetic basis of *Neisseria gonorrhoeae* lipooligosaccharide antigenic variation. J Bacteriol 177(24):7275–7279

Davies JK, Harrison PF, Lin YH, Bartley S, Khoo CA, Seemann T, Ryan CS, Kahler CM, Hill SA (2014) The use of high-throughput DNA sequencing in the investigation of antigenic variation: application to *Neisseria* species. PLoS One 9(1):e86704. https://doi.org/10.1371/journal.pone.0086704

Davis J, Smith AL, Hughes WR, Golomb M (2001) Evolution of an autotransporter: domain shuffling and lateral transfer from pathogenic *Haemophilus* to *Neisseria*. J Bacteriol 183 (15):4626–4635. https://doi.org/10.1128/JB.183.15.000-000.2001

de Vries FP, Cole R, Dankert J, Frosch M, van Putten JP (1998) *Neisseria meningitidis* producing the Opc adhesin binds epithelial cell proteoglycan receptors. Mol Microbiol 27(6):1203–1212

Deasy AM, Guccione E, Dale AP, Andrews N, Evans CM, Bennett JS, Bratcher HB, Maiden MC, Gorringe AR, Read RC (2015) Nasal inoculation of the commensal *Neisseria lactamica* inhibits carriage of *Neisseria meningitidis* by young adults: a controlled human infection study. Clin Infect Dis 60(10):1512–1520. https://doi.org/10.1093/cid/civ098

Del Tordello E, Vacca I, Ram S, Rappuoli R, Serruto D (2014) *Neisseria meningitidis* NalP cleaves human complement C3, facilitating degradation of C3b and survival in human serum. Proc Natl Acad Sci U S A 111(1):427–432. https://doi.org/10.1073/pnas.1321556111

Derrick JP, Urwin R, Suker J, Feavers IM, Maiden MC (1999) Structural and evolutionary inference from molecular variation in *Neisseria* porins. Infect Immun 67(5):2406–2413

Didelot X, Walker AS, Peto TE, Crook DW, Wilson DJ (2016) Within-host evolution of bacterial pathogens. Nat Rev Microbiol 14(3):150–162. https://doi.org/10.1038/nrmicro.2015.13

Dillard JP, Seifert HS (2001) A variable genetic island specific for *Neisseria gonorrhoeae* is involved in providing DNA for natural transformation and is found more often in disseminated infection isolates. Mol Microbiol 41(1):263–277

Donati C, Zolfo M, Albanese D, Tin Truong D, Asnicar F, Iebba V, Cavalieri D, Jousson O, De Filippo C, Huttenhower C, Segata N (2016) Uncovering oral *Neisseria* tropism and persistence using metagenomic sequencing. Nat Microbiol 1(7):16070. https://doi.org/10.1038/nmicrobiol.2016.70

Dossett JH, Appelbaum PC, Knapp JS, Totten PA (1985) Proctitis associated with *Neisseria cinerea* misidentified as *Neisseria gonorrhoeae* in a child. J Clin Microbiol 21(4):575–577

Edwards JL, Apicella MA (2004) The molecular mechanisms used by *Neisseria gonorrhoeae* to initiate infection differ between men and women. Clin Microbiol Rev 17 (4):965–981, Table of contents. https://doi.org/10.1128/CMR.17.4.965-981.2004

Edwards JL, Butler EK (2011) The pathobiology of *Neisseria gonorrhoeae* lower female genital tract infection. Front Microbiol 2:102. https://doi.org/10.3389/fmicb.2011.00102

Elkins C, Thomas CE, Seifert HS, Sparling PF (1991) Species-specific uptake of DNA by gonococci is mediated by a 10-base-pair sequence. J Bacteriol 173(12):3911–3913

Evans NJ, Harrison OB, Clow K, Derrick JP, Feavers IM, Maiden MC (2010) Variation and molecular evolution of HmbR, the *Neisseria meningitidis* haemoglobin receptor. Microbiology 156(Pt 5):1384–1393. https://doi.org/10.1099/mic.0.036475-0

Fairley CK, Hocking JS, Zhang L, Chow EP (2017) Frequent transmission of gonorrhea in men who have sex with men. Emerg Infect Dis 23(1):102–104. https://doi.org/10.3201/eid2301.161205

Feavers IM, Maiden MC (1998) A gonococcal porA pseudogene: implications for understanding the evolution and pathogenicity of *Neisseria gonorrhoeae*. Mol Microbiol 30(3):647–656

Finlay BB, Falkow S (1997) Common themes in microbial pathogenicity revisited. Microbiol Mol Biol Rev 61(2):136–169

Frasch CE, Zollinger WD, Poolman JT (1985) Serotype antigens of *Neisseria meningitidis* and a proposed scheme for designation of serotypes. Rev Infect Dis 7(4):504–510

Frasch CE, Preziosi MP, LaForce FM (2012) Development of a group A meningococcal conjugate vaccine, MenAfriVac(TM). Hum Vaccin Immunother 8(6):715–724. https://doi.org/10.4161/hv.19619

Garcia DL, Dillard JP (2008) Mutations in ampG or ampD affect peptidoglycan fragment release from *Neisseria gonorrhoeae*. J Bacteriol 190(11):3799–3807. https://doi.org/10.1128/JB.01194-07

Gault J, Ferber M, Machata S, Imhaus AF, Malosse C, Charles-Orszag A, Millien C, Bouvier G, Bardiaux B, Pehau-Arnaudet G, Klinge K, Podglajen I, Ploy MC, Seifert HS, Nilges M, Chamot-Rooke J, Dumenil G (2015) *Neisseria meningitidis* type IV pili composed of sequence invariable pilins are masked by multisite glycosylation. PLoS Pathog 11(9):e1005162. https://doi.org/10.1371/journal.ppat.1005162

Girardin SE, Boneca IG, Carneiro LA, Antignac A, Jehanno M, Viala J, Tedin K, Taha MK, Labigne A, Zahringer U, Coyle AJ, DiStefano PS, Bertin J, Sansonetti PJ, Philpott DJ (2003) Nod1 detects a unique muropeptide from gram-negative bacterial peptidoglycan. Science 300 (5625):1584–1587. https://doi.org/10.1126/science.1084677

Goldschneider I, Gotschlich EC, Artenstein MS (1969) Human immunity to the meningococcus. I. The role of humoral antibodies. J Exp Med 129(6):1307–1326

Goodman SD, Scocca JJ (1988) Identification and arrangement of the DNA sequence recognized in specific transformation of *Neisseria gonorrhoeae*. Proc Natl Acad Sci U S A 85(18):6982–6986

Gotschlich EC (1994) Genetic locus for the biosynthesis of the variable portion of *Neisseria gonorrhoeae* lipooligosaccharide. J Exp Med 180(6):2181–2190

Gray-Owen SD, Lorenzen DR, Haude A, Meyer TF, Dehio C (1997) Differential Opa specificities for CD66 receptors influence tissue interactions and cellular response to *Neisseria gonorrhoeae*. Mol Microbiol 26(5):971–980

Grissa I, Bouchon P, Pourcel C, Vergnaud G (2008) On-line resources for bacterial micro-evolution studies using MLVA or CRISPR typing. Biochimie 90(4):660–668. https://doi.org/10.1016/j.biochi.2007.07.014

Haas R, Meyer TF (1986) The repertoire of silent pilus genes in *Neisseria gonorrhoeae*: evidence for gene conversion. Cell 44:107–115

Hacker J, Kaper JB (2000) Pathogenicity islands and the evolution of microbes. Annu Rev Microbiol 54:641–679. https://doi.org/10.1146/annurev.micro.54.1.641

Hagblom P, Segal E, Billyard E, So M (1985) Intragenic recombination leads to pilus antigenic variation in *Neisseria gonorrhoeae*. Nature 315(6015):156–158

Hagen TA, Cornelissen CN (2006) *Neisseria gonorrhoeae* requires expression of TonB and the putative transporter TdfF to replicate within cervical epithelial cells. Mol Microbiol 62 (4):1144–1157. https://doi.org/10.1111/j.1365-2958.2006.05429.x

Hagman M, Forslin L, Moi H, Danielsson D (1991) *Neisseria meningitidis* in specimens from urogenital sites. Is increased awareness necessary? Sex Transm Dis 18(4):228–232

Haines KA, Yeh L, Blake MS, Cristello P, Korchak H, Weissmann G (1988) Protein I, a translocatable ion channel from *Neisseria gonorrhoeae*, selectively inhibits exocytosis from human neutrophils without inhibiting O_2-generation. J Biol Chem 263(2):945–951

Hamilton HL, Dominguez NM, Schwartz KJ, Hackett KT, Dillard JP (2005) *Neisseria gonorrhoeae* secretes chromosomal DNA via a novel type IV secretion system. Mol Microbiol 55 (6):1704–1721. https://doi.org/10.1111/j.1365-2958.2005.04521.x

Hammarstrom S (1999) The carcinoembryonic antigen (CEA) family: structures, suggested functions and expression in normal and malignant tissues. Semin Cancer Biol 9(2):67–81. https://doi.org/10.1006/scbi.1998.0119

Hammerschmidt S, Muller A, Sillmann H, Muhlenhoff M, Borrow R, Fox A, van Putten J, Zollinger WD, Gerardy-Schahn R, Frosch M (1996) Capsule phase variation in *Neisseria meningitidis* serogroup B by slipped-strand mispairing in the polysialyltransferase gene (siaD): correlation with bacterial invasion and the outbreak of meningococcal disease. Mol Microbiol 20(6):1211–1220

Hamrick TS, Dempsey JA, Cohen MS, Cannon JG (2001) Antigenic variation of gonococcal pilin expression in vivo: analysis of the strain FA1090 pilin repertoire and identification of the *pilS* gene copies recombining with *pilE* during experimental human infection. Microbiology 147 (Pt 4):839–849

Hansen MT (1978) Multiplicity of genome equivalents in the radiation-resistant bacterium *Micrococcus radiodurans*. J Bacteriol 134(1):71–75

Harrison OB, Evans NJ, Blair JM, Grimes HS, Tinsley CR, Nassif X, Kriz P, Ure R, Gray SJ, Derrick JP, Maiden MC, Feavers IM (2009) Epidemiological evidence for the role of the hemoglobin receptor, *hmbR*, in meningococcal virulence. J Infect Dis 200(1):94–98. https://doi.org/10.1086/599377

Harrison OB, Bennett JS, Derrick JP, Maiden MC, Bayliss CD (2013a) Distribution and diversity of the haemoglobin-haptoglobin iron-acquisition systems in pathogenic and non-pathogenic *Neisseria*. Microbiology 159(Pt 9):1920–1930. https://doi.org/10.1099/mic.0.068874-0

Harrison OB, Claus H, Jiang Y, Bennett JS, Bratcher HB, Jolley KA, Corton C, Care R, Poolman JT, Zollinger WD, Frasch CE, Stephens DS, Feavers I, Frosch M, Parkhill J, Vogel U, Quail MA, Bentley SD, Maiden MC (2013b) Description and nomenclature of *Neisseria meningitidis* capsule locus. Emerg Infect Dis 19(4):566–573. https://doi.org/10.3201/eid1904.111799

Helm RA, Seifert HS (2010) Frequency and rate of pilin antigenic variation of *Neisseria meningitidis*. J Bacteriol 192(14):3822–3823. https://doi.org/10.1128/JB.00280-10

Howell-Adams B, Seifert HS (2000) Molecular models accounting for the gene conversion reactions mediating gonococcal pilin antigenic variation. Mol Microbiol 37(5):1146–1158

Jamet A, Nassif X (2015) New players in the toxin field: polymorphic toxin systems in bacteria. MBio 6(3):e00285–e00215. https://doi.org/10.1128/mBio.00285-15

Jamet A, Jousset AB, Euphrasie D, Mukorako P, Boucharlat A, Ducousso A, Charbit A, Nassif X (2015) A new family of secreted toxins in pathogenic *Neisseria* species. PLoS Pathog 11(1): e1004592. https://doi.org/10.1371/journal.ppat.1004592

Jansen R, Embden JD, Gaastra W, Schouls LM (2002) Identification of genes that are associated with DNA repeats in prokaryotes. Mol Microbiol 43(6):1565–1575

Jarva H, Ram S, Vogel U, Blom AM, Meri S (2005) Binding of the complement inhibitor C4bp to serogroup B *Neisseria meningitidis*. J Immunol 174(10):6299–6307

Jennings MP, Srikhanta YN, Moxon ER, Kramer M, Poolman JT, Kuipers B, van der Ley P (1999) The genetic basis of the phase variation repertoire of lipopolysaccharide immunotypes in *Neisseria meningitidis*. Microbiology 145(Pt 11):3013–3021. https://doi.org/10.1099/00221287-145-11-3013

Jennings MP, Jen FE, Roddam LF, Apicella MA, Edwards JL (2011) *Neisseria gonorrhoeae* pilin glycan contributes to CR3 activation during challenge of primary cervical epithelial cells. Cell Microbiol 13(6):885–896. https://doi.org/10.1111/j.1462-5822.2011.01586.x

Johannsen DB, Johnston DM, Koymen HO, Cohen MS, Cannon JG (1999) A *Neisseria gonorrhoeae* immunoglobulin A1 protease mutant is infectious in the human challenge model of urethral infection. Infect Immun 67(6):3009–3013

John CM, Liu M, Phillips NJ, Yang Z, Funk CR, Zimmerman LI, Griffiss JM, Stein DC, Jarvis GA (2012) Lack of lipid A pyrophosphorylation and functional *lptA* reduces inflammation by *Neisseria* commensals. Infect Immun 80(11):4014–4026. https://doi.org/10.1128/IAI.00506-12

Johswich KO, McCaw SE, Islam E, Sintsova A, Gu A, Shively JE, Gray-Owen SD (2013) In vivo adaptation and persistence of *Neisseria meningitidis* within the nasopharyngeal mucosa. PLoS Pathog 9(7):e1003509. https://doi.org/10.1371/journal.ppat.1003509

Jonsson AB, Nyberg G, Normark S (1991) Phase variation of gonococcal pili by frameshift mutation in *pilC*, a novel gene for pilus assembly. EMBO J 10(2):477–488

Jordan PW, Snyder LA, Saunders NJ (2005) Strain-specific differences in *Neisseria gonorrhoeae* associated with the phase variable gene repertoire. BMC Microbiol 5:21. https://doi.org/10.1186/1471-2180-5-21

Judson FN (1990) Gonorrhea. Med Clin North Am 74(6):1353–1366

Kahler CM, Blum E, Miller YK, Ryan D, Popovic T, Stephens DS (2001) *exl*, an exchangeable genetic island in *Neisseria meningitidis*. Infect Immun 69(3):1687–1696. https://doi.org/10.1128/IAI.69.3.1687-1696.2001

Kent CK, Chaw JK, Wong W, Liska S, Gibson S, Hubbard G, Klausner JD (2005) Prevalence of rectal, urethral, and pharyngeal chlamydia and gonorrhea detected in 2 clinical settings among men who have sex with men: San Francisco, California, 2003. Clin Infect Dis 41(1):67–74. https://doi.org/10.1086/430704

King GJ, Swanson J (1978) Studies on gonococcus infection. XV. Identification of surface proteins of *Neisseria gonorrhoeae* correlated with leukocyte association. Infect Immun 21:575–584

Kinghorn G (2010) Pharyngeal gonorrhoea: a silent cause for concern. Sex Transm Infect 86 (6):413–414. https://doi.org/10.1136/sti.2010.043349

Kitten T, Barbour AG (1992) The relapsing fever agent *Borrelia hermsii* has multiple copies of its chromosome and linear plasmids. Genetics 132(2):311–324

Kline KA, Seifert HS (2005) Role of the Rep helicase gene in homologous recombination in *Neisseria gonorrhoeae*. J Bacteriol 187(8):2903–2907

Klughammer J, Dittrich M, Blom J, Mitesser V, Vogel U, Frosch M, Goesmann A, Muller T, Schoen C (2017) Comparative genome sequencing reveals within-host genetic changes in *Neisseria meningitidis* during invasive disease. PLoS One 12(1):e0169892. https://doi.org/10.1371/journal.pone.0169892

Knapp JS (1988) Historical perspectives and identification of *Neisseria* and related species. Clin Microbiol Rev 1(4):415–431

Komaki K, Ishikawa H (1999) Intracellular bacterial symbionts of aphids possess many genomic copies per bacterium. J Mol Evol 48(6):717–722

Koomey M, Gotschlich EC, Robbins K, Bergstrom S, Swanson J (1987) Effects of *recA* mutations on pilus antigenic variation and phase transitions in *Neisseria gonorrhoeae*. Genetics 117 (3):391–398

Kuryavyi V, Cahoon LA, Seifert HS, Patel DJ (2012) RecA-binding *pilE* G4 sequence essential for pilin antigenic variation forms monomeric and 5′ end-stacked dimeric parallel G-quadruplexes. Structure 20(12):2090–2102. https://doi.org/10.1016/j.str.2012.09.013

Lee MH, Walker GC (1996) Interactions of *Escherichia coli* UmuD with activated RecA analyzed by cross-linking UmuD monocysteine derivatives. J Bacteriol 178(24):7285–7294

Leighton MP, Kelly DJ, Williamson MP, Shaw JG (2001) An NMR and enzyme study of the carbon metabolism of *Neisseria meningitidis*. Microbiology 147(Pt 6):1473–1482. https://doi.org/10.1099/00221287-147-6-1473

Leo JC, Grin I, Linke D (2012) Type V secretion: mechanism(s) of autotransport through the bacterial outer membrane. Philos Trans R Soc Lond Ser B Biol Sci 367(1592):1088–1101. https://doi.org/10.1098/rstb.2011.0208

Lewis LA, Ram S (2014) Meningococcal disease and the complement system. Virulence 5 (1):98–126. https://doi.org/10.4161/viru.26515

Liguori A, Malito E, Lo Surdo P, Fagnocchi L, Cantini F, Haag AF, Brier S, Pizza M, Delany I, Bottomley MJ (2016) Molecular basis of ligand-dependent regulation of NadR, the transcriptional repressor of meningococcal virulence factor NadA. PLoS Pathog 12(4):e1005557. https://doi.org/10.1371/journal.ppat.1005557

Lin L, Ayala P, Larson J, Mulks M, Fukuda M, Carlsson SR, Enns C, So M (1997) The *Neisseria* type 2 IgA1 protease cleaves LAMP1 and promotes survival of bacteria within epithelial cells. Mol Microbiol 24(5):1083–1094

Liu G, Tang CM, Exley RM (2015) Non-pathogenic *Neisseria*: members of an abundant, multi-habitat, diverse genus. Microbiology 161(7):1297–1312. https://doi.org/10.1099/mic.0.000086

Loh E, Kugelberg E, Tracy A, Zhang Q, Gollan B, Ewles H, Chalmers R, Pelicic V, Tang CM (2013) Temperature triggers immune evasion by *Neisseria meningitidis*. Nature 502 (7470):237–240. https://doi.org/10.1038/nature12616

Louwen R, Staals RH, Endtz HP, van Baarlen P, van der Oost J (2014) The role of CRISPR-Cas systems in virulence of pathogenic bacteria. Microbiol Mol Biol Rev 78(1):74–88. https://doi.org/10.1128/MMBR.00039-13

Mackey WC, Immerman RS (2003) A proposed feedback loop of sexually transmitted diseases and sexual behavior: the Red Queen's Dilemma. Soc Biol 50(3–4):281–299

MacNeil J, Cohn A (2011) Meningococcal disease. In: Roush LMB SW (ed) Vaccine preventable diseases surveillance manual, 6th edn. Centers for Disease Control and Prevention, Atlanta, GA

Maizels N, Gray LT (2013) The G4 genome. PLoS Genet 9(4):e1003468. https://doi.org/10.1371/journal.pgen.1003468

Marraffini LA (2015) CRISPR-Cas immunity in prokaryotes. Nature 526(7571):55–61. https://doi.org/10.1038/nature15386

Marri PR, Paniscus M, Weyand NJ, Rendon MA, Calton CM, Hernandez DR, Higashi DL, Sodergren E, Weinstock GM, Rounsley SD, So M (2010) Genome sequencing reveals widespread virulence gene exchange among human *Neisseria* species. PLoS One 5(7):e11835. https://doi.org/10.1371/journal.pone.0011835

Massari P, Henneke P, Ho Y, Latz E, Golenbock DT, Wetzler LM (2002) Cutting edge: immune stimulation by Neisserial porins is toll-like receptor 2 and MyD88 dependent. J Immunol 168 (4):1533–1537

Massari P, Ram S, Macleod H, Wetzler LM (2003) The role of porins in Neisserial pathogenesis and immunity. Trends Microbiol 11(2):87–93

McGee ZA, Johnson AP, Taylor-Robinson D (1981) Pathogenic mechanisms of *Neisseria gonorrhoeae*: observations on damage to human fallopian tubes in organ culture by gonococci of colony type 1 or type 4. J Infect Dis 143(3):413–422

McKenna JG, Fallon RJ, Moyes A, Young H (1993) Anogenital non-gonococcal Neisseriae: prevalence and clinical significance. Int J STD AIDS 4(1):8–12. https://doi.org/10.1177/095646249300400103

Mehr IJ, Seifert HS (1997) Random shuttle mutagenesis: gonococcal mutants deficient in pilin antigenic variation. Mol Microbiol 23(6):1121–1131

Mehr IJ, Seifert HS (1998) Differential roles of homologous recombination pathways in *Neisseria gonorrhoeae* pilin antigenic variation, DNA transformation and DNA repair. Mol Microbiol 30 (4):697–710

Melly MA, McGee ZA, Rosenthal RS (1984) Ability of monomeric peptidoglycan fragments from *Neisseria gonorrhoeae* to damage human fallopian-tube mucosa. J Infect Dis 149(3):378–386

Metruccio MM, Pigozzi E, Roncarati D, Berlanda Scorza F, Norais N, Hill SA, Scarlato V, Delany I (2009) A novel phase variation mechanism in the meningococcus driven by a ligand-responsive repressor and differential spacing of distal promoter elements. PLoS Pathog 5(12):e1000710. https://doi.org/10.1371/journal.ppat.1000710

Meyer TF, Billyard E, Haas R, Storzbach S, So M (1984) Pilus genes of *Neisseria gonorrheae*: chromosomal organization and DNA sequence. Proc Natl Acad Sci U S A 81:6110–6114

Meyer J, Brissac T, Frapy E, Omer H, Euphrasie D, Bonavita A, Nassif X, Bille E (2016) Characterization of MDAPhi, a temperate filamentous bacteriophage of *Neisseria meningitidis*. Microbiology 162(2):268–282. https://doi.org/10.1099/mic.0.000215

Miller F, Phan G, Brissac T, Bouchiat C, Lioux G, Nassif X, Coureuil M (2014) The hypervariable region of meningococcal major pilin PilE controls the host cell response via antigenic variation. MBio 5(1):e01024–e01013. https://doi.org/10.1128/mBio.01024-13

Minton KW (1994) DNA repair in the extremely radioresistant bacterium *Deinococcus radiodurans*. Mol Microbiol 13(1):9–15

Miyoshi D, Nakao A, Sugimoto N (2003) Structural transition from antiparallel to parallel G-quadruplex of d(G4T4G4) induced by Ca2+. Nucleic Acids Res 31(4):1156–1163

Moore J, Bailey SE, Benmechernene Z, Tzitzilonis C, Griffiths NJ, Virji M, Derrick JP (2005) Recognition of saccharides by the OpcA, OpaD, and OpaB outer membrane proteins from *Neisseria meningitidis*. J Biol Chem 280(36):31489–31497. https://doi.org/10.1074/jbc.M506354200

Moxon R, Bayliss C, Hood D (2006) Bacterial contingency loci: the role of simple sequence DNA repeats in bacterial adaptation. Annu Rev Genet 40:307–333. https://doi.org/10.1146/annurev.genet.40.110405.090442

Mulks MH, Shoberg RJ (1994) Bacterial immunoglobulin A1 proteases. Methods Enzymol 235:543–554

Murphy GL, Connell TD, Barritt DS, Koomey M, Cannon JG (1989) Phase variation of gonococcal protein II: regulation of gene expression by slipped-strand mispairing of a repetitive DNA sequence. Cell 56(4):539–547

Muzzi A, Mora M, Pizza M, Rappuoli R, Donati C (2013) Conservation of meningococcal antigens in the genus *Neisseria*. MBio 4(3):e00163–e00113. https://doi.org/10.1128/mBio.00163-13

Nagpal P, Jafri S, Reddy MA, Das HK (1989) Multiple chromosomes of *Azotobacter vinelandii*. J Bacteriol 171(6):3133–3138

Neil RB, Apicella MA (2009) Role of HrpA in biofilm formation of *Neisseria meningitidis* and regulation of the *hrpBAS* transcripts. Infect Immun 77(6):2285–2293. https://doi.org/10.1128/IAI.01502-08

Ngampasutadol J, Ram S, Gulati S, Agarwal S, Li C, Visintin A, Monks B, Madico G, Rice PA (2008) Human factor H interacts selectively with *Neisseria gonorrhoeae* and results in species-specific complement evasion. J Immunol 180(5):3426–3435

Obergfell KP, Seifert HS (2015) Mobile DNA in the pathogenic *Neisseria*. Microbiol Spectr 3(1). MDNA3-0015-2014. https://doi.org/10.1128/microbiolspec.MDNA3-0015-2014

Park HS, Wolfgang M, van Putten JP, Dorward D, Hayes SF, Koomey M (2001) Structural alterations in a type IV pilus subunit protein result in concurrent defects in multicellular behaviour and adherence to host tissue. Mol Microbiol 42(2):293–307

Pecoraro V, Zerulla K, Lange C, Soppa J (2011) Quantification of ploidy in proteobacteria revealed the existence of monoploid, (mero-)oligoploid and polyploid species. PLoS One 6(1):e16392. https://doi.org/10.1371/journal.pone.0016392

Perkins-Balding D, Ratliff-Griffin M, Stojiljkovic I (2004) Iron transport systems in *Neisseria meningitidis*. Microbiol Mol Biol Rev 68(1):154–171

Pirofski LA, Casadevall A (2012) Q and A: what is a pathogen? A question that begs the point. BMC Biol 10:6. https://doi.org/10.1186/1741-7007-10-6

Plaut AG, Gilbert JV, Artenstein MS, Capra JD (1975) *Neisseria gonorrhoeae* and *neisseria meningitidis*: extracellular enzyme cleaves human immunoglobulin A. Science 190 (4219):1103–1105

Pohlner J, Halter R, Beyreuther K, Meyer TF (1987) Gene structure and extracellular secretion of *Neisseria gonorrhoeae* IgA protease. Nature 325(6103):458–462. https://doi.org/10.1038/325458a0

Poole SJ, Diner EJ, Aoki SK, Braaten BA, t'Kint de Roodenbeke C, Low DA, Hayes CS (2011) Identification of functional toxin/immunity genes linked to contact-dependent growth inhibition (CDI) and rearrangement hotspot (Rhs) systems. PLoS Genet 7(8):e1002217. https://doi.org/10.1371/journal.pgen.1002217

Power PM, Roddam LF, Rutter K, Fitzpatrick SZ, Srikhanta YN, Jennings MP (2003) Genetic characterization of pilin glycosylation and phase variation in *Neisseria meningitidis*. Mol Microbiol 49(3):833–847

Price AA, Sampson TR, Ratner HK, Grakoui A, Weiss DS (2015) Cas9-mediated targeting of viral RNA in eukaryotic cells. Proc Natl Acad Sci U S A 112(19):6164–6169. https://doi.org/10.1073/pnas.1422340112

Rhodes D, Lipps HJ (2015) G-quadruplexes and their regulatory roles in biology. Nucleic Acids Res 43(18):8627–8637. https://doi.org/10.1093/nar/gkv862

Rice PA, Shafer WM, Ram S, Jerse AE (2017) *Neisseria gonorrhoeae*: drug resistance, mouse models, and vaccine development. Annu Rev Microbiol 71:665–686. https://doi.org/10.1146/annurev-micro-090816-093530

Robson RL, Chesshyre JA, Wheeler C, Jones R, Woodley PR, Postgate JR (1984) Genome size and complexity in *Azotobacter chroococcum*. J Gen Microbiol 130(7):1603–1612

Rotman E, Seifert HS (2014) The genetics of *Neisseria* species. Annu Rev Genet 48:405–431. https://doi.org/10.1146/annurev-genet-120213-092007

Rotman E, Webber DM, Seifert HS (2016) Analyzing *Neisseria gonorrhoeae* pilin antigenic variation using 454 sequencing technology. J Bacteriol 198:2470–2482. https://doi.org/10.1128/JB.00330-16

Rouphael NG, Stephens DS (2012) *Neisseria meningitidis:* biology, microbiology, and epidemiology. Methods Mol Biol 799:1–20. https://doi.org/10.1007/978-1-61779-346-2_1

Rudel T, van Putten JP, Gibbs CP, Haas R, Meyer TF (1992) Interaction of two variable proteins (PilE and PilC) required for pilus-mediated adherence of *Neisseria gonorrhoeae* to human epithelial cells. Mol Microbiol 6(22):3439–3450

Sadarangani M, Pollard AJ (2010) Serogroup B meningococcal vaccines-an unfinished story. Lancet Infect Dis 10(2):112–124. https://doi.org/10.1016/S1473-3099(09)70324-X

Sadarangani M, Pollard AJ, Gray-Owen SD (2011) Opa proteins and CEACAMs: pathways of immune engagement for pathogenic *Neisseria*. FEMS Microbiol Rev 35(3):498–514. https://doi.org/10.1111/j.1574-6976.2010.00260.x

Saleem M, Prince SM, Rigby SE, Imran M, Patel H, Chan H, Sanders H, Maiden MC, Feavers IM, Derrick JP (2013) Use of a molecular decoy to segregate transport from antigenicity in the FrpB iron transporter from *Neisseria meningitidis*. PLoS One 8(2):e56746. https://doi.org/10.1371/journal.pone.0056746

Sarantis H, Gray-Owen SD (2012) Defining the roles of human carcinoembryonic antigen-related cellular adhesion molecules during neutrophil responses to *Neisseria gonorrhoeae*. Infect Immun 80(1):345–358. https://doi.org/10.1128/IAI.05702-11

Sarkari J, Pandit N, Moxon ER, Achtman M (1994) Variable expression of the Opc outer membrane protein in *Neisseria meningitidis* is caused by size variation of a promoter containing poly-cytidine. Mol Microbiol 13(2):207–217

Saunders NJ (2000) Repeat-associated phase variable genes in the complete genome sequence of *Neisseria meningitidis* strain MC58. Mol Microbiol 37(1):207–215

Schielke S, Huebner C, Spatz C, Nagele V, Ackermann N, Frosch M, Kurzai O, Schubert-Unkmeir A (2009) Expression of the meningococcal adhesin NadA is controlled by a transcriptional regulator of the MarR family. Mol Microbiol 72(4):1054–1067. https://doi.org/10.1111/j.1365-2958.2009.06710.x

Schmidt H, Hensel M (2004) Pathogenicity islands in bacterial pathogenesis. Clin Microbiol Rev 17 (1):14–56

Schmitt C, Turner D, Boesl M, Abele M, Frosch M, Kurzai O (2007) A functional two-partner secretion system contributes to adhesion of *Neisseria meningitidis* to epithelial cells. J Bacteriol 189(22):7968–7976. https://doi.org/10.1128/JB.00851-07

Schoen C, Blom J, Claus H, Schramm-Gluck A, Brandt P, Muller T, Goesmann A, Joseph B, Konietzny S, Kurzai O, Schmitt C, Friedrich T, Linke B, Vogel U, Frosch M (2008) Whole-genome comparison of disease and carriage strains provides insights into virulence evolution in *Neisseria meningitidis*. Proc Natl Acad Sci U S A 105(9):3473–3478. https://doi.org/10.1073/pnas.0800151105

Scholten RJ, Kuipers B, Valkenburg HA, Dankert J, Zollinger WD, Poolman JT (1994) Lipo-oligosaccharide immunotyping of *Neisseria meningitidis* by a whole-cell ELISA with monoclonal antibodies. J Med Microbiol 41(4):236–243

Schryvers AB, Stojiljkovic I (1999) Iron acquisition systems in the pathogenic *Neisseria*. Mol Microbiol 32(6):1117–1123

Sechman EV, Rohrer MS, Seifert HS (2005) A genetic screen identifies genes and sites involved in pilin antigenic variation in *Neisseria gonorrhoeae*. Mol Microbiol 57(2):468–483. https://doi.org/10.1111/j.1365-2958.2005.04657.x

Serino L, Virji M (2000) Phosphorylcholine decoration of lipopolysaccharide differentiates commensal *Neisseriae* from pathogenic strains: identification of licA-type genes in commensal *Neisseriae*. Mol Microbiol 35(6):1550–1559

Siena E, D'Aurizio R, Riley D, Tettelin H, Guidotti S, Torricelli G, Moxon ER, Medini D (2016) In-silico prediction and deep-DNA sequencing validation indicate phase variation in 115 *Neisseria meningitidis* genes. BMC Genomics 17(1):843. https://doi.org/10.1186/s12864-016-3185-1

Sigurlásdóttir S, Engman J, Eriksson OS, Saroj SD, Zguna N, Lloris-Garcerá P, Ilag LL, Jonsson A-B, Tang C (2017) Host cell-derived lactate functions as an effector molecule in *Neisseria meningitidis* microcolony dispersal. PLoS Pathog 13(4):e1006251

Simon MC, Schmidt HJ (2007) Antigenic variation in ciliates: antigen structure, function, expression. J Eukaryot Microbiol 54(1):1–7. https://doi.org/10.1111/j.1550-7408.2006.00226.x

Simons MP, Nauseef WM, Apicella MA (2005) Interactions of *Neisseria gonorrhoeae* with adherent polymorphonuclear leukocytes. Infect Immun 73(4):1971–1977. https://doi.org/10.1128/IAI.73.4.1971-1977.2005

Skaar EP, Lazio MP, Seifert HS (2002) Roles of the *recJ* and *recN* genes in homologous recombination and DNA repair pathways of *Neisseria gonorrhoeae*. J Bacteriol 184(4):919–927

Smith H, Yates EA, Cole JA, Parsons NJ (2001) Lactate stimulation of gonococcal metabolism in media containing glucose: mechanism, impact on pathogenicity, and wider implications for other pathogens. Infect Immun 69(11):6565–6572. https://doi.org/10.1128/IAI.69.11.6565-6572.2001

Snyder LA, Butcher SA, Saunders NJ (2001) Comparative whole-genome analyses reveal over 100 putative phase-variable genes in the pathogenic *Neisseria* spp. Microbiology 147 (Pt 8):2321–2332

Sparling PF (1966) Genetic transformation of *Neisseria gonorrhoea*e to streptomycin resistance. J Bacteriol 92:1364–1371

Spinosa MR, Progida C, Tala A, Cogli L, Alifano P, Bucci C (2007) The *Neisseria meningitidis* capsule is important for intracellular survival in human cells. Infect Immun 75(7):3594–3603. https://doi.org/10.1128/IAI.01945-06

Stein DC, Gunn JS, Radlinska M, Piekarowicz A (1995) Restriction and modification systems of *Neisseria gonorrhoeae*. Gene 157(1–2):19–22

Stein DC, Miller CJ, Bhoopalan SV, Sommer DD (2011) Sequence-based predictions of lipooligosaccharide diversity in the *Neisseriaceae* and their implication in pathogenicity. PLoS One 6(4):e18923. https://doi.org/10.1371/journal.pone.0018923

Stephens DS (2009) Biology and pathogenesis of the evolutionarily successful, obligate human bacterium *Neisseria meningitidis*. Vaccine 27(Suppl 2):B71–B77. https://doi.org/10.1016/j.vaccine.2009.04.070

Stephens DS, Spellman PA, Swartley JS (1993) Effect of the (alpha 2-->8)-linked polysialic acid capsule on adherence of *Neisseria meningitidis* to human mucosal cells. J Infect Dis 167 (2):475–479

Stern A, Brown M, Nickel P, Meyer TF (1986) Opacity genes in *Neisseria gonorrhoeae*: control of phase and antigenic variation. Cell 47(1):61–71

Stohl EA, Dale EM, Criss AK, Seifert HS (2013) *Neisseria gonorrhoeae* metalloprotease NGO1686 is required for full piliation, and piliation is required for resistance to H2O2- and neutrophil-mediated killing. MBio 4(4):e00399-13. https://doi.org/10.1128/mBio.00399-13

Swanson J (1978) Studies on gonococcus infection. XII. Colony color and opacity variants of gonococci. Infect Immun 19:320–331

Swartley JS, Marfin AA, Edupuganti S, Liu LJ, Cieslak P, Perkins B, Wenger JD, Stephens DS (1997) Capsule switching of *Neisseria meningitidis*. Proc Natl Acad Sci U S A 94(1):271–276

Taha MK, Claus H, Lappann M, Veyrier FJ, Otto A, Becher D, Deghmane AE, Frosch M, Hellenbrand W, Hong E, Parent du Chatelet I, Prior K, Harmsen D, Vogel U (2016) Evolutionary events associated with an outbreak of meningococcal disease in men who have sex with men. PLoS One 11(5):e0154047. https://doi.org/10.1371/journal.pone.0154047

Takahashi H, Yanagisawa T, Kim KS, Yokoyama S, Ohnishi M (2012) Meningococcal PilV potentiates *Neisseria meningitidis* type IV pilus-mediated internalization into human endothelial and epithelial cells. Infect Immun 80(12):4154–4166. https://doi.org/10.1128/IAI.00423-12

Tala A, Progida C, De Stefano M, Cogli L, Spinosa MR, Bucci C, Alifano P (2008) The HrpB-HrpA two-partner secretion system is essential for intracellular survival of *Neisseria meningitidis*. Cell Microbiol 10(12):2461–2482. https://doi.org/10.1111/j.1462-5822.2008.01222.x

Tettelin H, Saunders NJ, Heidelberg J, Jeffries AC, Nelson KE, Eisen JA, Ketchum KA, Hood DW, Peden JF, Dodson RJ, Nelson WC, Gwinn ML, DeBoy R, Peterson JD, Hickey EK, Haft DH, Salzberg SL, White O, Fleischmann RD, Dougherty BA, Mason T, Ciecko A, Parksey DS, Blair E, Cittone H, Clark EB, Cotton MD, Utterback TR, Khouri H, Qin H, Vamathevan J, Gill J, Scarlato V, Masignani V, Pizza M, Grandi G, Sun L, Smith HO, Fraser CM, Moxon ER, Rappuoli R, Venter JC (2000) Complete genome sequence of *Neisseria meningitidis* serogroup B strain MC58. Science 287(5459):1809–1815

Tobiason DM, Seifert HS (2006a) The obligate human pathogen, *Neisseria gonorrhoeae*, is polyploid. PLoS Biol 4(6):e185. https://doi.org/10.1371/journal.pbio.0040185

Tobiason DM, Seifert HS (2006b) The obligate human pathogen, *Neisseria gonorrhoeae,* is polyploid. PLoS Biol 4(6):1069–1078

Tobiason DM, Seifert HS (2010) Genomic content of *Neisseria* species. J Bacteriol 192 (8):2160–2168. https://doi.org/10.1128/JB.01593-09

Toleman M, Aho E, Virji M (2001) Expression of pathogen-like Opa adhesins in commensal *Neisseria*: genetic and functional analysis. Cell Microbiol 3(1):33–44

Tommassen J, Vermeij P, Struyve M, Benz R, Poolman JT (1990) Isolation of *Neisseria meningitidis* mutants deficient in class 1 (*porA)* and class 3 (*porB*) outer membrane proteins. Infect Immun 58(5):1355–1359

Tonjum T (2005) Genus I. *Neisseria*. In: Garrity G, Brenner DJ, Krieg NR, Staley JR (eds) Bergey's manual of systematic bacteriology. Springer, New York, pp 777–798

Toussi DN, Carraway M, Wetzler LM, Lewis LA, Liu X, Massari P (2012) The amino acid sequence of *Neisseria lactamica* PorB surface-exposed loops influences Toll-like receptor 2-dependent cell activation. Infect Immun 80(10):3417–3428. https://doi.org/10.1128/IAI.00683-12

Turner DP, Wooldridge KG, Ala'Aldeen DA (2002) Autotransported serine protease A of *Neisseria meningitidis*: an immunogenic, surface-exposed outer membrane, and secreted protein. Infect Immun 70(8):4447–4461

Tzeng YL, Datta A, Ambrose K, Lo M, Davies JK, Carlson RW, Stephens DS, Kahler CM (2004) The MisR/MisS two-component regulatory system influences inner core structure and immunotype of lipooligosaccharide in *Neisseria meningitidis*. J Biol Chem 279 (33):35053–35062. https://doi.org/10.1074/jbc.M401433200

Tzeng YL, Thomas J, Stephens DS (2016) Regulation of capsule in *Neisseria meningitidis*. Crit Rev Microbiol 42(5):759–772. https://doi.org/10.3109/1040841X.2015.1022507

Tzeng YL, Bazan JA, Turner AN, Wang X, Retchless AC, Read TD, Toh E, Nelson DE, Del Rio C, Stephens DS (2017) Emergence of a new *Neisseria meningitidis* clonal complex 11 lineage 11.2 clade as an effective urogenital pathogen. Proc Natl Acad Sci U S A 114(16):4237–4242. https://doi.org/10.1073/pnas.1620971114

Uehara T, Park JT (2007) An anhydro-N-acetylmuramyl-L-alanine amidase with broad specificity tethered to the outer membrane of *Escherichia coli*. J Bacteriol 189(15):5634–5641. https://doi.org/10.1128/JB.00446-07

Unkmeir A, Latsch K, Dietrich G, Wintermeyer E, Schinke B, Schwender S, Kim KS, Eigenthaler M, Frosch M (2002) Fibronectin mediates Opc-dependent internalization of *Neisseria meningitidis* in human brain microvascular endothelial cells. Mol Microbiol 46 (4):933–946

Urra E, Alkorta M, Sota M, Alcala B, Martinez I, Barron J, Cisterna R (2005) Orogenital transmission of *Neisseria meningitidis* serogroup C confirmed by genotyping techniques. Eur J Clin Microbiol Infect Dis 24(1):51–53. https://doi.org/10.1007/s10096-004-1257-7

van der Ende A, Hopman CT, Dankert J (2000) Multiple mechanisms of phase variation of PorA in *Neisseria meningitidis*. Infect Immun 68(12):6685–6690

van der Woude MW (2011) Phase variation: how to create and coordinate population diversity. Curr Opin Microbiol 14(2):205–211. https://doi.org/10.1016/j.mib.2011.01.002

van Ulsen P, van Alphen L, ten Hove J, Fransen F, van der Ley P, Tommassen J (2003) A *Neisserial* autotransporter NalP modulating the processing of other autotransporters. Mol Microbiol 50 (3):1017–1030

van Vliet SJ, Steeghs L, Bruijns SC, Vaezirad MM, Snijders Blok C, Arenas Busto JA, Deken M, van Putten JP, van Kooyk Y (2009) Variation of *Neisseria gonorrhoeae* lipooligosaccharide directs dendritic cell-induced T helper responses. PLoS Pathog 5(10):e1000625. https://doi.org/10.1371/journal.ppat.1000625

Vaughan AT, Gorringe A, Davenport V, Williams NA, Heyderman RS (2009) Absence of mucosal immunity in the human upper respiratory tract to the commensal bacteria *Neisseria lactamica* but not pathogenic *Neisseria meningitidis* during the peak age of nasopharyngeal carriage. J Immunol 182(4):2231–2240. https://doi.org/10.4049/jimmunol.0802531

Viala J, Chaput C, Boneca IG, Cardona A, Girardin SE, Moran AP, Athman R, Memet S, Huerre MR, Coyle AJ, DiStefano PS, Sansonetti PJ, Labigne A, Bertin J, Philpott DJ, Ferrero RL (2004) Nod1 responds to peptidoglycan delivered by the *Helicobacter pylori* cag pathogenicity island. Nat Immunol 5(11):1166–1174. https://doi.org/10.1038/ni1131

Virji M (2009) Pathogenic Neisseriae: surface modulation, pathogenesis and infection control. Nat Rev Microbiol 7(4):274–286. https://doi.org/10.1038/nrmicro2097

Virji M, Makepeace K, Ferguson DJ, Achtman M, Sarkari J, Moxon ER (1992) Expression of the Opc protein correlates with invasion of epithelial and endothelial cells by *Neisseria meningitidis*. Mol Microbiol 6(19):2785–2795

Virji M, Makepeace K, Ferguson DJ, Achtman M, Moxon ER (1993a) Meningococcal Opa and Opc proteins: their role in colonization and invasion of human epithelial and endothelial cells. Mol Microbiol 10(3):499–510

Virji M, Saunders JR, Sims G, Makepeace K, Maskell D, Ferguson DJ (1993b) Pilus-facilitated adherence of *Neisseria meningitidis* to human epithelial and endothelial cells: modulation of adherence phenotype occurs concurrently with changes in primary amino acid sequence and the glycosylation status of pilin. Mol Microbiol 10(5):1013–1028

Virji M, Makepeace K, Ferguson DJ, Watt SM (1996) Carcinoembryonic antigens (CD66) on epithelial cells and neutrophils are receptors for Opa proteins of pathogenic Neisseriae. Mol Microbiol 22(5):941–950

Virji M, Evans D, Hadfield A, Grunert F, Teixeira AM, Watt SM (1999) Critical determinants of host receptor targeting by *Neisseria meningitidis* and *Neisseria gonorrhoeae*: identification of Opa adhesiotopes on the N-domain of CD66 molecules. Mol Microbiol 34(3):538–551

Walker CK, Sweet RL (2011) Gonorrhea infection in women: prevalence, effects, screening, and management. Int J Womens Health 3:197–206. https://doi.org/10.2147/IJWH.S13427

Weyand NJ, Ma M, Phifer-Rixey M, Taku NA, Rendon MA, Hockenberry AM, Kim WJ, Agellon AB, Biais N, Suzuki TA, Sait LG, Harrison OB, Bratcher HB, Nachman MW, Maiden MC, So M (2016) Isolation and characterization of a new species of *Neisseria, Neisseria musculi,* from the wild house mouse. Int J Syst Evol Microbiol 66(9):3585–3593. https://doi.org/10.1099/ijsem.0.001237

WHO (2012) Global incidence and prevalence of selected curable sexually transmitted infections – 2008. World Health Organization, Geneva, Switzerland

Wiesner PJ, Thompson SE 3rd (1980) Gonococcal diseases. Dis Mon 26(5):1–44

Wolf DM, Vazirani VV, Arkin AP (2005) A microbial modified prisoner's dilemma game: how frequency-dependent selection can lead to random phase variation. J Theor Biol 234 (2):255–262. https://doi.org/10.1016/j.jtbi.2004.11.021

Wolfgang M, Lauer P, Park HS, Brossay L, Hebert J, Koomey M (1998) PilT mutations lead to simultaneous defects in competence for natural transformation and twitching motility in piliated *Neisseria gonorrhoeae*. Mol Microbiol 29(1):321–330

Woodhams KL, Benet ZL, Blonsky SE, Hackett KT, Dillard JP (2012) Prevalence and detailed mapping of the gonococcal genetic island in *Neisseria meningitidis*. J Bacteriol 194 (9):2275–2285. https://doi.org/10.1128/JB.00094-12

Woodhams KL, Chan JM, Lenz JD, Hackett KT, Dillard JP (2013) Peptidoglycan fragment release from *Neisseria meningitidis*. Infect Immun 81(9):3490–3498. https://doi.org/10.1128/iai.00279-13

Workowski KA (2015) Centers for disease control and prevention sexually transmitted diseases treatment guidelines. Clin Infect Dis 61(Suppl 8):S759–S762. https://doi.org/10.1093/cid/civ771

Workowski KA, Bolan GA, Centers for Disease C, Prevention (2015) Sexually transmitted diseases treatment guidelines, 2015. MMWR Recomm Rep 64(RR-03):1–137

Wormann ME, Horien CL, Bennett JS, Jolley KA, Maiden MC, Tang CM, Aho EL, Exley RM (2014) Sequence, distribution and chromosomal context of class I and class II pilin genes of *Neisseria meningitidis* identified in whole genome sequences. BMC Genomics 15:253. https://doi.org/10.1186/1471-2164-15-253

Yang QL, Gotschlich EC (1996) Variation of gonococcal lipooligosaccharide structure is due to alterations in poly-G tracts in lgt genes encoding glycosyl transferases. J Exp Med 183 (1):323–327

Yazdankhah SP, Kriz P, Tzanakaki G, Kremastinou J, Kalmusova J, Musilek M, Alvestad T, Jolley KA, Wilson DJ, McCarthy ND, Caugant DA, Maiden MC (2004) Distribution of serogroups and genotypes among disease-associated and carried isolates of *Neisseria meningitidis* from the Czech Republic, Greece, and Norway. J Clin Microbiol 42(11):5146–5153. https://doi.org/10.1128/JCM.42.11.5146-5153.2004

Zhang Y (2017) The CRISPR-Cas9 system in *Neisseria* spp. Pathog Dis 75(4). https://doi.org/10.1093/femspd/ftx036

Zhang D, de Souza RF, Anantharaman V, Iyer LM, Aravind L (2012) Polymorphic toxin systems: comprehensive characterization of trafficking modes, processing, mechanisms of action, immunity and ecology using comparative genomics. Biol Direct 7:18. https://doi.org/10.1186/1745-6150-7-18

Zhang Y, Heidrich N, Ampattu BJ, Gunderson CW, Seifert HS, Schoen C, Vogel J, Sontheimer EJ (2013) Processing-independent CRISPR RNAs limit natural transformation in *Neisseria meningitidis*. Mol Cell 50(4):488–503. https://doi.org/10.1016/j.molcel.2013.05.001
Zimmer SM, Stephens DS (2006) Serogroup B meningococcal vaccines. Curr Opin Investig Drugs 7(8):733–739
Zughaier SM, Tzeng YL, Zimmer SM, Datta A, Carlson RW, Stephens DS (2004) *Neisseria meningitidis* lipooligosaccharide structure-dependent activation of the macrophage CD14/ Toll-like receptor 4 pathway. Infect Immun 72(1):371–380

Chapter 14
Sulfur Assimilation and Trafficking in Methanogens

John J. Perona, Benjamin Julius Rauch, and Camden M. Driggers

1 Introduction

Life on Earth originated under conditions in which the atmosphere lacked significant amounts of molecular oxygen and was consequently much more reducing than it is today (Holland 2006). Other aspects of ancient global biogeochemistry reflected these oxygen-lean conditions: in particular, reduced forms of sulfur predominated over the sulfate anion that is abundant in the present-day marine environment (Fig. 14.1) (Lyons and Gill 2010). Consistent with these observations, contemporary anaerobic microorganisms that originated prior to the Great Oxygenation Event some 2.4 billion years ago (2.4 Ga) are able to assimilate sulfur in the reduced sulfide form (Canfield et al. 2000). In contrast, almost all known present-day organisms conducting O_2-based respiration instead assimilate sulfur as sulfate (Mueller 2006; Hidese et al. 2011; Kessler 2006). While the enzymes and metabolic pathways for sulfate assimilation and subsequent sulfur incorporation are fairly well understood for these aerobes, until recently little had been known of the corresponding sulfide-based pathways in anaerobes present on the early Earth.

J. J. Perona (✉)
Department of Chemistry, Portland State University, Portland, OR, USA

Department of Biochemistry and Molecular Biology, Oregon Health and Science University, Portland, OR, USA
e-mail: perona@pdx.edu

B. J. Rauch
Department of Microbiology and Immunology, University of California San Francisco, San Francisco, CA, USA

C. M. Driggers
Department of Chemistry, Portland State University, Portland, OR, USA

P. H. Rampelotto (ed.), *Molecular Mechanisms of Microbial Evolution*, Grand Challenges in Biology and Biotechnology,
https://doi.org/10.1007/978-3-319-69078-0_14

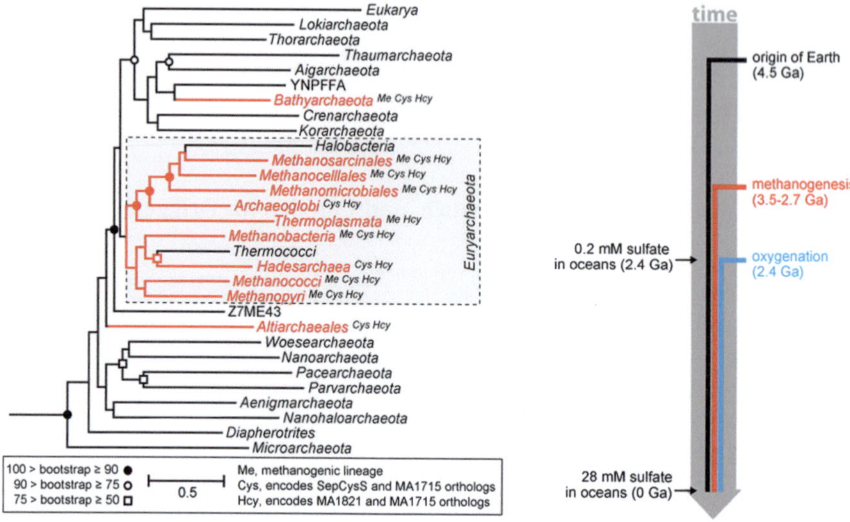

Fig. 14.1 Left: Evolutionary tree based on 16S rRNA sequences, using concatenated ribosomal protein and rRNA alignments generated by Banfield and colleagues (Hug et al. 2016). Lineages containing methanogens and unique proteins for Cys and Hcy biosynthesis are highlighted. The Altiarchaeales are uncultivated anaerobes found in deep groundwater environments. (Probst and Moissl-Eichinger 2015) Right: Timeline of major relevant events in Earth history

Methanogens are among the best-studied anaerobes and are models for elucidating sulfide-based assimilatory sulfur metabolism (Liu and Whitman 2008; Liu et al. 2012c). They are presently known to exist in two archaeal phyla: the *Euryarchaeota* and *Bathyarchaeota* (Fig. 14.1) (Evans et al. 2015). Notwithstanding the detection of aerobic methane emissions from plants (Fraser et al. 2015), the ancient methanogenesis pathway found in these two archaeal phyla remains the only known source of biogenic methane and provides the basis for a cellular metabolism that operates under chronic energy stress and close to thermodynamic limits. Numerous molecular adaptations underlie the successful low-energy lifestyle of methanogens, including the use of low and variable stoichiometries of ion pumping to support chemiosmotic ATP synthesis, and maintenance of many of the individual methanogenesis reactions at near equilibrium (Valentine 2007).

Methanogens are capable of utilizing a variety of growth substrates, including H_2/CO_2, formate, methanol, methylamines, acetate, and methoxylated aromatic compounds (Liu and Whitman 2008; Mayumi et al. 2016). In *Methanothermobacter*, over 200 genes are required for methane formation just from H_2 and CO_2, including some that are involved in synthesis of the coenzymes and prosthetic groups required in the central, eight-step methanogenesis pathway (Kaster et al. 2011). Biosynthesis of these compounds provides an important connection to sulfur acquisition and trafficking pathways. For example, the unique sulfur-containing cofactors coenzyme M and coenzyme B participate directly in methane formation (Graham 2011), and many core and peripheral methanogenesis enzymes require iron-sulfur clusters for

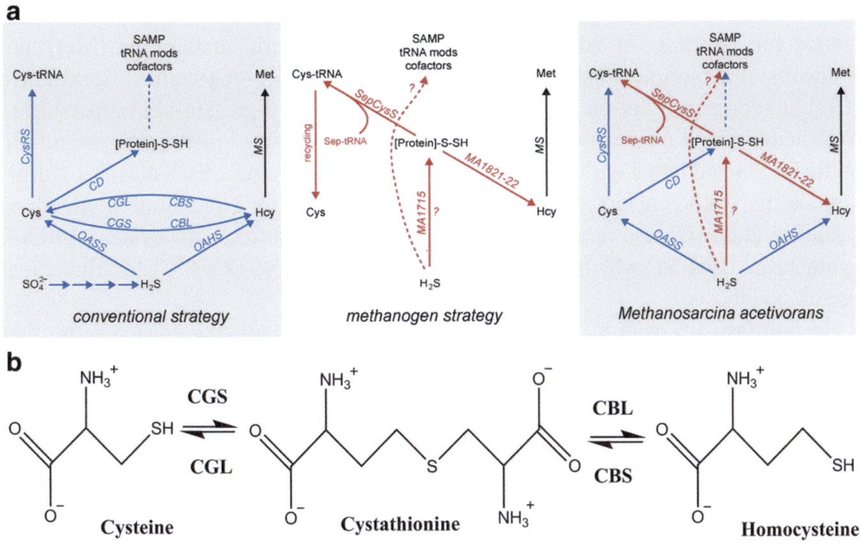

Fig. 14.2 (**a**) Sulfur assimilation and trafficking patterns. Left: Pathways present in most aerobic organisms that assimilate sulfur as sulfate. The four-step ATP-dependent of sulfate to sulfide is depicted at bottom. The dotted line depicts many distinct CD-dependent transpersulfidation pathways that deliver sulfur in the S^0 state to cofactors and other metabolites. OASS, O-acetylserine sulfhydrylase; OAHS, O-acetylhomoserine sulfhydrylase; CD, cysteine desulfurase; CysRS, cysteinyl-tRNA synthetase; MS, methionine synthase; CGS, cystathionine γ-synthase; CBL, cystathionine β-lyase; CBS, cystathionine β-synthase; CGL, cystathionine γ-lyase. Center: Pathways present in many hydrogenotrophic methanogens, exemplified by *M. jannaschii*. Enzymes depicted in blue at left are missing from this organism. MA1715 and MA1821-22 are the *M. acetivorans* homologs of the *M. jannaschii* proteins. Cys is produced by recycling of cellular proteins and (probably) from deacylation of Cys-tRNA^{Cys}. The assignment of MA1715 as a persulfide-generating protein is based on indirect physiological data and on mass spectrometry of SepCysS and MA1821 proteins (see the text). Proposed direct assimilation of H_2S for biosynthesis of cofactors is depicted by the dotted line and has yet to be well demonstrated in most cases. Right: Pathways present in *M. acetivorans* demonstrating substantial metabolic redundancies in this later-evolving methanogen. (**b**) Interconversion of Cys and Hcy by "transsulfuration" pathways involving the cystathionine intermediate. Abbreviations are defined in the legend to panel **a**

function (Thauer 1998; Major et al. 2004). Ancient mechanisms of sulfur assimilation and trafficking are thus likely to be preserved, together with the methanogenesis pathway, in contemporary methanogens that inhabit anaerobic, sulfidic environments (Blank 2009a). Further, recent comparative genomics of archaeal and bacterial species suggests that the gene repertoire of the common, anaerobic prokaryotic ancestor is presently most abundant among methanogens and clostridia (Sousa et al. 2016). This offers additional impetus for the study of contemporary methanogens as models for understanding how life arose.

In bacteria and eukaryotes, sulfate is taken up and converted to sulfide in four steps by an ATP-dependent 8-electron reduction process involving several sulfurated adenosine derivatives (Fig. 14.2a) (Sekowska et al. 2000). Sulfide then combines with O-acetylserine to form cysteine (Cys), which acts as a central sulfur source for synthesis of homocysteine (Hcy) and, in turn, methionine (Met) and S-adenosylmethionine (Mino and Ishikawa 2003). Cysteine desulfurase (CD) enzymes also act on Cys to produce alanine and a protein-bound persulfide intermediate on a conserved Cys in the enzyme active site. In turn, this persulfidic sulfur is donated into a variety of anabolic pathways to produce Fe-S clusters, sulfur-containing RNA nucleosides, and other essential sulfur-containing metabolites (Mueller 2006; Hidese et al. 2011). Cys is also the substrate for cysteinyl-tRNA synthetase (CysRS), which catalyzes biosynthesis of Cys-tRNA^{Cys} for ribosomal protein synthesis.

In contrast, the genome sequence of *Methanocaldococcus jannaschii*, the first Archaeon to be sequenced, suggested that many of these enzymatic activities are not present (Fig. 14.2a) (Bult et al. 1996). *M. jannaschii* lacks not only the ATP-dependent pathway for sulfide biosynthesis from sulfate, as expected, but also homologs of the conserved genes responsible for biosynthesis of Cys (O-acetylserine sulfhydrylase; OASS), Hcy (O-acetylhomoserine sulfhydrylase; OAHS), persulfide groups (cysteine desulfurase; CD), and cysteinyl-tRNA^{Cys} (cysteinyl-tRNA synthetase; CysRS) (Bult et al. 1996; Liu et al. 2012c). Many other methanogens also lack some or all of these genes (Rauch et al. 2014; Liu et al. 2012c), making discovery of the new pathways of broad significance to metabolomics studies in these organisms. It is worth noting that *M. jannaschii* is an ancient and obligate hydrogenotroph that inhabits deep hydrothermal vent environments and grows only on H_2/CO_2 (Liu et al. 2012c). It possesses a streamlined genome 1.66 Mbp in size (Bult et al. 1996). Other methanogens that are able to grow on methanol, methylamines, or methyl sulfides (methylotrophic methanogens) emerged more recently (Blank 2009b) and often have acquired many bacterial genes by horizontal transfer (Costa and Leigh 2014; Galagan et al. 2002). While the later-developing methanogens, including *Methanosarcina acetivorans*, possess many of the canonical sulfur assimilation genes, they also retain at least some of the ancestral pathways (Sauerwald et al. 2005; Rauch and Perona 2016; Rauch et al. 2014). However, the selective pressures and benefits of this redundancy are not well understood.

This article reviews new advances in understanding the assimilation and trafficking of sulfide in methanogens. One major focus is on sulfide uptake and incorporation into Cys, Hcy, methanogenesis coenzymes, and protein persulfide groups. The discovery of new assimilatory proteins that fill key metabolic gaps has answered some of the outstanding questions posed by the *M. jannaschii* genome (Table 14.1). However, further efforts to better elucidate the biochemical mechanisms of these enzymes are warranted, and this endeavor is also likely to yield novel insights into the mechanisms of sulfur transfer in enzymes generally. Little is known in many other areas, especially with respect to how sulfur is incorporated into tRNA, iron-sulfur clusters, and other cofactors and how sulfide assimilation and trafficking are regulated (see Box 14.1 for a summary of outstanding questions). A recently emerging theme that emerges from the study of these downstream processes is that

Table 14.1 Conserved sulfur assimilation proteins in methanogens

Protein	Function	Representative[a]	Citation
Phosphoseryl (Sep)-tRNA synthetase (SepRS)	Sep-tRNA^{Cys} biosynthesis	MJ1660	Sauerwald et al. (2005)
Sep-tRNA:Cys-tRNA synthase (SepCysS)	Cys-tRNA^{Cys} biosynthesis; Cys biosynthesis	MJ1678	Sauerwald et al. (2005)
COG1900a/CBS	Hcy biosynthesis from aspartate semialdehyde	MA1821	Rauch et al. (2014), Allen et al. (2015)
NIL-Fer	Hcy biosynthesis from aspartate semialdehyde	MA1822	Rauch et al. (2014), Allen et al. (2015)
COG1900d	Coenzyme M and coenzyme B biosynthesis (proposed)	MA3299	Rauch et al. (2014)
COG2122	Sulfide assimilation; persulfide formation (proposed)	MA1715	Rauch and Perona (2016)

[a]MJ, *Methanocaldococcus jannaschii*; MA, *Methanosarcina acetivorans*

sulfide may be taken up directly by individual enzymes, in place of (or in addition to) the involvement of sulfur relay chains involving persulfides (Mueller 2006).

2 Biosynthesis of Cysteine and Cys-tRNA^{Cys}

2.1 *Two-Step Pathway to Cys-tRNA^{Cys} in Methanogens and* Archaeoglobi

Many methanogens, especially members of the *Methanococcales*, lack the gene encoding CysRS. By taking a biochemical approach, Söll and colleagues identified two enzymes from *M. jannaschii* cell-free extracts that together are capable of biosynthesizing Cys-tRNA^{Cys}, thus bypassing the need for CysRS (Fig. 14.3) (Sauerwald et al. 2005). The first enzyme, O-phosphoseryl-tRNA synthetase (SepRS), is a class II aminoacyl-tRNA synthetase that possesses canonical structural and sequence signatures and is most closely related to phenylalanyl-tRNA synthetase (PheRS) (Figs. 14.3 and 14.4) (Sauerwald et al. 2005). SepRS catalyzes the archetypical two-step reaction of this enzyme family, by which phosphoserine (Sep) is first activated with ATP to yield phosphoseryl adenylate, followed by transfer of Sep to the 3′-end of tRNA^{Cys} to yield Sep-tRNA^{Cys} (Perona and Hadd 2012). The enzyme is an α_4 homotetramer that binds two tRNAs and exhibits half-of-the-sites reactivity for phosphoseryl adenylate synthesis (Fig. 14.4) (Hauenstein et al. 2008; Fukunaga and Yokoyama 2007a; Kamtekar et al. 2007). Studies of phosphoserylation by *M. jannaschii* SepRS revealed several interesting features, including the use of the posttranscriptionally modified base m^{1}G37 in the anticodon loop as a positive determinant for aminoacylation (Zhang et al. 2008). Mutations of

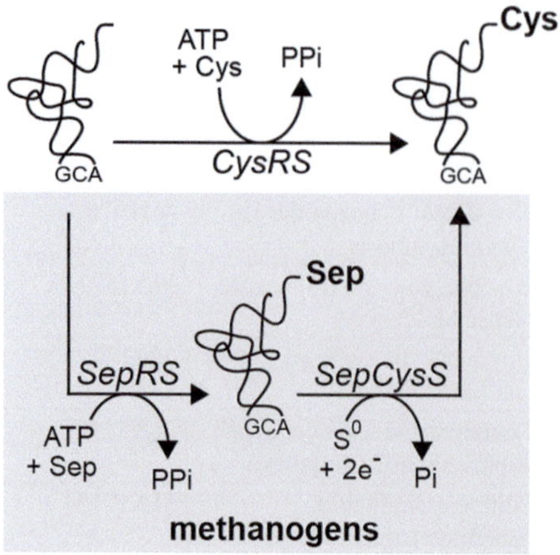

Fig. 14.3 Pathways for biosynthesis of Cys-tRNA^{Cys}. The shaded region depicts the two-step pathway present in nearly all methanogens. Known species lacking the two-step pathway are limited to those inhabiting the intestinal tract, which is not a sulfide-rich environment

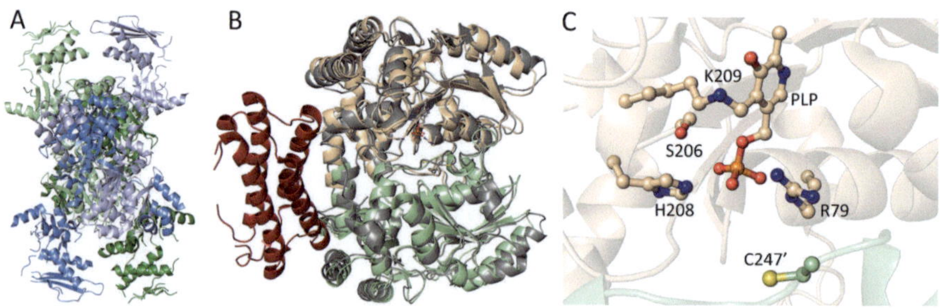

Fig. 14.4 (**a**) Structure of *Methanococcus maripaludis* SepRS with each of the four subunits of the tetramer shown (PDB code 2odr) (Kamtekar et al. 2007). (**b**) Structure of SepCysS from *Archaeoglobus fulgidus* (green/tan carbons PDB code 2e7j) (Fukunaga and Yokoyama 2007b), aligned on the SepCysS-SepCysE complex from *Methanocaldococcus jannaschii* (PDB code 3kwr) (Liu et al. 2014b), with the dimer of SepCysE shown (red carbons). (**c**) Closeup in the same orientation of the SepCysS active site that forms at the dimer interface, with Cys247 on the opposing subunit from PLP

conserved tRNA^{Cys} nucleotides had smaller effects on catalytic efficiency than is typical for other aminoacyl-tRNA synthetases (Zhang et al. 2008), including CysRS. However, a deeper understanding of the reaction mechanism is limited by the SepRS-tRNA^{Cys} cocrystal structures, which reveal tRNA bound solely through its anticodon loop rather than in a productive complex with the enzyme (Fukunaga and Yokoyama 2007a).

The misacylated Sep-tRNA^{Cys} product of the SepRS reaction is converted to Cys-tRNA^{Cys} through the action of the pyridoxal phosphate (PLP)-dependent enzyme Sep-tRNA:Cys-tRNA synthase (SepCysS) (Figs. 14.3 and 14.4) (Sauerwald et al. 2005; Fukunaga and Yokoyama 2007b). The SepCysS dimer possesses two domains. The larger domain forms a seven-stranded parallel β-sheet and is structurally homologous to CDs, which also function by a PLP-dependent mechanism (Hidese et al. 2011). Comparisons among SepCysS sequences reveal that the large CD-like domain possesses both a conserved active-site lysine that makes a covalent linkage with PLP and three conserved cysteines. However, detailed comparison of the structures of *Archaeoglobus fulgidus* SepCysS with the *E. coli* CD CsdB showed that the structures of the active site clefts diverge significantly, with a much broader and deeper cleft in SepCysS to accommodate the large tRNA substrate (Fukunaga and Yokoyama 2007b). While CDs also possess a conserved Cys that acquires a persulfide group in the reaction, the locations of the conserved Cys residues in SepCysS and CsdB (or other CDs) do not coincide (Fukunaga and Yokoyama 2007b). SepCysS does not possess CD-like activity and does not use Cys as a sulfur source, as the SepRS-SepCysS pathway is also a source of free Cys (see below) (Sauerwald et al. 2005). Phylogenetic analysis showed that SepCysS sequences have substantially diverged from CDs to form a distinct monophyletic group, which also predicts a distinct function (O'Donoghue et al. 2005).

The reaction mechanism of SepCysS is proposed to be similar to that of the PLP-dependent Sep-tRNA:Sec-tRNA synthase (SepSecS) (Fig. 14.5). In archaea

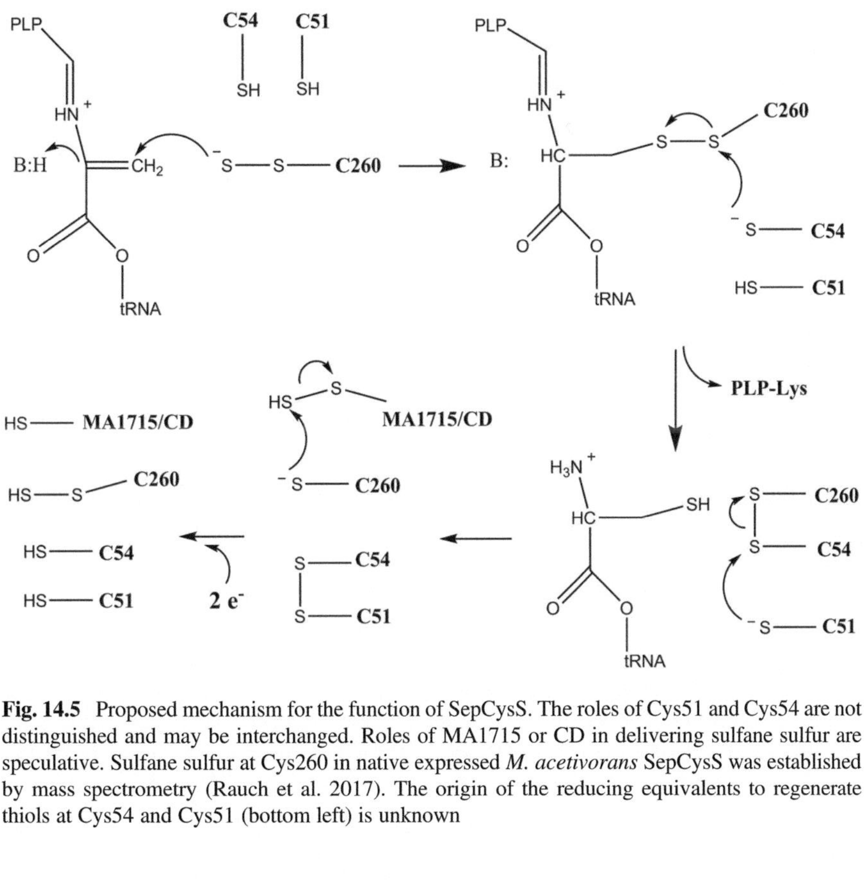

Fig. 14.5 Proposed mechanism for the function of SepCysS. The roles of Cys51 and Cys54 are not distinguished and may be interchanged. Roles of MA1715 or CD in delivering sulfane sulfur are speculative. Sulfane sulfur at Cys260 in native expressed *M. acetivorans* SepCysS was established by mass spectrometry (Rauch et al. 2017). The origin of the reducing equivalents to regenerate thiols at Cys54 and Cys51 (bottom left) is unknown

and eukaryotes, selenocysteine (Sec) is inserted into proteins in response to certain amber (UGA) codons in a three-step pathway involving (1) aminoacylation of the unusual amber-decoding tRNA^{Sec} species with serine by seryl-tRNA synthetase (SerRS), (2) phosphorylation of the tRNA^{Sec}-linked serine by the kinase PSTK to form Sep-tRNA^{Sec}, and (3) conversion of Sep-tRNA^{Sec} to Sec-tRNA^{Sec} by SepSecS, using selenophosphate as the selenium donor (Carlson et al. 2004; Yuan et al. 2006; Xu et al. 2007). SepCysS and SepSecS thus each displace phosphate from Sep-tRNA, while delivering sulfur or selenium, respectively, to form Cys or Sec on the tRNA 3′-end. Using an in vivo assay for formate dehydrogenase activity, which requires either an active-site Sec or Cys for function, it was established that SepCysS can convert Sep-tRNA^{Sec} to Cys-tRNA^{Sec} in vivo—although, unlike SepSecS, it is not able to use selenophosphate as a selenium donor to achieve Sec-tRNA^{Sec} biosynthesis (Yuan et al. 2010). Using this assay, it was then shown that a number of amino acids in SepCysS, including the lysine that forms the external aldimine with PLP and all three conserved active-site Cys residues, are critical for tRNA-dependent Cys formation (Helgadottir et al. 2012; Liu et al. 2012a).

SepCysS is able to function in vitro using sulfide, thiophosphate, or Cys as sulfur donors (Hauenstein and Perona 2008; Sauerwald et al. 2005). However, the rate of Cys-tRNA^{Cys} formation with sulfide was about 500-fold lower than the rate of Sep-tRNA^{Cys} formation by SepRS, suggesting that sulfide is unlikely to function as the sulfur donor in vivo (Hauenstein and Perona 2008). Instead, by analogy with other well-characterized enzymes, it was thought likely that SepCysS inserts sulfur into Sep-tRNA^{Cys} using an active-site persulfide group previously formed on one of the three conserved cysteines (Hauenstein and Perona 2008; Fukunaga and Yokoyama 2007b). PLP-dependent formation of Sec-tRNA^{Cys} from Ser-tRNA^{Cys} by bacterial selenocysteine synthase (a reaction that bypasses the two-step PSTK/SepSecS pathway in archaea and eukaryotes) has provided the leading model (Fig. 14.5) (Forchhammer and Bock 1991). In the proposed mechanism, productive binding of Sep-tRNA^{Cys} to SepCysS is followed by formation of a Schiff base between the α-amino group of Sep and the formyl group of PLP, resulting in elimination of the phosphate group and formation of covalently linked aminoacrylyl-tRNA^{Cys}. The outer sulfur of an active-site persulfide (the *sulfane* sulfur) then adds to the double bond, and the S-S linkage is resolved by disulfide formation on the enzyme with one of the other active-site cysteines. Regeneration of active enzyme then requires both a source of exogenous reductant to reduce the enzyme disulfide to thiols and delivery of sulfur to reform the persulfide (Fig. 14.5) (Hauenstein and Perona 2008; Liu et al. 2012a).

This proposed mechanism is supported by mass spectrometry and biochemical data. When expressed heterologously in *E. coli*, mass spectrometry showed that *M. jannaschii* SepCysS is the target of persulfidation by the CD IscS on either Cys64 or Cys67 (Liu et al. 2012a). Several chemical tests also showed that a disulfide bond is formed between Cys64 and Cys67 in *E. coli*-expressed SepCysS (Liu et al. 2012a). The Cys64-Cys67 disulfide was confirmed by mass spectrometry of tryptic peptides derived from *M. acetivorans* SepCysS expressed in native *M. acetivorans* cells, but this study did not reveal a persulfide on the equivalent residues (Rauch et al. 2017). However, it was shown that the third conserved active-site Cys, Cys260 (Cys272 in the *M. jannaschii* enzyme), carries persulfide when SepCysS is expressed in *M. acetivorans* (Rauch et al. 2017). The functional role of the persulfide on Cys260 was demonstrated in experiments showing that SepCysS from this source can form Cys-tRNA^{Cys} in vitro in the absence of any exogenous sulfur source. Therefore, the sulfane sulfur located on Cys260 of SepCysS is the source of the cysteine sulfur for protein synthesis in methanogens (Rauch et al. 2017). This suggests a three-Cys mechanism involving disulfide formation between Cys51 and Cys54 (equivalent to Cys64 and Cys67 of *M. jannaschii* SepCysS) and persulfide formation on Cys260 (Cys272 of *M. jannaschii* SepCysS) (Fig. 14.5). Interestingly, the SepCysS dimer structure shows that these conserved Cys residues are in proximity to the PLP binding site on the opposing subunit (Fig. 14.5) (Fukunaga and Yokoyama 2007b). Further, in one of two independent structures of the SepCysS dimer, PLP binds in just one of the two subunits. Together, these data suggest that, like SepRS, SepCysS may exhibit half-of-the-sites activity.

The proposed SepCysS mechanism does not yet resolve several outstanding questions. Most obviously, the means of delivering the sulfane sulfur to Cys260 remains unknown (see below). The identity of the reductant that regenerates thiols on Cys64 and Cys67 is also unresolved. However, it has recently been shown that, when expressed in *E. coli*, *M. jannaschii* SepCysS binds a [3Fe-4S] iron-sulfur cluster through its three conserved cysteines (Liu et al. 2016). Reduction of cystines by Fe-S clusters has precedent in ferredoxin:thioredoxin reductase and heterodisulfide reductase (Walters et al. 2009; Duin et al. 2002), suggesting that SepCysS in native methanogen cells may use the cluster to regenerate thiols at Cys64 and Cys67 in the last step of the mechanism (Fig. 14.5).

Finally, in a subset of methanogens, SepCysS forms a large complex with a 25 kDa protein, SepCysE, and a ternary complex with both SepRS and SepCysE (Liu et al. 2014b). (There is conflicting information regarding the ability of *M. jannaschii* SepRS and SepCysS to form a binary complex in the absence of SepCysE) (Liu et al. 2014b; Zhang et al. 2008). The SepCysS-SepCysE, SepRS-SepCysE, and SepCysS-SepCysE-SepRS complexes are composed of four subunits of each protein, with the ternary complex exhibiting a molecular weight of about 600,000 daltons. Complex formation with SepCysE substantially improves the tRNA affinities of SepRS and SepCysS, the rate of Sep-tRNA^{Cys} synthesis, and the conversion of Sep-tRNA^{Cys} to Cys-tRNA^{Cys} (Liu et al. 2014b).

A crystal structure of the *M. jannaschii* SepCysS-SepCysE complex revealed that an N-terminal 67 amino acid domain of SepCysE forms antiparallel helix bundles that bind SepCysS, while the C-terminal domain, although playing a role in enhancing tRNA binding and reaction rates, is disordered in the structure and not required for ternary complex formation (Fig. 14.4). Molecular docking of the *A. fulgidus* SepRS-tRNA^{Cys} complex with the SepCysS-SepCysE complex, based on aligning the symmetry axes of these two complexes, suggests that the complex facilitates substrate channeling: movement of tRNA between the enzymes without dissociation of the Sep-tRNA^{Cys} product (Liu et al. 2014b). However, it is important to note that no crystal structure is yet available of SepCysS bound to either tRNA or SepRS.

Such a structure is expected to offer substantial insight into the mechanisms of both enzymes.

2.2 Cysteine Biosynthesis from Cys-tRNA^{Cys}

Some methanogens lack the genes encoding common Cys biosynthetic pathways found in most microorganisms and eukaryotes (White 2003). The missing enzymes include O-acetylserine sulfhydrylase (OASS), which catalyzes Cys biosynthesis from sulfide and O-acetylserine (Fig. 14.2; Table 14.2) (Borup and Ferry 2000). Also often missing are cystathionine β-synthase and L-cystathionine cysteine lyase, which catalyze conversion of Hcy to Cys through the intermediary metabolite cystathionine (Fig. 14.2) (White 2003) (interestingly, cystathionine is present in a variety of methanogens irrespective of whether the genes are detected by homology search; RH White, personal communication). In *Methanococcus maripaludis*, which possesses both CysRS and SepRS pathways but lacks the *oass* gene, it was shown that deletion of the gene encoding SepRS in *M. maripalu*dis gave rise to a Cys auxotroph (Sauerwald et al. 2005). This demonstrated that the SepRS/SepCysS pathway provides free Cys, either through deacylation of Cys-tRNA^{Cys}, protein turnover, or both. In double knockout strains lacking SepRS and SepCysE, both proteins were required in *trans* to relieve Cys auxotrophy, showing that SepCysE is essential in vivo even though it is absent in many methanogens (Liu et al. 2014b). Interestingly, an analogous metabolic strategy involving tRNA-dependent amino acid biosynthesis is present in microorganisms lacking asparagine biosynthesis genes, where Asn-tRNA^{Asn} is the source of free asparagine (Min et al. 2002).

The SepRS pathway may be an inefficient source of Cys as compared to canonical biosynthetic pathways (Fig. 14.2), as the concentration of free Cys in *M. maripaludis* is only about 20 μM, five to tenfold lower than commonly observed in bacteria (Liu et al. 2010). However, the concentration of Cys is also lowered in *M. maripaludis* by the enzyme L-cysteine desulfidase, which catalyzes Cys breakdown into pyruvate, ammonia, and sulfide (Tchong et al. 2005). This enzyme is broadly distributed in anaerobes. The low concentration of Cys is presumably adaptive in methanogens that also lack CDs and/or CysRS; indeed, in these organisms the only other reaction that requires free Cys as substrate is the biosynthesis of coenzyme A (Spry et al. 2008). Metabolic labeling experiments also showed that sulfide rather than Cys is the sulfur source for both iron-cluster and methionine biosynthesis in *M. maripaludis*, reflecting the presence of alternative, novel pathways for biosynthesis of these metabolites (Liu et al. 2010).

Table 14.2 Occurrence of sulfur assimilation proteins in methanogens and related anaerobes

	COG1900a NIL-Fer	COG2122	COG1900d	SepRS SepCysS	$tRNA^{Cys}$	CysRS	OASS CBS/CGL	CD	OAHS CGS/CBL	
M. barkeri	+	+	+	+	4	+	+	+	+	*Methanomicrobia*
M. acetivorans	+	+	+	+	3	+	+	+	+	
M. mazei	+	+	+	+	3	+	–	+	–	
M. psychrophilus	+	+	+	+	5	+	+	+	+	
M. hollandica	+	+	+	+	4	+	+	+	+	
M. burtonii	+	+	+	+	3	–	+	+	+	
M. zhilinae	+	+	+	+	3	+	+	+	+	
M. evestigatum	+	+	+	+	3	+	–	+	+	
M. concilii	+	+	+	+	1	–	–	+	–	
M. thermophila	+	+	+	+	2	–	–	+	–	
M. arvoryzae	+	+	+	+	5	+	–	+	–	
M. conradii	+	+	+	+	3	+	–	+	+	
M. paludicola	+	+	+	+	2	+	–	+	+	
M. labreanum	+	+	+	+	4	–	+	+	+	
M. marisnigri	+	+	+	+	1	–	–	+	+	
M. hungatei	+	+	+	+	2	+	+	+	+	
M. palustris	+	+	+	+	8	+	+	+	+	
M. boonei	+	+	+	+	4	+	+	+	+	
A. sulfaticallidus	+	+	–	+	1	+	–	–	–	*Archaeog*
A. veneficus	+	+	–	+	1	+	–	–	–	
A. fulgidus	+	+	–	+	1	+	–	+	–	
F. placidus	+	+	–	+	1	+	–	+	–	
A. profundus	+	+	–	+	1	+	–	–	–	

(continued)

Table 14.2 (continued)

	COG1900a NIL-Fer	COG2122	COG1900d	SepRS SepCysS	$tRNA^{Cys}$	CysRS	OASS CBS/CGL	CD	OAHS CGS/CBL	
M. fervidus	+	+	+	+	1	–	–	+	–	*Methanobacteria*
M. marburgensis	+	+	+	+	1	–	–	+	–	
M. therm.	+	+	+	+	1	–	–	+	–	
M. lacus	+	+	+	+	2	+	+	+	+	
M. paludis	+	+	+	+	2	+	+	+	+	
M. stadtmanae	–	+	+	–	1	+	+	+	+	
M. ruminantium	–	–	+	–	1	+	+	+	+	
M. smithii	–	+	+	–	1	+	+	+	+	
M. fervens	+	+	+	+	1	–	–	–	–	*Methanococci*
M. jannaschii	+	+	+	+	1	–	–	–	–	
M. vulcanius	+	+	+	+	1	–	–	–	–	
M. infernus	+	+	+	+	1	–	–	–	–	
M. igneus	+	+	+	+	1	–	–	–	–	
M. okinawensis	+	+	+	+	1	+	–	–	–	
M. aeolicus	+	+	+	+	1	+	–	–	–	
M. votae	+	+	+	+	1	+	–	–	–	
M. maripaludis	+	+	+	+	1	+	–	–	–	
M. vannielii	+	+	+	+	1	+	+	+	+	
M. kandleri	+	+	+	+	1	–	–	+	–	[a]

Archaeog, Archaeoglobales
therm., thermautotrophicus
[a]*Methanopyri*

2.3 Evolution of Cys-tRNA^{Cys} Biosynthesis Pathways

The two-step pathway for Cys-tRNA^{Cys} biosynthesis is confined to most methanogens and some closely related *Archaeoglobus* species, which conserve energy by dissimilatory sulfate reduction to sulfide and possess most enzymes in the methanogenesis pathway (Klenk et al. 1997). Some methanogens, including species of *Methanococcales* and thermophilic *Methanobacteriales*, possess the SepRS/SepCysS pathway as the sole route to Cys-tRNA^{Cys} (Bult et al. 1996). However, other methanogens, including many species of *Methanosarcinales* and *Methanomicrobiales*, possess both the ancient two-step and the modern Cys-tRNA^{Cys} biosynthesis pathways (Table 14.2) (Galagan et al. 2002). Many methanogens also retain canonical eukaryotic or bacterial enzymes for Cys biosynthesis together with SepRS/SepCysS (Liu et al. 2012c). The selective pressures for retaining these redundancies are not well understood, although it is known that the Cys content of proteins in methanogens is nearly double the level in other archaea (Klipcan et al. 2008). This is at least in part due to the great abundance of Fe-S clusters in these organisms (Major et al. 2004). Both SepRS/SepCysS and Fe-S assembly pathways require net addition of sulfide, suggesting that they may be linked in a common metabolic framework for sulfide assimilation (see below). The only known methanogens that possess CysRS instead of SepRS/SepCysS are those that inhabit specialized anaerobic environments that are not sulfide-rich (Table 14.2), such as the human intestinal tract (Dridi et al. 2011; Liu et al. 2012c).

Comparative phylogenetic analyses of the CysRS and SepRS/SepCysS routes to Cys-tRNA^{Cys} biosynthesis showed that the SepRS/SepCysS pathway is ancient and was present at the time of the last universal common ancestral state (LUCAS) (O'Donoghue et al. 2005). It was later shown that SepCysE also originated at this early stage; all three enzymes show phylogenetic patterns that are equivalent to those exhibited by methanogenesis enzymes (Liu et al. 2014b). Interestingly, CysRS is equally ancient and was also present at LUCAS, although it was conserved then only in species that went on to differentiate into bacterial lineages. In contemporary methanogens, CysRS was acquired later by horizontal transfer from bacteria (O'Donoghue et al. 2005).

In addition to redundant pathways for Cys-tRNA^{Cys} biosynthesis, analysis of methanogen genomes also reveals a remarkable expansion in the number of tRNA^{Cys} genes, with up to eight found in some organisms. The tRNA gene expansions primarily occur in the methanogens that also possess both pathways for Cys-tRNA^{Cys} biosynthesis (Table 14.2), suggesting that some selective pressure exists. In *Methanosarcina mazei*, which possesses both enzyme pathways and three tRNA^{Cys} genes, biochemical studies using in vitro transcripts modified to contain m^{1}G37 (an important recognition element for both enzymes) showed that CysRS and SepRS prefer distinct tRNAs as substrates (Hauenstein and Perona 2008). Two of the three *M. mazei* tRNA^{Cys} possess canonical structures and are preferred four- to fivefold by SepRS. The third *M. mazei* tRNA^{Cys} species is unusual: it possesses a rare purine at position 33 in the anticodon loop and a rare A65-C49 mismatch together with a GU wobble pair in the T-stem. This tRNA is preferred five- to ninefold by CysRS (Hauenstein and Perona 2008). Swapping six divergent nucleotides in the globular core regions of these tRNA^{Cys} species is sufficient to interconvert aminoacylation preferences. These experiments suggested that, in methanogens with multiple tRNA^{Cys} species, the tRNAs may be preferentially dedicated to CysRS or SepRS based on their structural differences.

All three *M. mazei* tRNA^{Cys} species possess the U73 nucleotide, an identity determinant for both CysRS and SepRS in methanogens and bacteria (Hohn et al. 2006; Komatsoulis and Abelson 1993). U73 is unusual in tRNAs and is found primarily in tRNA^{Cys} from all three domains of life. However, in methanogens containing both CysRS and SepRS, some tRNA^{Cys} in the expanded set of these species possess other nucleotides at this position. Remarkably, these particular tRNAs possessing A73, C73, or G73 are usually found adjacent to CysRS in the chromosome. Moreover, the identity of the nucleotide at position 73 covaries with the identities of several amino acids located in the tRNA acceptor-stem binding domain of CysRS, near the active site (amino acids Tyr152 and Asn237 of *E. coli* CysRS) (Fig. 14.6). These observations support the notion that particular tRNA^{Cys}

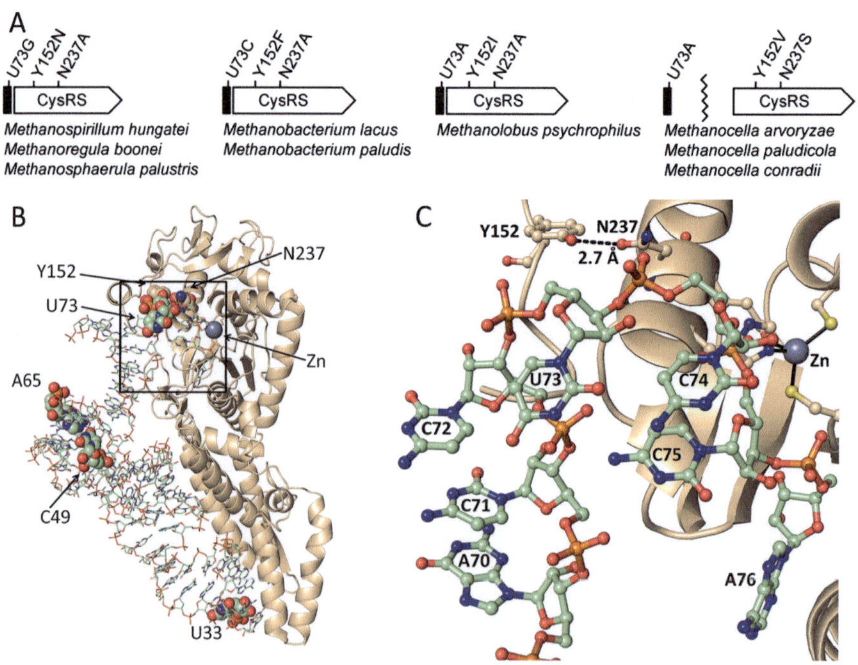

Fig. 14.6 (**a**) Top: Covariations in the sequences of methanogen tRNA^{Cys} species at position 73 near the 3′-terminus of the molecule, with amino acids in the tRNA acceptor stem-binding domain of CysRS. (**b**) Bottom left: structure of the *E. coli* CysRS-tRNA^{Cys} complex (pdb 1u0b). The relative positions of U73 in the tRNA and the covarying amino acids Tyr152 and Asn237 are shown. (**c**) Closeup of 3′-tRNA end binding in the CysRS active site

species are preferentially dedicated for aminoacylation by either SepRS or CysRS. However, the evolutionary rationale for retaining both pathways remains obscure.

3 Biosynthesis of Homocysteine, Coenzyme M, and Coenzyme B

3.1 *Homocysteine Biosynthesis from Aspartate Semialdehyde and Hydrogen Sulfide*

Many methanogens lack the widely distributed enzymes responsible for Hcy biosynthesis in microorganisms and eukaryotes (Fig. 14.2; Table 14.2) (White 2003; Liu et al. 2012c). Hcy is often biosynthesized from activated derivatives of homoserine (O-succinyl-L-homoserine, O-phospho-L-homoserine, or O-acetyl-L-homoserine) or by an alternative two-step conversion ("transsulfuration") from Cys through cystathionine, the reverse of the pathway to Cys biosynthesis from Hcy (Fig. 14.2). Novel proteins that are essential to Hcy biosynthesis in methanogens lacking all these enzymes were recently discovered through a bioinformatics approach (Rauch et al. 2014). This was accomplished by occurrence profiling using SepCysS as the focal point. Given that SepCysS is found in all methanogens except those not inhabiting sulfidic environments, and is retained regardless of whether CysRS is also present, it was reasoned that the enzyme is closely linked to other activities associated with sulfide assimilation. Comparative genomics of about 100 archaeal genomes identified candidate genes present in organisms containing SepCysS and absent in other organisms. This resulted in the identification of three previously uncharacterized proteins, corresponding to MA1821, MA1822, and MA1715 in *M. acetivorans* (Fig. 14.7) (Rauch et al. 2014).

MA1821 is a 500 aa protein consisting of a 350 aa N-terminal domain that is associated with cluster of orthologous genes (COG) 1900. COG1900 is of unknown structure and function and possesses two highly conserved Cys residues. At the

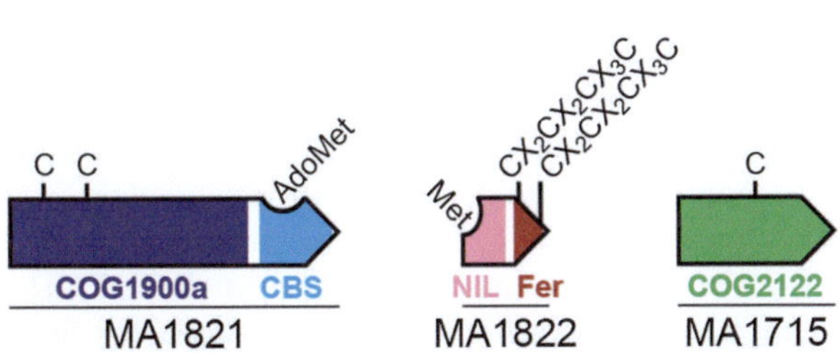

Fig. 14.7 Cartoon representation of proteins encoded by MA1821, MA1822, and MA1715 in *M. acetivorans*. CBS represents a regulatory domain first identified in cystathionine β-synthase. NIL is homologous to the intracellular domain of a methionine transport protein, and Fer refers to the domain containing two Fe_4S_4 clusters. Positions of conserved Cys residues are indicated. Proposed binding sites for AdoMet and Met are based on homology to known protein domains

C-terminus, MA1821 and its homologs, including the *M. jannaschii* MJ0100 protein, possess tandem copies of a ~60 amino acid CBS cystathionine β-synthase (CBS) domain. These tandem CBS domains from MJ0100 are structurally characterized and possess a regulatory adenosine binding site in the interdomain cleft, which bind either S-adenosyl-L-methionine (SAM) or S-methyl-5′-thioadenosine (MTA) and undergo conformational changes that plausibly regulate activity (Lucas et al. 2010). Tandem CBS domains possess a regulatory adenosine binding site in the interdomain cleft (Lucas et al. 2010) and are usually fused to partner domains involved in energy and sulfur metabolism. MA1822 homologs in both archaea and bacteria are almost always encoded immediately downstream of MA1821, strongly suggesting a common function. MA1822 is a 130 amino acid protein with eight conserved Cys residues and is predicted to contain two 4Fe-4S clusters near its C-terminus (Fig. 14.7). Sequences at the N-terminus of MA1822 are homologous to the intracellular domain of the methionine ABC transporter protein (NIL) (Kadaba et al. 2008).

The genes encoding MA1821 and MA1822 were deleted from the chromosome of *M. acetivorans*, which also possess the *oahs* gene for Hcy biosynthesis from O-acetylhomoserine (Fig. 14.2) (Rauch et al. 2014). While deletion of either *oahs* or *ma1821-22* did not affect growth phenotypes under conditions where sulfide was provided as the sole sulfur source, the multiple deletion *oahs/ma1821-22* was not viable unless Hcy was provided to the culture medium. This triple deletion strain was not a Cys auxotroph, suggesting that the MA1821 and MA1822 proteins are dedicated to Hcy biosynthesis alone. These experiments established that MA1821 and MA1822 are essential to Hcy biosynthesis when *oahs* is deleted and thus constitute part of the ancient sulfide assimilation pathway in hydrogenotrophic methanogens lacking the other pathways for Hcy biosynthesis. Addbacks of plasmid-borne wild-type and mutant *ma1821-22* genes into the Hcy auxotrophic strain showed that the COG1900 domain of MA1821 and the iron-sulfur cluster binding domain of MA1822 are essential to activity, while the CBS and NIL domains are dispensable (Fig. 14.7) (Rauch et al. 2014).

Isotope labeling experiments using cell extracts from *M. jannaschii* and *M. acetivorans* established that the precursor for Hcy biosynthesis is aspartate semialdehyde (Asa), and that the MA1821-22 proteins participate in the reaction (Allen et al. 2015). Asa is a common intermediate in lysine, threonine, and methionine biosynthesis (Fig. 14.8). The experiments also showed that sulfide is the sulfur source, although its direct binding to MA1821-22 was not demonstrated.

Using mass spectrometry, a persulfide group was identified on conserved Cys131 in the COG1900 domain of MA1821 when the protein is expressed in native methanogen cells under anaerobic conditions (Rauch et al. 2017). This is consistent with a proposed mechanism in which a sulfane sulfur attacks the aldehyde carbon of Asa to yield a disulfide hemiacetal adduct. The thiolate of another Cys then reacts with the disulfide to yield an enzyme-bound thioaldehyde intermediate and enzyme disulfide, with release of water (Fig. 14.8) (Allen et al. 2015). Electron donation by the iron-sulfur cluster domain of MA1822 is then capable of reducing the enzyme disulfide to thiols. As in SepCysS, an exogenous source of electrons is required to

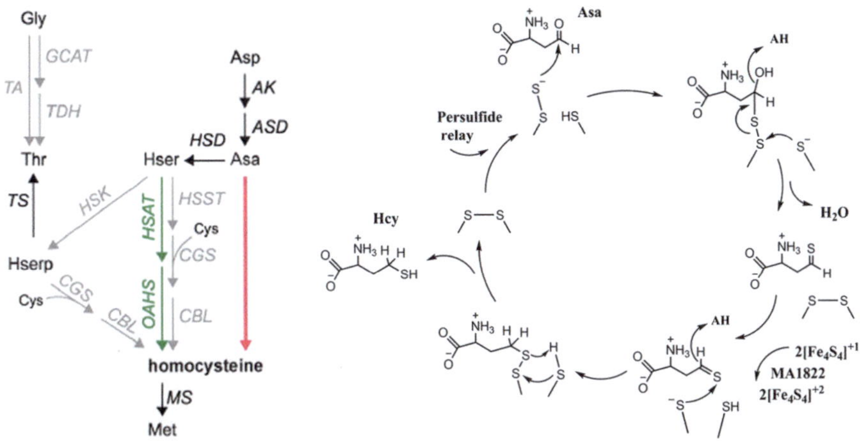

Fig. 14.8 Left: Biosynthetic pathways for homocysteine (Hcy). The novel transformation in methanogens and some other anaerobes is depicted by the pink arrow. Reactions depicted in black are highly conserved in all domains of life; those depicted in gray are typically present in aerobes but missing from *M. acetivorans*. *Hserp* Phosphohomoserine; *Hser* Homoserine; *Asa* Aspartate semialdehyde; *AK* Aspartate kinase; *ASD* Aspartate semialdehyde dehydrogenase; *HSD* Homoserine dehydrogenase; *HSK* Homoserine kinase; *GCAT* Glycine C-acetyltransferase; *TDH* Threonine dehydrogenase; *TS* Threonine synthase; *HSST* Homoserine succinyltransferase. Other enzymes are defined in the legend to Fig. 14.2. Right: Proposed mechanism for biosynthesis of Hcy from aspartate-semialdehyde (Asa). After Allen et al. (2015)

regenerate the reduced form of the iron-sulfur cluster for the next round of catalysis. Interestingly, while conserved Cys54 of MA1821 was identified as essential in the Hcy auxotrophy rescue experiments, Cys131 was found to be dispensable (Rauch et al. 2014). Gel filtration experiments showed that MA1821 forms a disulfide-linked dimer, suggesting that an intersubunit disulfide between Cys54 of each subunit may be an integral feature of the reaction (Rauch et al. 2017). The mechanism by which sulfane sulfur is delivered to MA1821 is unknown, but a role for Cys131 in this process could help reconcile the finding that this residue carries a persulfide when the protein is expressed in native cells (see below).

3.2 COG1900 Subfamilies and Possible Roles in Coenzyme Biosynthesis

Homologs of MA1821 and MA1822 have been identified in all archaea possessing the SepCysS enzyme, which include the great majority of methanogens and some species of *Archaeoglobi*. Homologs were also identified in over 50 bacterial genera, all of which are anaerobic (Rauch et al. 2014). The sulfide-dependent conversion of Asa to Hcy is thus a broadly conserved process in anaerobic microorganisms.

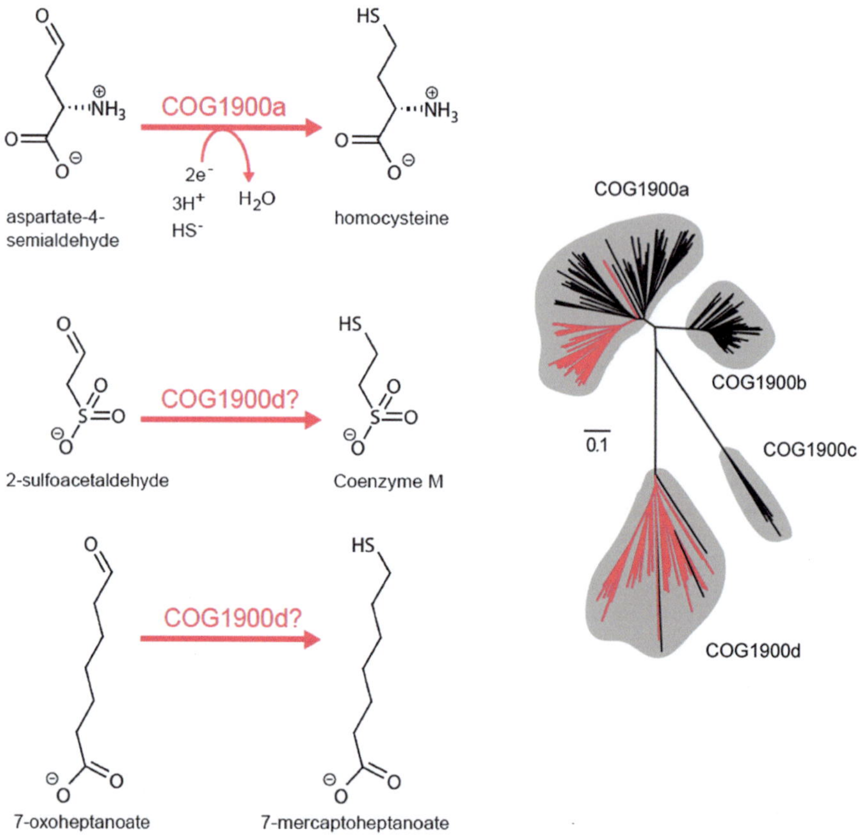

Fig. 14.9 Left: Depiction of the reaction catalyzed by COG 1900a proteins and plausible similar reactions that may be catalyzed by COG 1900d proteins (Graham and White 2002). 7-mercaptoheptanoate is intermediate in the coenzyme B pathway. Right: Condensed phylogenetic reconstruction of all members of COG 1900, with methanogen lineages indicated in pink (Rauch et al. 2014)

Phylogenetic analysis of MA1821, MA1822, and the third protein that co-occurs with SepCysS, MA1715, showed congruence with evolutionary trees built from organismal phylogenies and from genes encoding enzymes in the methanogenesis pathways. Therefore, all three proteins were vertically inherited, together with the methanogenesis pathway, from the ancestral euryarchaeote (Rauch et al. 2014).

Additional phylogenetic analysis of COG1900 revealed that it is separable into four subdivisions, termed COG1900a-d (Fig. 14.9; the orthologous group containing MA1821 was assigned as COG1900a) (Rauch et al. 2014). COG1900b and COG1900c occur solely in cyanobacteria, while the COG1900d proteins are found in all methanogens, including those lacking SepCysS (Table 14.2). COG1900d proteins possess the large catalytic domain found in COG1900a but lack both conserved cysteines (corresponding to Cys54 and Cys131 of MA1821). The CBS domain is not present, but is replaced by a 4Fe-4S cluster domain similar to that found in MA1822. Therefore, with the exception of Cys54, COG1900d proteins contain all the catalytic elements essential for Hcy biosynthesis. As COG1900a and COG1900d coexist in methanogen genomes, each in single copy, they presumably function as paralogs.

A plausible function for COG1900d, consistent with its phylogenetic distribution solely in methanogens, is insertion of sulfur into the metabolic precursors of coenzyme M and coenzyme B. These cofactors are required for both methanogenesis and anaerobic methane oxidation ("reverse methanogenesis") by anaerobic methanotrophic archaea (ANME) (Scheller et al. 2010). In both the coenzyme M and coenzyme B pathways, the enzyme responsible for inserting sulfur remains unidentified (Graham 2011). Strikingly, in each case, the precursor compound contains an aldehyde group that must be converted to the thiol of each coenzyme, identical to the reaction chemistry that produces Hcy from Asa (Fig. 14.9) (Graham and White 2002). The congruence of phylogenetic distribution and reaction chemistry strongly implicates COG1900d proteins in coenzyme M and coenzyme B biosynthesis. Experiments to investigate the physiological role of COG1900d and the nature of the catalytic element that presumably replaces the essential Cys54 of MA1821 are needed to evaluate this hypothesis.

Coenzyme M biosynthesis also implicates sulfur assimilation pathways for incorporation of sulfur into phosphoenolpyruvate to form (R)-phosphosulfolactate, the first intermediate in the pathway found in the *Methanococcales* (Graham and White 2002; Graham 2011; Graham et al. 2002). An alternative pathway initiating from phosphoserine and leading to sulfopyruvate is found in other methanogens (Graham et al. 2009; Liu et al. 2012d). In both cases the enzymes responsible incorporate the sulfur in the form of sulfite. Sulfite is an inhibitor of methanogenesis, and its availability to these enzymes must be carefully controlled (Susanti and Mukhopadhyay 2012). Despite the toxicity, some methanogens are able to grow on sulfite as the sole sulfur source (Daniels et al. 1986; Johnson and Mukhopadhyay 2005), and this is enabled by the presence of sulfite reductases that convert the sulfite to sulfide. Several classes of sulfite reductases are known; the best characterized is the Fsr enzyme isolated from *M. jannaschii*, which uses the cofactor F420 as the reductant (Johnson and Mukhopadhyay 2005, 2008). Sulfite reductases were also essential for the development of dissimilatory sulfate reduction in the early archaean era (Canfield et al. 2000; Susanti and Mukhopadhyay 2012). The source of the required sulfite in methanogens grown with sulfide as the sole sulfur source is not yet clear, although it has been suggested that Fsr or a homologous dissimilatory sulfate reductase may produce it from sulfide (Susanti and Mukhopadhyay 2012).

Another metabolic pathway that involves the same chemical transformation of an aldehyde to a sulfhydryl group has recently been identified in *M. jannaschii* (Allen and White 2016). This pathway leads to the biosynthesis of 3-mercaptopropionic acid (MPA). Analysis of cell extracts showed that L-malate semialdehyde (MSA) is converted to the thiol-containing 2-hydroxy-4-mercaptobutyric acid (L-HMBA) in a sulfide-dependent reaction catalyzed by an as-yet unidentified enzyme. The biosynthetic origin of the MSA is also unknown. Three further steps are necessary to convert L-HMBA to MPA. MPA is a prevalent natural product in marine environments, and the discovery of a biosynthetic route to this compound in *M. jannaschii* suggests that it plays a previously unanticipated biological role (Allen and White 2016).

4 Sulfide Assimilation and Formation of Protein Persulfides

Biochemical precedents suggest two possible routes to the formation of protein persulfides in methanogens. First, the modification can be generated via the activity of cysteine desulfurases (CDs), which utilize a PLP-dependent mechanism to deliver sulfur from a substrate Cys to a Cys residue in the enzyme active site, generating alanine as the second product (Hidese et al. 2011) (Fig. 14.10). Second, the sulfurtransferase rhodanese transfers sulfur from a donor species to a Cys residue in its active site, again generating a protein persulfide (Fig. 14.10) (Aird et al. 1987; Mueller 2006). Proteins with domains homologous to rhodanese are also able to receive sulfur from CDs via interprotein transpersulfidation (Mueller 2006; Mishanina et al. 2015); a well-known example is persulfide transfer from the *E. coli* CD IscS to the rhodanese-like sulfurtransferase ThiI, which in turn donates the sulfane sulfur to form s^4U in tRNA (see below) (Kambampati and Lauhon 1999; Liu et al. 2012b). Persulfides have been detected on conserved cysteines of the MA1821 and SepCysS proteins partially purified from *M. acetivorans* cells (Figs. 14.5 and 14.8) (Rauch et al. 2017) and were also present on nearly 100 peptides derived from contaminating *M. acetivorans* proteins in those preparations. Cysteine persulfides were also detected on several *M. maripaludis* and *M. jannaschii* proteins expressed in *E. coli* (Liu et al. 2012a, b, 2014a). Persulfides, annotated as S-mercaptocysteine (CSS), have further been found in nearly 100 protein X-ray structures deposited to the PDB, attesting to the widespread nature of this modification (Berman et al. 2000).

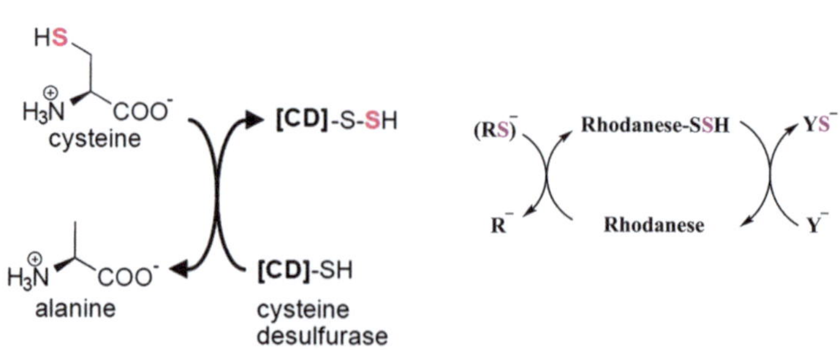

Fig. 14.10 Left: General mechanism for persulfide generation by cysteine desulfurases. Right: Double displacement reaction catalyzed by rhodanese domains. Rhodanese is able to catalyze sulfur transfers among a variety of donor (RS-) and acceptor (Y-) molecules. After Westley (1973)

The prevalence of protein persulfides on native proteins in methanogens lacking identifiable CDs has not yet been investigated. However, supplementation of *M. maripaludis* cell-free extracts with Cys and PLP did reveal a weak CD-like activity despite the apparent absence of CDs encoded in the genome (Liu et al. 2010). This suggests that a distinct structural class of CD in some methanogens apparently lacking these enzymes might be present. However, Cys cannot be the source of persulfide sulfur on SepCysS given that the SepCysS pathway is itself the source of free intracellular Cys (Sauerwald et al. 2005). Another possibility is that persulfides may be generated directly by rhodanese homology domain proteins (Su et al. 2012; Liu et al. 2012b). Rhodanese superfamily proteins are best known for catalyzing the reaction between thiosulfate and cyanide, which proceeds by a double displacement mechanism and involves a persulfide intermediate on the protein (Westley 1973). The reaction yields thiocyanate and sulfite and appears to play an important role in cyanide detoxification in both mammals and bacteria (Cipollone et al. 2007). However, in general rhodanese proteins are able to accept sulfur from a variety of donors, including thiosulfonates and polysulfides (Westley 1973). Therefore, a role for rhodanese in interprotein persulfidation and/or generation of sulfur-containing metabolites in methanogens is plausible.

It is possible that the recently identified *M. acetivorans* MA1715 protein may play a role in persulfide formation (Fig. 14.7) (Rauch and Perona 2016). MA1715 homologs are conserved in all methanogens together with SepCysS and the COG1900 proteins. An *M. acetivorans* strain in which the *ma1715* gene is deleted grows very poorly at sub-millimolar levels of sulfide as the sole sulfur source, while wild-type cells grow at sulfide concentrations as low as 50 μM and attain optimal growth by 200 μM (Rauch and Perona 2016). Thus, MA1715 appears to function as a sulfide biosensor. Further physiological experiments implicated MA1715 in delivery of sulfur to MA1821 and SepCysS for Hcy and Cys biosynthesis, respectively. Both MA1821 and SepCysS carry persulfide at conserved active-site cysteines in native cells, suggesting that MA1715 is involved in generating sulfane sulfur on these proteins (Rauch et al. 2017). Further, the very strong co-conservation of MA1715 with COG1900d proteins, including in methanogens that do not inhabit sulfidic environments, suggests that the two proteins could be jointly responsible for delivery of the thiol sulfur to precursors of coenzyme M and coenzyme B (Fig. 14.9; Table 14.2) (Graham and White 2002).

The biochemical activity of MA1715, however, is unknown. The protein belongs to the uncharacterized protein domain COG2122 and is a distant homolog of the ApbE enzyme (Fig. 14.7). ApbE catalyzes the Mg^{2+}- or Mn^{2+}-dependent hydrolysis of flavin-adenine dinucleotide (FAD) to flavin mononucleotide (FMN) and adenosine monophosphate (AMP) (Boyd et al. 2011; Deka et al. 2013). ApbE has a unique bimetal catalytic center, with two metal-binding residues (Thr and Asp) strictly conserved among Cog2122 and ApbE. However, the portion of ApbE that binds FAD is missing in the MA1715 protein, suggesting a distinct function (Figs. 14.7 and 14.11) (Rauch and Perona 2016). Since MA1715 facilitates sulfide mobilization at low ambient concentrations, it is possible that the protein may catalyze a reaction in which sulfide binding results in formation of a persulfide on the enzyme. While

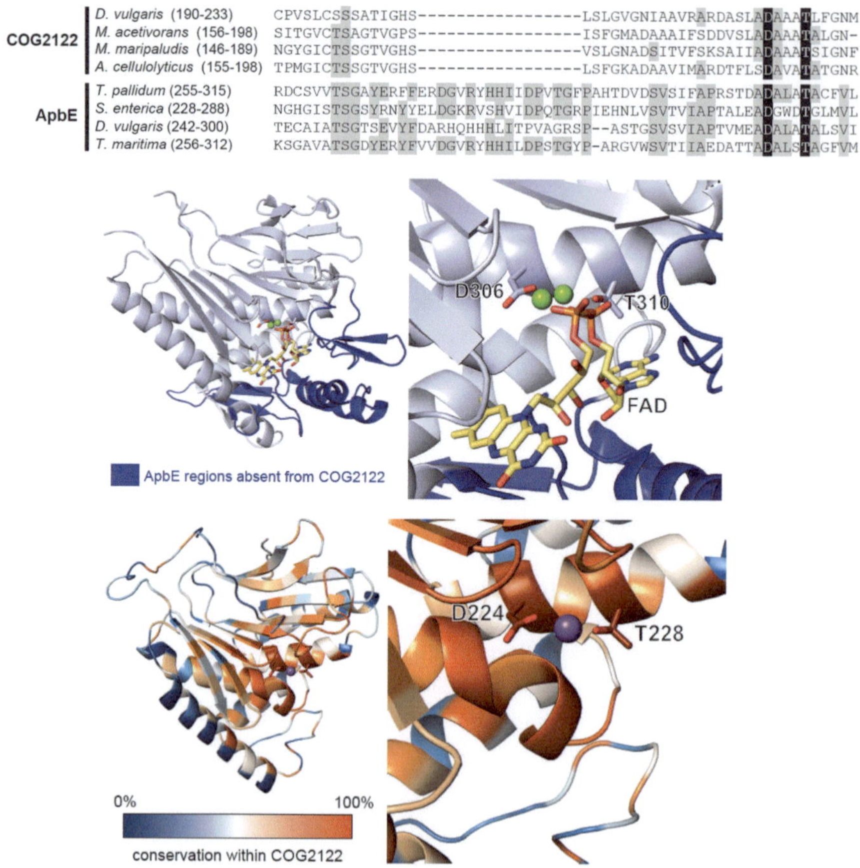

Fig. 14.11 Top: Structure-based sequence alignment of COG2122 and ApbE with consensus ApbE identities shaded (gray) and strictly conserved metal-binding residues indicated (black). Center: Structure of *Treponema pallidum* ApbE (PDB code 4ifx) (Deka et al. 2013) with regions absent from COG2122 shown in dark blue. Shown are the FAD substrate (yellow carbons), Mg^{2+} ions (green atoms), and strictly conserved metal-binding residues (gray carbons; D306 and T310). Bottom: Structure of *Desulfovibrio vulgaris* COG2122 (PDB code 2o34) colored from blue to red by conservation among all COG2122 members. Sticks are shown for the strictly conserved metal-binding residues (D224, T228)

direct reaction of hydrogen sulfide with Cys thiols is not favored in reducing environments, the reaction would be stimulated if oxidized forms of Cys, such as cysteine sulfenic acid, are present (Park et al. 2015; Mishanina et al. 2015). Such an MA1715 persulfide could be long-lived because there is no adjacent Cys to eliminate it. Another possibility is that an intermolecular disulfide-linked dimer formed via the conserved Cys on each of two MA1715 monomers could form persulfide upon direct reaction with hydrogen sulfide. This would resemble the activities of cystathionine β-synthase and cystathionine γ-lyase (Fig. 14.2), which synthesize cysteine persulfide from cystine (Ida et al. 2014). Another precedent is sulfide quinone oxidoreductase, which oxidizes sulfide to persulfide as part of the human mitochondrial H_2S oxidation pathway (Jackson et al. 2012; Libiad et al. 2014).

We have previously suggested that MA1715 might facilitate the nucleophilic attack of sulfide on a different metabolite, such as a high-energy phosphate compound that might yield thiophosphate (Rauch and Perona 2016). Thiophosphate could then function as a sulfur source for SepCysS, analogous to the role of selenophosphate as selenium donor for conversion of Sep-tRNA^{Sec} to Sec-tRNA^{Sec} (Itoh et al. 2009). However, the lower intrinsic reactivity of thiophosphate compared to selenophosphate argues against this mechanism (Reich and Hondal 2016). The proposed SepSecS mechanism, based on crystal structures, also does not invoke conserved Cys (or selenoCys) residues, in contrast to SepCysS (Fig. 14.5) (Palioura et al. 2009).

5 Sulfur Incorporation into Fe-S Clusters, tRNAs, and Other Metabolites

5.1 *Fe-S Cluster Assembly*

In many bacteria and eukaryotes, transpersulfidation initiated by CDs is responsible for delivery of sulfur to key cellular metabolites (Fig. 14.10), including Fe-S clusters and a subset of tRNA-modifying enzymes (Roche et al. 2013; Shigi 2014). The CDs are differentiated into two widely distributed classes based on detailed differences in local active-site structure and reactivity: the *E. coli* IscS/*A. vinelandii* NifS class (class I) and the SufS/CsdA class (class II) (Black and Dos Santos 2015). For the prototypical class I *E. coli* IscS, sulfane sulfur acquired from substrate Cys is transferred to a scaffold protein, IscU, which also binds Fe^{2+} (Schwartz et al. 2000). Fe-S clusters are assembled on IscU and then released for targeting to apoproteins in a process that involves a number of chaperones (Roche et al. 2013; Kim et al. 2015). In turn, some of the recipient Fe-S cluster proteins provide sulfur for biosynthesis of key metabolites, including certain thionucleosides in tRNA, biotin, and lipoic acid (Figs. 14.12 and 14.13) (Shigi 2014; Black and Dos Santos 2015). Although orthologs of class I and class II CDs are found in most organisms, including some archaea, both classes are conspicuously absent from a subset of methanogens, especially the obligate hydrogenotrophs (such as *M. jannaschii*) that originated at the earliest evolutionary stages of life (Table 14.2) (Blank 2009b).

The pathway of Fe-S biosynthesis in methanogens lacking CDs is still largely unknown. A few homologs of scaffold and assembly proteins from the SUF Fe-S biogenesis system (including SufB and SufC) are found in methanogens, but their roles have not been investigated (Liu et al. 2012d). Given the key importance of Fe-S clusters in energy conservation and in providing the sulfur source for downstream cofactors, this represents a major gap in our understanding of the metabolic map in these organisms. In *M. maripaludis*, which lacks CDs, metabolic labeling studies

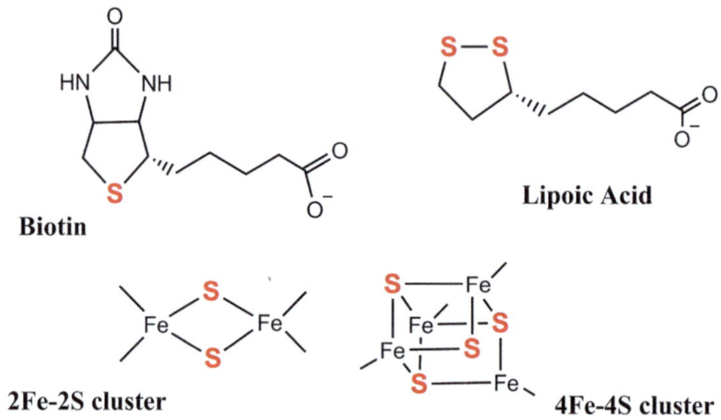

Fig. 14.12 Structures of biotin, lipoic acid, and several common iron-sulfur clusters

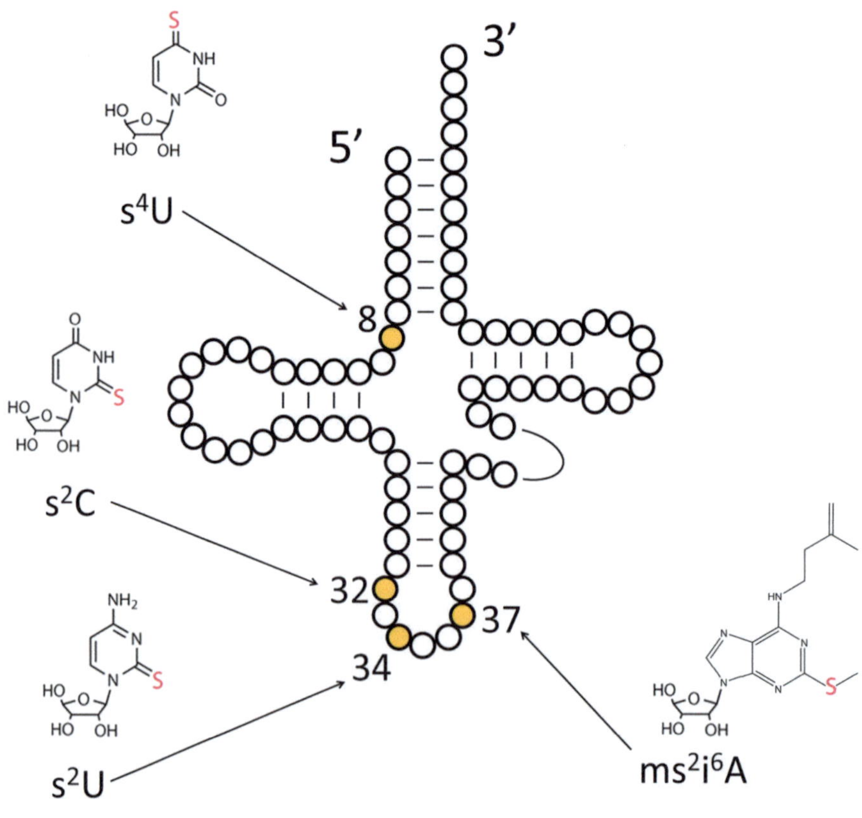

Fig. 14.13 Depiction of the secondary structure of tRNA with the positions and structures of sulfur-modified nucleotides indicated

confirmed that Cys is not the sulfur source for Fe-S biosynthesis (Liu et al. 2010). Instead, the sulfur originates from exogenous sulfide (Liu et al. 2010). In *M. acetivorans*, a triple deletion strain lacking the sulfide biosensor protein MA1715 and the bacterial sulfide uptake proteins OAHS and OASS (for synthesis of Hcy and Cys, respectively; Fig. 14.2) is viable when grown on sulfide as the sole sulfur source. This strain is also unaffected in its ability to biosynthesize either Fe-S clusters or thiolated tRNAs (Rauch and Perona 2016). Together, these studies suggest that abundant exogenous sulfide plays a central role in the biosynthesis of Fe-S clusters, the downstream recipients of Fe-S sulfur, tRNA nucleosides, and perhaps also other sulfur-containing metabolites in methanogens.

The mechanism by which sulfide may directly participate in Fe-S cluster assembly is not known. However, a recent report on an apparently inactive IscS homolog in *A. fulgidus* may offer some insight (Pagnier et al. 2015). The protein lacks canonical activity as a CD because it is missing the essential active-site lysine that links to PLP, but it nonetheless forms a complex with IscU and contributes a conserved Cys side-chain to the binding of a nascent Fe_2S_2 cluster on IscU (Marinoni et al. 2012). The conserved Cys accepts sulfane sulfur from NifS in vitro, but no Fe-S assembly on IscU (in the IscS-IscU complex) occurs unless DTT is added, generating sulfide in the active site. Further, exogenous sulfide alone is able to promote cluster assembly on IscU, when added to an IscS-IscU complex in which the conserved Cys on IscS is mutated (Pagnier et al. 2015). Based on this precedent, it is not unreasonable to speculate that SufB Fe-S scaffold protein homologs in methanogens may also be able to bind exogenous sulfide for direct Fe-S assembly, bypassing the need for CDs.

Sulfur is also incorporated into the highly specific FeMoCo prosthetic group found in nitrogenase. The metabolic capacity for nitrogen fixation is widespread in methanogens (Leigh 2000), and the presence of key *nif* genes in many species strongly suggests that it is evolutionarily related to the bacterial process. The process of FeMoCo assembly is complex and not yet fully understood, but it is known that in bacteria, the origin of the sulfur is the CD NifS, which subsequently donates its sulfane sulfur to NifU for initial Fe-S cluster formation that occurs prior to further processing and molybdenum incorporation (Hu and Ribbe 2011). Hence, sulfur incorporation into FeMoCo in methanogens lacking CDs is a specialized case of incorporation into Fe-S clusters generally and may also proceed by direct insertion of sulfide. Nitrogen fixation has been reported in *M. maripaludis*, which lacks identifiable CDs while possessing other genes necessary for the process (Blank et al. 1995).

5.2 Transfer of Sulfur from Fe-S Scaffolds to Biotin and Lipoic Acid

Sulfur insertion into biotin and lipoic acid has been best studied in *E. coli*. In each case the origin of the sulfurs in the cofactor is a bound Fe-S cluster that has been delivered to the biosynthetic enzymes BioB (biotin) and LipA (lipoate) (Cronan 2016). Mobilizable sulfurs within the cluster are transferred to precursors of biotin and lipoate by a radical SAM mechanism involving two Fe-S clusters with mechanistically distinct roles (Atta et al. 2012). Biotin-dependent carboxylases are widespread in methanogens (Lombard and Moreira 2011), as are genes for the biotin synthase BioB. Sulfur insertion into biotin in methanogens is thus predicted to proceed similarly to *E. coli*. Interestingly, however, analysis of archaeal genomes showed that the genes for lipoic acid biosynthesis have been largely lost in anaerobes, including most of the methanogens (Borziak et al. 2014). Analysis of cellular extracts from several methanogens by sensitive GC-MS approaches did not reveal the presence of lipoic acid, substantiating this finding (RH White, personal communication). Retention of LipA in *Methanocellales* may be related to the capacity of these organisms to withstand oxygen stress, rather than a role for lipoic acid in energy metabolism.

5.3 Incorporation of Sulfur into Transfer RNAs

Over 100 posttranscriptional modifications have been found in tRNAs, more than 20 of which contain sulfur (Machnicka et al. 2013). The sulfur-modified nucleotides are of four types—4-thiouridine (s^4U), 2-thiocytidine (s^2C), 2-thiouridine (s^2U), and methylthioadenosine ($ms^2(i/t)^6A$)—with the latter two comprising families of modifications that contain a variety of substitutions at other positions on the base (Fig. 14.13) (Shigi 2014; Black and Dos Santos 2015; Kimura and Suzuki 2015). Isolation and characterization of tRNAs from a variety of *Methanococcus* species by mass spectrometry showed that all four types of thionucleotides are present (McCloskey et al. 2001). The enzymes and pathways of tRNA thiomodification are diverse and share intermediates with pathways of sulfur incorporation into the other cofactors. The tRNA thiomodification enzymes in methanogens are generally present in all three domains of life and are not limited to the domain archaea.

The s^2C and $ms^2(i/t)^6A$ modifications are biosynthesized by enzymes containing Fe-S clusters (Fig. 14.12). The five known methylthio-modified bases feature the methylthio group at the 2-position of the adenine ring (designated ms^2) paired with a different moieties at N6, including N6-threonylcarbamoyl adenosine (ms^2t^6A) and N6-isopentenyladenosine (ms^2i^6A) (Machnicka et al. 2013). This modification is found only at position 37 in a subset of tRNAs, directly 3′ to the anticodon (Fig. 14.13) (Shigi 2014). Sulfur is incorporated into ms^2t^6A and ms^2i^6A by the radical SAM superfamily enzymes MtaB and MiaB, respectively (Atta et al. 2012).

The mechanism of sulfuration is thought to be similar to BioB and LipA. A subfamily of MtaB enzymes exists in archaea (Atta et al. 2012), but no MiaB-like or MtaB-like tRNA modifying enzyme has yet been clearly identified from a methanogen. However, radical SAM enzymes are widespread, accounting for 2% of the *M. jannaschii* genome (Miller et al. 2014).

s^2C is found at the 5′-end of the anticodon loop in a limited number of tRNA species (Fig. 14.13), where it increases the rate of A-site selection for the rare $tRNA^{Arg}{}_{AGG}$ species and decreases the rate of translation by $tRNA^{Arg}{}_{ICG}$ in a codon-specific manner (Jager et al. 2004). This modification is incorporated by the enzyme TtcA, which contains a redox-active [4Fe-4S] cluster that is assembled on a conserved Cys-X1-X2-Cys motif of the protein (Bouvier et al. 2014; Jager et al. 2004). The presence of the Fe-S cluster and three of six conserved Cys residues in the protein is essential for catalysis, although the role of the cluster in sulfur transfer, if any, has not been established (Bouvier et al. 2014). TtcA is not a member of the radical SAM enzyme superfamily, suggesting that its mechanism of sulfur incorporation does not resemble that for biotin, lipoic acid, or ms^2i^6A. Instead, the enzyme contains a motif characteristic of the PP-loop in the ATPase superfamily, which is also present in MnmA and ThiI proteins involved in s^2U and s^4U biosynthesis, respectively (see below) (Kambampati and Lauhon 2003; Mueller and Palenchar 1999). The TtcA enzyme is broadly distributed in all three domains of life and is annotated in many methanogen genomes.

In most organisms, biosynthesis of s^4U and s^2U in tRNA occurs by transpersulfidation initiated from sulfane sulfur carried on a CD (Fig. 14.10), so that Cys is the ultimate sulfur donor (Shigi 2014; Black and Dos Santos 2015). s^4U occurs exclusively at position 8 in tRNA and functions as a UV photosensor that fosters decreased cell growth and division under stress conditions (Ramabhadran and Jagger 1976). In *E. coli*, the CD IscS first transfers sulfane sulfur to a conserved Cys on the rhodanese domain of the protein ThiI, which in turn inserts the sulfur into tRNA (Mueller et al. 2001; Veerareddygari et al. 2016; Neumann et al. 2014). As part of the mechanism, ThiI catalyzes ATP-dependent adenylation of U8 prior to sulfur transfer. However, methanogens that lack CDs cannot biosynthesize s^4U8 by this pathway; moreover, ThiI homologs in these organisms also lack the rhodanese domain (Liu et al. 2012b). Mass spectrometry, metabolite labeling, and in vivo complementation studies of wild-type and mutant *M. maripaludis* ThiI proteins expressed in *E. coli* showed that two Cys residues on a conserved CXXC motif are essential for s^4U biosynthesis in vivo and that these two cysteines can be modified via disulfide and persulfide formation (Liu et al. 2012b). Moreover, in vitro formation of s^4U occurs with sulfide as the sulfur donor under conditions commensurate with physiological sulfide levels in the millimolar range (Liu et al. 2010, 2012b). *M. maripaludis* ThiI offers a good example of a methanogen-specific mechanism for sulfur incorporation, involving direct uptake of sulfide that bypasses the need for a CD. While persulfide is found on *E. coli* expressed ThiI, whether the modification occurs in *M. maripaludis* cells or is obligately involved in the mechanism remain open questions.

The s^2U modification is present as the common feature of a large family of modified uracils found at the wobble position of the anticodon, where it plays important roles in aminoacylation specificity and ribosome fidelity (Fig. 14.13) (Machnicka et al. 2013; Rodriguez-Hernandez et al. 2013; Sylvers et al. 1993). Diverse mechanisms for s^2U formation are known (Black and Dos Santos 2015; Shigi 2014). In *E. coli* and in mitochondria, the pathway proceeds via persulfide formation on IscS followed by a series of transpersulfidation reactions in which the sulfur atom is transferred through several persulfide carriers. In *E. coli*, the sulfane sulfur is relayed from IscS to TusA, TusBCD, and finally TusE. TusE transfers persulfide to the protein MnmA, which incorporates the sulfur to form s^2U by an ATP-dependent mechanism involving adenylate formation at the C2 position of the uracil ring (Numata et al. 2006a, b). This mechanism resembles adenylation of U8 by ThiI (Mueller et al. 2001). As TusBCD and TusE are not well conserved in bacteria (Kotera et al. 2010), the process presumably is capable of functioning efficiently when these proteins are bypassed.

The distinct Ncs6/Urm1 pathway of s^2U formation is present in the cytosol of eukaryotes (Shigi 2014). This pathway also begins with persulfide formation on a CD (Nfs1) followed by transfer of sulfane sulfur to the rhodanese domain proteins Tum1 and Uba4. Sulfane sulfur is then transferred from Uba4 to the C-terminus of the protein Urm1, which is previously activated by ATP-dependent adenylation. Finally, the sulfur in the C-terminal thiocarboxylate of Urm1 is incorporated to form s^2U34 through the action of the tRNA modification complex Ncs2/Ncs6; Ncs6 contains the ATP-dependent adenylation motif also found in ThiI and MnmA (Noma et al. 2009). Interestingly, the C-terminal thiocarboxylate of Urm1 is also essential in coupling it to other proteins to target them for degradation. In this context, Uba4 functions as an activating enzyme (E1) of the ubiquitin pathway, and Urm1 is a ubiquitin-like protein (Ubl), thus linking the s^2U incorporation and protein degradation machineries. Similar coupling is found in archaea (Miranda et al. 2011), where the Ubl proteins are termed SAMPs (Maupin-Furlow 2013), and via the TtuA/TtuB pathway in the thermophilic bacterium *T. thermophilus* (Maupin-Furlow 2014; Hepowit et al. 2016; Shigi 2014).

Like the s^4U8 biosynthesis pathway, the pathway for s^2U formation in methanogens lacking CDs is also distinctive. In addition to the absence of the CD Nfs1, the rhodanese protein Tum1 is also missing in many species, while the Uba4 homolog lacks the rhodanese domain (Liu et al. 2014a). However, an Ncs6 homolog protein is present. When expressed in *E. coli*, the protein acquires sulfane sulfur from IscS, suggesting that it may serve as the sulfur donor for s^2U formation. When expressed in *M. maripaludis* cells, the Ncs6 homolog copurifies with a SAMP protein that is likely to be co-transcribed, further supporting the notion that key aspects of the methanogen s^2U34 biosynthesis pathway resemble those in the eukaryotic cytosol and in other archaea. However, the mechanism of sulfur acquisition by Ncs6 in methanogens that lack CDs remains unresolved.

5.4 Incorporation of Sulfur into Thiamine and Molybdopterin

The enzymes and pathways for sulfur incorporation into thiamine and molybdopterin substantially overlap those discussed above (Fig. 14.14). Like s^4U8 biosynthesis in tRNA, the thiamine pathway in *B. subtilis* also initiates from persulfided IscS (Begley et al. 2012). The sulfane sulfur is then mobilized through the action of the protein ThiF, resulting in formation of a C-terminal thiocarboxylate on a second key protein, ThiS. Further well-described steps result in biosynthesis of the thiamin thiazole ring and its condensation with the pyrimidine portion of the cofactor (Begley et al. 2012).

A second pathway has been discovered in yeast and fungi that requires only one protein, thiazole synthase (Thi4p), for synthesis of the thiazole ring. Thi4p catalyzes a single-turnover, iron-dependent reaction in which the sulfur of an active-site Cys is transferred to the nascent thiazole, with formation of dehydroalanine and irreversible inactivation of the enzyme (Begley et al. 2012). Many archaea, including methanogens, lack the bacterial thiazole biosynthesis pathway but possess orthologs of Thi4p that are lacking the Cys sulfur donor. Biochemical studies showed that the *M. jannaschii* ortholog (MjThi4) binds Fe(II) or Fe(III) and can catalyze multiple rounds of thiazole formation with sulfide as the donor (Eser et al. 2016). The immediate product of the reaction is an adenylated derivative of the thiazole in a

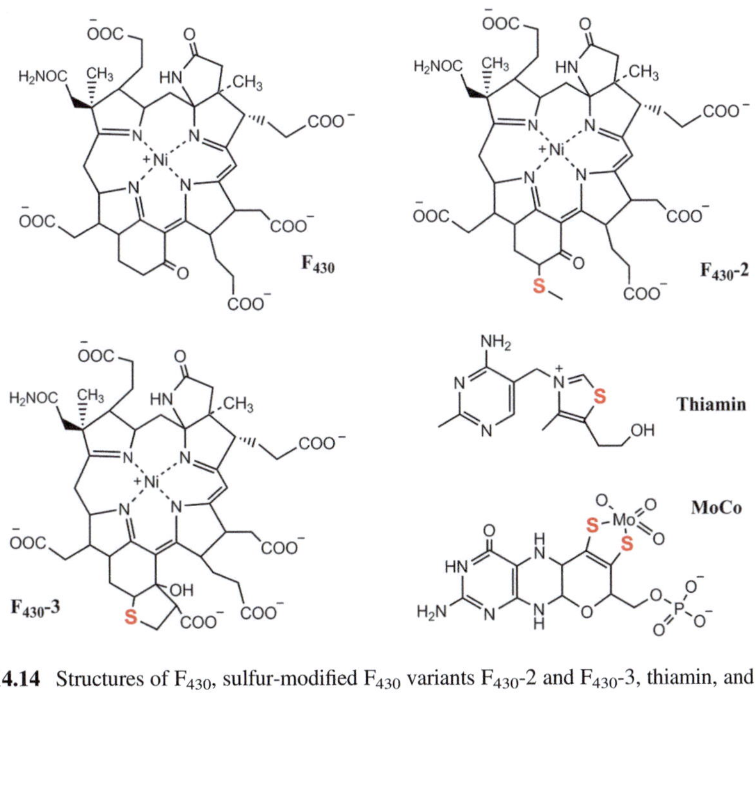

Fig. 14.14 Structures of F_{430}, sulfur-modified F_{430} variants F_{430}-2 and F_{430}-3, thiamin, and Moco

distinct tautomeric form. A crystal structure of the related *M. igneus* Thi4 that includes an active-site iron atom suggests an iron-dependent mechanism for sulfide incorporation that features direct iron-sulfur ligation (Zhang et al. 2016; Eser et al. 2016). This mechanistic proposal demonstrates a plausible means for direct sulfide incorporation into thiazole that does not depend on an enzyme persulfide.

Molybdenum has been identified as a component of over 50 enzymes, usually as a component of the so-called Moco cofactor (Fig. 14.14) (Mendel and Leimkuhler 2015). In Moco, the molybdenum is ligated to two sulfurs in the dithiolene portion of the pterin molybdopterin (MPT). The sulfur atoms in MPT are delivered from a C-terminal thiocarboxylate found on the protein MoaD, which in turn is generated in a reaction catalyzed by MoeB (Mendel and Leimkuhler 2015; Black and Dos Santos 2015). MoeB does not carry the sulfur but facilitates its transfer from a relay system composed of the CD IscS and the protein TusA; as discussed above, TusA also participates in sulfur relay to tRNA s^2U34 (Numata et al. 2006b). In archaea, work in *H. volcanii* has led to a model in which the source of sulfur for Moco is a C-terminal thiocarboxylate on an ubiquitin-like SAMP protein (Maupin-Furlow 2013). This is similar to the proposal made for s^2U formation, which also involves a C-terminal thiocarboxylate on a SAMP protein. However, in both cases, the nature of the sulfur donor in methanogens lacking CDs has not been investigated.

5.5 *Incorporation of Sulfur into Modified F_{430} Coenzymes*

F_{430} is a specialized nickel-containing cofactor found in the methanogen enzyme methyl-CoM reductase (MCR), which catalyzes formation of methane from methyl-coenzyme M—the terminal step in methanogenesis (Ermler et al. 1997). Recent analysis of extracts from several methanogens and anaerobic methane oxidizers (anaerobic methanotrophic archaea; ANME) has revealed the presence of nine F_{430} variants, four of which contain sulfur (Fig. 14.14) (Allen et al. 2014; Mayr et al. 2008). The best studied of these in methanogens, F_{430}-3, features the addition of either 2- or 3-mercaptopropionate to the F_{430} structure. Mass spectrometry and spectroscopic measurements suggested that F_{430}-3 is biosynthesized by addition of a methylthio group to the F_{430} skeleton, as observed for the F_{430}-2 modification that is found in ANME isolates but not in methanogens (Mayr et al. 2008). The functional roles of the modified F_{430} cofactors are not yet known. Biosynthesis of both F_{430}-2 and F_{430}-3 may be catalyzed by radical SAM superfamily methylthiotransferase enzymes, functioning similarly to the MiaB and MtaB enzymes that incorporate ms^2i^6A and ms^2t^6A into tRNA (Allen et al. 2014).

6 Conclusion

Over 20 years ago, the genome sequence of *M. jannaschii* opened a new era for investigating the metabolism of methanogenesis. Since unique sulfur-containing cofactors are essential to this process, and the organism thrives in an extreme anaerobic and highly sulfidic environment, it is not surprising (at least in hindsight) that the sulfur assimilation pathways found in aerobes are absent, and that many other obligate hydrogenotrophs share this property. Today the genes responsible for Cys-tRNA^{Cys}, Cys, and Hcy biosynthesis in these and other methanogens have been elucidated, and biochemical studies of the enzyme mechanisms are underway. Testable hypotheses have also been generated regarding the identities of the genes responsible for introducing thiols into coenzyme M and coenzyme B and for persulfide formation on the Cys and Hcy biosynthetic enzymes (Table 14.1). Less is known of the downstream pathways by which sulfur is incorporated into iron-sulfur clusters, tRNAs, and other metabolites, especially in organisms where neither of the two known classes of CDs are present. However, in several instances—incorporation of s^{4}U into tRNA, assembly of Fe-S clusters, and biosynthesis of the thiazole ring in thiamine—recent evidence suggests that the sulfur can be incorporated directly as sulfide rather than arriving via interprotein persulfide relay from CDs or other persulfide-bearing proteins. The enzymatic mechanisms of these proteins and the extent to which direct sulfide uptake represents a common theme for sulfur assimilation in methanogens are now among the most pressing questions for future research (see Box 14.1 for a summary of outstanding questions).

Box 14.1 Outstanding Questions

- How is sulfane sulfur delivered to conserved active-site cysteines of SepCysS and COG1900a/CBS in *Methanosarcina acetivorans*?
- Does a [3Fe-4S] cluster serve as the external reductant in SepCysS? In general, how are iron-sulfur clusters biosynthesized in methanogens that lack CDs?
- What is the biochemical reaction catalyzed by the conserved MA1715 protein, which appears to function as a sulfide biosensor in vivo?
- Does the conserved COG1900d protein catalyze thiol group formation in coenzyme M and coenzyme B?
- What is the biological significance, if any, of the expansion in the number tRNA^{Cys} genes in some methanogens? Why is CysRS retained in some methanogens despite the presence of the SepRS-SepCysS pathway?
- Do protein persulfides and interprotein persulfide relay occur in methanogens without evident cysteine desulfurases? If so, to what extent? Do methanogens with cysteine desulfurases have redundant pathways to incorporate sulfur into some cofactors?
- Is the direct uptake of sulfide by biosynthetic enzymes a general feature of cofactor and metabolite formation in methanogens?

References

Aird BA, Heinrikson RL, Westley J (1987) Isolation and characterization of a prokaryotic sulfurtransferase. J Biol Chem 262:17327–17335

Allen KD, White RH (2016) Occurrence and biosynthesis of 3-mercaptopropionic acid in Methanocaldococcus jannaschii. FEMS Microbiol Lett 363

Allen KD, Wegener G, White RH (2014) Discovery of multiple modified F(430) coenzymes in methanogens and anaerobic methanotrophic archaea suggests possible new roles for F(430) in nature. Appl Environ Microbiol 80:6403–6412

Allen KD, Miller DV, Rauch BJ, Perona JJ, White RH (2015) Homocysteine is biosynthesized from aspartate semialdehyde and hydrogen sulfide in methanogenic archaea. Biochemistry 54:3129–3132

Atta M, Arragain S, Fontecave M, Mulliez E, Hunt JF, Luff JD, Forouhar F (2012) The methylthiolation reaction mediated by the radical-SAM enzymes. Biochim Biophys Acta 1824:1223–1230

Begley TP, Ealick SE, McLafferty FW (2012) Thiamin biosynthesis: still yielding fascinating biological chemistry. Biochem Soc Trans 40:555–560

Berman HM, Westbrook J, Feng Z, Gilliland G, Bhat TN, Weissig H, Shindyalov IN, Bourne PE (2000) The protein data bank. Nucleic Acids Res 28:235–242

Black KA, Dos Santos PC (2015) Shared-intermediates in the biosynthesis of thio-cofactors: mechanism and functions of cysteine desulfurases and sulfur acceptors. Biochim Biophys Acta 1853:1470–1480

Blank CE (2009a) Phylogenomic dating–a method of constraining the age of microbial taxa that lack a conventional fossil record. Astrobiology 9:173–191

Blank CE (2009b) Phylogenomic dating–the relative antiquity of archaeal metabolic and physiological traits. Astrobiology 9:193–219

Blank CE, Kessler PS, Leigh JA (1995) Genetics in methanogens: transposon insertion mutagenesis of a Methanococcus maripaludis nifH gene. J Bacteriol 177:5773–5777

Borup B, Ferry JG (2000) Cysteine biosynthesis in the Archaea: Methanosarcina thermophila utilizes O-acetylserine sulfhydrylase. FEMS Microbiol Lett 189:205–210

Borziak K, Posner MG, Upadhyay A, Danson MJ, Bagby S, Dorus S (2014) Comparative genomic analysis reveals 2-oxoacid dehydrogenase complex lipoylation correlation with aerobiosis in archaea. PLoS One 9:e87063

Bouvier D, Labessan N, Clemancey M, Latour JM, Ravanat JL, Fontecave M, Atta M (2014) TtcA a new tRNA-thioltransferase with an Fe-S cluster. Nucleic Acids Res 42:7960–7970

Boyd JM, Endrizzi JA, Hamilton TL, Christopherson MR, Mulder DW, Downs DM, Peters JW (2011) FAD binding by ApbE protein from Salmonella enterica: a new class of FAD-binding proteins. J Bacteriol 193:887–895

Bult CJ, White O, Olsen GJ, Zhou L, Fleischmann RD, Sutton GG, Blake JA, FitzGerald LM, Clayton RA, Gocayne JD, Kerlavage AR, Dougherty BA, Tomb JF, Adams MD, Reich CI, Overbeek R, Kirkness EF, Weinstock KG, Merrick JM, Glodek A, Scott JL, Geoghagen NS, Venter JC (1996) Complete genome sequence of the methanogenic archaeon, Methanococcus jannaschii. Science 273:1058–1073

Canfield DE, Habicht KS, Thamdrup B (2000) The Archean sulfur cycle and the early history of atmospheric oxygen. Science 288:658–661

Carlson BA, Xu XM, Kryukov GV, Rao M, Berry MJ, Gladyshev VN, Hatfield DL (2004) Identification and characterization of phosphoseryl-tRNA[Ser]Sec kinase. Proc Natl Acad Sci U S A 101:12848–12853

Cipollone R, Ascenzi P, Visca P (2007) Common themes and variations in the rhodanese superfamily. IUBMB Life 59:51–59

Costa KC, Leigh JA (2014) Metabolic versatility in methanogens. Curr Opin Biotechnol 29:70–75

Cronan JE (2016) Assembly of lipoic acid on its cognate enzymes: an extraordinary and essential biosynthetic pathway. Microbiol Mol Biol Rev 80:429–450

Daniels L, Belay N, Rajagopal BS (1986) Assimilatory reduction of sulfate and sulfite by methanogenic bacteria. Appl Environ Microbiol 51:703–709

Deka RK, Brautigam CA, Liu WZ, Tomchick DR, Norgard MV (2013) The TP0796 lipoprotein of Treponema pallidum is a bimetal-dependent FAD pyrophosphatase with a potential role in flavin homeostasis. J Biol Chem 288:11106–11121

Dridi B, Raoult D, Drancourt M (2011) Archaea as emerging organisms in complex human microbiomes. Anaerobe 17:56–63

Duin EC, Madadi-Kahkesh S, Hedderich R, Clay MD, Johnson MK (2002) Heterodisulfide reductase from Methanothermobacter marburgensis contains an active-site [4Fe-4S] cluster that is directly involved in mediating heterodisulfide reduction. FEBS Lett 512:263–268

Ermler U, Grabarse W, Shima S, Goubeaud M, Thauer RK (1997) Crystal structure of methyl-coenzyme M reductase: the key enzyme of biological methane formation. Science 278:1457–1462

Eser BE, Zhang X, Chanani PK, Begley TP, Ealick SE (2016) From suicide enzyme to catalyst: the iron-dependent sulfide transfer in Methanococcus jannaschii thiamin thiazole biosynthesis. J Am Chem Soc 138:3639–3642

Evans PN, Parks DH, Chadwick GL, Robbins SJ, Orphan VJ, Golding SD, Tyson GW (2015) Methane metabolism in the archaeal phylum Bathyarchaeota revealed by genome-centric metagenomics. Science 350:434–438

Forchhammer K, Bock A (1991) Selenocysteine synthase from Escherichia coli. Analysis of the reaction sequence. J Biol Chem 266:6324–6328

Fraser WT, Blei E, Fry SC, Newman MF, Reay DS, Smith KA, McLeod AR (2015) Emission of methane, carbon monoxide, carbon dioxide and short-chain hydrocarbons from vegetation foliage under ultraviolet irradiation. Plant Cell Environ 38:980–989

Fukunaga R, Yokoyama S (2007a) Structural insights into the first step of RNA-dependent cysteine biosynthesis in archaea. Nat Struct Mol Biol 14:272–279

Fukunaga R, Yokoyama S (2007b) Structural insights into the second step of RNA-dependent cysteine biosynthesis in archaea: crystal structure of Sep-tRNA:Cys-tRNA synthase from Archaeoglobus fulgidus. J Mol Biol 370:128–141

Galagan JE, Nusbaum C, Roy A, Endrizzi MG, Macdonald P, FitzHugh W, Calvo S, Engels R, Smirnov S, Atnoor D, Brown A, Allen N, Naylor J, Stange-Thomann N, DeArellano K, Johnson R, Linton L, McEwan P, McKernan K, Talamas J, Tirrell A, Ye W, Zimmer A, Barber RD, Cann I, Graham DE, Grahame DA, Guss AM, Hedderich R, Ingram-Smith C, Kuettner HC, Krzycki JA, Leigh JA, Li W, Liu J, Mukhopadhyay B, Reeve JN, Smith K, Springer TA, Umayam LA, White O, White RH, Conway de Macario E, Ferry JG, Jarrell KF, Jing H, Macario AJ, Paulsen I, Pritchett M, Sowers KR, Swanson RV, Zinder SH, Lander E, Metcalf WW, Birren B (2002) The genome of M. acetivorans reveals extensive metabolic and physiological diversity. Genome Res 12:532–542

Graham DE (2011) 2-oxoacid metabolism in methanogenic CoM and CoB biosynthesis. Methods Enzymol 494:301–326

Graham DE, White RH (2002) Elucidation of methanogenic coenzyme biosyntheses: from spectroscopy to genomics. Nat Prod Rep 19:133–147

Graham DE, Xu H, White RH (2002) Identification of coenzyme M biosynthetic phosphosulfolactate synthase: a new family of sulfonate-biosynthesizing enzymes. J Biol Chem 277:13421–13429

Graham DE, Taylor SM, Wolf RZ, Namboori SC (2009) Convergent evolution of coenzyme M biosynthesis in the Methanosarcinales: cysteate synthase evolved from an ancestral threonine synthase. Biochem J 424:467–478

Hauenstein SI, Perona JJ (2008) Redundant synthesis of cysteinyl-tRNACys in Methanosarcina mazei. J Biol Chem 283:22007–22017

Hauenstein SI, Hou YM, Perona JJ (2008) The homotetrameric phosphoseryl-tRNA synthetase from Methanosarcina mazei exhibits half-of-the-sites activity. J Biol Chem 283:21997–22006

Helgadottir S, Sinapah S, Soll D, Ling J (2012) Mutational analysis of Sep-tRNA:Cys-tRNA synthase reveals critical residues for tRNA-dependent cysteine formation. FEBS Lett 586:60–63
Hepowit NL, de Vera IM, Cao S, Fu X, Wu Y, Uthandi S, Chavarria NE, Englert M, Su D, Sll D, Kojetin DJ, Maupin-Furlow JA (2016) Mechanistic insight into protein modification and sulfur mobilization activities of noncanonical E1 and associated ubiquitin-like proteins of Archaea. FEBS J 283:3567–3586
Hidese R, Mihara H, Esaki N (2011) Bacterial cysteine desulfurases: versatile key players in biosynthetic pathways of sulfur-containing biofactors. Appl Microbiol Biotechnol 91:47–61
Hohn MJ, Park HS, O'Donoghue P, Schnitzbauer M, Soll D (2006) Emergence of the universal genetic code imprinted in an RNA record. Proc Natl Acad Sci U S A 103:18095–18100
Holland HD (2006) The oxygenation of the atmosphere and oceans. Philos Trans R Soc Lond Ser B Biol Sci 361:903–915
Hu Y, Ribbe MW (2011) Biosynthesis of nitrogenase FeMoco. Coord Chem Rev 255:1218–1224
Hug LA, Baker BJ, Anantharaman K, Brown CT, Probst AJ, Castelle CJ, Butterfield CN, Hernsdorf AW, Amano Y, Ise K, Suzuki Y, Dudek N, Relman DA, Finstad KM, Amundson R, Thomas BC, Banfield JF (2016) A new view of the tree of life. Nat Microbiol 1:16048
Ida T, Sawa T, Ihara H, Tsuchiya Y, Watanabe Y, Kumagai Y, Suematsu M, Motohashi H, Fujii S, Matsunaga T, Yamamoto M, Ono K, Devarie-Baez NO, Xian M, Fukuto JM, Akaike T (2014) Reactive cysteine persulfides and S-polythiolation regulate oxidative stress and redox signaling. Proc Natl Acad Sci U S A 111:7606–7611
Itoh Y, Sekine S, Matsumoto E, Akasaka R, Takemoto C, Shirouzu M, Yokoyama S (2009) Structure of selenophosphate synthetase essential for selenium incorporation into proteins and RNAs. J Mol Biol 385:1456–1469
Jackson MR, Melideo SL, Jorns MS (2012) Human sulfide:quinone oxidoreductase catalyzes the first step in hydrogen sulfide metabolism and produces a sulfane sulfur metabolite. Biochemistry 51:6804–6815
Jager G, Leipuviene R, Pollard MG, Qian Q, Bjork GR (2004) The conserved Cys-X1-X2-Cys motif present in the TtcA protein is required for the thiolation of cytidine in position 32 of tRNA from Salmonella enterica serovar Typhimurium. J Bacteriol 186:750–757
Johnson EF, Mukhopadhyay B (2005) A new type of sulfite reductase, a novel coenzyme F420-dependent enzyme, from the methanarchaeon Methanocaldococcus jannaschii. J Biol Chem 280:38776–38786
Johnson EF, Mukhopadhyay B (2008) Coenzyme F420-dependent sulfite reductase-enabled sulfite detoxification and use of sulfite as a sole sulfur source by Methanococcus maripaludis. Appl Environ Microbiol 74:3591–3595
Kadaba NS, Kaiser JT, Johnson E, Lee A, Rees DC (2008) The high-affinity E. coli methionine ABC transporter: structure and allosteric regulation. Science 321:250–253
Kambampati R, Lauhon CT (1999) IscS is a sulfurtransferase for the in vitro biosynthesis of 4-thiouridine in Escherichia coli tRNA. Biochemistry 38:16561–16568
Kambampati R, Lauhon CT (2003) MnmA and IscS are required for in vitro 2-thiouridine biosynthesis in Escherichia coli. Biochemistry 42:1109–1117
Kamtekar S, Hohn MJ, Park HS, Schnitzbauer M, Sauerwald A, Soll D, Steitz TA (2007) Toward understanding phosphoseryl-tRNACys formation: the crystal structure of Methanococcus maripaludis phosphoseryl-tRNA synthetase. Proc Natl Acad Sci U S A 104:2620–2625
Kaster AK, Goenrich M, Seedorf H, Liesegang H, Wollherr A, Gottschalk G, Thauer RK (2011) More than 200 genes required for methane formation from H(2) and CO(2) and energy conservation are present in Methanothermobacter marburgensis and Methanothermobacter thermautotrophicus. Archaea 2011:973848
Kessler D (2006) Enzymatic activation of sulfur for incorporation into biomolecules in prokaryotes. FEMS Microbiol Rev 30:825–840
Kim JH, Bothe JR, Alderson TR, Markley JL (2015) Tangled web of interactions among proteins involved in iron-sulfur cluster assembly as unraveled by NMR, SAXS, chemical crosslinking, and functional studies. Biochim Biophys Acta 1853:1416–1428

Kimura S, Suzuki T (2015) Iron-sulfur proteins responsible for RNA modifications. Biochim Biophys Acta 1853:1272–1283

Klenk HP, Clayton RA, Tomb JF, White O, Nelson KE, Ketchum KA, Dodson RJ, Gwinn M, Hickey EK, Peterson JD, Richardson DL, Kerlavage AR, Graham DE, Kyrpides NC, Fleischmann RD, Quackenbush J, Lee NH, Sutton GG, Gill S, Kirkness EF, Dougherty BA, McKenney K, Adams MD, Loftus B, Peterson S, Reich CI, McNeil LK, Badger JH, Glodek A, Zhou L, Overbeek R, Gocayne JD, Weidman JF, McDonald L, Utterback T, Cotton MD, Spriggs T, Artiach P, Kaine BP, Sykes SM, Sadow PW, D'Andrea KP, Bowman C, Fujii C, Garland SA, Mason TM, Olsen GJ, Fraser CM, Smith HO, Woese CR, Venter JC (1997) The complete genome sequence of the hyperthermophilic, sulphate-reducing archaeon Archaeoglobus fulgidus. Nature 390:364–370

Klipcan L, Frenkel-Morgenstern M, Safro MG (2008) Presence of tRNA-dependent pathways correlates with high cysteine content in methanogenic Archaea. Trends Genet 24:59–63

Komatsoulis GA, Abelson J (1993) Recognition of tRNA(Cys) by Escherichia coli cysteinyl-tRNA synthetase. Biochemistry 32:7435–7444

Kotera M, Bayashi T, Hattori M, Tokimatsu T, Goto S, Mihara H, Kanehisa M (2010) Comprehensive genomic analysis of sulfur-relay pathway genes. Genome Inform 24:104–115

Leigh JA (2000) Nitrogen fixation in methanogens: the archaeal perspective. Curr Issues Mol Biol 2:125–131

Libiad M, Yadav PK, Vitvitsky V, Martinov M, Banerjee R (2014) Organization of the human mitochondrial hydrogen sulfide oxidation pathway. J Biol Chem 289:30901–30910

Liu Y, Whitman WB (2008) Metabolic, phylogenetic, and ecological diversity of the methanogenic archaea. Ann N Y Acad Sci 1125:171–189

Liu Y, Sieprawska-Lupa M, Whitman WB, White RH (2010) Cysteine is not the sulfur source for iron-sulfur cluster and methionine biosynthesis in the methanogenic archaeon Methanococcus maripaludis. J Biol Chem 285:31923–31929

Liu Y, Dos Santos PC, Zhu X, Orlando R, Dean DR, Soll D, Yuan J (2012a) Catalytic mechanism of Sep-tRNA:Cys-tRNA synthase: sulfur transfer is mediated by disulfide and persulfide. J Biol Chem 287:5426–5433

Liu Y, Zhu X, Nakamura A, Orlando R, Soll D, Whitman WB (2012b) Biosynthesis of 4-thiouridine in tRNA in the methanogenic archaeon Methanococcus maripaludis. J Biol Chem 287(44):36683–36692

Liu Y, Beer LL, Whitman WB (2012c) Methanogens: a window into ancient sulfur metabolism. Trends Microbiol 20:251–258

Liu Y, Beer LL, Whitman WB (2012d) Sulfur metabolism in archaea reveals novel processes. Environ Microbiol 14:2632–2644

Liu Y, Long F, Wang L, Soll D, Whitman WB (2014a) The putative tRNA 2-thiouridine synthetase Ncs6 is an essential sulfur carrier in Methanococcus maripaludis. FEBS Lett 588:873–877

Liu Y, Nakamura A, Nakazawa Y, Asano N, Ford KA, Hohn MJ, Tanaka I, Yao M, Soll D (2014b) Ancient translation factor is essential for tRNA-dependent cysteine biosynthesis in methanogenic archaea. Proc Natl Acad Sci U S A 111(29):10520–10525

Liu Y, Vinyard DJ, Reesbeck ME, Suzuki T, Manakongtreecheep K, Holland PL, Brudvig GW, Soll D (2016) A [3Fe-4S] cluster is required for tRNA thiolation in archaea and eukaryotes. Proc Natl Acad Sci U S A 113(45):12703–12708

Lombard J, Moreira D (2011) Early evolution of the biotin-dependent carboxylase family. BMC Evol Biol 11:232

Lucas M, Encinar JA, Arribas EA, Oyenarte I, Garcia IG, Kortazar D, Fernandez JA, Mato JM, Martinez-Chantar ML, Martinez-Cruz LA (2010) Binding of S-methyl-5'-thioadenosine and S-adenosyl-L-methionine to protein MJ0100 triggers an open-to-closed conformational change in its CBS motif pair. J Mol Biol 396:800–820

Lyons TW, Gill BC (2010) Ancient sulfur cycling and oxygenation of the early biosphere. Elements 6:93–99

Machnicka MA, Milanowska K, Osman Oglou O, Purta E, Kurkowska M, Olchowik A, Januszewski W, Kalinowski S, Dunin-Horkawicz S, Rother KM, Helm M, Bujnicki JM, Grosjean H (2013) MODOMICS: a database of RNA modification pathways--2013 update. Nucleic Acids Res 41:D262–D267

Major TA, Burd H, Whitman WB (2004) Abundance of 4Fe-4S motifs in the genomes of methanogens and other prokaryotes. FEMS Microbiol Lett 239:117–123

Marinoni EN, de Oliveira JS, Nicolet Y, Raulfs EC, Amara P, Dean DR, Fontecilla-Camps JC (2012) (IscS-IscU)2 complex structures provide insights into Fe2S2 biogenesis and transfer. Angew Chem Int Ed Eng 51:5439–5442

Maupin-Furlow JA (2013) Ubiquitin-like proteins and their roles in archaea. Trends Microbiol 21:31–38

Maupin-Furlow JA (2014) Prokaryotic ubiquitin-like protein modification. Annu Rev Microbiol 68:155–175

Mayr S, Latkoczy C, Kruger M, Gunther D, Shima S, Thauer RK, Widdel F, Jaun B (2008) Structure of an F430 variant from archaea associated with anaerobic oxidation of methane. J Am Chem Soc 130:10758–10767

Mayumi D, Mochimaru H, Tamaki H, Yamamoto K, Yoshioka H, Suzuki Y, Kamagata Y, Sakata S (2016) Methane production from coal by a single methanogen. Science 354:222–225

McCloskey JA, Graham DE, Zhou S, Crain PF, Ibba M, Konisky J, Soll D, Olsen GJ (2001) Post-transcriptional modification in archaeal tRNAs: identities and phylogenetic relations of nucleotides from mesophilic and hyperthermophilic Methanococcales. Nucleic Acids Res 29:4699–4706

Mendel RR, Leimkuhler S (2015) The biosynthesis of the molybdenum cofactors. J Biol Inorg Chem 20:337–347

Miller D, O'Brien K, Xu H, White RH (2014) Identification of a 5'-deoxyadenosine deaminase in Methanocaldococcus jannaschii and its possible role in recycling the radical S-adenosylmethionine enzyme reaction product 5'-deoxyadenosine. J Bacteriol 196:1064–1072

Min B, Pelaschier JT, Graham DE, Tumbula-Hansen D, Soll D (2002) Transfer RNA-dependent amino acid biosynthesis: an essential route to asparagine formation. Proc Natl Acad Sci U S A 99:2678–2683

Mino K, Ishikawa K (2003) A novel O-phospho-L-serine sulfhydrylation reaction catalyzed by O-acetylserine sulfhydrylase from Aeropyrum pernix K1. FEBS Lett 551:133–138

Miranda HV, Nembhard N, Su D, Hepowit N, Krause DJ, Pritz JR, Phillips C, Soll D, Maupin-Furlow JA (2011) E1- and ubiquitin-like proteins provide a direct link between protein conjugation and sulfur transfer in archaea. Proc Natl Acad Sci U S A 108:4417–4422

Mishanina TV, Libiad M, Banerjee R (2015) Biogenesis of reactive sulfur species for signaling by hydrogen sulfide oxidation pathways. Nat Chem Biol 11:457–464

Mueller EG (2006) Trafficking in persulfides: delivering sulfur in biosynthetic pathways. Nat Chem Biol 2:185–194

Mueller EG, Palenchar PM (1999) Using genomic information to investigate the function of ThiI, an enzyme shared between thiamin and 4-thiouridine biosynthesis. Protein Sci 8:2424–2427

Mueller EG, Palenchar PM, Buck CJ (2001) The role of the cysteine residues of ThiI in the generation of 4-thiouridine in tRNA. J Biol Chem 276:33588–33595

Neumann P, Lakomek K, Naumann PT, Erwin WM, Lauhon CT, Ficner R (2014) Crystal structure of a 4-thiouridine synthetase-RNA complex reveals specificity of tRNA U8 modification. Nucleic Acids Res 42:6673–6685

Noma A, Sakaguchi Y, Suzuki T (2009) Mechanistic characterization of the sulfur-relay system for eukaryotic 2-thiouridine biogenesis at tRNA wobble positions. Nucleic Acids Res 37:1335–1352

Numata T, Fukai S, Ikeuchi Y, Suzuki T, Nureki O (2006a) Structural basis for sulfur relay to RNA mediated by heterohexameric TusBCD complex. Structure 14:357–366

Numata T, Ikeuchi Y, Fukai S, Suzuki T, Nureki O (2006b) Snapshots of tRNA sulphuration via an adenylated intermediate. Nature 442:419–424

O'Donoghue P, Sethi A, Woese CR, Luthey-Schulten ZA (2005) The evolutionary history of Cys-tRNACys formation. Proc Natl Acad Sci U S A 102:19003–19008

Pagnier A, Nicolet Y, Fontecilla-Camps JC (2015) IscS from Archaeoglobus fulgidus has no desulfurase activity but may provide a cysteine ligand for [Fe2S2] cluster assembly. Biochim Biophys Acta 1853:1457–1463

Palioura S, Sherrer RL, Steitz TA, Soll D, Simonovic M (2009) The human SepSecS-tRNASec complex reveals the mechanism of selenocysteine formation. Science 325:321–325

Park CM, Weerasinghe L, Day JJ, Fukuto JM, Xian M (2015) Persulfides: current knowledge and challenges in chemistry and chemical biology. Mol BioSyst 11:1775–1785

Perona JJ, Hadd A (2012) Structural diversity and protein engineering of the aminoacyl-tRNA synthetases. Biochemistry 51:8705–8729

Probst AJ, Moissl-Eichinger C (2015) "Altiarchaeales": uncultivated archaea from the subsurface. Life (Basel) 5:1381–1395

Ramabhadran TV, Jagger J (1976) Mechanism of growth delay induced in Escherichia coli by near ultraviolet radiation. Proc Natl Acad Sci U S A 73:59–63

Rauch BJ, Perona JJ (2016) Efficient sulfide assimilation in Methanosarcina acetivorans is mediated by the MA1715 protein. J Bacteriol 198:1974–1983

Rauch BJ, Gustafson A, Perona JJ (2014) Novel proteins for homocysteine biosynthesis in anaerobic microorganisms. Mol Microbiol 94:1330–1342

Rauch BJ, Klimek J, David L, Perona JJ (2017) Persulfide formation mediates cysteine and homocysteine biosynthesis in Methanosarcina acetivorans. Biochemistry 56(8):1051–1061

Reich HJ, Hondal RJ (2016) Why Nature Chose Selenium. ACS Chem Biol 11:821–841

Roche B, Aussel L, Ezraty B, Mandin P, Py B, Barras F (2013) Iron/sulfur proteins biogenesis in prokaryotes: formation, regulation and diversity. Biochim Biophys Acta 1827:455–469

Rodriguez-Hernandez A, Spears JL, Gaston KW, Limbach PA, Gamper H, Hou YM, Kaiser R, Agris PF, Perona JJ (2013) Structural and mechanistic basis for enhanced translational efficiency by 2-thiouridine at the tRNA anticodon wobble position. J Mol Biol 425(20):3888–3906

Sauerwald A, Zhu W, Major TA, Roy H, Palioura S, Jahn D, Whitman WB, Yates JR, Ibba M 3rd, Soll D (2005) RNA-dependent cysteine biosynthesis in archaea. Science 307:1969–1972

Scheller S, Goenrich M, Boecher R, Thauer RK, Jaun B (2010) The key nickel enzyme of methanogenesis catalyses the anaerobic oxidation of methane. Nature 465:606–608

Schwartz CJ, Djaman O, Imlay JA, Kiley PJ (2000) The cysteine desulfurase, IscS, has a major role in in vivo Fe-S cluster formation in Escherichia coli. Proc Natl Acad Sci U S A 97:9009–9014

Sekowska A, Kung HF, Danchin A (2000) Sulfur metabolism in Escherichia coli and related bacteria: facts and fiction. J Mol Microbiol Biotechnol 2:145–177

Shigi N (2014) Biosynthesis and functions of sulfur modifications in tRNA. Front Genet 5:67

Sousa FL, Nelson-Sathi S, Martin WF (2016) One step beyond a ribosome: the ancient anaerobic core. Biochim Biophys Acta 1857:1027–1038

Spry C, Kirk K, Saliba KJ (2008) Coenzyme a biosynthesis: an antimicrobial drug target. FEMS Microbiol Rev 32:56–106

Su D, Ojo TT, Soll D, Hohn MJ (2012) Selenomodification of tRNA in archaea requires a bipartite rhodanese enzyme. FEBS Lett 586:717–721

Susanti D, Mukhopadhyay B (2012) An intertwined evolutionary history of methanogenic archaea and sulfate reduction. PLoS One 7:e45313

Sylvers LA, Rogers KC, Shimizu M, Ohtsuka E, Soll D (1993) A 2-thiouridine derivative in tRNAGlu is a positive determinant for aminoacylation by Escherichia coli glutamyl-tRNA synthetase. Biochemistry 32:3836–3841

Tchong SI, Xu H, White RH (2005) L-cysteine desulfidase: an [4Fe-4S] enzyme isolated from Methanocaldococcus jannaschii that catalyzes the breakdown of L-cysteine into pyruvate, ammonia, and sulfide. Biochemistry 44:1659–1670

Thauer RK (1998) Biochemistry of methanogenesis: a tribute to Marjory Stephenson. 1998 Marjory Stephenson Prize Lecture. Microbiology 144(Pt 9):2377–2406

Valentine DL (2007) Adaptations to energy stress dictate the ecology and evolution of the Archaea. Nat Rev Microbiol 5:316–323

Veerareddygari GR, Klusman TC, Mueller EG (2016) Characterization of the catalytic disulfide bond in E. coli 4-thiouridine synthetase to elucidate its functional quaternary structure. Protein Sci 25:1737–1743

Walters EM, Garcia-Serres R, Naik SG, Bourquin F, Glauser DA, Schurmann P, Huynh BH, Johnson MK (2009) Role of histidine-86 in the catalytic mechanism of ferredoxin:thioredoxin reductase. Biochemistry 48:1016–1024

Westley J (1973) Rhodanese. Adv Enzymol Relat Areas Mol Biol 39:327–368

White RH (2003) The biosynthesis of cysteine and homocysteine in Methanococcus jannaschii. Biochim Biophys Acta 1624:46–53

Xu XM, Carlson BA, Mix H, Zhang Y, Saira K, Glass RS, Berry MJ, Gladyshev VN, Hatfield DL (2007) Biosynthesis of selenocysteine on its tRNA in eukaryotes. PLoS Biol 5:e4

Yuan J, Palioura S, Salazar JC, Su D, O'Donoghue P, Hohn MJ, Cardoso AM, Whitman WB, Soll D (2006) RNA-dependent conversion of phosphoserine forms selenocysteine in eukaryotes and archaea. Proc Natl Acad Sci U S A 103:18923–18927

Yuan J, Hohn MJ, Sherrer RL, Palioura S, Su D, Soll D (2010) A tRNA-dependent cysteine biosynthesis enzyme recognizes the selenocysteine-specific tRNA in Escherichia coli. FEBS Lett 584:2857–2861

Zhang CM, Liu C, Slater S, Hou YM (2008) Aminoacylation of tRNA with phosphoserine for synthesis of cysteinyl-tRNA(Cys). Nat Struct Mol Biol 15:507–514

Zhang X, Eser BE, Chanani PK, Begley TP, Ealick SE (2016) Structural basis for iron-mediated sulfur transfer in archael and yeast thiazole synthases. Biochemistry 55:1826–1838

Chapter 15
Molecular Mechanisms of Fungal Adaptive Evolution

Yongjie Zhang and Jianping Xu

1 Introduction

Traditionally, fungi refer to the group of heterotrophic eukaryotes with thick cell walls made of chitin and cellulose, containing mitochondria but not chloroplasts. With the broad application of DNA sequencing in the systematics and taxonomy of cellular organisms over the last two decades, our understanding of what a fungus is has changed significantly and continues to be refined and debated. While most fungi as revealed based on DNA sequencing still have the traditional structural and genetic traits, some of the "old fungi" such as the structurally similar myxomycetes (slime molds) and oomycetes (water molds that include the agent responsible for the Irish Potato Famine) are no longer considered true fungi (Kirk et al. 2008; Alexopoulos et al. 1996). Because some scientists studying slime molds and oomycetes still identify them as mycologists, when appropriate, we will include examples of these organisms in this chapter. Conversely, many newly discovered fungi exhibit variations in cell wall structures and in whether they contain functional mitochondrial genomes. The emergence of these novel fungal lineages discovered through genomic and metagenomic sequencing has significantly influenced not only our understanding of the evolutionary history of fungi but also their potential novel mechanisms for survival, reproduction, and adaptation (Xu 2016).

Fungal ecological niches are very diverse and include not only those commonly associated with humans and human activities such as agricultural soil, forests, fresh

Y. Zhang
School of Life Sciences, Shanxi University, Taiyuan, Shanxi, China
e-mail: zhangyj2008@sxu.edu.cn

J. Xu (✉)
Department of Biology, McMaster University, Hamilton, ON, Canada
e-mail: jpxu@mcmaster.ca

P. H. Rampelotto (ed.), *Molecular Mechanisms of Microbial Evolution*, Grand Challenges in Biology and Biotechnology,
https://doi.org/10.1007/978-3-319-69078-0_15

and marine water, indoor and outdoor air, households, and workplaces but also niches in extreme environments such as deserts, deep-sea sediments, hydrothermal vents, and areas with high salt concentrations or ionizing radiations (e.g., Dadachova et al. 2007; Raghukumar and Raghukumar 1998; Le Calvez et al. 2009). Indeed, the applications of genomic and metagenomic tools to analyze fungi, especially fungal extremophiles, are bringing researches on fungal adaptation into an exciting new era (Xu 2016).

Though traditionally studied in Botany departments instead of Zoology departments, evolutionarily, the true fungi (Eumycota) are more closely related to animals than to plants. Collectively, fungi share a fundamental feature with animals in that they are unable to photosynthesize but acquire their carbohydrates by absorbing dissolved energy-rich organic molecules. This method of nutrient acquisition means that wherever energy-rich organic compounds are present, there is a possibility that some fungi may live there. This method of nutrient acquisition is also found in water molds (the oomycetes), a group of filamentous protists previously considered as fungi. However, unlike true fungi, the oomycetes lack chitin in their cell walls but contain a mixture of cellulosic compounds and glycan.

Having an efficient means to obtain nutrients is essential for the survival and reproduction of any organism, including fungi. However, fungi differ from animals in several aspects that can impact their nutrient acquisition, survival, and reproduction. In animals, the food is ingested either actively or passively before being digested to release the absorbable forms that are further taken up by cells. In contrast, fungi typically obtain their food by secreting digestive enzymes into their environments to degrade complex organic compounds into monomeric units such as monosaccharides and amino acids before they can be absorbed. Thus, the fungal nutrient-acquisition pathway is more similar to those of heterotrophic prokaryotes than to animals, and the feast or famine condition is likely the norm for most environmental fungi. Another major difference between animals and fungi is in their dispersal abilities. In animals, the somatic body is typically mobile, allowing the animals actively forage for nutrients. In contrast, fungi are typically not mobile except by hyphal growth or spore dispersal. The limited mobility of fungal hyphae means that each fungus often develops specific adaptations to the dominant, local ecological niche that the fungus occupies.

Ecologically, fungi can be broadly classified into four types based on their interaction patterns with other organisms: (a) saprophytes where the fungi feed on dead organic matter and interactions with other living organisms are not required components of their regular life cycle in nature; (b) mutualists where the fungi form mutually beneficial symbiotic relationships with other organisms such as plants, animals (including humans), and other microbes (including other fungi), obtaining organic compounds from and contributing nutrients to their interacting partners; (c) commensals where the fungi benefit from being associated with a biotic partner (plants, animals, and other microbes) for nutrient access while exerting little or no detrimental effect to the partner; and (d) parasites where the fungi benefit from the interaction at the expense of their interacting partners. While some of the above associations are obligate, many are loose, with the fungal partner capable of living in more than one type of ecological niches. Furthermore, in these diverse situations, the ecological conditions can vary widely in their physical/chemical/biological parameters. Aside from growing in most of the "normal" ecological niches, fungal niches also include those with extreme temperatures, high and low water activities, high salinity, extreme pH values, antifungal drugs, lack of oxygen, biotic competitors, and host defense mechanisms (for fungal parasites) (e.g., Brem and Lips 2008). While the fundamental issue is the same for the fungi (i.e., to obtain sufficient nutrients to survive and reproduce), living in these extreme environments would require the fungi to have specific adaptations for each of the extreme factors.

Their ubiquitous distributions and abilities to digest a variety of organic compounds mean that ecologically, fungi play essential roles in nutrient cycling, environmental protection, plant and animal health, and human welfare such as issues relating to food security and infectious diseases. Thus, understanding the factors involved in their adaptations will have tremendous implications for managing these fungi in many fields including the conservation and sustainable use of biodiversity, ecological monitoring, and the prevention and control of fungal pathogens, including human fungal pathogens.

Compared to plants and animals, fungi are often considered simple. However, fungi exhibit great diversities not only ecologically as mentioned above but also in morphology and life cycles (Kirk et al. 2008; Alexopoulos et al. 1996). Morphologically, fungi can range from microscopic single-celled yeasts to filamentous molds and macroscopic multicellular mushrooms. Some fungi can switch morphological forms depending on environmental condition or in response to specific environmental cues. Reproductively, they can propagate both asexually and/or sexually in a variety of structures such as conidiophores and mushroom fruiting bodies, often producing sexual and/or asexual spores that can survive extreme environmental stresses and disperse long distances (Kirk et al. 2008; Alexopoulos et al. 1996; Xu 2005). Indeed, sporulation in fungi represents a major adaptation in the face of environmental adversity, and fungal spores are among the most stress-resistant biological forms.

In this chapter, we first review the basic concepts of adaptation and adaptive evolution as well as the approaches that have been used to study fungal adaptation and adaptative evolution. This is then followed by the specific examples on how fungi adapt to several environmental stresses, including extreme temperatures, drought/low water activity, antifungal drugs, and host defense mechanisms. We draw examples from all four major ecological groups (saprophytes, commensals, mutualists, and parasites), with a special emphasis on the *Basidiomycota* and *Ascomycota*, the two dominant phyla in the fungal kingdom (Kirk et al. 2008; Alexopoulos et al. 1996; Xu 2005, 2016).

2 Adaptive Evolution and Approaches to Study Them in Fungi

In biological studies, the term "adaptation" typically refers to genetic and/or physiological changes leading to a population's increased survival and/or reproduction in a particular environment. The population's genetic changes could be due to the changing frequency of existing alleles and/or the emergence of new alleles through mutations. Genetic mutations are the ultimate sources of phenotypic variations we see among organisms within and among species in Earth's biosphere. Mutations can occur every time DNA replicates and transcribes, with the rate related to adjacent base pairs, transcription levels, replication timing, chromatin states, meiotic crossover rates, and GC content (Chen et al. 2017). Those mutations that cause protein sequence changes can influence protein's structure and function, potentially contributing to organisms' survival and reproduction in the specific environment. If the particular environmental pressure persists, this can lead to fixation of the allele(s) for that trait in the population. In contrast to genetic changes that can permanently change the genetic makeup of the population, physiological adaptations due to epigenetic modifications are temporary and may include DNA methylation/demethylation, increased/decreased expression of specific genes, alternative splicing of messenger RNA transcript(s), and modification of proteins through phosphorylation/dephosphorylation and/or acetylation/deacetylation.

While adaptation refers to genetic and/or physiological changes that benefit the population's survival and reproduction, evolution is generally a neutral term and commonly defined as the change in allele frequencies in a population over time. Such allelic frequency changes are produced by mutation, genetic drift, migration, and selection. Consequently, several factors can influence the direction and speed of allele frequency changes in a population, including mutation rate, population size, gene flow, mating system, and the types and intensities of selection pressure. Selection pressure is often classified into three categories: natural selection, sexual selection, and artificial selection imposed by humans, either intentionally or unintentionally. These allelic frequency changes may be beneficial, neutral, or even (slightly) deleterious. In this chapter, we refer the genetic changes associated with increased survival and reproduction in a population over time as adaptive evolution. Some of these changes may be historical, happened in the distant past but are present in all individuals of a population/species but absent in other populations of the same species or sister species.

To fully understand the mechanism(s) of the adaptation of a fungus to a specific environmental factor, several approaches and methods may be necessary (Singh et al. 2012). Below we briefly describe the general approaches that have been used to elucidate the potential mechanisms underlying fungal adaptations in a diversity of traits. The applications of these specific methods for understanding fungal adaptations will be further described in Sects. 3, 4, 5, 6, and 7 using specific examples.

For evolutionary and population biologists, one common approach to investigate the molecular mechanism of adaptation is to compare the genotypes and phenotypes among strains/species that have adapted to different environments. Depending on the research objectives, the specific investigative methods may differ. For example, if the study is on comparing strains and populations within an individual species, a sufficient number of strains that represent the different phenotypes will be needed. These strains and populations are then subjected to genotyping at candidate loci or even at the whole-genome level. The relationships between genotypes and phenotypes can be statistically investigated to determine whether certain genetic polymorphisms are associated with specific phenotypic values. In humans, genome-wide association studies have helped reveal many candidate genes or genomic regions associated with diseases. While this approach can be applied to any organism, it is especially useful for those where genetic crosses and genetic manipulations can't be performed due to ethical and/or biological constraints.

For organisms where experimental crosses and genetic manipulations can be performed, both the forward and reverse genetics approaches could be applied to understand the genetic basis of phenotypic variations and adaptation. Briefly, the forward genetics (or a forward genetic screen) is an approach used to identify genes (or sets of genes) responsible for a particular phenotype of an organism. This approach typically involves crossing individuals with divergent phenotypic trait values, obtaining and genotyping sexual offspring, constructing a genetic linkage map, and identifying the genetic marker(s) or genomic region(s) that co-segregate with phenotypic trait values (Vogan et al. 2016). Depending on the genetic basis of a specific trait, a large progeny population (e.g., hundreds to thousands) may be required to identify the specific gene(s) controlling the phenotypic trait. In contrast, reverse genetics (or a reverse genetic screen) analyzes the phenotype of an organism following the disruption or targeted modification of a known gene. The CRISPR-cas9 system has significantly facilitated the reverse genetic screening in a diversity of organisms, including fungi (Krappmann 2017).

For organisms that can be grown in lab conditions, there is another approach for investigating the genetic basis of adaptation. In this approach, the investigator can directly create experimentally adapted populations from non-adapted ones and compare the genes and genomes between the derived strains and the ancestral strain(s) (Kohn and Anderson 2014). Using this approach to investigate fungal adaptation, the fungal isolates are typically placed in a specific environmental condition (e.g., high temperature or high drug concentration environment) to clonally select for mutants capable of growing in such an environment. Because of their known and extremely close genetic relatedness between the ancestral and evolved strains, any novel genetic or genomic differences between the original non-adapted and the derived adapted strains would represent candidate loci potentially contributing to their phenotypic differences.

In recent years, comparative transcriptomics, proteomics, and metabolomics have become an increasingly common approach for investigating the differences between fungal strains grown under different stress conditions and to help identify the genes, proteins, and metabolites involved in stress response. The increasing availability and affordability of whole-genome sequencing services make this approach extremely attractive for organisms with relatively small genomes and fast rates of reproduction under laboratory conditions (e.g., prokaryotes and fungi). Below we provide examples on our understanding of the genetic bases related to fungal adaptation to extreme temperatures, low water activity, antifungal drugs, and host immune system attacks.

3 Adaptation to High and Low Temperatures

While most known fungi grow at temperatures from 15 to 35 °C, some have adapted to high-temperature habitats (e.g., human bodies and compost) and others have adapted to low-temperature habitats (e.g., glaciers and snow). Fungi growing in such diverse temperatures have evolved different mechanisms of adaptation. Below we separately summarize our current knowledge with regard to molecular mechanisms of fungal adaptation to high and low temperatures (Table 15.1).

The heat-shock (HS) response is a common mechanism that organisms use in their adaptation to elevated temperatures. Elevated temperatures can cause substantial changes in the composition of cellular membranes, proteins, and soluble carbohydrates. To protect the cellular macromolecules, thermophilic organisms have evolved mechanisms of persistent thermotolerance. During thermal stress response, the synthesis of heat-shock proteins (HSPs) and their elevated expressions have been well documented in multiple fungi. At present, five major families of HSPs are recognized: HSP100, HSP90, HSP70, HSP60, and small HSP (sHSP), and these proteins are broadly distributed in many fungi. However, the number of HSP proteins and the relative importance of each HSP family in stress tolerance can vary among organisms. For example, sHSPs are important in the acquisition of thermotolerance in the ectomycorrhizal fungus *Pisolithus* sp. under thermal stress (Ferreira et al. 2005), while HSP60 orchestrates the adaptation of the human pathogenic fungus *Histoplasma capsulatum* to high-temperature stress (Guimaraes et al. 2011).

Aside from HSPs, other genes and proteins have also been reported as associated with cold and heat-shock responses and that the responses can differ among fungi. For example, using 2D gel protein electrophoresis, Tesei et al. (2012) found that rock-inhabiting black fungi likely used a different strategy to cope with nonoptimal temperature compared with the cosmopolitan and mesophilic hyphomycete *Penicillium chrysogenum*. In this study, three black rock-inhabiting fungi were chosen, *Exophiala jeanselmei*, *Coniosporium perforans*, and *Friedmanniomyces endolithicus*, and they were incubated at different temperatures, ranging from 1 °C, 15 °C, and 28 °C to 40 °C. 2D protein gel electrophoresis patterns revealed that *P. chrysogenum* expressed the highest number of proteins at 40 °C, whereas when exposed to temperatures far above their growth optimum, black fungi decreased the number of expressed proteins, suggesting a downregulation of their metabolism and the lack of a heat-shock response at the protein level. In contrast, at the low temperature of 1 °C, there was an increased number of expressed proteins in all fungi, with the exception of *P. chrysogenum*. At present, the specific genes and proteins that cause such expression differences are not known.

Table 15.1 Examples of molecular mechanisms of fungal adaptation to extreme environments

Environmental stress	Molecular mechanisms	Reference
High temperature	Increased transcription of heat-shock genes	Steen et al. (2002)
	Elevated expression of heat-shock proteins	Ferreira et al. (2005), Guimaraes et al. (2011)
	Decreased expression of nonessential proteins	Tesei et al. (2012)
	Increased trehalose concentration	Oberson et al. (1999)
	More saturated fatty acids	Oberson et al. (1999)
Low temperature	Expansion of gene families coding for antifreeze proteins	Hu et al. (2013)
	Decreased sterol and glycolipids	Tereshina and Memorskaya (2005)
	Increased non-saturated fatty acids	Tereshina and Memorskaya (2005)
	Increased glycerol	Tereshina and Memorskaya (2005)
	Increased arabitol and trehalose	Tereshina and Memorskaya (2005)
	Increased metabolites from TCA cycles	Tsuji (2016)
	Expression of antifreeze proteins	Hoshino et al. (2009)
	Cytoskeleton rearrangements	Blasi et al. (2015)
	Increased expression of nonessential proteins	Tesei et al. (2012)
Drought/low water activity	Increased intracellular concentration of glycerol and erythritol	Pettersson and Leong (2001)
	Production of walleminol and walleminone	Jancic et al. (2016a, b)
	Loss of gene clusters involved in secondary metabolite production	Leong et al. (2015)
	Increased expression of genes involved in fatty acid oxidation and the glyoxylate cycle	Singh et al. (2005)
	Increased catalase expression	Franca et al. (2005)
	Increased autophagy	Ratnakumar et al. (2011)
	Mycelial compartmentalization and interconnectivity	Guhr et al. (2015)
	Overall metabolic modulation	Wang et al. (2015)
Antifungal drug—flucytosine	Mutations in genes Fcy1, Fcy2, and Fur1 that either render the cell unable to take flucytosine up or unable to convert flucytosine to its toxic form	Espinel-Ingroff (2008)
Antifungal drug—azoles	Synthesis of alternative sterols	Pemán et al. (2009)
	Elevated expression of the target gene ERG11/CYP51	Pemán et al. (2009)
	Mutation in the target gene ERG11/CYP51	Pemán et al. (2009)
	Increased expression of efflux pumps	Pemán et al. (2009)

(continued)

Table 15.1 (continued)

Environmental stress	Molecular mechanisms	Reference
Antifungal drug—echinocandins	Mutations in FKS1 and FKS2, the target genes of echinocandins	Perlin (2007)
Antifungal drug—polyenes	Low ergosterol content	Pemán et al. (2009)
	Defects in genes involved in ergosterol biosynthesis	Pemán et al. (2009)
	Enhanced catalase activity	Pemán et al. (2009), Blum et al. (2008)
Plant host defense	Effectors	Dong et al. (2015), Kemen et al. (2011)
	Plant cell-wall degrading enzymes	Ma et al. (2017)
Animal host defense	Chromatin remodeling	O'Meara et al. (2010)
	Hydrolytic enzymes such as extracellular proteases	Ortiz-Urquiza and Keyhani (2013)
	Adhesins and specialized adhesive structures	Ortiz-Urquiza and Keyhani (2013)
	Metabolites facilitating infection	Ortiz-Urquiza and Keyhani (2013)
	Acquisition of host genes involved in immune response	Wang et al. (2016)
	Reduction/expansion of gene families	Wichadakul et al. (2015)
	Melanin and capsule	Alspaugh (2015)
Anoxia	Expression of genes involved in sterol and glycerol transport	Snoek and Steensma (2007)
	Genes for posttranslational fucosylation	Youssef et al. (2013)
	Lack of mitochondria but presence of hydrogenosome	Youssef et al. (2013)
	Expansion of gene families coding for glycoside hydrolases	Youssef et al. (2013)
Heavy metal	Increased concentration of intracellular malondialdehyde, intracellular thiol, and proline	Mukherjee et al. (2010)
	Increased ROS scavenging ability	Mukherjee et al. (2010)
High salt	Expression of the pentose phosphate pathway	Kashyap et al. (2016)
	Elevated expression of heat-shock proteins	Kashyap et al. (2016)
	Increased HOG expression	Padamsee et al. (2012)
	Increased expression of Na^+ efflux proteins	Ma et al. (2015)
Domestication	Selective enrichment of specific genes related to human-created environments	Xiao et al. (2016)
	Expansion of gene families related to substrate transport and utilization	Ropars et al. (2015)
UV irradiation	Melanin pigment	Wang and Casadevall (1994)

Aside from changes at the protein level, changes in metabolites have been intensively investigated for their roles in acquired thermotolerance in fungi (i.e., thermotolerance acquired after being exposed to HS). The accumulation of trehalose (a membrane-stabilizing cytosolic carbohydrate) and changes in membrane composition are often observed and that differences may be observed among closely related fungi. For example, a comparative study of the response to HS between two closely related fungi, the mesophilic *Chaetomium brasiliense* and thermophilic *Chaetomium thermophilum* var. *thermophilum*, revealed that fatty acids of the thermophilic fungus were more saturated than those of the mesophilic fungus. However, under optimal conditions, both fungi synthesized comparable amounts of trehalose, and in response to HS, both fungi increased similar amounts of trehalose (Oberson et al. 1999). In a different study of two thermophilic fungi *Rhizomucor tauricus* and *Myceliophthora thermophila*, in response to HS, the proportions of phosphatidic acids and sterols increased, while the amounts of phosphatidylcholines and phosphatidylethanolamines decreased (Ianutsevich et al. 2016). However, there was no increase in the degree of fatty acid saturation in the major phospholipids under HS in these two species. Furthermore, these two fungi did not show any "acquired" thermotolerance as a result of the HS probably due to their inability to further increase the synthesis of trehalose, already at 8–10% of dry weight at their optimal growth temperatures (Ianutsevich et al. 2016).

Similar to those observed at HS, changes in the profiles of proteins, lipids, carbohydrates, and other metabolites have also been observed in fungal response to cold stress. For example, changes in membrane lipids were observed during the adaptation of the white-rot fungus *Flammulina velutipes* to hypothermia (5 to −5 °C) in natural environments (Tereshina and Memorskaya 2005). Specifically, the levels of sterols and glycolipids decreased, and the proportion of phospholipids with a high degree of non-saturation (2.2) increased. Similarly, glycerol, known to have antifreeze properties, accumulated in the cell cytosol along with arabitol and trehalose (Tereshina and Memorskaya 2005). By analyzing two strains of the Antarctic basidiomycetous yeast *Mrakia blollopis* that exhibited distinct growth characteristics under subzero conditions, Tsuji (2016) found that these two strains also showed different cold adaptation mechanisms. In response to cold shock, strain SK-4, which grew well under subzero temperatures, accumulated high levels of TCA-cycle metabolites as well as lactic acid, aromatic amino acids, and polyamines. In contrast, in strain TKG1-2, which did not grow as efficiently under subzero temperatures, cold stress strongly induced the TCA cycle, but other metabolites did not show significantly increased accumulation within the cells (Tsuji 2016). These results suggest that closely related species and even different strains within the same species can have very different metabolic responses to cold stresses.

For fungi that are historically adapted to cold environments, their genetic and physiological features associated with cold adaptation have also been investigated and reviewed (e.g., Hoshino et al. 2009). Different strategies have been found in different taxa of snow molds. For example, basidiomycetous snow molds produce extracellular antifreeze proteins to keep the immediate extracellular environment ice-free. However, the psychrophilic ascomycete *Sclerotinia borealis* does not produce extracellular antifreeze proteins but instead increased its osmotic stress tolerance in order to grow at subzero temperatures. Like most psychrophilic fungi, *S. borealis* grows faster under cold/frozen conditions than under normal unfrozen conditions.

Aside from differences in cold-resistant mechanisms between strains and species, the cell's developmental states can also exert effects on the types of physiological response to low temperatures. For example, Cheawchanlertfa et al. (2011) showed that after being similarly acclimated to low temperatures, mycelia and phenethyl alcohol-induced yeast cultures of the dimorphic fungus *Mucor rouxii* had different fatty acid profiles due to different fatty acid desaturations through cooperative upregulation of the desaturase genes.

In addition to the above-described physiological mechanisms, results from genomic, transcriptomic, and proteomic analyses are also increasing our understanding on fungal adaptation to high/low temperatures. For example, Hu et al. (2013) found that the caterpillar fungus *Ophiocordyceps sinensis*, which is endemic to high altitudes on the Tibetan Plateau and adapted to extreme cold, had enriched gene families encoding putative antifreeze proteins and mechanisms for increasing lipid accumulation and fatty acid unsaturation. In the opportunistic pathogen *Cryptococcus neoformans* that causes meningitis in humans and other animals, growth at a lower temperature (25 °C) increased transcript levels for histone-encoding genes, indicating a general influence of temperature on chromatin structure. At a higher temperature of 37 °C, there were elevated transcript levels for several genes encoding heat-shock proteins and translation machinery (Steen et al. 2002). Transcriptome analysis of the pathogenic oomycete *Pythium insidiosum*, which can infect both humans and animals, revealed a total of 1074 genes either significantly upregulated (625 genes) or downregulated (449 genes) at body temperature (37 °C), in comparison with those grown at room temperature (28 °C) (Krajaejun et al. 2014). These 1074 genes can be divided into 309 gene product groups, with the biggest group consisting of 408 genes. Murata et al. (2006) studied the genome-wide transcriptional response in *Saccharomyces cerevisiae* in the presence of a cold shock at 4 °C. They found that genes related to energy production, metabolism, cell rescue, defense, and virulence were upregulated, while those related to protein synthesis were generally downregulated. In the truffle fungus *Tuber melanosporum*, compared to those at 25 °C, a total of 423 genes were differentially expressed (>2.5-fold; P value <0.05) when the mycelia were exposed to cold (7 days at 4 °C) (Zampieri et al. 2011). Among these 423 genes, 187 were upregulated, while 236 were downregulated. Sixty-six and fifty-one percent, respectively, of the up- or downregulated transcripts had no KOG classification.

Through an integrated analysis using genomic, transcriptomic, and proteomic tools, Su et al. (2016) analyzed the mechanisms of both cold adaptation and inability to grow at above 20 °C in the obligate psychrophilic fungus *Mrakia psychrophila*. They found that several strategies used by *M. psychrophila* are shared with other psychrophiles, including the upregulation at 4 °C of desaturase and glycerol 3-phosphate dehydrogenase, which are involved in biosynthesis of unsaturated fatty acid and glycerol, respectively. The lack of growth of the fungus at above

20 °C was at least partially due to the accumulation of unfolded proteins in the endoplasmic reticulum. However, differences with other psychrophiles were also observed, including codon usage bias and alternative splicing events that were unique to *M. psychrophila*. Codon usage bias and alternative splicing events might contribute to the cold adaptation of *M. psychrophila*.

Blasi et al. (2015) presented the functional analysis of the transcriptional response of the black fungus *Exophiala dermatitidis*, a human pathogen, at 1 °C, 37 °C, and 45 °C at two different exposure time points. At 1 °C, *E. dermatitidis* activated several mechanisms to acclimatize, such as lipid membrane fluidization, trehalose production, or cytoskeleton rearrangement, and allowed the fungus to remain metabolically active. At 45 °C, the fungus drifts into a replicative state and increases the activity of the Golgi apparatus. In addition to expression differences in protein-coding genes, this study also found differential expressions in noncoding RNAs, circular RNAs, as well as fusion transcripts among the temperature treatments, suggesting potentially novel mechanisms involved in low- and high-temperature adaptations.

In the model microbial eukaryote *Neurospora crassa*, Ellison et al. (2011) discovered two cryptic and recently diverged populations, one in the tropical Caribbean and the other endemic to subtropical Louisiana, USA. Comparison of the genomes of the two populations revealed two "islands of differentiation." The subtropical Louisiana population has a higher fitness at low temperature (10 °C), and several of the genes within these distinct regions have functions related to low-temperature adaptation. These results suggest the divergent genomic islands may be the result of local adaptation to the 9 °C temperature difference in the average yearly minimum temperature between these two populations.

4 Adaptation to Drought/Low Water Activity

Drought is a common phenomenon in nature. In microbial ecology, including fungal ecology, drought typically refers to an environment with an insufficient water activity to sustain microbial growth. Quantitatively, water activity (a_w) is defined as the ratio between the partial vapor pressure of water in a substance (e.g., soil) over the partial vapor pressure of pure water, everything else being equal. Water activity or the availability of usable water is one of the most significant variables for fungi in natural ecosystems. In typical terrestrial environments, seasonal, monthly, or even daily changes of water activity are common, often with short periods of high a_w interspersed with long periods of low a_w, including complete desiccation in certain environments such as the desserts under midday sun. While the presence of water typically facilitates the growths of fungi, the absence of water can elicit a diversity of physiological and/or life cycle responses. Thus, adaptation to low water activity, including desiccation, represents a common type of physiological response in fungi. Indeed, in combination with other factors, low water activity often leads to vegetative growth arrests and the initiation of sexual reproduction in fungi. Below we briefly describe the molecular mechanisms of fungi living under drought stress (Table 15.1).

Fungi can be divided into two groups based on their growth abilities around water activity of 0.85 a_w: those that can't grow at or below this water activity are called xerophobic fungi and those that can grow are commonly called xerophilic fungi. Under low water activity condition, xerophilic fungi can synthesize compatible solutes (e.g., sugar alcohols, especially glycerol and erythritol) to balance the internal water activity with the outside and enable their enzyme systems to function (Pettersson and Leong 2001). As a group widely spread on the fungal tree of life, xerophiles are extremely important in the spoilage of many processed foods and stored commodities and in indoor environments (Micheluz et al. 2015; Oetari et al. 2016; Jancic et al. 2016a, b; Skrinjar et al. 2012; Vytrasova et al. 2002). Some xerophiles have a preference for salt or sugar substrates, whereas other species can be isolated from both jam and salterns. The most xerophilic fungi include *Xeromyces bisporus*, *Aspergillus penicillioides*, and *Wallemia* species. The latter species also produce secondary metabolites (e.g., walleminol, walleminone) which may enhance their competitive advantages in environments where there is insufficient water but rich sugar or salt (Jancic et al. 2016a, b). Several other genera such as *Eurotium* and *Penicillium* also include xerophilic species.

Among the xerophilic fungi, *Xeromyces bisporus* is regarded as the most xerophilic to date, capable of growing on sugary substrates down to an extremely low a_w of 0.61 (Leong et al. 2011). Genome sequencing of the fungus revealed the apparent loss of all gene clusters to produce secondary metabolites, key molecules for competition, and interaction with other organisms. Transcriptomes at optimal (approximate to 0.89) versus low a_w (0.68) revealed differential expression of only a few stress-related genes; among these, certain (not all) steps for glycerol synthesis were upregulated (Leong et al. 2015).

Singh et al. (2005) analyzed the transcriptional response of the budding yeast *Saccharomyces cerevisiae* to desiccation and rehydration under glucose-limiting conditions. They found that expression of genes involved in fatty acid oxidation and the glyoxylate cycle increased during drying and remained in this state during the rehydration phase. Franca et al. (2005) found that cytoplasmic catalase of *S. cerevisiae* plays a role in the maintenance of the intracellular redox balance during dehydration and, therefore, in tolerance against a water stress.

Saccharomyces cerevisiae is more desiccation tolerant during stationary phase (one in five cells surviving desiccation) than exponential phase (only one in a million cells) (Calahan et al. 2011). Welch et al. (2013) exploited the desiccation sensitivity of exponentially dividing cells to understand the stresses imposed by desiccation and their stress response pathways. They found that a transient heat shock induced a 5000-fold increase in desiccation tolerance in desiccation-sensitive, exponential-phase cells, whereas hyper-ionic, -reductive, -oxidative, or -osmotic stresses induced much less. They provided evidence that the Sch9p-regulated branch of the TOR (target of rapamycin) and Ras-cAMP signaling pathway inhibited desiccation tolerance by inhibiting Gis1p, Msn2p, and Msn4p (transcription factors critical for heat stress response) and by activating Sfp1p (a ribosome biogenesis transcription factor).

Among the 41 mutants defective in ribosome biogenesis, a subset defective in the 60S subunit showed a dramatic increase in desiccation tolerance independent of growth rate. Their results suggest that reduction of a specific intermediate in 60S biogenesis, resulting from conditions such as heat shock and nutrient deprivation, increases desiccation tolerance.

Another survival analysis of a mixture of approximately 4800 mutant strains of *S. cerevisiae* subjected to desiccation, each deleted for a different nonessential gene, suggested that about 653 genes (constituting about 14% of nonessential genes in the yeast and about 10% of its genome) may be important for desiccation tolerance of cells growing after diauxic shift (Ratnakumar et al. 2011). Desiccation of yeast cells in the post-diauxic phase of growth induced changes in transcription of 12% (814 genes) of the yeast genome, activating expression of 484 genes (7%) and downregulating 330 (5%). Autophagy processes were significantly overrepresented, indicating the importance of the clearance of protein aggregates/damaged organelles and the recycling of nutrients for the survival of desiccation in yeast (Ratnakumar et al. 2011).

Lichens are well known to survive severe drought conditions (Sancho et al. 2007). In drought resistance, the mycobiont partner in lichens seems to be the main contributor (Zhang and Wei 2011), probably due to the fact that within lichens, the photobionts are housed by the fungal cells, providing protection for their photosynthetic partners from drought environments. Comparative transcriptome analysis of the lichen-forming fungus *Endocarpon pusillum* showed that a total of 1781 genes were differentially expressed between samples cultured under normal and PEG (polyethylene glycol)-induced drought stress conditions. Among these 1781 genes, 1004 were significantly upregulated, while 777 were downregulated. A large number of differentially expressed genes were classified into metabolism process, which suggests that *E. pusillum* had active metabolism under PEG-induced drought stress. This phenomenon is different from several other drought-resistant organisms, where metabolic processes were largely suppressed during desiccation including PEG-induced stress and suggests that *E. pusillum* is intrinsically adapted to water limitations and drought (Wang et al. 2015).

The desiccation of upper soil horizons is a common phenomenon, leading to a decrease in soil microbial activity and mineralization. In general, fungal communities and fungal-based food webs are less sensitive and more resilient to soil desiccation than bacterial communities and bacterial-based food webs (de Vries et al. 2012; Six 2012). Part of the reason for the greater resilience of fungal communities might be related to the presence of mycelial networks where the loss of water in part of the colony could be compensated for by interconnected mycelia in another part of the colony. Indeed, a recent study on a saprotrophic fungus *Agaricus bisporus* demonstrates that hydraulic redistribution can partly compensate water deficiency if water is available in other zones of the mycelia network (Guhr et al. 2015).

5 Adaptation to Antifungal Drugs

In natural environments such as soil, antibiotic compounds are produced by a diversity of microbes. These antibiotics can inhibit the growth of other microbes and help the producers gain an advantage in their competition against others for nutrients. In return, the nonproducing microbes can develop resistant mechanisms to overcome their competitive disadvantage. Different microbes may produce different compounds or use different strategies against others. Indeed, such arms races among microbes have long existed and are widespread in natural environments (Pawlowski et al. 2016). However, it was not until the last 50 years that antibiotic resistance started attracting worldwide attention. According to data from the World Health Organization, antibiotic resistance is among the most urgent issues facing public health in the world.

While most of our attention on antibiotic resistance has focused on bacteria, there is increasing indication that antifungal drug resistance is also on the rise, threatening the continued use of many of the existing antifungal drugs (Pfaller 2012). Compounding this problem is the increasing spectrum of fungal pathogens capable of causing diseases in humans (Köhler et al. 2015). Among the diversities of human fungal pathogens, two yeast species *Candida glabrata* and *Candida krusei* are causing increasing proportions of invasive fungal infections. These two species are intrinsically more resistant than others to two common groups of antifungal drugs, the triazoles and amphotericin B (Pfaller 2012). Similarly, over the last 10 years, multiple triazole-resistant strains of *Aspergillus fumigatus* have been reported from many countries, including the Netherlands, Britain, India, China, and the USA (Ashu et al. 2017). Infections caused by drug-resistant strains are typically associated with high morbidity and mortality. In order to help control the rapid emergence of antifungal resistance, it's essential to understand the mechanisms of resistance. Below we focus on the main types of antifungal drugs and summarize the main resistance mechanisms (Table 15.1). Most of the molecular mechanisms of resistance described below were identified by comparing drug-susceptible and drug-resistant isolates from the same patients over the course of infection time using genetic, transcriptomic, proteomic, and/or metabolomic approaches. However, a number of studies have also employed QTL mapping of progeny from genetic crosses and laboratory experimental evolution analyses, followed by genome comparisons (e.g., Vogan et al. 2016; Kohn and Anderson, 2014).

There are four major groups of systemic antifungal drugs used in clinics: (a) antimetabolites such as flucytosine, (b) azoles such as fluconazole and voriconazole, (c) echinocandins such as caspofungin and micafungin, and (d) polyenes such as amphotericin B and nystatin.

Flucytosine is a pyrimidine analog that can inhibit the synthesis of DNA and RNA, causing cell death. The molecular mechanisms of flucytosine resistance among fungal pathogens are commonly associated with mutations in three genes (Espinel-Ingroff 2008). The first is the Fcy2 gene, encoding the purine-cytosine permease that is also responsible for the uptake of flucytosine into the cell. The second is the Fcy1 gene, encoding the cytosine deaminase enzyme responsible for converting flucytosine to 5-flurouracil. The third is the Fur1 gene, encoding the kinase responsible for converting 5-flurouracil to 5-flurouridine monophosphate. Mutations in these three genes can either render the cell unable to take flucytosine up or unable to convert flucytosine to its toxic form inside the cells, resulting in flucytosine resistance.

The second group of antifungal drugs is the azoles. Azole drugs inhibit fungal growth by interfering with the biosynthesis of ergosterol, the key fungal cell membrane sterol. The specific target of azole drugs is the enzyme lanosterol-14-α-demethylase encoded by ERG11 or CYP51, and the binding of azole drugs to this enzyme leads to failed conversion of lanosterol into ergosterol, resulting in reduced ergosterol content, altered membrane structure and functions, and ultimately inability to grow. The high solubility and low side effects of azole drugs on humans make these drugs the first-line antibiotic for treating most of the invasive fungal infection. However, like many other categories of antibiotics, the use of azoles has also resulted in frequent azole resistance among fungal pathogens. Based on our current understanding, the mechanisms of resistance can be grouped into four major types; however, more than one mechanism may be present in a given resistant strain, and these mutations can act either additively or synergistically to enhance the strains' resistance level (Pemán et al. 2009; Oliver et al. 2005; Kohn and Anderson 2014). Briefly, the first major type of azole resistance mechanism involves the development of alternate pathways and the synthesis of alternative sterols to negate the membrane-disruptive effects of azole drugs on the depletion of ergosterol. The second common azole resistance mechanism is through elevated expression of the target enzyme. The third major mechanism involves the induction of efflux pumps that reduce intracellular drug concentration. The dominant efflux pumps involved in azole resistance are those encoded by either the major facilitator gene MDR1 or the ATP-binding cassette transporter genes CDR1 and CDR2. It should be noted that the genes and their specific involvements might differ among species and azole drugs. For example, upregulation of the MDR1 gene is predominantly related to fluconazole resistance, while the upregulation of CDR1 and to a lesser extent the CDR2 gene is commonly found in strains resistant to multiple triazoles (Pemán et al. 2009; Oliver et al. 2005). The fourth type of azole resistance mechanism is due to non-synonymous substitutions in the gene encoding for the target enzyme (ERG11 or CYP51). A diversity of mutations have been found, and such mutations can either completely eliminate or reduce the drugs' affinity with the enzymes and maintain its enzymatic activity in the presence of the drugs (Pemán et al. 2009; Chang et al. 2016; Oliver et al. 2005; Vogan et al. 2016).

The third major class of antifungal drugs is the echinocandins represented by anidulafungin, caspofungin, and micafungin. These drugs target the 1,3-β-D-glucan synthetase, a fungal-specific enzyme involved in the synthesis of the fungal cell wall. The inhibiting of cell wall synthesis by echinocandins leads to cell rupture and/or aberrant hyphal growth. As with the azoles and flucytosine, reduced susceptibility to echinocandins has been observed in several *Candida* species, including *C. glabrata*, *C. krusei*, *C. tropicalis*, and *C. dubliniensis*. The dominant resistant mechanisms have been linked to point mutations in the presumed catalytic domains of genes FKS1 and FKS2, the genes encoding the major subunits of 1,3-β-D-glucan synthase (Perlin 2007). However, different from azole resistance mechanisms, overexpression of the MDR and CDR genes seemed to have little effect on the susceptibility of fungal pathogens to echinocandins, suggesting that echinocandins are probably not good substrates for these efflux pumps (Pemán et al. 2009).

The fourth major group of clinically important systemic antifungal drugs is the polyenes, including nystatin and amphotericin B. Polyenes inhibit fungal growth by intercalating with each other and with membrane ergosterol to form aqueous pores, leading to leakage of cytoplasmic materials. Though resistance to amphotericin B is low, the analyses of target strains have found generally lower levels of ergosterol in cell membrane and the presence of alternative, probably less effective sterols to maintain fungal growths. Indeed, defects in genes (e.g., *ERG3*, *ERG6*, and *ERG11*) involved in ergosterol biosynthesis are among the major mechanisms of amphotericin B resistance in human fungal pathogens (Pfaller 2012; Pemán et al. 2009). It should be noted that in such strains, due to the absence or limitation of their preferred sterol (i.e., the ergosterol), their growths are typically impaired compared to the wild-type strains in the absence of amphotericin B. In addition, aside from intercalating with ergosterol, amphotericin B can also cause oxidative damage to cell membrane through generating reactive oxygen species. Thus, fungal strains with high levels of resistance to amphotericin B typically also have enhanced catalase activity to counter the oxidative stress. Indeed, in *Aspergillus terreus*, amphotericin B resistance was not due to reduced ergosterol content but to significantly upregulated catalase activity (Blum et al. 2008).

As shown above, the development of antifungal resistance is predominantly through mutation of existing genes and is quite different from those in bacterial pathogens where antibiotics resistance genes are often acquired by horizontal gene transfers. However, the impacts of the emergence of drug resistance on our healthcare systems are very similar among microbial pathogens, from viruses to bacteria, protozoa, and fungi. Understanding the mechanisms of resistance and their epidemiological patterns will have significant practical implications in helping develop rapid diagnosis and targeted treatments for patients with fungal infections.

6 Adaptation to Host Defenses

Fungi are common pathogens of plants and animals, including humans. However, both plants and animals possess a diversity of protective and defense mechanisms. In order for pathogenic fungi to successfully colonize and invade host tissues, they need specific adaptations to overcome such host defenses. Below we highlight a few examples in this broad area (Table 15.1).

As one of the major sources of organic compounds, plant materials, both living and dead, are among the favored substrates for fungal growth. Indeed, each year, fungal pathogens cause tens of billions dollars worth of crop loss. In order for fungal pathogens to cause plant diseases, the pathogens have to colonize and invade plant tissues and overcome plant defense mechanisms. Effectors are specific types of proteins secreted into plants by phytopathogens to interfere with plant defenses. By applying genomic and transcriptomic analytical tools, 134 candidate effectors were identified in the rice blast pathogen *Magnaporthe oryzae*; overexpression of two effectors, Iug6 and Iug9, suppressed defense-related gene expression in rice (Dong et al. 2015). Studies on the obligate biotroph white rust pathogen (*Albugo laibachii*, Oomycota) of *Arabidopsis* found biotrophy also requires "effectors" to suppress host defense. Two effectors, RXLR and Crinkler, in *A. laibachii* were found to be shared with other oomycetes. In addition, a novel class of effectors that share a CHXC motif within 50 amino acids of the signal peptide cleavage site was discovered and experimentally verified (Kemen et al. 2011).

A recent study by Ma et al. (2017) revealed that during the soybean-oomycete pathogen *Phytophthora sojae* interaction, an intriguing molecular war takes place at the plant's extracellular space, the apoplast. In this interaction, the pathogen releases the virulence factor, the apoplastic xyloglucan-specific endoglucanase PsXEG1. The soybean counters by producing a PsXEG1 inhibitor, GmGIP1, that binds to PsXEG1 and abolishes its function. However, the pathogen also secretes a PsXEG1-like protein, PsXLP1, as a decoy. PsXLP1 does not have an enzymatic activity but has a highly binding affinity to GmGIP1 than PsXEG1, thus freeing PsXEG1 to degrade host plant tissue and release nutrients for their growth. Similar to the RXLR and Crinkler effectors, homologs of the PsXEG1 and PsXLP1 pair are widely distributed in *Phytophthora* species.

The basidiomycete yeasts *Cryptococcus neoformans* species complex and *Cryptococcus gattii* species complex are a group of opportunistic human and other animal pathogens (Kwon-Chung et al. 2017). They are broadly distributed in nature, including soil, bird excreta, and plant debris. Through inhalation of propagules from the environment by the host, *C. neoformans* can survive and replicate within macrophages in vivo. Two cellular components are critically for *C. neoformans*' survival within host, capsule and melanin, and many genes have been found influencing their syntheses and production (Steenbergen et al. 2001; Alspaugh 2015). To better understand the origin of *C. neoformans* virulence, Derengowski et al. (2013) compared the transcriptional profiles of *C. neoformans* 6 h after phagocytosis by the amoeba *Acanthamoeba castellanii* and by murine macrophages. The results revealed 656 and 293 genes in *C. neoformans* ingested by amoebae and macrophages, respectively, whose expression changed at least twofold relative to non-phagocytosed cells. Despite their differences on the number of modulated genes, the overall categorization of the protein-coding genes revealed a very similar gene expression profile of the fungus inside either phagocyte. Genes related to nutrient transport, general metabolism, and oxidative stress response showed increased expression, while genes involved in transcription, translation, and ergosterol biosynthesis were suppressed. Overall, these results are consistent with the view that cryptococcal virulence for mammals originated from fungus-protozoan interactions in the environment. Another study in *C. neoformans* demonstrates that chromatin remodeling by the conserved histone acetyltransferase Gcn5 is important in regulating the expression of specific genes that allow *C. neoformans* to respond appropriately to the human host (O'Meara et al. 2010).

Entomogenous fungi are able to parasitize susceptible hosts via direct penetration of the cuticle with the initial and potentially determining interaction occurring between the fungal spore and the insect epicuticle. Entomogenous fungi have evolved mechanisms for adhesion and recognition of host surface cues that help direct an adaptive response that includes the production of: (a) hydrolytic, assimilatory, and/or detoxifying enzymes including lipase/esterases, catalases, cytochrome P450s, proteases, and chitinases; (b) specialized infectious structures, e.g., appressoria or penetrant tubes; and (c) secondary and other metabolites that facilitate infection (Ortiz-Urquiza and Keyhani 2013).

Wang et al. (2016) sequenced the genome of *Zancudomyces culisetae*, formerly known as *Smittium culisetae*. The species has been shown to benefit the in vivo development of infested mosquito larvae under specific conditions (Horn and Lichtwardt 1981). In contrast, *Z. culisetae* can also lead to the death of mosquito larvae, in situations where the host's hindgut becomes overgrown with this fungus (Williams 2001). An insect-like polyubiquitin chain was encoded by the fungus. Ubiquitin and ubiquitin-like proteins are universally involved in protein degradation and regulation of immune response in eukaryotic organisms. Multiple lines of evidence support this polyubiquitin gene in *Z. culisetae* was obtained via a horizontal gene transfer event from the host. The acquired polyubiquitin gene in *Z. culisetae* may be useful during the invasive processes of the fungus, to induce the hosts' ubiquitin-proteasome systems by labeling and degrading host cell membrane proteins. *Z. culisetae* may also use it as a defense against bacteria, viruses, or other microbes that coexist in the insect guts, whether for its own competitive advantage or as an ally of the host.

Wichadakul et al. (2015) sequenced the genome of *Ophiocordyceps polyrhachis-furcata*, a species in the *Ophiocordyceps unilateralis* species complex specialized in colonizing the ant *Polyrhachis furcata*, and performed a comparative genomic analysis of insect fungi. They found evidence for genome contractions for species with narrow host ranges. Interestingly, the sizes of several gene families, including cuticle-degrading genes (proteases, carbohydrate esterases) and some families of pathogen-host interaction (PHI) genes, were reduced for specialized obligate fungal parasites. However, two gene families also showed evidence of expansions: (1) the genes involved in the production of bacteria-like toxins in *O. polyrhachis-furcata*, compared with other entomopathogenic fungi, and (2) retrotransposable elements. The loss of various genes involved in the pathogenesis for *O. unilateralis* would result in a reduced capacity to exploit larger ranges of hosts and therefore in the different level of host specificity. In contrast, the expansions of other gene families suggest an adaptation to particular environments with unexpected strategies like oral toxicity to its host insects, through the production of bacteria-like toxins, or sophisticated mechanisms underlying pathogenicity mediated by genes within or mobilized by retrotransposons.

7 Molecular Mechanisms of Adaptation to Other Stressors

Aside from our understanding of fungal adaptations to stressors mentioned above, studies have also analyzed the molecular mechanisms of fungal adaptations to other environmental factors, including anoxic environments, heavy metal contaminations, high-salt condition, and human domestications. Below we briefly review examples of our current understanding of these adaptations (Table 15.1).

Anoxia The ascomycete yeast *Saccharomyces cerevisiae* is a model eukaryote for research and is commonly used by the baking and fermentation industry. This unicellular fungus can grow under both aerobic and anaerobic conditions. Our understanding of fungal adaptation to anaerobic environments has mainly come from research conducted using this yeast. Different molecules have to be transported into and out of the cell using pathways not commonly expressed under aerobic conditions (Snoek and Steensma 2007). In *S. cerevisiae*, this adaptation is mainly controlled at the transcriptional level. About 500 genes showed differential expression when transcriptomes from aerobic and anaerobic cultures are compared (ter Linde et al. 1999; Kwast et al. 2002; Tai et al. 2005). Among these, 23 genes were expressed only under anaerobic conditions. Apart from *ARV1* (functioning in sterol metabolism/transport), *NPT1* (nicotinate phosphoribosyl transferase), and *GUP1* (glycerol transporter), the 20 other genes have no obvious function in anaerobic metabolism (Snoek and Steensma 2007). In addition, posttranscriptional regulation also contributed to the adaptation of *S. cerevisiae* to anaerobic growth (Bruckmann et al. 2009).

Anaerobic gut fungi represent a distinct early-branching fungal phylum (Neocallimastigomycota) and reside in the rumen, hindgut, and feces of ruminant and nonruminant herbivores. Youssef et al. (2013) sequenced the genome of an anaerobic fungal isolate, *Orpinomyces* sp. strain C1A. Comparative genomic analysis identified multiple genes and pathways that are absent in Dikarya genomes but present in early-branching fungal lineages and/or nonfungal Opisthokonta. These included genes for posttranslational fucosylation, the production of specific intramembrane proteases and extracellular protease inhibitors, the formation of a complete axoneme and intraflagellar trafficking machinery, and a near-complete focal adhesion machinery. The mitochondrial reductive evolution to a hydrogenosome, the apparent replacement of ergosterol with tetrahymanol in the cell membrane, and the sole dependence on a mixed-acid fermentation pathway for pyruvate metabolism and energy production in strain C1A are clear adaptations to anaerobiosis. The development of cellulosomes and the acquisition of many glycoside hydrolases could be viewed as an adaptation to improve the access, speed, and efficacy of biomass degradation.

Heavy Metals The toxicity of heavy metals to microorganisms has attracted considerable research attention in recent years. Mukherjee et al. (2010) reported an *Aspergillus niger* strain that was able to thrive even in a medium with 100 mg/L arsenate, an unusually high concentration for most other living organisms to survive.

To understand the possible cellular strategy toward tolerance of arsenate-induced toxicity, the responses evoked to counter arsenate toxicity were analyzed by assaying alterations of certain enzymes and several biomolecules. MDA (malondialdehyde), intracellular thiol, and proline contents increased up to a certain level. Activities of GR (glutathione reductase), SOD (superoxide dismutase), and CAT (catalase) declined following a rise at low concentrations; SDH (succinate dehydrogenase) activity decreased gradually with increased arsenate stress. These results showed that cells of *A. niger* are equipped with an elaborate network of anti-oxidative enzymes, which are involved in scavenging ROS (reactive oxygen species) and other oxidative products generated by arsenate insult. By sequencing cDNA libraries of the aquatic fungus *Blastocladiella emersonii* submitted to heat shock and cadmium stress, Georg et al. (2009) found that environmental stresses, particularly cadmium treatment, inhibit intron processing (2.9% ESTs containing introns from stress library vs. 0.2% from the unstressed library), revealing a new adaptive response to cellular exposure to this heavy metal.

High-Salt Environments Kashyap et al. (2016) isolated a *Penicilliopsis clavariiformis* culture AP from mangrove in India. The fungus is salt tolerant, being able to tolerate up to 10% (w/v) NaCl. To understand the mechanism of adaptation to high salinity, activities of the key enzymes regulating glycolysis, pentose phosphate pathway, and tricarboxylic acid cycle were investigated under normal (0% NaCl) and saline stress environment (10% NaCl). The results revealed a rerouting of carbon metabolism away from glycolysis to the pentose phosphate pathway, a common pathway related to saline stress tolerance in fungi. In addition, several other genes such as *Hsp98*, *Hsp60*, *HTB*, and *RHO* were significantly upregulated under saline stress, suggesting that they likely play significant roles under such conditions.

The ability to tolerate environments with reduced water activity and high salt concentrations are rare in Basidiomycota. However, species of the basidiomycetous genus *Wallemia*, most commonly found as food contaminants, have been isolated from hypersaline environments. The diverse habitats from which strains of *Wallemia sebi* have been isolated (e.g., jam, dried fish, marine sponges, and house dust) suggest that it can adjust its physiology to adapt to different environments. Recently, the genome of *W. sebi* was sequenced in order to understand its adaptations for surviving in osmotically challenging environments. *W. sebi* has a compact genome (9.8 Mb), with few repeats and the largest fraction of genes with functional domains compared with other Basidiomycota. In silico analyses identified 93 putative osmotic stress proteins; homology searches showed the HOG (high-osmolarity glycerol) pathway to be mostly conserved. Despite the highly reduced genome size, several gene family expansions (esp., HSP20, Dabb, and AA_trans) and a high number of transporters (549) were found that also provide clues to the ability of *W. sebi* to colonize harsh environments (Padamsee et al. 2012).

In fungi, the gene encoding ENA ATPase (ENA is from "exitus natru: exit of sodium") is phylogenetically broadly distributed, and this enzyme plays a central role in Na^+ efflux and Na^+ tolerance. Like in all cellular organisms, the K^+ and Na^+

concentrations within fungal cells are strictly regulated to maintain constant concentrations, relatively high for K^+ and low for Na^+. However, in high saline environments, the high Na^+ concentrations outside of the cell will create a high Na^+ influx and elevate the intracellular Na^+ concentration. To counter such an influx, Na^+ effluxes need to be activated. Indeed, Ma et al. (2015) demonstrated that the deletion of the ENA ATPase gene from the entomopathogenic fungus *Metarhizium acridum* (Δ*MaENA1*) resulted in reduced tolerance to high salt concentration. In addition, the deletion strain was also less tolerant to other stressors such as heat and UV radiation than its wild-type counterpart. Transcriptome profiling showed a large number of differentially expressed genes between the WT and Δ*MaENA1* strains, including 6 cytochrome P450 superfamily genes, 35 oxidoreductase genes, 24 ion-binding genes, 7 DNA repair genes, and 8 genes involved in 5 stress response pathways (the Ras-cAMP PKA pathway, the RIM101 pathway, the Ca^{2+}/calmodulin pathway, the TOR pathway, and the HOG/Spcl/Styl/JNK pathway). These results are consistent with *MaENA1* playing a very important role in the adaptation and survival of this entomopathogenic fungi in stressful conditions.

Comparative genomics was conducted to analyze the genomes of eight *Aspergillus* spp. (*A. nidulans*, *A. clavatus*, *A. flavus*, *A. fumigatus*, *A. niger*, *A. oryzae*, *A. terreus*, and *Neosartorya fischeri*) to identify stress response proteins. All genomes harbored elements of the SskA-HogA/SakA stress signaling pathway, suggesting the importance of SskA-HogA/SakA signaling in different types of stress responses (e.g., responses to osmotic, oxidative, starvation, and even heat stress in germinating conidia) in the aspergilli. The abundance of annotated histidine kinases, MAPKs (HogA/SakA, MpkC), response regulators (two SskAs in *A. flavus*), and transcriptional regulators, e.g., AtfA, AtfB, NapA (AfYap1), MsnA, and RpdA (two orthologues in *A. flavus*), may be indicative of a complex and robust stress defense system controlled by a high-complexity regulatory network in these filamentous fungi (Miskei et al. 2009).

Melanin as a Common Stress Response Factor Melanin is a pigment produced by laccase, a phenoloxydase enzyme. Several experiments suggest that melanin serves a protective role against a variety of harmful stimuli in a diversity of fungi. For example, melanin can protect *Cryptococcus neoformans* against ultraviolet light (Wang and Casadevall 1994), suggesting a role in protection against solar radiation. Melanized *C. neoformans* cells are usually less susceptible to antimicrobial drugs and oxidant agents; in addition they are able to trick the host immune system by inactivating the drugs normally used on therapeutics (Mauch et al. 2013). Melanization also affects susceptibility of *C. neoformans* to heat and cold, with melanized cells being less susceptible than non-melanized cells (Rosas and Casadevall 1997).

Domestication Some fungi were domesticated to produce useful products. Domestication is an excellent model for studies of adaptation because it involves recent and strong selection on a few identified traits. By comparing the genomes of 10 *Penicillium* species, Ropars et al. (2015) reported that adaptation to cheese was associated with multiple recent horizontal transfers of large genomic regions carrying crucial metabolic genes. They identified seven horizontally transferred regions (HTRs) spanning more than 10 kb each, flanked by specific transposable elements, and displaying nearly 100% identity between distant *Penicillium* species. Two HTRs carried genes with functions involved in the utilization of cheese nutrients or competition and were found nearly identical in multiple strains and species of cheese-associated *Penicillium* fungi, indicating recent selective sweeps; they were experimentally associated with faster growth and greater competitiveness on cheese and contained genes highly expressed in the early stage of cheese maturation.

The effect of human domestication on genomic variation was also revealed in the cultivated edible mushroom *Lentinula edodes*. In the study by Xiao et al. (2016), they compared the genome sequences of 39 wild and 21 cultivated strains and identified three distinct genetic groups in the Chinese *L. edodes* population with the majority of the cultivated strains in one genetic cluster. Interestingly, this genetic cluster had enriched non-synonymous nucleotide substitutions in genes related to stress response and in fruiting body formations. These genes include those encoding a protein kinase Pbs2-like MAPKK protein, a cofactor (DnaJ) of heat-shock protein HSP70, a zinc-finger DNA binding protein PriB correlated with mushroom fruiting, a cyclopropane fatty acid synthase that triggers reproductive shift from vegetative to sexual reproduction, and a DEAD-box ATP-dependent RNA helicase related to cold stress response.

8 Conclusions and Perspectives

The fungal kingdom is phylogenetically very ancient and ecologically broadly distributed. Fungi can survive and reproduce in a diversity of environments and have evolved a variety of mechanisms to cope with environmental stresses. In this chapter, we selectively reviewed the molecular mechanisms of fungal adaptation to several common stresses such as extreme temperatures, desiccation, antifungal drugs, host defenses, and osmotic stress such as high salt concentrations. Our surveys identified that some of the genes and pathways are involved in responses to multiple stresses (e.g., the efflux pumps such as the ENA ATPase), and others may be unique to a specific stressor (e.g., a mutation in a drug target). While significant progresses have been made over the last couple of decades, much remains unknown. Increasingly, comparative genomics have been used to identify the genomic differences among species living under different environments to infer the potential signatures of adaptation. Similarly, information from transcriptome profiling for the same species/strains but under different conditions is used to infer the regulatory mechanisms of a variety of adaptations. Together, such data are generating abundant hypotheses for further experimental tests using targeted approaches.

Acknowledgments The research was supported by the Natural Sciences and Engineering Research Council (NSERC) of Canada (J.X.), the Natural Science Foundation of Shanxi Province (201601D011065; Y.Z.), and the Fund Program for the Scientific Activities of Selected Returned Overseas Professionals in Shanxi Province (Y.Z.).

References

Alexopoulos CJ, Mims CW, Blackwell MM (1996) Introductory mycology, 4th edn. Wiley, New York

Alspaugh JA (2015) Virulence mechanisms and *Cryptococcus neoformans* pathogenesis. Fungal Genet Biol 78:55–58

Ashu EE, Hagen F, Chowdhary A, Meis JF, Xu J (2017) Global population genetic analysis of *Aspergillus fumigatus*. mSphere 2(1):e00019–e00017

Blasi B, Tafer H, Tesei D, Sterflinger K (2015) From glacier to sauna: RNA-Seq of the human pathogen black fungus *Exophiala dermatitidis* under varying temperature conditions exhibits common and novel fungal response. PLoS One 10(6):e0127103

Blum G, Perkhofer S, Haas H, Schrettl M, Wurzner R, Dierich MP, Lass-Florl C (2008) Potential basis for amphotericin B resistance in *Aspergillus terreus*. Antimicrob Agents Chemother 52 (4):1553–1555

Brem FM, Lips KR (2008) *Batrachochytrium dendrobatidis* infection patterns among Panamanian amphibian species, habitats and elevations during epizootic and enzootic stages. Dis Aquat Org 81(3):189–202

Bruckmann A, Hensbergen PJ, Balog CIA, Deelder AM, Brandt R, Snoek ISI, Steensma HY, van Heusden GPH (2009) Proteome analysis of aerobically and anaerobically grown *Saccharomyces cerevisiae* cells. J Proteomics 71(6):662–669

Calahan D, Dunham M, DeSevo C, Koshland DE (2011) Genetic analysis of desiccation tolerance in *Saccharomyces cerevisiae*. Genetics 189(2):507–519

Chang H, Ashu E, Sharma C, Kathuria S, Chowdhary A, Xu J (2016) Diversity and origins of Indian multi-triazole resistant strains of *Aspergillus fumigatus*. Mycoses 59(7):450–466

Cheawchanlertfa P, Cheevadhanarak S, Tanticharoen M, Maresca B, Laoteng K (2011) Up-regulated expression of desaturase genes of *Mucor rouxii* in response to low temperature associates with pre-existing cellular fatty acid constituents. Mol Biol Rep 38(5):3455–3462

Chen C, Qi HJ, Shen YF, Pickrell J, Przeworski M (2017) Contrasting determinants of mutation rates in germline and soma. Genetics 207(1):255–267

Dadachova E, Bryan RA, Huang X, Moadel T, Schweitzer AD, Aisen P, Nosanchuk JD, Casadevall A (2007) Ionizing radiation changes the electronic properties of melanin and enhances the growth of melanized fungi. PloS One 2(5):e457

de Vries FT, Liiri ME, Bjørnlund L, Bowker MA, Christensen S, Setälä HM, Bardgett RD (2012) Land use alters the resistance and resilience of soil food webs to drought. Nat Clim Change 2 (4):276–280

Derengowski LS, Paes HC, Albuquerque P, Tavares AHFP, Fernandes L, Silva-Pereira I, Casadevall A (2013) The transcriptional response of *Cryptococcus neoformans* to ingestion by *Acanthamoeba castellanii* and macrophages provides insights into the evolutionary adaptation to the mammalian host. Eukaryot Cell 12(5):761–774

Dong Y, Li Y, Zhao M, Jing M, Liu X, Liu M, Guo X, Zhang X, Chen Y, Liu Y, Liu Y, Ye W, Zhang H, Wang Y, Zheng X, Wang P, Zhang Z (2015) Global genome and transcriptome analyses of *Magnaporthe oryzae* epidemic isolate 98-06 uncover novel effectors and pathogenicity-related genes, revealing gene gain and lose dynamics in genome evolution. PLoS Pathog 11(4):e1004801

Ellison CE, Hall C, Kowbel D, Welch J, Brem RB, Glass NL, Taylor JW (2011) Population genomics and local adaptation in wild isolates of a model microbial eukaryote. Proc Natl Acad Sci USA 108(7):2831–2836

Espinel-Ingroff A (2008) Mechanisms of resistance to antifungal agents: yeasts and filamentous fungi. Rev Iberoam Micol 25(2):101–106

Ferreira AS, Totola MR, Kasuya MCM, Araujo EF, Borges AC (2005) Small heat shock proteins in the development of thermotolerance in *Pisolithus* sp. J Therm Biol 30(8):595–602

Franca MB, Panek AD, Eleutherio ECA (2005) The role of cytoplasmic catalase in dehydration tolerance of *Saccharomyces cerevisiae*. Cell Stress Chaperon 10(3):167–170

Georg RC, Stefani RMP, Gomes SL (2009) Environmental stresses inhibit splicing in the aquatic fungus *Blastocladiella emersonii*. BMC Microbiol 9:231

Guhr A, Borken W, Spohn M, Matzner E (2015) Redistribution of soil water by a saprotrophic fungus enhances carbon mineralization. Proc Natl Acad Sci USA 112(47):14647–14651

Guimaraes AJ, Nakayasu ES, Sobreira TJP, Cordero RJB, Nimrichter L, Almeida IC, Nosanchuk JD (2011) *Histoplasma capsulatum* heat-shock 60 orchestrates the adaptation of the fungus to temperature stress. PLoS One 6(2):e14660

Horn BW, Lichtwardt RW (1981) Studies on the nutritional relationship of larval *Aedes aegypti* (Diptera: Culicidae) with *Smittium culisetae* (Trichomycetes). Mycologia 73:724–740

Hoshino T, Xiao N, Tkachenko OB (2009) Cold adaptation in the phytopathogenic fungi causing snow molds. Mycoscience 50(1):26–38

Hu X, Zhang YJ, Xiao GH, Zheng P, Xia YL, Zhang XY, St Leger RJ, Zhong LX, Shu WC (2013) Genome survey uncovers the secrets of sex and lifestyle in caterpillar fungus. Chinese Sci Bull 58(23):2846–2854

Ianutsevich EA, Danilova OA, Groza NV, Kotlova ER, Tereshina VM (2016) Heat shock response of thermophilic fungi: membrane lipids and soluble carbohydrates under elevated temperatures. Microbiology 162:989–999

Jancic S, Frisvad JC, Kocev D, Gostincar C, Dzeroski S, Gunde-Cimerman N (2016a) Production of secondary metabolites in extreme environments: food- and airborne *Wallemia* spp. produce toxic metabolites at hypersaline conditions. PLoS One 11(12):e0169116

Jancic S, Zalar P, Kocev D, Schroers H-J, Dzeroski S, Gunde-Cimerman N (2016b) Halophily reloaded: new insights into the extremophilic life-style of *Wallemia* with the description of *Wallemia hederae* sp nov. Fungal Divers 76(1):97–118

Kashyap PL, Rai A, Singh R, Chakdar H, Kumar S, Srivastava AK (2016) Deciphering the salinity adaptation mechanism in *Penicilliopsis clavariiformis* AP, a rare salt tolerant fungus from mangrove. J Basic Microb 56(7):779–791

Kemen E, Gardiner A, Schultz-Larsen T, Kemen AC, Balmuth AL, Robert-Seilaniantz A, Bailey K, Holub E, Studholme DJ, MacLean D, Jones JDG (2011) Gene gain and loss during evolution of obligate parasitism in the white rust pathogen of *Arabidopsis thaliana*. PLoS Biol 9(7): e1001094

Kirk PM, Cannon PF, Minter DW, Stalpers JA (2008) Ainsworth & Bisby's dictionary of the fungi, 10th edn. CAB International, Wallingford

Köhler JR, Casadevall A, Perfect J (2015) The spectrum of fungi that infects humans. CSH Perspect Med 5(1):a019273

Kohn LM, Anderson JB (2014) The underlying structure of adaptation under strong selection in 12 experimental yeast populations. Eukaryot Cell 13(9):1200–1206

Krajaejun T, Lerksuthirat T, Garg G, Lowhnoo T, Yingyong W, Khositnithikul R, Tangphatsornruang S, Suriyaphol P, Ranganathan S, Sullivan TD (2014) Transcriptome analysis reveals pathogenicity and evolutionary history of the pathogenic oomycete *Pythium insidiosum*. Fungal Biol 118(7):640–653

Krappmann S (2017) CRISPR-Cas9, the new kid on the block of fungal molecular biology. Med Mycol 55(1):16–23

Kwast KE, Lai L-C, Menda N, James DT, Aref S, Burke PV (2002) Genomic analyses of anaerobically induced genes in *Saccharomyces cerevisiae*: functional roles of Rox1 and other factors in mediating the anoxic response. J Bacteriol 184(1):250–265

Kwon-Chung KJ, Bennett JE, Wickes BL, Meyer W, Cuomo CA, Wollenburg KR, Bicanic TA, Castaneda E, Chang YC, Chen J, Cogliati M, Dromer F, Ellis D, Filler SG, Fisher MC, Harrison TS, Holland SM, Kohno S, Kronstad JW, Lazera M, Levitz SM, Lionakis MS, May RC, Ngamskulrongroj P, Pappas PG, Perfect JR, Rickerts V, Sorrell TC, Walsh TJ, Williamson PR, Xu J, Zelazny AM, Casadevall A (2017) The case for adopting the "species complex" nomenclature for the etiologic agents of *Cryptococcosis*. mSphere 2(1):e00357-16

Le Calvez T, Burgaud G, Mahé S, Barbier G, Vandenkoornhuyse P (2009) Fungal diversity in deep-sea hydrothermal ecosystems. Appl Environ Microbiol 75(20):6415–6421

Leong S-lL, Pettersson OV, Rice T, Hocking AD, Schnurer J (2011) The extreme xerophilic mould *Xeromyces bisporus* - Growth and competition at various water activities. Int J Food Microbiol 145(1):57–63

Leong S-lL, Lantz H, Pettersson OV, Frisvad JC, Thrane U, Heipieper HJ, Dijksterhuis J, Grabherr M, Pettersson M, Tellgren-Roth C, Schnurer J (2015) Genome and physiology of the ascomycete filamentous fungus *Xeromyces bisporus*, the most xerophilic organism isolated to date. Environ Microbiol 17(2):496–513

Ma Q, Jin K, Peng G, Xia Y (2015) An ENA ATPase, *MaENA1*, of *Metarhizium acridum* influences the Na^+-, thermo- and UV-tolerances of conidia and is involved in multiple mechanisms of stress tolerance. Fungal Genet Biol 83:68–77

Ma Z, Zhu L, Song T, Wang Y, Zhang Q, Xia Y, Qiu M, Lin Y, Li H, Kong L, Fang Y, Ye W, Wang Y, Dong S, Zheng X, Tyler BM, Wang Y (2017) A paralogous decoy protects *Phytophthora sojae* apoplastic effector PsXEG1 from a host inhibitor. Science 335 (6326):710–714

Mauch RM, Cunha VO, Dias ALT (2013) The copper interference with the melanogenesis of *Cryptococcus neoformans*. Rev Inst Med Trop São Paulo 55(2):117–120

Micheluz A, Manente S, Tigini V, Prigione V, Pinzari F, Ravagnan G, Varese GC (2015) The extreme environment of a library: xerophilic fungi inhabiting indoor niches. Int Biodeter Biodegr 99:1–7

Miskei M, Karanyi Z, Pocsi I (2009) Annotation of stress-response proteins in the aspergilli. Fungal Genet Biol 46(Suppl 1):S105–S120

Mukherjee A, Das D, Mondal SK, Biswas R, Das TK, Boujedaini N, Khuda-Bukhsh AR (2010) Tolerance of arsenate-induced stress in *Aspergillus niger*, a possible candidate for bioremediation. Ecotox Environ Safe 73(2):172–182

Murata Y, Homma T, Kitagawa E, Momose Y, Sato MS, Odani M, Shimizu H, Hasegawa-Mizusawa M, Matsumoto R, Mizukami S, Fujita K, Parveen M, Komatsu Y, Iwahashi H (2006) Genome-wide expression analysis of yeast response during exposure to 4°C. Extremophiles 10(2):117–128

O'Meara TR, Hay C, Price MS, Giles S, Alspaugh JA (2010) *Cryptococcus neoformans* histone acetyltransferase Gcn5 regulates fungal adaptation to the host. Eukaryot Cell 9(8):1193–1202

Oberson J, Rawyler A, Brändle R, Canevascini G (1999) Analysis of the heat-shock response displayed by two *Chaetomium* species originating from different thermal environments. Fungal Genet Biol 26(3):178–189

Oetari A, Susetyo-Salim T, Sjamsuridzal W, Suherman EA, Monica M, Wongso R, Fitri R, Nurlaili DG, Ayu DC, Teja TP (2016) Occurrence of fungi on deteriorated old dluwang manuscripts from Indonesia. Int Biodeter Biodegr 114:94–103

Oliver BG, Silver PM, White TC (2005) Evolution of drug resistance in pathogenic fungi. In: Xu JP (ed) Evolutionary genetics of fungi. Horizon Bioscience, Norfolk, pp 253–288

Ortiz-Urquiza A, Keyhani NO (2013) Action on the surface: entomopathogenic fungi versus the insect cuticle. Insects 4(3):357–374

Padamsee M, Kumar TKA, Riley R, Binder M, Boyd A, Calvo AM, Furukawa K, Hesse C, Hohmann S, James TY, LaButti K, Lapidus A, Lindquist E, Lucas S, Miller K, Shantappa S, Grigoriev IV, Hibbett DS, McLaughlin DJ, Spatafora JW, Aime MC (2012) The genome of the xerotolerant mold *Wallemia sebi* reveals adaptations to osmotic stress and suggests cryptic sexual reproduction. Fungal Genet Biol 49(3):217–226

Pawlowski AC, Wang W, Koteva K, Barton HA, McArthur AG, Wright GD (2016) A diverse intrinsic antibiotic resistome from a cave bacterium. Nat Commun 7:13803

Pemán J, Cantón E, Espinel-Ingroff A (2009) Antifungal drug resistance mechanisms. Expert Rev Anti Infect Ther 7(4):453–460

Perlin DS (2007) Resistance to echinocandin-class antifungal drugs. Drug Resist Update 10 (3):121–130

Pettersson OV, Leong S-lL (2001) Fungal Xerophiles (Osmophiles). In: eLS (Encyclopaedia of Life Sciences). Wiley, Chichester

Pfaller MA (2012) Antifungal drug resistance: mechanisms, epidemiology, and consequences for treatment. Am J Med 125(1 Suppl):S3–S13

Raghukumar C, Raghukumar S (1998) Barotolerance of fungi isolated from deep-sea sediments of the Indian Ocean. Aquatic Microbial Ecology 15(2):153–163

Ratnakumar S, Hesketh A, Gkargkas K, Wilson M, Rash BM, Hayes A, Tunnacliffe A, Oliver SG (2011) Phenomic and transcriptomic analyses reveal that autophagy plays a major role in desiccation tolerance in *Saccharomyces cerevisiae*. Mol BioSyst 7(1):139–149

Ropars J, de la Vega RCR, Lopez-Villavicencio M, Gouzy J, Sallet E, Dumas E, Lacoste S, Debuchy R, Dupont J, Branca A, Giraud T (2015) Adaptive horizontal gene transfers between multiple cheese-associated fungi. Curr Biol 25(19):2562–2569

Rosas AL, Casadevall A (1997) Melanization affects susceptibility of *Cryptococcus neoformans* to heat and cold. FEMS Microbiol Lett 153(2):265–272

Sancho LG, de la Torre R, Horneck G, Ascaso C, de Los Rios A, Pintado A, Wierzchos J, Schuster M (2007) Lichens survive in space: results from the 2005 LICHENS experiment. Astrobiology. 7(3):443–454

Singh J, Kumar D, Ramakrishnan N, Singhal V, Jervis J, Garst JF, Slaughter SM, DeSantis AM, Potts M, Helm RF (2005) Transcriptional response of *Saccharomyces cerevisiae* to desiccation and rehydration. Appl Environ Microbiol 71(12):8752–8763

Singh RS, Xu JP, Kulathinal R (2012) Evolution in the fast lane: rapid evolution of genes and genetic systems. Oxford University Press, Oxford

Six J (2012) Soil science: fungal friends against drought. Nat Clim Change 2(4):234–235

Skrinjar M, Blagojev N, Petrovic L, Soso V, Veskovic-Moracanin S, Skaljac S (2012) Diversity of moulds on the *Petrovska klobasa* raw materials, casings and in the processing unit environment. Rom Biotech Lett 17(6):7726–7736

Snoek ISI, Steensma HY (2007) Factors involved in anaerobic growth of *Saccharomyces cerevisiae*. Yeast 24(1):1–10

Steen BR, Lian T, Zuyderduyn S, MacDonald WK, Marra M, Jones SJM, Kronstad JW (2002) Temperature-regulated transcription in the pathogenic fungus *Cryptococcus neoformans*. Genome Res 12(9):1386–1400

Steenbergen JN, Shuman HA, Casadevall A (2001) *Cryptococcus neoformans* interactions with amoebae suggest an explanation for its virulence and intracellular pathogenic strategy in macrophages. Proc Natl Acad Sci USA 98(26):15245–15250

Su Y, Jiang X, Wu W, Wang M, Hamid MI, Xiang M, Liu X (2016) Genomic, transcriptomic and proteomic analysis provide insights into the cold adaptation mechanism of the obligate psychrophilic fungus *Mrakia psychrophila*. G3 6(11):3603–3613

Tai SL, Boer VM, Daran-Lapujade P, Walsh MC, de Winde JH, Daran JM, Pronk JT (2005) Two-dimensional transcriptome analysis in chemostat cultures. Combinatorial effects of oxygen availability and macronutrient limitation in *Saccharomyces cerevisiae*. J Biol Chem 280 (1):437–447

ter Linde JJ, Liang H, Davis RW, Steensma HY, van Dijken JP, Pronk JT (1999) Genome-wide transcriptional analysis of aerobic and anaerobic chemostat cultures of *Saccharomyces cerevisiae*. J Bacteriol 181(24):7409–7413

Tereshina VM, Memorskaya AS (2005) Adaptation of *Flammulina velutipes* to hypothermia tip in natural environments: the role of lipids and carbohydrates. Microbiology 74(3):279–283

Tesei D, Marzban G, Zakharova K, Isola D, Selbmann L, Sterflinger K (2012) Alteration of protein patterns in black rock inhabiting fungi as a response to different temperatures. Fungal Biol 116 (8):932–940

Tsuji M (2016) Cold-stress responses in the Antarctic basidiomycetous yeast *Mrakia blollopis*. R Soc Open Sci 3(7):160106

Vogan AA, Khankhet J, Samarasinghe H, Xu J (2016) Identification of QTLs associated with virulence related traits and drug resistance in *Cryptococcus neoformans*. G3 6(9):2745–2759

Vytrasova J, Pribanova P, Marvanova L (2002) Occurrence of xerophilic fungi in bakery gingerbread production. Int J Food Microbiol 72(1-2):91–96

Wang Y, Casadevall A (1994) Decreased susceptibility of melanized *Cryptococcus neoformans* to UV light. Appl Environ Microbiol 60:3864–3866

Wang Y, Zhang X, Zhou Q, Zhang X, Wei J (2015) Comparative transcriptome analysis of the lichen-forming fungus *Endocarpon pusillum* elucidates its drought adaptation mechanisms. Sci China Life Sci 58(1):89–100

Wang Y, White MM, Kvist S, Moncalvo J-M (2016) Genome-wide survey of gut fungi (Harpellales) reveals the first horizontally transferred ubiquitin gene from a mosquito host. Mol Biol Evol 33(10):2544–2554

Welch AZ, Gibney PA, Botstein D, Koshland DE (2013) TOR and RAS pathways regulate desiccation tolerance in *Saccharomyces cerevisiae*. Mol Biol Cell 24(2):115–128

Wichadakul D, Kobmoo N, Ingsriswang S, Tangphatsornruang S, Chantasingh D, Luangsa-ard JJ, Eurwilaichitr L (2015) Insights from the genome of *Ophiocordyceps polyrhachis-furcata* to pathogenicity and host specificity in insect fungi. BMC Genomics 16:881

Williams MC (2001) Trichomycetes a brief review of research. In: Misra JK, Horn B (eds) Trichomycetes and other fungal groups. Science Publishers, Enfield, NH, p 19

Xiao Y, Cheng X, Liu J, Li C, Nong W, Bian Y, Cheung MK, Kwan HS (2016) Population genomic analysis uncovers environmental stress-driven selection and adaptation of *Lentinula edodes* population in China. Sci Rep 6:36789

Xu J (2005) Evolutionary genetics of fungi. Horizon Bioscience, Norfolk

Xu J (2016) Fungal DNA barcoding. Genome 59(11):913–932

Youssef NH, Couger MB, Struchtemeyer CG, Liggenstoffer AS, Prade RA, Najar FZ, Atiyeh HK, Wilkins MR, Elshahed MS (2013) The genome of the anaerobic fungus *Orpinomyces* sp. strain C1A reveals the unique evolutionary history of a remarkable plant biomass degrader. Appl Environ Microbiol 79(15):4620–4634

Zampieri E, Balestrini R, Kohler A, Abba S, Martin F, Bonfante P (2011) The Perigord black truffle responds to cold temperature with an extensive reprogramming of its transcriptional activity. Fungal Genet Biol 48(6):585–591

Zhang T, Wei J (2011) Survival analyses of symbionts isolated from *Endocarpon pusillum* Hedwig to desiccation and starvation stress. Sci China Life Sci 54(5):480–489

Index

P. H. Rampelotto (ed.), *Molecular Mechanisms of Microbial Evolution*, Grand Challenges in Biology and Biotechnology,
https://doi.org/10.1007/978-3-319-69078-0

Q

R